Flexible Electronics, Volume 2

Thin-film transistors

Flexible Electronics, Volume 2

Thin-film transistors

Vinod Kumar Khanna

CSIR-Central Electronics Engineering Research Institute (CSIR-CEERI), Pilani, India

IOP Publishing, Bristol, UK

ISBN   978-0-7503-2453-3 (ebook)
ISBN   978-0-7503-2450-2 (print)
ISBN   978-0-7503-2452-6 (mobi)

DOI   10.1088/2053-2563/ab0d18

Version: 20190701

IOP Expanding Physics
ISSN 2053-2563 (online)
ISSN 2054-7315 (print)

British Library Cataloguing-in-Publication Data: A catalogue record for this book is available from the British Library.

Published by IOP Publishing, wholly owned by The Institute of Physics, London

IOP Publishing, Temple Circus, Temple Way, Bristol, BS1 6HG, UK

US Office: IOP Publishing, Inc., 190 North Independence Mall West, Suite 601, Philadelphia, PA 19106, USA

To my late mother Smt Pushpa Khanna,
my sweetest and ever-lasting memory,
as a humble tribute of my thankfulness
for her endless love, affection and gentleness,
her sacrifices in my upbringing
and enthusiastic mentoring!

Contents

Preface

Electronics has created many wonders. Communication, computing, control, sensing/actuation and power systems have gone through dramatic changes due to advances in electronics. Automobiles, aerospace, healthcare, information, instrumentation and entertainment systems have been totally transformed over the years.

Vacuum tubes, transistors, solid-state devices, very large scale and ultra large scale integrated circuits (VLSI and ULSI), ... it is an unending success story of unabated progress towards smaller geometries and minimum feature sizes. Dimensions continued shrinking to the nanoscale and the miniaturization wave still continues. Then electronic scientists and engineers realized that a major portion of the silicon substrate on which the devices and circuits are built was of little use during the operation of the device except for providing mechanical support. It was an unnecessary burden and added to the weight of the electronic equipment. In particular, the equipment and photovoltaics used on space missions could be made lightweight by thinning down the processed wafer after the chip fabrication was completed. Soon, it was also realized that the redundant wafer thickness was an obstacle to the flexibility of the chip. Thinning of the wafer increased its ability to withstand bending. Thus, silicon wafers could be thinned down post-processing and the thin chips mounted on plastic substrates.

Gradually, scientists thought that they should develop capabilities for chip fabrication directly on plastic substrates. This would lead to process simplification and cost reduction. But this was a difficult decision. Plastic substrates cannot be exposed to high processing temperatures but most silicon fabrication processes require elevated temperatures. So processes were divided into two classes, namely those which could be implemented at low temperatures and those for which the high temperature route was mandatory. Lowering the temperature means compromising quality. Either the desired product will not be formed or its properties will not be up to the requisite standard. A strategy was evolved in which the high temperature steps were completed on silicon, glass or some other hard material and the low-temperature steps were done after the structure was transferred to a polymeric substrate. This paved the way towards a strategy of dividing the process into parts to be implemented on hard and soft substrates.

A scheme of momentarily exposing the organic substrates to high temperatures also received much attention, e.g. amorphous silicon could be crystallized by shining bursts of laser energy at them which would locally heat the material for a time duration short enough not to cause any harm to the underlying plastic substrate. In this way, rapid thermal annealing-based processes were also formulated.

Another line of thinking looked at organic and printed electronics. These processes are comparatively cheap. They do not require the expensive infrastructure of clean rooms. Moreover, these processes are scalable to large-area roll-to-roll manufacturing, promising high throughput. Thus, electronic circuits can be contemplated coming out of a printer in which inks for conducting, semiconducting and insulating materials are used in place of the inks of different colors for printing letters and pictures on paper. Noteworthy applications of this type include the large-area display and solar cell panels where emphasis shifts from microelectronics to macroelectronics. These applications require low-cost manufacturing using solution-processable printable polymers.

In yet another alternative, the unique properties of nanostructured materials, notably graphene, 2D transition metal dichalcogenides (TMDs), carbon nanotubes (CNTs), silicon nanowires (Si NWs) and nanoribbons are utilized to fabricate flexible chips.

With the vast spectrum of opportunities and options available, flexible electronics has made great strides. Already, a wide assortment of thin film transistors (TFTs) exist. Starting from amorphous silicon (a-Si) TFTs, we have polysilicon TFTs and single crystal silicon TFTs on the inorganic side. On the organic side, we have small molecule, polymeric and organic single crystal TFT varieties. Flexible energy devices like solar cells, batteries, supercapacitors and energy harvesters have all made great progress. The list of flexible electronic applications is very large. Displays, sensors, memories and radio frequency identification (RFID) tags are the hot topics in flexible electronics.

Flexible electronics is blossoming like a flower. It is growing and reaching diverse aspects of human life and activities. It is maturing into a very interesting discipline that is bound to touch our entire lives. Many flexible electronics products can be worn like clothing. Wearable and stretchable electronics are sister branches of flexible electronics wherein the flexing capabilities are further augmented by providing extendability and conformability features.

With all the good things that flexible electronics has brought to humankind, the field has become a completely new discipline in itself and this book is a humble attempt to bring the reader a taste of the groundbreaking research and break-throughs in this field over the last few decades. We shall start from the very basic concepts of mechanical bending and take the reader on an entertaining journey where there are stations called materials for flexible electronics, the manufacturing processes for flexible electronic components, the thin film transistors which are the ubiquitous amplifying and switching devise in flexible electronic circuits, the sources of energy that power the flexible electronics gadgetry, and the service to mankind in the form of multifaceted manifold applications of flexible electronics. We promise that this joyride will be very captivating, informative and enjoyable. We wish that readers will find this a very useful and pleasurable experience.

Acknowledgements

First and foremost, I am thankful with all my heart to Almighty God with whose grace, blessings and kindness all tasks are accomplished.

I am highly indebted to the various authors/editors of books, journal articles and web pages whose work forms the subject matter of this book. Most of these names appear in the references at the end of respective chapters. Accolades to these pioneer workers. Errors and omissions exempted, please!

I feel greatly obliged to my editors/editorial assistants at IOP Publishing for their overwhelming support, kind co-operation and valuable encouragement throughout the project.

Last but not least, extreme gratitude is extended to my family members. They lent me their unfailing support and exercised great patience in tolerating my hectic work schedule for this project. They always boosted my confidence and encouraged me with full zeal and gusto to put my best foot forward.

My profound thanks to all the above!

Vinod Kumar Khanna
Chandigarh, INDIA

About the book

Flexible Electronics is a book on an interesting subject in which electronics, a science and technology initially developed on brittle and hard silicon substrates gradually evolved to transform into a softer, bendable and shape-conformable discipline smoothly riding on pliable, rollable and foldable plastic substrates. The book traces the course of progress from the brittle silicon era to the bendable polymeric era. Within the pages of this book are crammed the endeavors of several researchers and pioneers in the field who passionately worked on the innovative translation-of-technology approaches day and night to achieve the flexibility goal. This upcoming electronics has become more adaptable and adjustable to the demands of the user, a technological breakthrough that has considerably enlarged its range of applications.

The book comprehensively surveys the basics of the field, the state of the art and the trends in research being pursued worldwide on flexible electronics. Themes covered include the mechanical theory, materials science aspects, fabrication technologies, devices and applications.

The subject matter is lucidly presented as 3 volumes, organized in 6 parts, and further divided into 35 chapters, all profusely illustrated with over 300 line diagrams to make it easily understandable and to work as a stepping stone for students as well as professional engineers and scientists, to follow original research papers in this field advancing at a breathtakingly fast pace.

Flexible Electronics, Volume 1 deals with the three foundation pillars of flexible electronics: mechanics, materials and manufacturing. It lays down the groundwork for the whole book. Contents include bending theory, stresses and strains in thin films, curvature and overlay alignment, controlled buckling, brittle and ductile film behaviors, permeation barriers; inorganic, organic and nanomaterials; printing and vacuum techniques; silicon IC/MEMS processes and packaging.

Flexible Electronics, Volume 2 provides a detailed treatment of thin film transistors which have the same ubiquity and occupy similar pivotal status in flexible electronics as the silicon MOSFET in mainstream electronics. The treatment covers a wide gamut of devices covering inorganic (amorphous silicon, polysilicon, single-crystal silicon, metal-oxide); organic (small organic molecule, polymer, organic single crystal, electrolyte-gated, organic electrochemical) to nanomaterial (2D material, CNT, nanowire) TFTs.

Flexible Electronics, Volume 3 is devoted to the especially designed energy sources for flexible electronics and the applications of flexible electronics in various fields. The energy infrastructure for flexible electronics comprises supercapacitors, batteries, energy harvesters and solar cells, all built on supple substrates. These elements constitute the muscles of flexible electronics. Applications of flexible electronics in displays, CNT field emitters, sensors, memories and RFID tags are highlighted.

Author biography

Vinod Kumar Khanna

Vinod Kumar Khanna (born 25 November 1952, Lucknow, India) is a former emeritus scientist, CSIR (Council of Scientific and Industrial Research) and emeritus professor, AcSIR (Academy of Scientific and Innovative Research), India. He is a retired Chief Scientist and Head, MEMS and Microsensors Group, CSIR-CEERI (CSIR-Central Electronics Engineering Research Institute), Pilani (Rajasthan) and Professor, AcSIR, India. During his tenure of work at CSIR-CEERI, Pilani spanning over 37 years, he worked on a large number of CSIR and sponsored projects on power semiconductor devices, notably a high-current and high-voltage rectifier, high-voltage transistor, fast-switching inverter grade thyristor, power DMOSFET and IGBT. Another key research area included the device and process design, and fabrication of micro- and nanoelectronics and MEMS technology-based sensors and dosimeters. His contributions focused primarily on the development of technology and characterization of moisture and microheater-embedded gas sensor, ion-sensitive field-effect transistor pH sensor, MEMS acoustic sensor and capacitive MEMS ultrasonic transducer; PIN diode neutron dosimeter, and PMOSFET gamma ray dosimeter.

He is widely traveled and has worked at Technische Universität Darmstadt, Germany, 1999 and Kurt-Schwabe-Institut für Mess- und Sensortechnik e.V., Meinsberg, Germany, 2008. He also visited the Institute of Chemical Physics, Novosibirsk, Russia, 2009 as a member of Indian delegation and Fondazione Bruno Kessler (FBK), Trento, Italy, 2011, under India-Trento Programme of Advanced Research (ITPAR). He participated and presented research papers at the IEEE-IAS Annual meeting at Denver, Colorado, USA, 1986.

Awarded a national scholarship by the Ministry of Education and Youth Services, Govt. of India in 1970, he received his MSc degree in Physics in 1975 from University of Lucknow, India and PhD (Physics) from Kurukshetra University, India in 1988.

Prior to the present book, he has authored 12 books. He is the author/co-author of 192 research papers in prestigious refereed journals and national/international conference proceedings; he has five patents to his credit, including two US patents. He is a life member of several leading professional societies, and fellow of the Institution of Electronics and Telecommunication Engineers (IETE), India.

After superannuating as Head, MEMS and Microsensors Group, CSIR-CEERI in November 2014, and completing his tenure as Emeritus Scientist, CSIR, in November 2017, he presently resides at Chandigarh, India. He is a passionate author, avidly reading and writing.

Abbreviations, acronyms, chemical symbols and formulae

A	Ampere
Å	Angstrom unit
AC	Alternating current
ACF	Anisotropic conductive film
A/D	Analog-to-digital (converter)
AFE	Analog front end
AFM	Atomic force microscope
Ag	Argentum (silver)
$AgNO_3$	Silver nitrate
Ag NWs	Silver nanowires
AgO; Ag_2O, Ag_2O_3	Silver (I, III) oxide (Ag_4O_4)
Ag_2O	Silver (I) oxide
a-IGZO	Amorphous indium gallium zinc oxide
Al	Aluminum
AlAs	Aluminum arsenide
$Al_2(CH_3)_6$	Trimethylaluminum (TMA) (exists as a dimer, molar mass 144.18 g mol^{-1})
ALD	Atomic layer deposition
AlGaAs	Aluminum gallium arsenide
AlGaN	Aluminum gallium nitride
Al:Nd	Aluminum–neodymium
AlN_x	Aluminum nitride (x unknown)
AlO_x	Aluminum oxide (x unknown)
AlO_x–Nd	Aluminum oxide (x unknown)–neodymium
Al_2O_3	Aluminum oxide
Alq_3	$C_{27}H_{18}AlN_3O_3$, tris(8-hydroxyquinolinato)aluminum, molar mass 459.43, yellow powder
Al:ZnO	Al-doped ZnO
AM1.5G	Air mass 1.5-global (illumination); 'global' means direct plus diffuse radiation
AMLCD	Active-matrix liquid crystal display
AMOLED	Active-matrix light emitting diode
APS	Ammonium persulphate, $(NH_4)_2S_2O_8$
APTES	(3-aminopropyl)triethoxysilane
Ar	Argon
ARC	Antireflection coating
Ar/CHF_3	Argon/fluoroform
As	Arsenic
AsF_5	Arsenic pentafluoride
AsH_3	Arsine
a-Si:H	Amorphous silicon: hydrogenated
$a\text{-}SiN_x$	Amorphous silicon nitride (x unknown)
at%	Atomic percent [= (number of one kind of atoms/ total number of atoms) × 100%]

a-TiO$_2$	Amorphous titanium dioxide
Au	Aurum (gold)
AuBe	Gold–beryllium
AuGe	Gold–germanium
AuGeNi	Gold–germanium–nickel
AZO	Aluminum-doped zinc oxide
B	Boron, byte
b	Bit
BaTiO$_3$	Barium titanate
BBr$_3$	Boron tribromide
BCB	Benzocyclobutene
BCE	Back channel etch
BCl$_3$	Boron trichloride
Be	Beryllium
BF$_3$	Boron trifluoride
B$_2$H$_6$	Diborane
Bi$_2$O$_3$	Bismuth (III) oxide
BL	Bit line
[(bmim)(Tf2N)]	1-butyl-3-methylimidazolium bis(trifluoromethyl-sulfonyl)imide, C$_{10}$H$_{15}$F$_6$N$_3$O$_4$S$_2$
BN	Boron nitride
B$_2$O$_3$	Boron trioxide
BOE	Buffered oxide etch
BOX	Buried oxide
BPDA	Biphenyl-tetracarboxylic dianhydride (C$_{16}$H$_6$O$_6$)
BPEA	9,10-bis(phenylethynyl)anthracene (C$_{30}$H$_{18}$)
B-PEH	Bimorph piezoelectric energy harvester
Bphen	4,7-diphenyl-1,10-phenanthroline (C$_{24}$H$_{16}$N$_2$)
BPM	Beats per minute
Br$_2$	Bromine
B/s	Byte per second (1 byte = 8 bits)
b/s	Bit per second
BSF	Back surface field
BSG	Borosilicate glass
BTBT	Benzothienobenzothiophene
C	Carbon
°C	Degree centigrade
C$_{60}$	Fullerene, Buckminsterfullerene
Ca	Calcium
CaCl$_2$	Calcium chloride
CAD	Computer-aided design
CaF$_2$	Calcium fluoride
Ca(NO$_3$)$_2$	Calcium nitrate
CBC	Cymbet Corporation
C$_4$Cl$_4$O$_2$S	Tetrachlorothiophene dioxide
C$_6$Cl$_4$O$_2$	o-chloranil
C$_{32}$CuF$_{16}$N$_8$	Copper hexadecafluorophthalocyanine (F$_{16}$CuPc)
cd/m^2	Candela/square meter
CD-ROM	Compact disc-read-only memory
CdS	Cadmium sulfide

$-CF_3$	Trifluoromethyl (functional group)
CF_4	Tetrafluoromethane
$(C_2F_4)_n$	Polytetrafluoroethylene (PTFE)
C_4F_8	Octafluorocyclobutane
C_6F_5SH	2,3,4,5,6-pentafluorothiophenol, pentafluorobenzenethiol (PFBT)
CG	Conversion gain, center of gravity
CGL	Charge generating layer
CH	Cysteamine hydrochloride
CH_4	Methane
C_2H_2	Acetylene
$(C_2H_2)_n$	Polyacetylene
C_2H_4	Ethylene, ethene
$(C_2H_4)_n$	Polyethylene
C_2H_6	Ethane
C_4H_8	Butylene, butene
C_6H_{14}	Hexane
C_7H_8	Toluene
$(C_8H_8)_n$	Polystyrene (PS)
C_8H_{18}	Octane
$C_{10}H_{12}$	Tetralin
$C_{18}H_{12}$	Tetracene
$C_{18}H_{22}$	1,1-diphenylhexane
$C_{22}H_{14}$	Pentacene
$C_{25}H_{34}$	9,9-dihexylfluorene
$C_{30}H_{18}$	Bis(phenylethynyl)anthracene
$C_{42}H_{28}$	Rubrene
$C_{27}H_{18}AlN_3O_3$	Alq$_3$, tris(8-hydroxyquinolinato)aluminum
CH/Au	Cysteamine hydrochloride/gold
$C_6H_4Br_2$	1, 2 dibromobenzene
$[CH_2C(CH_3)(CO_2CH_3)]_n$	Poly(methyl methacrylate) (PMMA)
$(-CH_2CF_2-)_x[-CF_2CF(CF_3)-]_y$	Poly(vinylidenefluoride-co-hexafluoropropene) (PVDF-HFP)
$C_6H_5CH_3$	Toluene
$[-CH_2CH = C(CH_3)CH_2-]_x$ $[-CH_2CH_2CH(CH_3)CH_2-]_y$	Poly(propylene-alt-ethylene), multi-arm
$CH_3CH_2C(CH_3)_2OH$	Tert-amyl alcohol
$C_6H_5CH = CH_2$	Styrene
$(C_6H_5CHCH_2)_n$	Polystyrene
$[CH_2CH(CH_3)]_n$	Polypropylene (PP)
$(CH_2CH = CHCH_2)_n$	Polybutadiene
$[CH_2CH(C_6H_5)]_x(CH_2CH = CHCH_2)_y$	Polystyrene-block-polybutadiene (PS-b-PB)
$CH_3(CH_2)_9CH[(CH_2)_7CH_3]CH_2OH$	Octyldodecanol
$[CH_2CH(CH_2NH_2 \cdot HCl)]_n$	PAH {poly(allylamine hydrochloride)}
$CH_3CH_2CH_2OH$	Propanol
$[CH_2CH(C_6H_4OH)]_n$	Poly-4-vinylphenol, polyvinylphenol (PVP)
$[CH_2CH(C_6H_4SO_3R)]_x[CH(CO_2R)$ $CH(CO_2R)]_y$, R = H or Na	PSSMA {Poly(4-styrenesulfonic acid-co-maleic acid} sodium salt
$C_6H_5CH_2CH_2SiCl_3$	Trichloro(phenethyl)silane (PETS)
$CH_3(CH_2)_{11}C_6H_4SO_3Na$	Sodium dodecylbenzenesulfonate (SDBS)

$CH_2 = CHCOOCH_2CH_2OH$	2-hydroxylethyl acrylate
$CH_2 = CHCOOH$	Acrylic acid
$CH_3(CH_2)_{15}N(Br)(CH_3)_3$	Cetyltrimethylammonium bromide (CTAB)
$(CH_2)_3CH_2O$	Tetrahydrofuran
$C_{10}H_7CHO$	2-naphthaldehyde
$(-CH_2CH_2O-)_n$	Polyethylene oxide (PEO)
$(C_2H_4)_n(C_4H_6O_2)_m$	Ethylene vinyl acetate (EVA)
$(C_8H_8)_n(C_3H_4O_2)_n$	Polystyrene-block-polyactide (PS-b-PLA)
$[-CH_2CHOH-]_n$	Polyvinylalcohol (PVA)
CH_3CH_2OH	Ethanol
$(CH_3)_2CHOH$	Isopropyl alcohol, 2-propanol
$CH_3(CH_2)_5OH$	Hexanol
$CH_3CHOHCH_3$	Isopropanol (isopropyl alcohol)
$CH_3(CH_2)_{11}OSO_3Na$	Sodium dodecyl sulfate (SDS)
$CH_3(CH_2)_{13}P(O)(OH)_2$	N-tetradecylphosphonic acid (TDPA)
$CH_3C_6H_4SO_3H$	p-toluenesulfonic acid (PTSA)
$CHCl_3$	Chloroform
$C_6H_2Cl_4$	1,2,4,5-tetetrachlorobenzene
$C_6H_4Cl_2$	1,2-dichlorobenzene
C_6H_5Cl	Chlorobenzene
$C_{16}H_{14}Cl_2$	Dichloro-[2.2]-paracyclophane
$C_9H_{24}ClNO_3Si$	N-[3-(trimethoxysilyl)propyl]-N,N,N-trimethyl-ammonium chloride
CH_3CN	Acetonitrile
$CH_3C(O)CH_2CH_3$	2-butanone, methyl ethyl ketone (MEK)
CH_3COCH_3	Acetone
$CH_3CO_2CH = CH_2$	Vinyl acetate
$CH_3CO_2CH(CH_3)CH_2OCH_3$	Propylene glycol monomethyl ether acetate (PGMEA)
$(C_8H_8)_n (C_5O_2H_8)_n$	Polystyrene-block-poly(methyl methacrylate) (PS-b-PMMA)
$CH_3CON(CH_3)_2$	DMAc or DMA (N,N-dimethylacetamide)
$(CH_3)_3COOC(CH_3)_3$	Tert-butyl peroxide (TBPO)
$CH_3COO(CH_2)_3CH_3$	n-butyl acetate
CH_3COOH	Acetic acid
$C_{32}H_{16}CuN_8$	Copper(II) phthalocyanine
CHF_3	Fluoroform
$-(C_2H_2F_2)_n-$	Polyvinylidene fluoride (PVDF)
$-(C_2HF_3)_n-$	Polytrifluoroethylene (PTrFE)
$C_{10}H_{10}Fe$	Ferrocene
$C_{10}H_{15}F_6N_3O_4S_2$	1-butyl-3-methylimidazolium bis(trifluoromethylsulfonyl)imide [(bmim)(Tf2N)]
$C_6H_2F_2N_2S$	5,6-difluoro-2,1,3-benzothiadiazole (FBT)
$C_8H_5F_3O_2S$	2-thenoyltrifluoroacetone
C_6HF_5S	Pentafluorobenzenethiol (PFBT)
$C_8H_{24}HfN_4$ or $[(CH_3)_2N]_4Hf$	Tetrakis(dimethylamino) hafnium
C_4H_9Li	n-butyllithium
C_2H_3N or CH_3CN	Acetonitrile
$C_3H_5N_2^+$	Imidazolium$^+$ ionic liquid-based electrolyte, 1H-imidazol-3-ium

$(CH_2)_6N_4$ or $C_6H_{12}N_4$	Methenamine
C_4H_5N	Pyrrole
$(C_4H_5N)_n$	Polypyrrole
$(C_6H_7N)_x$	Polyaniline (PANI)
$(C_7H_7N)_n$	Poly(vinylpyridine) (PVP)
$C_8H_{11}N_5$	1-ethyl-3-methylimidazolium dicyanamide (EMIMDCA)
$C_8H_{17}N_3.HCl$	N-(3-dimethylaminopropyl)-N'-ethylcarbodiimide hydrochloride (EDC)
$C_{12}H_4N_4$	7,7,8,8-tetracyanoquinodimethane (TCNQ)
$C_{24}H_{16}N_2$	4,7-diphenyl-1,10-phenanthroline (Bphen)
$C_{44}H_{32}N_2$	N, N'-bis(naphthalen-1-yl)-N, N'-bis(phenyl)benzidine (NPB)N, N'-di(napthalen-1-yl)-N, N'-diphenyl-benzidine (NPD)
$C_{46}H_{36}N_2$	2,2'-dimethyl-N, N'-di-[(1-naphthyl)-N, N'-diphenyl]−1,1'-biphenyl-4,4'-diamine (α-NPD)
$(C_3H_3NaO_2)_n$	Sodium polyacrylate
$C_8H_{16}NaO_8$	Sodium carboxymethyl cellulose (CMC), molecular weight 263.198 g mol^{-1}
$C_3H_7NaO_3S_2$	3-mercapto-1-propanesulfonic acid sodium salt (MPS), Sodium 3-mercaptopropanesulphonate
$[C_6H_4N(C_6H_2(CH_3)_3)C_6H_4]_n$	Poly[bis(4-phenyl)(2,4,6-trimethylphenyl)amine (PTAA)
$C_6H_4(NH_2)_2$ or $C_6H_8N_2$	p-phenylenediamine
$C_4H_5NO_3$	NHS (N-hydroxysuccinimide)
C_5H_9NO	N-methyl-2-pyrrolidone (NMP), 1-methyl-2-pyrrolidinone
$(C_4H_8N_6O)_n$	Poly(melamine-co-formaldehyde) methylated
$(C_6H_9NO)_n$	Poly(vinyl pyrrolidone), PVP
$C_{40}H_{26}N_2O_6$	N,N'-bis(4-methoxybenzyl)perylene-3,4:9,10-bis(dicarboximide)
$C_{40}H_{36}N_2O_4$	N, N, N', N'-tetrakis(4-methoxyphenyl)-benzidine (MeO-TPD)
$C_{41}H_{22}N_4O_{11}$	Polyimide
$(C_{12}H_{12}N_2O.C_{10}H_2O_6)_x$	Polyamic acid
C_5H_3NS	2-thiophenecarbonitrile
$(C_2H_4O)_x$	Polyvinylalcohol (PVA)
$C_2H_6O_2$	Glycol, ethylene glycol
$C_{2n}H_{4n+2}O_{n+1}$	Polyethylene glycol
$(C_3H_4O_2)_n$	Polyacrylic acid (PAA)
$C_3H_2O_3$	Vinylene carbonate
$C_3H_4O_3$	Ethylene carbonate (EC)
C_3H_6O	Acetone
C_3H_8O	Isopropyl alcohol
$C_3H_8O_2$	2-methoxyethanol
$C_4H_6O_3$	Propylene carbonate
$C_4H_6O_4$	Succinic acid
$C_6H_{10}O$	Cyclohexanone
$C_7H_5O_2^-$	Benzoate
$C_7H_5O_2$–Eu complex	P6FBEu, Eu-complexed benzoate

$C_7H_{10}O_3$	Glycidyl methacrylate
$C_8H_{16}O_8$	Carboxymethyl cellulose
$C_{10}H_8O_4$	Ethylene terephthalate
$(C_{10}H_8O_4)_n$	Polyethylene terephthalate (PET)
$C_{10}H_{18}O$	α-terpineol
$C_{13}H_{20}O_2$	Isobornyl acrylate
$(C_{14}H_{10}O_4)_n$	Polyethylene napthalate (PEN)
$C_{16}H_{18}O_5$	Bisphenol A polycarbonate (PC)
$C_{18}H_{32}O_{16}$	Dextran
CH_3OCF_2R	Methoxyperfluorobutane, R = $-CF_2CF_2CF_3$ or $-CF(CF_3)_2$
$(C_3H_6O)_n(C_3H_6O)_n(C_3H_6O)_nC_{15}H_{20}O_6$	Trimethylolpropane polypropylene glycol triacrylate
$(CH_3O)_2CO$	Dimethyl carbonate (DMC)
CH_3OH	Methanol
C_2H_5OH	Ethyl alcohol, ethanol
$C_4H_{11}O_3P$	Butylphosphonic acid
$C_{18}H_{39}O_3P$	Octadecyl-phosphonic acid (ODPA)
$(C_2H_3OR)_n$	Polyvinylalcohol, R = H or $COCH_3$
C_2H_6OS	Dimethyl sulphoxide
$(C_6H_4O_2S)_n$	PEDOT {poly(3, 4-ethylenedioxythiophene)}
$C_7H_8O_3S$	Toluene-4-sulfonic acid
$C_7H_8O_3S. H_2O$	Toluene-4-sulfonic acid monohydrate
$(C_8H_8O_3S)_n$	Poly(4-styrenesulfonic acid) (PSSH)
$(C_{12}H_8O_3S)_n$	Polyethersulfone (PES)
$C_9H_{18}O_3Si_3$	1,3,5-trimethyl-1,3,5-trivinyl cyclotrisiloxane (V3D3)
$C_9H_{20}O_5Si$	3-glyci-doxypropyltrimethoxysilane (Silquest)
$(C_{10}H_{14}S)_n$	Poly(3-hexylthiophene-2,5-diyl) (P3HT)
$C_{16}H_{10}S_4$	2,2′:5′,2″:5″,2‴-quaterthiophene (Th$_4$)
$C_{22}H_{12}S_2$	Dinaphtho-[2, 3-b:2′,3′-f]thieno[3,2-b]thiophene (DNTT)
$(C_{42}H_{62}S_4)_n$	Poly[2,5-bis(3-tetradecylthiophen-2-yl)thieno[3,2-b]thiophene] (pBTTT-C14)
$C_{44}H_{54}Si_2$	6,13-bis(triisopropylsilylethynyl)pentacene (TIPS-pentacene)
$(CH_3)_3SiNHSi(CH_3)_3$	Hexamethyldisilazane (HMDS)
$[(CH_3)_2SiO]_n$	Polydimethylsiloxane (PDMS)
$(CH_3)_2SO$	Dimethyl sulfoxide (DMSO)
$(C_8H_7SO_3{}^-)_n$	PSS {poly(styrenesulfonate)}
$(C_2H_5)_2Zn$	Diethylzinc (DEZn)
CIF	Chip-in flex
CIGS	Cadmium indium gallium selenide
CIJ	Continuous inkjet
CISC	Complex instruction set computing
Cl_2	Chlorine
$ClCH = CCl_2$ or C_2HCl_3	Trichloroethylene
$ClCH_2CH_2Cl$ or $C_2H_4Cl_2$	1, 2-dichloroethane (DCE)
cm	Centimeter
CMC	Carboxymethyl cellulose ($C_8H_{16}O_8$)

CMOS	Complementary metal-oxide-semiconductor (field-effect transistor)
CMP	Chemical mechanical polishing
$C_{18}N_{12}$	1,4,5,8,9,11-hexaazatriphenylenehexacarbonitrile (HAT-CN6)
CNT	Carbon nanotube
CO	Carbon monoxide
Co	Cobalt
CO_2	Carbon dioxide
–COOH–	Carboxyl group
CP	Clear plastic
cP	Centipoise
CPS	Cyclopentasilane
CPU	Central processing unit
Cr	Chromium
C–S	Contacting and separation (effects)
Cs	Ceasium
cs, cSt	CentiStokes
C545T	2,3,6,7-tetrahydro-1,1,7,7,-tetramethyl-1H,5H,11H-10(2-benzothiazolyl)quinolizine-[9,9a,1gh]coumarin
CTAB	Cetyltrimethylammonium bromide, $CH_3(CH_2)_{15}N(Br)(CH_3)_3$
CTE	Coefficient of thermal expansion
CTM	Charge transfer memory
Cu	Copper
CuCl	Copper (I) chloride
$CuCl_2$	Copper (II) chloride
$CuGaSe_2$	Copper gallium selenide (CGS)
$CuIn_xGa_{(1-x)}Se_2$	Copper indium gallium diselenide (x: 0 to 1)
$CuInSe_2$	Copper indium selenide (CIS)
CuPc	Copper(II) phthalocyanine
Cu/Ti	Copper/titanium
CV	Cyclic voltammetry
CVD	Chemical vapor deposition
0D	Zero-dimensional
1D	One-dimensional
2D, 2-D	Two-dimensional
3D	Three-dimensional
D–A	Donor–acceptor
D/A	Digital-to-analog (converter)
Da	Dalton
dB	Decibel
DC	Direct current
DCE	1, 2-dichloroethane ($ClCH_2CH_2Cl$)
DEP	Dielectrophoresis
DEZn	Diethylzinc
DI	Deionized (water)
DL	Diffuse layer
DMAc	N,N-dimethylacetamide

DMC	Dimethyl carbonate {$(CH_3O)_2CO$} or $C_3H_6O_3$
DMF	N,N-dimethylformamide {$HCON(CH_3)_2$} or C_3H_7NO
DMSO	Dimethyl sulfoxide, $(CH_3)_2SO$ or C_2H_6OS
DNQ	Diazonaphthoquinone
DNTT	Dinaphtho-[2,3-b:2′,3′-f]thieno[3,2-b]thiophene, $C_{22}H_{12}S_2$
DoD	Droplet-on-demand
DPI	Dots per inch
DPP	Diketopyrrolopyrrole
DRAM	Dynamic random access memory
DRIE	Deep reactive ion etching
DRL	Dynamic releasing layer, die release layer
DTBDT-C_6	Dithieno[2,3-d; 2′,3′-d′]benzo[1.2-b;4,5-b′] dithiophene
DVD	Digital video disc or digital versatile disc
DWCNT	Double-walled carbon nanotube
E-beam	Electron beam (e.g. evaporation, lithography, resist)
EBL	Electron beam lithography
EC	Ethylene carbonate ($C_3H_4O_3$)
EC/DMC	Ethylene carbonate/dimethyl carbonate
ECR	Electron cyclotron resonance
ECR-CVD	Electron cyclotron resonance-chemical vapor deposition
EDC	N-(3-dimethylaminopropyl)-N'-ethylcarbodiimide hydrochloride, $C_8H_{17}N_3.HCl$
EDL	Electrical double layer
EDP	Ethylene diamine pyrocatechol
EEPROM or E2PROM	Electrically erasable programmable read-only memory
EFD	Electro-fluid dynamics
E-Fiber	Electronic fiber
EGOFET	Electrolyte-gated organic FET
EHD	Electrohydrodynamics
EIL	Electron injection layer
E-ink	Electronic ink (display)
EL	Emissive layer
ELA	Excimer laser annealing
ELC	Excimer laser crystallization
ELML	End-loaded meander-line
ELO	Epitaxial lift-off
EMIM DCA	1-Ethyl-3-methylimidazolium dicyanamide ($C_8H_{11}N_5$)
E-paper	Electronic paper
EPC	Electronic product code
EPD	Electronic paper display
EPDM	Ethylene propylene diene monomer
EPROM	Erasable programmable read-only memory

erfc	Complementary error function
ES	Etch stop
E-skin	Electronic skin
ESL	Etch stop layer
E-textile	Electronic textile
ETL	Electron transport layer
eV	Electron volt
EVA	Ethylene vinyl acetate
EWC-SPUDT	Electrode-width-controlled single-phase unidirectional interdigital transducer
exp	Exponential function
F	Fluorine
FBT	5,6-difluoro-2,1,3-benzothiadiazole
FBT-Th$_4$(1,4)	5,6-difluoro-2,1,3-benzothiadiazole-quarterthiophene ($C_6H_2F_2N_2S$–$C_{16}H_{10}S_4$)
FC-CVD	Floating catalyst-based CVD
f-CTM TFT	Floating charge-trap-type memory thin film transistor
F$_{16}$CuPc	Copper hexadecafluorophthalocyanine
Fe	Ferrum (iron)
FeCl$_2$	Ferrous chloride
FeCl$_3$	Ferric chloride
Fe(CN)$_6^{3-}$/Fe(CN)$_6^{4-}$	(Potassium) ferri/ferrocyanide
Fe$_2$O$_3$	Ferric oxide
FET	Field-effect transistor
FF	Fill factor (of a solar cell)
FN	Fowler–Nordheim (theory)
FOC	Foil on a carrier
FS	Fluorosurfactant
g	Gram
Ga	Gallium
GaAs	Gallium arsenide
GaInP	Gallium indium phosphide
GaN	Gallium nitride
Ga(NO$_3$)$_3$.xH$_2$O	Gallium nitrate hydrate
Ga$_2$O$_3$	Gallium oxide
Ge	Germanium
GeH$_4$	Germane
Gen 2	Generation 2
GHz	Gigahertz
GILD	Gas immersion laser doping
g/mol	Gram/mole
GO	Graphene oxide
GPa	Giga Pascal
GPC	Gas permeation chromatography
GZO	Gallium-doped zinc oxide
h	Hour
H$_2$	Hydrogen
HAT-CN6	1,4,5,8,9,11-hexaazatriphenylenehexacarbonitrile ($C_{18}N_{12}$)

h-BN	Hexagonal boron nitride
HBr	Hydrobromide
$H_2C = C(CH_3)COOH$ or $C_4H_6O_2$	Methacylic acid
$H_2C = CHCO_2C_{10}H_{21}$ or $C_{13}H_{24}O_2$	Isodecyl acrylate
$H_2C = CH(OCH_2CH_2)_nOCH = CH_2$	Poly(ethylene glycol) divinyl ether
HCl	Hydrochloric acid
$HClO_4$	Perchloric acid
$HCON(CH_3)_2$ or C_3H_7NO	N,N-dimethylformamide (DMF)
He	Helium
HF	Hydrofluoric acid
HfO_2	Hafnium oxide
HIL	Hole injection layer
HL	Helmholtz layer
HMDS	Hexamethyldisilazane
HMTA	Hexamethylenetetramine
HNA	(Hydrofluoric acid + nitric acid + acetic acid)-based silicon etchant
$H_2N(CH_2)_3Si(OCH_3)_3$	3-aminopropyl trimethoxysilane
$H_2N(CH_2)_3Si(OC_2H_5)_3$ or $C_9H_{23}NO_3Si$	3-(triethoxysilyl)propylamine
HNO_3	Nitric acid
H_2O	Water
H_2O_2	Hydrogen peroxide
$HOC(COOH)(CH_2COOH)_2$ or $C_6H_8O_7$	Citric acid
$HOCH_2CH_2OH$ or $C_2H_6O_2$	Ethylene glycol
$H(OCH_2CH_2)_nOH$	Polyethylene glycol
HOMO	Highest occupied molecular orbital
HPL	Hot-press lamination
H_3PO_4	Phosphoric acid
HRS	High resistance state
$HSCH_2CH_2NH_2 \cdot HCl$	Cysteamine hydrochloride (CH)
$HSCH_2CO_2CH_3$	Methyl thioglycoate
H_2SO_4	Sulfuric acid
HTL	Hole transport layer
HV	High vacuum
HW-CVD	Hot wire-chemical vapor deposition
Hz	Hertz
I-a-Si	Intrinsic amorphous silicon
IC	Integrated circuit
iCVD	Initiated chemical vapor deposition
ICP	Inductively-coupled plasma
ICP-CVD	Inductively-coupled plasma-chemical vapor deposition
ICP-RIE	Inductively-coupled plasma-reactive ion etching
IDEs	Interdigitated electrodes
IGBT	Insulated gate bipolar transistor
IGZO	Indium gallium zinc oxide
In	Indium
InAs	Indium arsenide

InGaAs	Indium gallium arsenide
InGaN	Indium gallium nitride
InGaP	Indium gallium phosphide
$InGaZnO_4$	Indium gallium zinc oxide (IGZO)
$In(NO_3)_3.xH_2O$	Indium (III) nitrate hydrate
In_2O_3	Indium oxide
I/O	Input/output
IPA	Isopropyl alcohol
ITO	Indium tin oxide
i-ZnO	Intrinsic zinc oxide
IZO	Indium zinc oxide
J	Joule
K	Kelvin
kB/s	Kilobyte/second
kb/s	Kilobit/second
kbyte	Kilobyte
K cell	Knudsen cell
kDa	Kilodalton
keV	Kiloelectron volt
kgf/cm^2	Kiologram force per cm^2
kHz	Kilohertz
km	Kilometer
$KMnO_4$	Potassium permanganate
KNO_3	Potassium nitrate
KOH	Potassium hydroxide
$k\Omega/sq$	kiloohm per square
KrF	Krypton fluoride
kS	Kilosiemen
kV	Kilovolt
L	Liter
LbL	Layer-by-layer
LbL-SA	Layer-by-layer self-assembly
LCD	Liquid crystal display
LCO	Lithium cobalt oxide ($LiCoO_2$)
LEAP	Laser-enabled advanced packaging
LED	Light emitting diode
LFP	Lithium iron phosphate ($LiFePO_4$)
Li	Lithium
LIB	Lithium ion battery
$LiBF_4$	Lithium tetrafluoroborate
LiCl	Lithium chloride
$LiClO_4$	Lithium perchlorate
$LiCoO_2$	Lithium cobalt oxide
LiF	Lithium fluoride
$LiFePO_4$	Lithium iron phosphate
LIFT	Laser-induced forward transfer
$LiMn_2O_4$	Lithium manganese oxide (LMO)
$LiNbO_3$	Lithium niobate
$LiNi_{0.8}Co_{0.2}O_2$	Lithium nickel cobalt oxide (LNCO)
$LiNi_xMn_yCo_{1-x-y}O_2$	Lithium nickel manganese cobalt oxide (NMC)

$LiPF_6$	Lithium hexafluorophosphate (lithium phosphorous fluoride)
LiPON	Lithium phosphorous oxynitride
LLO	Laser lift-off
LMO	Lithium manganese oxide ($LiMn_2O_4$)
lm W^{-1}	Lumen per watt
LNCO	Lithium nickel cobalt oxide ($LiNi_{0.8}Co_{0.2}O_2$)
LNMC	Lithium nickel manganese cobalt oxide ($LiNi_xMn_yCo_{1-x-y}O_2$)
LoD	Limit of detection
LPCVD	Low pressure chemical vapor deposition
LRS	Low resistance state
LT-GaN	Low-temperature GaN
LTPS	Low-temperature polysilicon
LUMO	Lowest unoccupied molecular orbital
M	Molar (unit)
mA	Milliampere
MacEtch	Metal-assisted chemical etching
mAh cm^{-1}	Milliamperehour per centimeter
mbar	Millibar
MBE	Molecular beam epitaxy
Mbyte	Megabyte
MCU	Microcontroller unit
2-ME	2-methoxyethanol
Me	Methyl group $-CH_3$
MEK	Methyl ethyl ketone, butanone, $CH_3CCH_2CH_3$ or C_4H_8O
MEMS	Microelectromechanical systems
MeO-TPD	N, N, N', N'-tetrakis(4-methoxyphenyl)-benzidine ($C_{40}H_{36}N_2O_4$)
mF	Millifarad
MFCs	Mass flow controllers
Mg	Magnesium
mg	Milligram
MgF_2	Magnesium fluoride
MgO	Magnesium oxide
MHz	Megahertz
MIM	Metal–insulator–metal
MIMCAPs	Metal/insulator/metal capacitors
min	Minute
MIPS	Million instructions per second
mJ	Millijoule
mL	Milliliter
mL min^{-1}	Milliliter per minute
mM	Millimolar
mm	Millimeter
mN	Millinewton
MnO_2	Manganese dioxide
Mn_2O_3	Manganese (III) oxide
Mo	Molybdenum

Mo–Cr	Molybdenum–chromium
MOCVD	Metal organic chemical vapor deposition
mol L^{-1}	Mole per liter
MoO_3	Molybdenum trioxide
MOS	Metal-oxide-semiconductor
MoS_2	Molybdenum disulfide
$MoSe_2$	Molybdenum diselenide
MOSFET	Metal-oxide-semiconductor field-effect transistor
MOVPE	Metal organic vapor phase epitaxy
MPa	Mega Pascal
MPB	Morphotropic phase boundary
MPE	Multiphoton emission (OLED)
MPS	3-Mercapto-1-propanesulfonic acid sodium salt ($C_3H_7NaO_3S_2$)
MQWs	Multiple quantum wells
ms	Millisecond
mTorr	Millitorr
MUX	Multiplexer
MV	Megavolt
mV	Millivolt
MVPE	Metalorganic vapor phase epitaxy (usually written as MOVPE)
mW	Milliwatt
MWCNT	Multi-walled carbon nanotube
(MWCNTs/PAH)$_n$	(Multi-walled carbon nanotubes/[{poly(allylamine hydrochloride)}]$_n$
mWh g^{-1}	Milliwatthour per gram
MX_2	A transition metal dichalcogenide (TMD) where M is a transition metal, e.g. Mo, W, etc, and X is a chalcogen atom, e.g. S, Se or Te
N	Newton, nitrogen
$NaBH_4$	Sodium borohydride or sodium tetrahydridoborate
NaCl	Sodium chloride
NaH_2PO_4	Sodium hydrogen phosphate
$NaNO_3$	Sodium nitrate or Chile saltpeter
NaOH	Sodium hydroxide
N-a-Si	N-type amorphous silicon
Na_2SO_4	Sodium sulfate
$Na_2S_2O_8$	Sodium persulfate
$Na_2WO_4.2H_2O$	Sodium tungstate dihydrate
Nb_2O_5	Niobium pentoxide
nc	Nanocrystalline (silicon)
Nd	Neodymium
Nd:YAG	Neodymium-doped yttrium–aluminum–garnet: garnets are silicate materials with the formula $\alpha_3\beta_2(SiO_4)_3$ where α and β are divalent and trivalent metals
nF cm^{-2}	Nanofarad per square centimeter
NG	Nanogenerator

–NH$_2$	Amino radical
NH$_3$	Ammonia
(NH$_4$)$_2$(MoO$_4$)	Ammonium molybdate (VI)
NHS	N-hydroxysuccinimide (C$_4$H$_5$NO$_3$)
(NH$_4$)$_2$S$_2$O$_8$	Ammonium persulfate
NI	Nano-imprinting
Ni	Nickel
NiCr	Nichrome
NiSO$_4$	Nickel (II) sulfate
nm	Nanometer
NMOSFET	N-channel metal-oxide-semiconductor field-effect transistor
NMP	N-methyl-2-pyrrolidone, C$_5$H$_9$NO
NMs	Nanomembranes
N$_2$O	Nitrous oxide
NO$_2$	Nitrogen dioxide
NPB	N, N'-bis(naphthalen-1-yl)-N, N'-bis(phenyl)benzidine (C$_{44}$H$_{32}$N$_2$)
NPD	N, N'-di(napthalen-1-yl)-N, N'-diphenyl-benzidine (C$_{44}$H$_{32}$N$_2$)
NPI	Nanoscale polyimide
NPs	Nanoparticles
ns	Nanosecond
NVM	Non-volatile memory
NWs	Nanowires
NZO	Ni-doped zinc oxide
O$_2$	Oxygen
OCV	Open circuit voltage
ODPA	Octadecyl-phosphonic acid (C$_{18}$H$_{39}$O$_3$P)
OECFET	Organic electrochemical FET
OFET	Organic field-effect transistor
Ohm/sq or Ω/sq	Ohm per square
OLED	Organic light emitting diode
OMBD	Organic molecular beam deposition
OMBE	Organic molecular beam epitaxy
OMVPE	Organometallic vapor phase epitaxy
OSC	Organic semiconductor compound
OTFT	Organic thin film transistor
OTR	Oxygen transmission rate
OTS	Octadecyltriethoxysilane
OVPD	Organic vapor phase deposition
P	Phosphorous
Pa	Pascal
pA	Picoampere
PAA	Poly(acrylic acid), (C$_3$H$_4$O$_2$)$_n$
PAC	Photoactive compound
PAc	Polyacetylene
PAH	{poly(allylamine hydrochloride)}, [CH$_2$CH (CH$_2$NH$_2$·HCl)]$_n$

PAH/PSSMA	{poly(allylamine hydrochloride)}/{poly(4-styrene-sulfonic acid-co-maleic acid} sodium salt; PAH is $[CH_2CH(CH_2NH_2.HCl)]_n$; PSSMA is $[CH_2CH(C_6H_4SO_3R)]_x[CH(CO_2R)CH(CO_2R)]_y$, R = H or Na
PANI	Polyaniline $(C_6H_7N)_x$
PAR	Polyarylate
Parylene C	Poly(2-chloro-para-xylylene) (PCPX), $C_{16}H_{14}Cl_2$
Pa s	Pascal second
P-a-SiC	P-type amorphous silicon carbide
Pb	Plumbum (lead)
$(1-x)[Pb(Mg_{1/3}Nb_{2/3})O_3].x[PbTiO_3]$	Lead magnesium niobate–lead titanate (PMN–PT)
PbO	Lead (II) oxide
pBTTT-C14	Poly(2,5-bis(3-tetradecylthiophen-2-yl)thienol [3,2-b]thiophene), $(C_{42}H_{62}S_4)_n$
$Pb[Zr_{(x)}Ti_{(1-x)}]O_3$	Lead zirconate titanate (PZT)
PC	Polycarbonate, personal computer
pc	Polycrystalline (silicon)
PCBM	Phenyl-C61-butyric acid methyl ester $(C_{72}H_{14}O_2)$
PcZ	Poly {9-(1-octylonoyl)−9H-carbazole-2,7-diyl}
Pd	Palladium
$PdCl_2$	Palladium (II) chloride
PDMS	Polydimethylsiloxane $(C_2H_6OSi)_n$
PDPP–PD	Poly(diketopyrrolopyrrole–phenylenediamine)
PDQT	Poly[3,6-bis(40-dodecyl[2,20]bithiophenyl-5-yl) −2,5-bis(2-hexyldecyl)−2,5-dihy-dropyrrolo[3,4-c]pyrrole-1,4-dione]
PEALD	Plasma-enhanced atomic layer deposition
PECVD	Plasma-enhanced chemical vapor deposition
PEDOT	Poly(3,4-ethylenedioxythiophene)–
PEDOT: PSS	Poly(3,4-ethylenedioxythiophene)–poly(styrenesulfonate)
PEEK	Polyether ether ketone, $[OC_6H_4OC_6H_4COC_6H_4]_n$
PEI	Polyetherimide, $(C_{37}H_{24}O_6N_2)_n$
PEN	Polyethylene naphthalate, $(C_{14}H_{10}O_4)_n$
PEO	Polyethylene oxide, $C_{2n}H_{4n+2}O_{n+1}$, $(-CH_2CH_2O)_n$, PECVD oxide
PEP	Poly(ethylene-alt-propylene)
PES	Polyethersulfone $(C_{12}H_8O_3S)_n$
PET	Polyethylene terephthalate, $(C_{10}H_8O_4)_n$
PETS	Phenethyltrichlorosilane or trichloro(phenethyl)silane, $C_6H_5CH_2CH_2SiCl_3$
pF	Picofarad
P6FBEu	Eu-complexed benzoate, $Eu-C_7H_5O_2$ complex
PFBT	Pentafluorobenzenethiol or 2,3,4,5,6-pentafluoro-thiophenol (C_6F_5SH)
PGE	Polymer gel electrolyte
PGMEA	Propylene glycol monomethyl ether acetate, $CH_3CO_2CH(CH_3)CH_2OCH_3$ or $C_6H_{12}O_3$

PH$_3$ — Phosphine (IUAC name: phosphane)
P3HT — Poly(3-hexylthiophene-2,5-diyl), (C$_{10}$H$_{14}$S)$_n$
PI — Polyimide, C$_{35}$H$_{28}$N$_2$O$_7$
PIC — Programmable intelligent computer
P-I-N — P type–intrinsic–N-type
PI-PR — Polyimide photoresist
PMGI — Polydimethylglutarimide
PMMA — Poly(methyl methacrylate)
PMN–PT — Lead magnesium niobate–lead titanate, $(1-x)$[Pb(Mg$_{1/3}$Nb$_{2/3}$)O$_3$].x[PbTiO$_3$]
0.72 PMN–0.28 PT — 0.72 lead magnesium niobate–0.28 lead titanate
PMOS TFT — P-channel metal-oxide semiconductor thin film transistor
P$_2$O$_5$ — Phosphorous pentoxide
POCl$_3$ — Phosphorous oxychloride
polySi — Polysilicon
[poly(ViEtIm)(Tf2N)] — poly(1-vinyl-3-methylimidazolium) bis(trifluoromethanesulfonimide)
PP — Polypropylene, [CH$_2$CH(CH$_3$)]$_n$ or (C$_3$H$_6$)$_n$, M_w ~ 250 000 (average)
ppb — Parts per billion
ppi — Pixels per inch
ppm — Parts per million
Ppy — Polypyrrole, (C$_4$H$_5$N)$_n$
PROM — Programmable read-only memory
PS — Polystyrene, (C$_8$H$_8$)$_n$
PS-b-PB — Polystyrene-block-polybutadiene, [CH$_2$CH(C$_6$H$_5$)]$_x$(CH$_2$CH = CHCH$_2$)$_y$
PS-b-PLA — Polystyrene-block-polyactide, (C$_8$H$_8$)$_n$(C$_3$H$_4$O$_2$)$_n$
PS-b-PMMA — Polystyrene-block-poly(methyl methacrylate), (C$_8$H$_8$)$_n$(C$_5$O$_2$H$_8$)$_n$
PSG — Phosphosilicate glass
PSI — Pounds per square inch
PSR — Pressure sensitive rubber
PSS — Poly(styrenesulfonate), C$_8$H$_8$O$_3$S
PSSH — Poly(4-styrenesulfonic acid) (C$_8$H$_8$O$_3$S)$_n$
PSSMA — {Poly(4-styrenesulfonic acid-co-maleic acid} sodium salt, [CH$_2$CH(C$_6$H$_4$SO$_3$R)]$_x$[CH(CO$_2$R)CH(CO$_2$R)]$_y$, R = H or Na
Pt — Platinum
PTAA — Poly(triaryl amine), poly[bis(4-phenyl)(2,4,6-trimethylphenyl)amine, (C$_{21}$H$_{19}$N)$_n$
PTFE — Polytetrafluoroethylene, (C$_2$F$_4$)$_n$
PTFE AF — PTFE amorphous fluoroplastic, poly[4,5-difluoro-2,2-bis(trifluoromethyl)–1,3-dioxole-co-tetrafluoroethylene]
PTrFE — Polytrifluoroethylene, –(C$_2$HF$_3$)$_n$–
PTSA — p-toluenesulfonic acid
PU — Polyurethane

PVA	Polyvinylalcohol
PVC	Polyvinyl chloride
PVDF	Polyvinylidene fluoride, $(C_2H_2F_2)_n$
PVDF-HFP	Poly(vinylidenefluoride-co-hexafluoropropene), $(-CH_2CF_2-)_x[-CF_2CF(CF_3)-]_y$
P(VDF-TrFE)	Poly[(vinylidenefluoride-co-trifluoroethylene]
PVOH	Polyvinylalcohol, $(C_2H_4O)_x$
PVP	Poly(4-vinylphenol), polyvinylpyrrolidone, $[CH_2CH(C_6H_4OH)]_n$ or $(C_8H_8O)_n$, $M_w \sim 25\,000$ (average)
PVP, P4VP	Poly(4-vinylpyridine), $(C_7H_7N)_n$, $M_w \sim 60\,000$ (average)
PZT	Lead zirconate titanate, $Pb[Zr_{(x)}Ti_{(1-x)}]O_3$
QD	Quantum dot
QE	Quasi-epitaxy
Q-factor	Quality factor
QW	Quantum well
RAM	Random-access memory
RER	Resist edge bead remover
RF	Radio frequency
RFCPU	Radio frequency central processing unit
RFIC	Radio frequency integrated circuit
RFID	Radio frequency identification integrated circuit
RFID IC	Radio frequency identification
RF-PECVD	Radio frequency chemical vapor deposition
rGO	Reduced graphene oxide
rGO–CH/Au	Reduced graphene oxide–cysteamine hydrochloride/Gold
RHEED	Reflection high-energy electron diffraction
RIE	Reactive ion etching
R–L–C	Resistance–inductance–capacitance
RMS	Root-mean-square
ROM	Read-only memory
RPM	Revolutions per minute
R2P	Roll-to-plate
RR	Read range
RRAM	Resistive random access memory
R2R	Roll-to-roll
RTA	Rapid thermal annealing
RTC	Real-time clock
RTP	Rapid thermal processing
S	Siemen, sulfur
s	Second
S11	Reflection coefficient or return loss of antenna
SAM	Self-assembled monolayer
SAW	Surface acoustic wave
Sb	Antimony (Stibium)
Sccm	Standard cubic centimeters per minute
S cm^{-1}	Siemen per cm

SDBS	Sodium dodecylbenzenesulfonate, $CH_3(CH_2)_{11}C_6H_4SO_3Na$
SDS	Sodium dodecyl sulfate
Se	Selenium
SEM	Scanning electron microscope
SF_6	Sulfur hexafluoride
Si	Silicon
SiF_4	Silicon tetrafluoride
SiGe	Silicon–germanium
SiH_4	Silane
Si_5H_{10}	CPS: cyclopentasilane
SiN	Silicon nitride
Si_3N_4	Silicon nitride
Si-NS	Silicon nanostructure
SiN_x	Silicon nitride (x unknown)
SiO	Silicon oxide
SiO_2	Silicon dioxide
$SiON_x$	Silicon oxynitride (x unknown)
SiO_x	Silicon oxide (x unknown)
SL	Source line
SLS	Sequential laser solidification
Sn	Stannum (tin)
SNAP	Superlattice nanowire pattern (transfer technique)
$SnCl_2$	Tin (II) chloride
SnO_2	Tin oxide
S/N ratio	Signal-to-noise ratio
SOI	Silicon-on-insulator (wafer)
SOM	Silicon-on-metal
SPC	Solid phase crystallization
SPI	Serial peripheral interface
SPOP	Stress-peel-off-process
SRAM	Static random access memory
SS	Subthreshold swing
SU-8	Epoxy-based, UV sensitive, negative photoresist
SUFTLA	Surface-free technology by laser annealing
SUV	Simultaneous thermal and UV
SWCNT	Single-walled carbon nanotube
TaN	Tantalum nitride
Ta_2O_5	Tantalum pentoxide
TBPO	Tert-butyl peroxide, $(CH_3)_3COOC(CH_3)_3$ or $C_8H_{18}O_2$
2T1C	2-transistor, 1-capacitor (cell)
TCNQ	7,7,8,8-tetracyanoquinodimethane, $C_{12}H_4N_4$
TCO	Transparent conducting oxide
Te	Tellurium
TEG	Triboelectric generator
TENG	Triboelectric nanogenerator
TFT	Thin film transistor
Th_4	Quaterthiophene ($C_{16}H_{10}S_4$)
THF	Tetrahydrofuran, C_4H_8O

Ti	Titanium
TiN	Titanium nitride
TiO_2	Titanium dioxide
$Ti[OCH(CH_3)_2]_4$ or $C_{12}H_{28}O_4Ti$	Titanium (IV) isopropoxide
TiO_x	Titanium oxide (x unknown)
TIPS-Pentacene	6,13-bis (triisopropylsilylethinyl)pentacene, $C_{44}H_{54}Si_2$
1T–1M	One transistor–one memristor
TMA	Trimethylaluminum, $C_6H_{18}Al_2$
TMAH	Tetra methyl ammonium hydroxide, $C_4H_{13}NO$
TMDs	Transition metal dichalcogenides
*tm*SLADT	Thermo-mechanical selective laser-assisted die transfer
TMSC	Trimethylsilyl cellulose
TRT	Thermal release tape
TTC	Through-thickness crack
TTIP	Titanium tetraisopropoxide, $Ti[OCH(CH_3)_2]_4$ or $C_{12}H_{28}O_4Ti$
UHF	Ultra-high frequency
UHF RFCPU	Ultra-high frequency radio frequency central processing unit
UHV	Ultra-high vacuum
UTMFs	Ultra thin metal films
UV	Ultraviolet
UVA	Ultraviolet adhesive
UVO	UV/ozone
UVR	UV-curable resin
V	Volt
V-CNTs	Vertically aligned carbon nanotubes
V3D3	1,3,5-trimethyl-1,3,5-trivinyl cyclotrisiloxane, $C_9H_{18}O_3Si_3$
VLS	Vapor–liquid–solid
vol%	Volume percent = (volume of solute/volume of solution) × 100%
V-SWCNTs	Vertical single-walled carbon nanotubes
v/v	Volume/volume
W	Watt, tungsten (Wolfram)
WL	Word line
WO_3	Tungsten trioxide
WORM	Write once read many times
WR/ER	Write (current)/erase (current)
WS_2	Tungsten disulfide
WSe_2	Tungsten diselenide
WSNs	Wireless sensor networks
wt%	Weight percent = percentage of solute in the solution = {mass of solute/mass of solution (solute + solvent)} × 100%
WVTR	Water vapor transmission rate
w/w	Weight/weight
XeCl	Xenon monochloride (laser)

XeF$_2$	Xenon difluoride
YAG	Yttrium–aluminum–garnet
Y$_2$O$_3$	Yttrium (III) oxide
Zn	Zinc
Zn(CH$_3$COO)$_2$.2H$_2$O	Zinc acetate dihydrate
Zn$_2$Ga$_2$O$_5$	Gallium zinc oxide (GZO)
Zn(NO$_3$)$_2$, 6H$_2$O	Zinc nitrate hexahydrate
ZnO	Zinc oxide
Zn(OH)$_2$	Zinc hydroxide
ZrO$_2$	Zirconium oxide
ZnS	Zinc sulfide

Greek letters

α-NPD	N, N'-di(napthalen-1-yl)-N,N'-diphenyl-benzidine (C$_{44}$H$_{32}$N$_2$)
α-terpineol	C$_{10}$H$_{18}$O
μA	Microampere
μAh cm^{-2}	Microampere-hour per square centimeter
μc	Microcrystalline
μCP	Micro-contact printing
μF	Microfarad
μg	Microgram
μm	Micrometer
μS	Microsiemen
μs	Microsecond
μW	Microwatt
π	Pi bond: A type of chemical bond formed by lateral or sideways overlapping of atomic orbitals
σ	Sigma bond: A type of chemical bond formed by axial or end-to-end overlapping of atomic orbitals
Ω	Ohm
Ω/sq	Ohm per square

Mathematical symbols and general notation

Roman alphabet

A	Area, ratio of out-of-plane deformation to thickness of the wafer
a, b, c	Lattice constants
A_o	Initial cross-sectional area, cross-sectional area of the film, a parameter dependent on the film material and the environment
$A/2$	Amplitude of bucking profile
A_c	Critical value of parameter A at which the circular wafer along with the film bifurcates and becomes geometrically unstable
A_f, A_s	Areas of the film and substrate
$2b$	Delamination width
C	Constant, capacitance
c	Distance, homogeneous strain, velocity of light
C_0	Impurity concentration at a distance $x = 0$, i.e. on the surface of the semiconductor (surface concentration)
$C(x, t)$	Impurity concentration at a distance x from the surface of a semiconductor at time t
$C(x, t = 0)$	Impurity concentration in a semiconductor at distance x from the surface at zero time
$C(x = \infty, t)$	Impurity concentration in a semiconductor at infinite distance from the surface at time t
D	Diameter of the substrate, the flexural rigidity of the ribbon, diffusion constant/diffusion coefficient/diffusivity of impurity
D_1	Diffusion constant at predeposition temperature
D_2	Diffusion constant at drive-in temperature
d	Atomic lattice spacing, thickness of beam, distance between plates, distance between the two electrodes
dA	Elementary area
$d_{Capping}$	Thickness of the capping layer
d_f	Film thickness
dL	Elementary length
ds	Element of length
$d_s, d_{Substrate}$	Substrate thickness
$d\theta$	Elementary angle
$dx_1 dy_1$	Areal element
dz	Element of length in the Z-direction
$\mathbf{E}$	Electric field vector
E	Plane-strain Young's modulus of the ribbon material, Young's modulus of a material, Young's modulus of SiN_x, applied electric field
E^*	Biaxial modulus of the ribbon material
e	Strain in an unrestricted stress-free material, shortening of the beam
E_f^*	Biaxial Young's modulus of the film
e_f	Strain in an unrestricted, stress-free film material
E_{film}	Elastic modulus of the film

$e_f(T)$	Value of e_f at temperature T
$E(k)$	Complete elliptic integral of the first kind
E_s^*	Biaxial Young's modulus of the substrate
e_s	Strain in an unrestricted, stress-free substrate material
$e_s(T)$	Value of e_s at temperature T
$E_{substrate}$, E_s	Elastic modulus of the substrate
E_{to}	Turn-on electric field
$\mathbf{F}$	Force vector
F	Axial compressive force
FF	Fill factor of a solar cell
F_f	Force acting on the film
f_{max}	Maximum oscillation frequency (of the TFT)
F_s	Force acting on the substrate
f_r	Resonance frequency of the microstrip patch antenna
f_T	Cut-off frequency (of the TFT)
$f(x)$	Function of x
G_0	Conductance in relaxed state
G	Shear modulus or modulus of rigidity of a material, energy release rate
G_1	Conductance in bent state
G_a	Shear modulus of the adhesive layer
g	(Conductance in bent state − conductance in relaxed state)/conductance in relaxed state
$g(\alpha, \beta)$	A function of the Dundurs' parameters α, β
$G_{Cracking}$	Energy liberation rate by cracking
$G_{Cracking}^{Critical}$	Critical fracture toughness
$G_{Delamination}^{Critical}$	Critical delamination toughness
$G_{Delamination}$	Energy release rate by delamination
g_m	Transconductance
g_m/W	Transconductance per unit gate width (normalized transconductance)
G_{Total}	Total energy release rate for steady-state delamination of the film followed by cracking
H	Height of an object
h	Thickness of the ribbon film
h_{film}, h_{Film}	Thickness of the film
$h_{substrate}$, $h_{Substrate}$	Thickness of the substrate
I, i	Electric current
I_{DS}	Drain–source current
$I_{DS(sat)}$	Saturation drain–source current
I_{off}	Off current
I_{on}	On current
I_{on}/I_{off}, I_{ON}/I_{OFF}	On current/off current
J	Flux of material
J_{SC}	Short-circuit current density
K	Stress intensity factor, Complex stress intensity factor $K_I + iK_{II}$
k	$\sin \theta/2$, curvature
K_I	Stress intensity factor
KI	Stress intensity factor for mode I

KIC	Critical stress intensity factor for mode I
K_{Ic}	Fracture toughness
KII	Stress intensity factor for mode II
KIIC	Critical stress intensity factor for mode II
KIII	Stress intensity factor for mode III
KIIIC	Critical stress intensity factor for mode III
$K(k)$	Complete elliptic integral of the second kind
L	Original length of the beam, initial length of the substrate, initial end-to-end substrate length, length of FET channel, length of the microstrip patch antenna
$L + \Delta L$	Length of the substrate after stretching
L_0	Original length of the Cu strip
L_1	Half of buckling wavelength
L_2	Half of post-relaxation sum of lengths of activated and inactivated regions
$L_{y=0}$	Length of the section of beam at position $y = 0$
$L_{y=y}$	Length of the section of beam at an arbitrary position y
L_f	Length of the film, length of an object
M	Internal bending moment of the structure about any axis, moment
M_f	Moment of the force F_f acting on the film
M_n	Number average molecular weight of a substance
M_s	Moment of the F_s acting on the substrate
M_V, M_v	Viscosity average molecular weight of a substance
M_W, M_w	Weight average molecular weight of a substance
M_{x1}	Uniformly distributed bending moment during the bending of a ribbon
n	A positive integer, a fatigue parameter
N_{x1}	Membrane force in the thin film
N_W, N_w	Number average molecular weight of a substance
P	Force $= F_f = -F_s$
$\mathbf{p}$	Moment of the induced dipole
P_{in}	Incident power on a solar cell
P, Q	Constants in Fowler–Nordheim equation
Q	Electric charge
q	Electric charge
R	Radius of curvature of the workpiece, the final resistance of the copper strip after straining, radius of curvature of an object, Bending radius
R_0	Initial resistance of the copper film, original resistance
R/R_0	Ratio of resistances of a material: final resistance/initial resistance
r	Ratio between critical cracking toughness and critical delamination toughness, rate of resistance change ($\Delta R/R_0$)
R_f	Final resistance of the sensor after gas exposure
R_i	Initial resistance of the sensor before exposure to gas
$R_{Nominal}$	Nominal bend radius at the center ($L/2$) of the length of the substrate
S	Scan rate in the C–V curve, sensitivity of gas sensor
s	Arc length
S11	Reflection coefficient of antenna, return loss
s, p, d, f	Atomic orbitals
SS	Subthreshold swing, subthreshold slope
$(S_{x,Centroid}, S_{y,Centroid})$	First moment of area about centroidal axis
(S_x, S_y)	First moment of area

T	Temperature
$\mathbf{T}$	A unit tangent vector
t	Time
t_1	Time of impurity predeposition
t_2	Time of impurity drive-in
t_b	Distance of a point along Z-direction from neutral line of strain
T_d	Temperature of deposition of the film
T_g	Glass transition temperature of a material
T_r	Room temperature
U	Total energy of the thin film/substrate system
u	A symbol for $(2\pi x_1/L_1)$
u_1	Small displacement of the middle point of the beam during bending (u_1 is in X-direction)
u_A	Axial displacement of point A
U_b	Bending energy for buckling of the thin film
u_B	Axial displacement of point B
U_m	Membrane energy in the film
U_s	Substrate energy
V	Voltage, potential difference
V_D	Drain voltage
V_DD	Positive supply voltage connected to the drain terminal of an FET device
V_DS	Drain–source voltage
V_EPD	Electric potential difference
V_FB	Flat-band voltage
V_G	Gate voltage
V_GS	Gate–source voltage
V_oc, V_OC or OCV	Open circuit voltage
V_Th	Threshold voltage of a device
W	Width of an object, width of the beam, width of the FET channel
w	Width of the composite film/substrate structure, small deflection of the ribbon in the Z-direction, deflection of the substrate in the vertical Z-direction
W_0	Original width of copper strip
w_0	Value of w at $x = L/2$, the center of the substrate
w_1	Out-of plane displacement corresponding to x_1 distance traversed along the X-direction, out-of-plane displacement between L_1 and L_2
W_Act	Width of the activated site
W_Inact	Width of the inactivated site
W/L ratio	Ratio of channel width to channel length of an FET device (channel aspect ratio)
x	Distance in X-direction
x_1	Distance traversed along the X-direction
x_Centroid	x co-ordinate of centroid
$(x_\mathrm{Centroid}, y_\mathrm{Centroid})$	(x, y) co-ordinates of centroid
Y	Young's modulus of a material
Y^*	Biaxial modulus of a material
y	Distance in Y-direction
y, y'	Initial and final ordinates of centroid

y_{Centroid}	y co-ordinate of centroid
Y_{f}	Young's modulus of the material of the film
Y_{f}^{*}	Biaxial strain modulus of the film
$y_{\text{f}}, y_{\text{s}}$	Ordinates of the centroids of the film and the substrate
Y_{s}	Young's modulus of the material of the substrate
Y_{s}^{*}	Biaxial strain modulus of the substrate
z	Thickness-through co-ordinate with arbitrarily located origin, the distance of fibers of stress σ_{B} from the chosen axis, distance along Z-axis

Greek letters

α	Coefficient of thermal expansion of a material, crack length, polarizability of the particle with respect to the fluid medium
α, β	Dundurs' parameters
α_{f}	CTE of the film
α_{s}	CTE of the substrate
β	Electric field enhancement factor
δ	Ratio = Stress (σ)/critical stress for bifurcation $(\sigma_{\text{c}}) = A/A_{\text{c}}$
$\Delta L, \Delta L_0$	Change in length L, change in length L_0
ΔG	Conductance in bent state − conductance in relaxed state
Δh_{Film}	Change in film thickness h_{Film}
ΔR	Change in resistance
$\Delta R/R_0$	Change in resistance/original resistance = relative change in resistance
ΔT	Change or difference in temperature
ΔV_{T}	Threshold voltage shift
ΔW_0	Change in width W_0
Δx	Distance AB between points A and B
ε	Strain produced, dielectric constant of a material, dielectric constant of the PDMS substrate
ε_0	Strain at the position $z = 0$, permittivity of air or free space (vacuum)
$\varepsilon_0 + \varepsilon_{\text{f}}$	Thermally-induced strain in the film
$\varepsilon_0 + \varepsilon_{\text{s}}$	Thermally-induced strain in the substrate
ε_{11}	Membrane strain
ε_{I}	First component of strain
ε_{II}	Second component of strain
ε_{III}	Third component of strain
ε_{A}	Strain in the axial direction
$(\varepsilon_{\text{A}})_{\text{Film}}, \varepsilon_{\text{Af}}$	Axial strain in the film
$\varepsilon_{\text{Assembly}}$	Strain of the whole assembly
$(\varepsilon_{\text{A}})_{\text{Substrate}}, \varepsilon_{\text{As}}$	Axial strain in the substrate
ε_{B}	Strain in the bending direction
ε_{b}	Bending strain
$(\varepsilon_{\text{B}})_{\text{Film}}, \varepsilon_{\text{Bf}}$	Bending strain in the film
ε_{bi}	Built-in strain
$(\varepsilon_{\text{B}})_{\text{Substrate}}, \varepsilon_{\text{Bs}}$	Bending strain in the substrate
$\varepsilon_{\text{c}}^{\text{Obs}}$	Observed critical strain
$\varepsilon_{\text{Critical}}, \varepsilon_{\text{c}}$	Critical strain at which the component fails, critical strain for buckling
$\varepsilon_{\text{External}}$	Externally applied strain (strain produced by external forces)

$\varepsilon_{Failure}$	Failure strain
ε_{Film}, ε_f	Strain in the film
$\varepsilon(T)$	Strain at temperature T
$\varepsilon_f(T)$	Strain in the film at temperature T
$\varepsilon_f(T_d)$	Strain in the film at the deposition temperature T_d
$\varepsilon_f(T_r)$	Strain in the film at room temperature T_r
ε_{Green}	Green strain
ε_{g, SiN_x}	Strain produced in SiN_x film during or after fabrication
$\varepsilon_{Infinitesimal}$	Infinitesimal strain
$\varepsilon_{Internal, SiN_x}$	Internal strain in SiN_x
ε_{ITO}	Strain in the ITO film
$\varepsilon_{maximum}$, ε_{max}	Maximum strain
$\varepsilon_{Nominal}$	Nominal bending strain
ε_{Pre}	Pre-strain produced in the substrate
ε_r	Dielectric constant or relative permittivity of a material
$\varepsilon_s(T)$	Strain in the substrate at temperature T
$\varepsilon_s(T_d)$	Strain in the substrate at the film deposition temperature T_d
$\varepsilon_s(T_r)$	Strain in the substrate at room temperature T_r
$\varepsilon_{Substrate}$	Strain in the substrate
$\varepsilon_{Surface}$	Surface strain
ε_{Top}	Strain on the top surface of the structure
ε_X	Linear strain along X-axis
ε_x	Strain in the X-direction
$\varepsilon_x(y)$	Strain in the X-direction at a distance y from the neutral axis
ε_y	Strain along the Y-direction
ε_Z	Bending strain along Z-axis
ε_z	Strain along the Z-direction
η	Ratio of film thickness (d_f)/substrate thickness (d_s), power conversion efficiency of a solar cell
θ	Angle, bending angle, diffraction cone angle
κ	Curvature
λ	Wavelength of x-ray/neutron wave, line tension in thin film, wavelength of the buckled structure, stress contraction ratio (e/L)
μ_0	Linear mobility before application of strain, the mobility in flat state of TFT
μ	Linear mobility after application of strain, mobility of carriers in a TFT
$\mu_{FET, HV}$	Mobility in FET at high voltage = mobility at $|V_{GS} - V_{FB}| = 60$ V
$\mu_{FET, LV}$	Mobility in FET at low voltage = mobility at $|V_{GS} - V_{FB}| = 1$ V
$\mu(R)$	Mobility when TFT is bent
μ_{sat}	Saturation mobility
ν	Poisson's ratio of a material, Poisson's ratio of the ribbon, crack growth velocity
ν_a	Poisson's ratio of the adhesive layer
ν_f	Poisson's ratio of film material
ν_s	Poisson's ratio of substrate material
π	Ratio of circumference of a circle to its diameter, a type of chemical bond
ρ	Radius of curvature, resistivity of film material
$\rho_{Critical}$	Critical radius of curvature
σ	Stress, interfacial peeling stress, applied macroscale stress, a type of chemical bond, charge density

σ_A	Stress in the axial direction
σ_B	Stress in the bending direction
σ_{bi}	Built-in stress
σ_c	Critical stress for bifurcation, tensile strength of the interface
σ_{Crack}	Cracking stress
σ_c^{Si}	Tensile strength of the Si film
σ_{Film}, σ_f	Stress in the film
$\sigma_{Internal, SiN_x}$	Internal stress in the SiN_x film
σ_{max}	Maximum peeling stress
$\sigma_{Substrate}, \sigma_s$	Stress in the substrate
σ_x	Stress in the direction along X-axis
σ_y	Stress in the direction along Y-axis
$\sigma(y)$	Stress at a distance y from the neutral axis
τ	Interfacial shear stress
τ_c	Shear strength of the interface
τ_{max}	Maximum interfacial shear stress
τ_{max}^{Beam}	Maximum interfacial shear stress for beam structure
τ_{max}^{Plate}	Maximum interfacial shear stress for plate structure
ϕ	Work function of a material
χ	Amount of low Young's modulus material/amount of high Young's modulus material, Ratio Y_f^*/Y_s^*
∇	Laplacian operator
$\nabla \mathbf{E}$	Electric field gradient

IOP Publishing

Flexible Electronics, Volume 2
Thin-film transistors
Vinod Kumar Khanna

Chapter 1

Amorphous Si TFT

The differences in fabrication between TFT and conventional FET devices are explained. TFT configurations and structures are introduced. Passivation of both sides of polyimide substrate with SiN_x film is essential for device processing as a protective layer for polyimide from chemical attack by reagents used, and as a barrier layer for polyimide out-diffusion. It also acts as the adhesion layer for construction of TFT structures and participates in the operation of the finished device as a gate dielectric. TFTs fabricated on 25 μm and 51 μm thick polyimide foils could be bent up to a radius of 0.5 mm without any impairment of TFT parameters (Gleskova *et al* 1998, 1999, 2001, Gleskova and Wagner 1999). Influence of cylindrical and spherical deformations on TFT characteristics are described (Gleskova *et al* 2000, 2002, Hsu *et al* 2002, 2004). TFTs on stainless steel substrates exhibit not only good flexing properties but also a high degree of unbreakability, as demonstrated by their continuing to work after falling from a height of four floors (Theiss and Wagner 1996, Hong *et al* 2006). TFTs on optically clear plastic substrates with temperature withstanding capability of up to 280 °C show stability like those on glass substrates (Long *et al* 2004, 2006a, 2006b, Cherenack *et al* 2007). Emphasis is laid on controlling the mechanical stress in surface passivation and TFT structural layers to achieve TFT reliability. Higher temperatures used during TFT processing are found to be conducive to reduction of threshold voltage shifts of TFTs, thereby providing operational stability of AMOLEDs (Hekmatshoar *et al* 2008a, 2008b).

1.1 Thin-film transistor (TFT)

A thin-film transistor (TFT) is a kind of field-effect transistor in which all the constituent strata, viz, the semiconducting, conducting and insulating layers, are deposited in the form of thin films on a substrate (Weimer 1962, le Comber *et al* 1979). The advantage gained by this fabrication method is that the transistor can be formed on a transparent glass or plastic substrate unlike the conventional transistor which is built on a silicon substrate. TFT is to the flexible electronics industry what MOSFET is to the silicon IC industry. A TFT is the flexible electronics counterpart of MOSFET in semiconductors ICs (figure 1.1). Learning about, and mastering, the concepts of TFT is as important in flexible electronics as knowing the basics of MOSFET in semiconductor electronics.

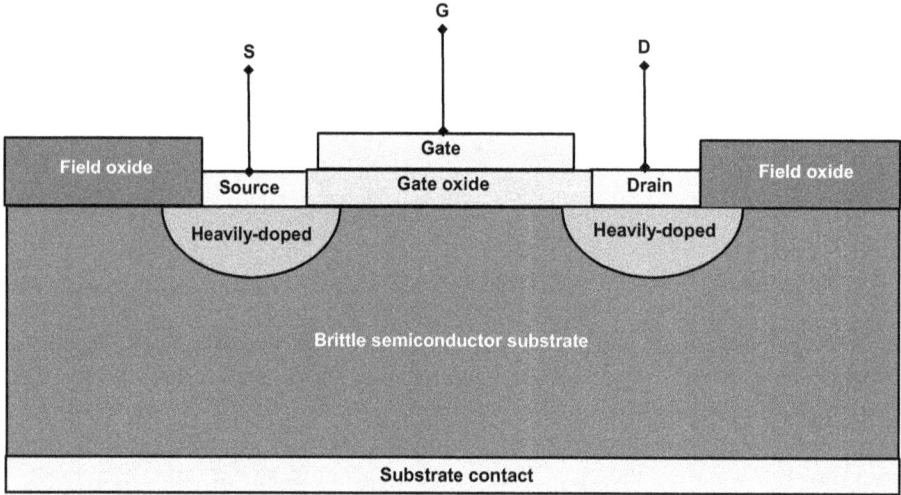

(a)

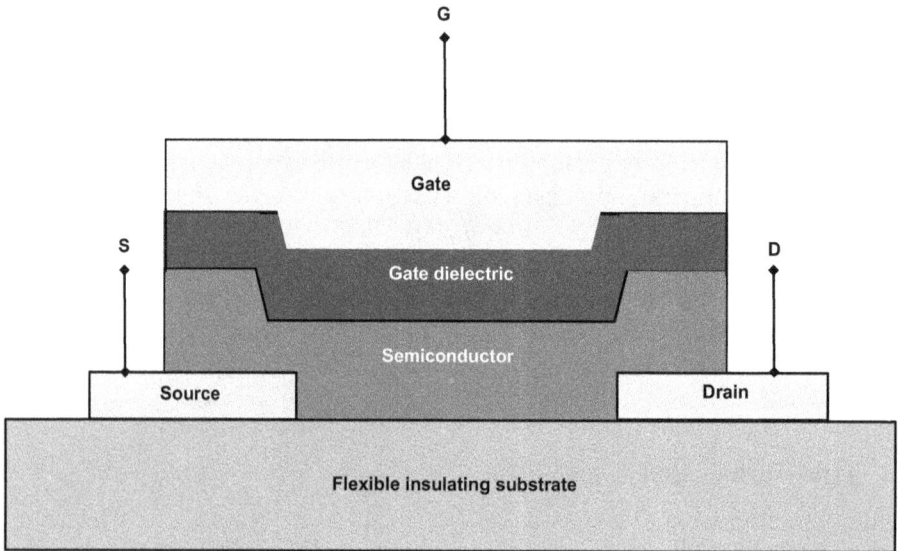

(b)

Figure 1.1. Comparison of (a) FET of semiconductor IC industry with (b) TFT of flexible electronics.

A cardinal application of TFTs lies in flat-screen liquid crystal displays (LCDs) used in computers and smart phones, which have an embedded TFT for each pixel (Brody *et al* 1973). The TFT can be quickly switched on and off enabling a smooth transition, e.g. when the cursor is moved across the TFT LCD screen with the help of the mouse, the cursor responds quickly enough in accordance with the movement of the mouse, whereas if a TFT is not used, the cursor temporarily disappears until the display is able to catch up. As the TFTs are entrenched within the screen, cross-talk among the pixels is reduced and the picture becomes more stable. Displays furnished with TFTs are called active matrix displays, distinct from those without TFTs, which are known as passive matrix displays.

1.2 TFT configurations and structures

There are two TFT configurations (Morita *et al* 2016): etch-stop (ES) and back-channel etch (BCE) (figure 1.2). In the ES TFT, an etch-stop layer protects the semiconductor thin film providing stable characteristics. The BCE TFT is not protected by such a layer but requires a smaller number of photolithographic steps. This is an advantageous feature from a manufacturability viewpoint leading to its wide adoption. Additionally, shorter channel lengths can be made. With resultant miniaturization, the parasitic capacitance and, therefore, delay in signal is reduced. But the susceptibility of BCE TFT to contamination necessitates stricter and uncompromising monitoring of the process.

Further, there are four TFT structures according to the relative positions of the constituent layers (Correia *et al* 2016). When the drain/source and gate are on opposite sides with respect to the semiconductor, the TFT structure is said to be staggered. When they are on the same side with respect to the semiconductor, the structure is known as coplanar. Looking at the location of the gate, the staggered or coplanar structure is further sub-classified either as top gate (figure 1.3) or bottom gate (figure 1.4). The staggered, bottom gate structure is the preferred choice when the dielectric layer requires high temperature processing. The coplanar, top gate structure is selected with the high temperature semiconductor, e.g. poly Si. Bottom gates are sometimes called inverted gates (Pappas *et al* 2009).

1.3 a-Si TFTs on polyimide foil substrates

1.3.1 TFTs on 51 μm and 25 μm thick polyimide foils

Gleskova *et al* (1998) reported the fabrication of high-performance a-Si TFTs on 51 μm thick polyimide foil substrates. During TFT fabrication, the gate SiN_x as well as undoped and N^+-doped a-Si layers are deposited by RF PECVD at a substrate temperature of 150 °C. The TFT has a BCE structure. Also, it has an inverted staggered gate. Here, the terms BCE, and 'inverted staggered gate' may be understood in accordance with discussions in section 1.2.

In a further course of work, Gleskova and Wagner (1999) made TFTs on 25 μm thick polyimide foil (Kapton E) substrates. The glass transition temperature T_g

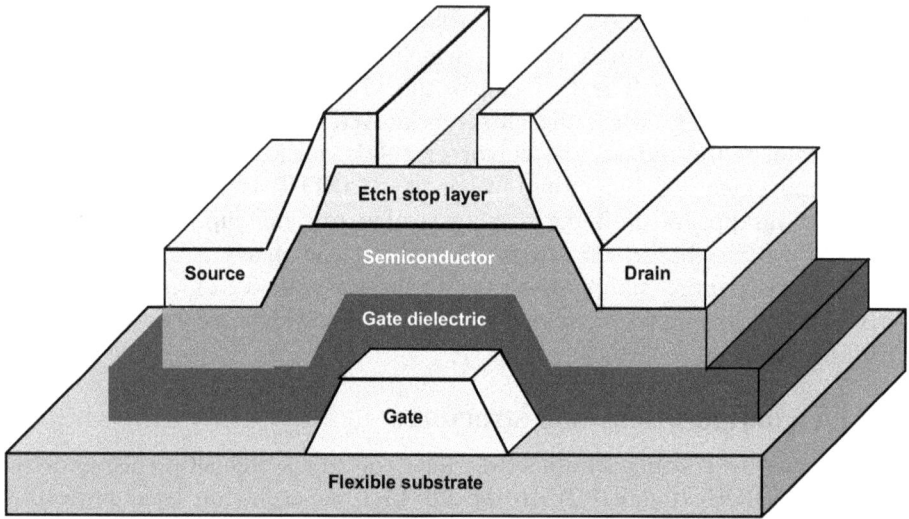

(a)

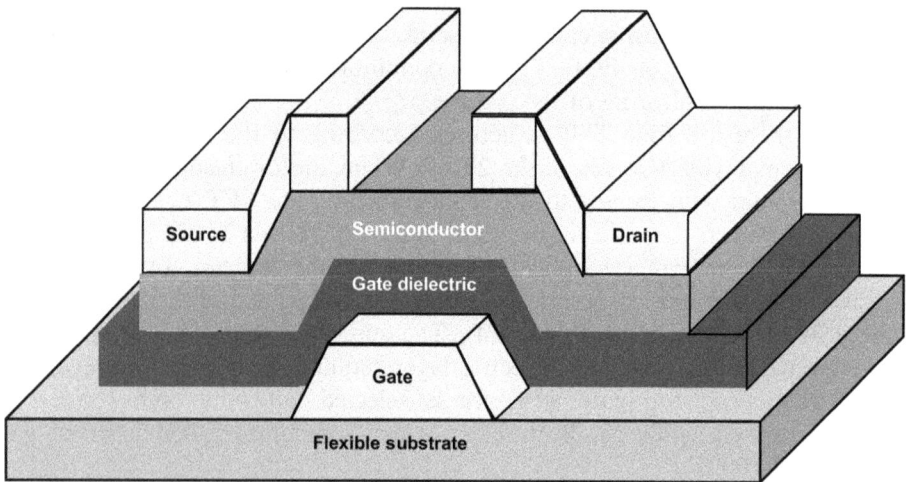

(b)

Figure 1.2. Schematic diagrams of TFT configurations: (a) etch-stop and (b) back-channel etch.

(the temperature of transition from a hard state to soft, rubbery material) of most plastics is <200 °C. It is $\ll T_g$ of commonly used glass substrates. The T_g of glass substrates is $\geqslant$ 600 °C. Hence, a temperature of 150 °C is chosen to carry out all fabrication process steps of TFT without sacrificing the desirable electrical properties. Retention of material properties called for a major re-optimization of process steps.

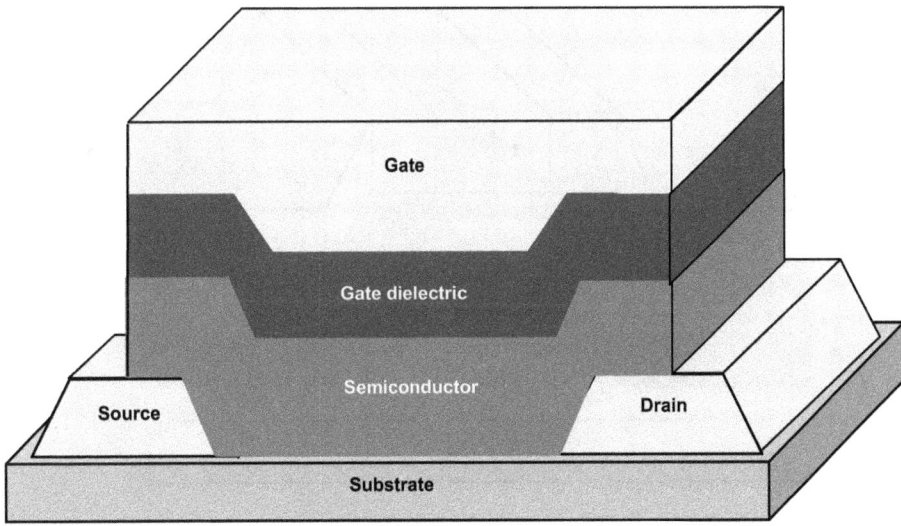

(a)

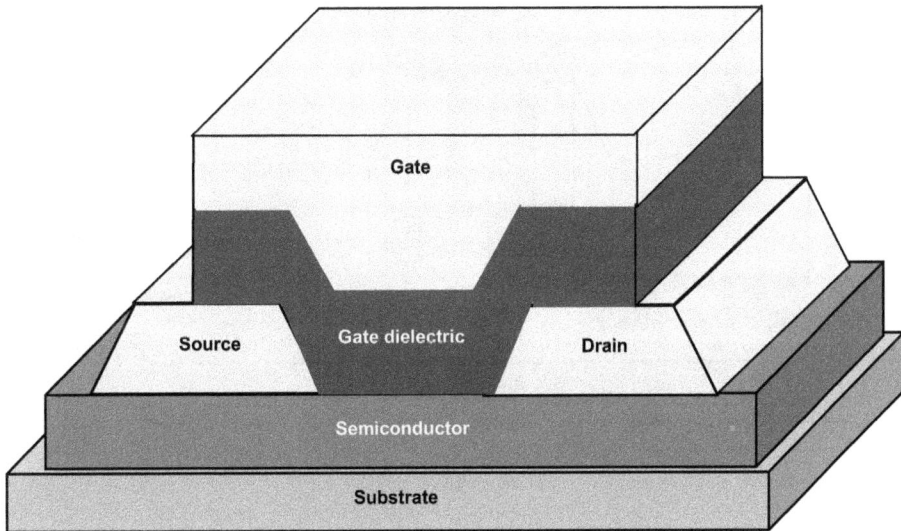

(b)

Figure 1.3. Top gate TFTs: (a) staggered top gate and (b) coplanar top gate.

1.3.2 Significance of SiN$_x$ passivation layer

An important process step is passivation of polyimide with a 0.5 μm thick SiN$_x$ layer on both sides (Gleskova and Wagner 1999). This SiN$_x$ layer protects the polyimide from the chemical attack of acids, bases and other reagents used in photolithography

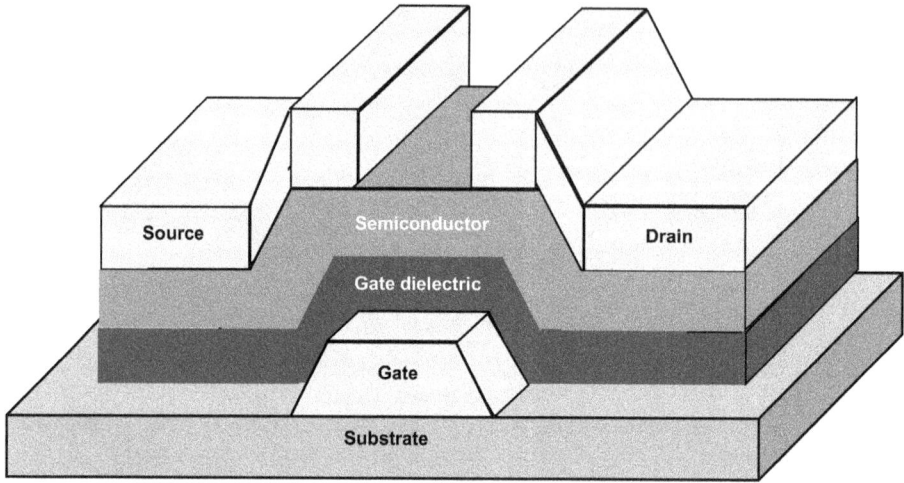

(a)

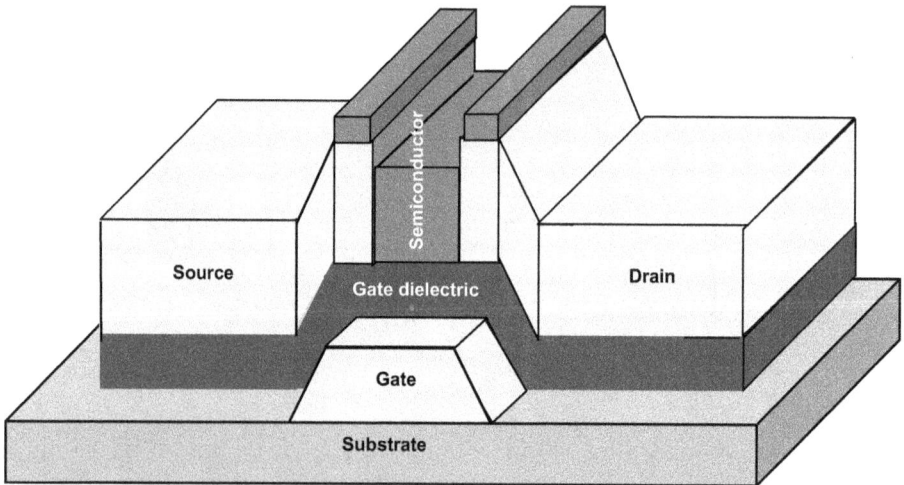

(b)

Figure 1.4. Bottom gate TFTs: (a) staggered bottom gate and (b) coplanar bottom gate.

and other process steps. It also serves as a barrier against gaseous out-diffusion of polyimide as well as release of water and oxygen from it. By completely sealing the polymer, the process engineer is concerned only with the chemistry of the passivating layer and not bothered by the chemistry of polyimide. The TFT function will degrade if the a-Si:H semiconducting film is contaminated. By sealing the polyimide within the passivating layers, such contamination is eliminated.

For building the TFT structure, the passivating layer is no less important. It is an adhesion layer over which the successive TFT layers are laid out, establishing a mechanical bond joining the substrate with the TFT layers, and also playing a vital role in device operation. It acts as the gate dielectric, and, therefore, must be a good insulator preventing the flow of any source–gate leakage current. Generally, the SiN_x layer is deposited at a temperature of 300 °C–350 °C. Therefore, lowering the deposition temperature of SiN_x to 150 °C must not compromise with its properties.

1.3.3 Optimization of deposition conditions of SiN_x

Gleskova et al (2001) carried out a systematic series of experiments for optimization of the deposition parameters of SiN_x at 150 °C keeping the leakage current as a quality control parameter. They reported the development of 150 °C technology for fabrication of a-Si TFTs. Using a 13.56 MHz PECVD reactor, they deposited SiN_x films at 500 mTorr by maintaining the SiH_4 and NH_3 flow rates constant while varying the H_2 flow rate from 55 to 220 sccm and RF power from 5 to 50 W. The SiH_4 flow rate is fixed at 5 sccm and NH_3 flow rate at 50 sccm. Growth rate, refractive index, leakage current between -100 V to $+100$ V, breakdown voltage, dielectric constant and etch rates of the films were measured. The leakage current was found to be 20–30 pA at 100 V for all the films.

Information gathered from SiN_x films deposited at 300 °C–350 °C revealed that the 150 °C film must be slightly rich in nitrogen and it must have low hydrogen content and its etching rate must be low. Breakdown voltage should be high. A refractive index around 1.85 along with high growth rate is required. From these considerations, Gleskova et al (2001) construed that the desirable qualities can be achieved in the 150 °C film with a gaseous mixture composition: SiH_4 (5 sccm), NH_3 (50 sccm) and H_2 (220 sccm). The RF power must be 20 W. The resulting film has an Si/N ratio of 0.67, film growth rate of 1.5 Å s^{-1}, H content of 2×10^{22} cm^{-3}, refractive index of 1.80, an etching rate of 61 Å s^{-1}, a dielectric constant of 7.46 and a breakdown voltage of >3.4 MV cm^{-1}.

1.3.4 TFT fabrication

Using the optimized SiN_x deposition parameters, the process starts with passivation of both sides of polyimide with SiN_x film of thickness 500 nm (figure 1.5), as already mentioned (Gleskova et al 2001). The next step is formation of a gate electrode by thermally evaporating a 100 nm thick Cr layer over SiN_x layer and patterning it. The patterned Cr gate electrode is covered with a 360–400 nm thick SiN_x layer. This SiN_x layer is the gate insulator. Over the SiN_x gate dielectric layer a 200 nm thick undoped a-Si:H layer from ($SiH_4 + H_2$) mixture is deposited. It is the semiconductor film in which the channel is formed. The undoped a-Si:H film is covered with a 50 nm thick N^+-doped a-Si:H film from ($SiH_4 + PH_3 + H_2$) mixture; here PH_3 is the phosphorous dopant source. The N^+-doped a-Si:H film is required for ohmic contact formation. Over the N^+-doped a-Si:H film, Cr is thermally evaporated. Source and drain contact regions are defined in Cr film by wet etching. The source and drain regions are separated in N^+-doped a-Si:H film by dry etching a separation window

**Passivation of both sides of
polyimide with SiN$_x$**

SiN$_x$

Polyimide

SiN$_x$

**Evaporating and patterning Cr
gate electrode**

Gate

Cr

SiN$_x$

Polyimide

SiN$_x$

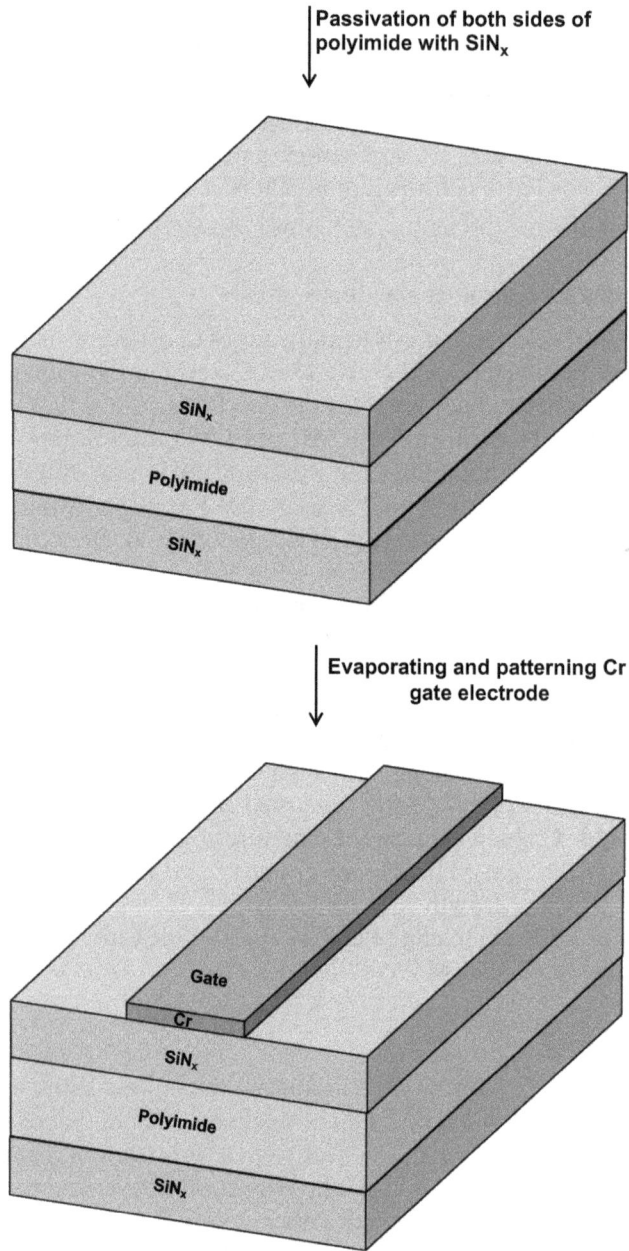

Figure 1.5. a-Si TFT on polyimide foil.

SiN$_x$ gate insulator film deposition

Undoped a-Si:H layer deposition

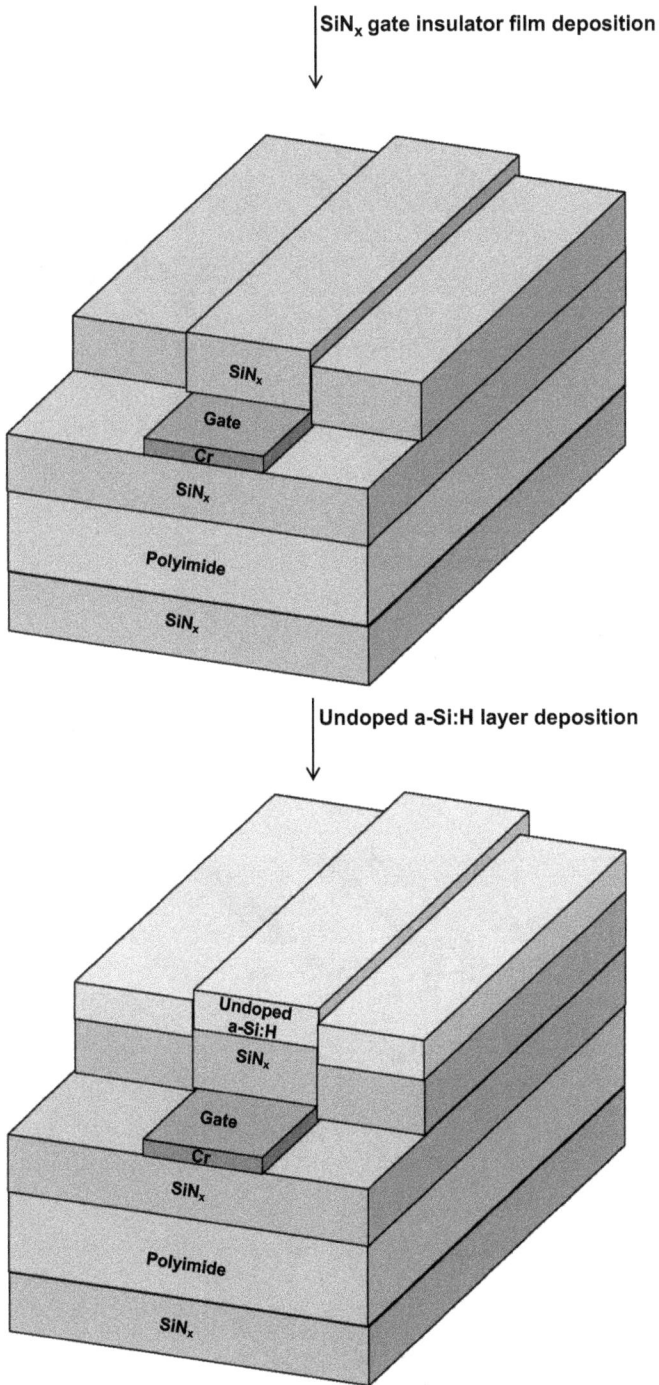

Figure 1.5. (Continued.)

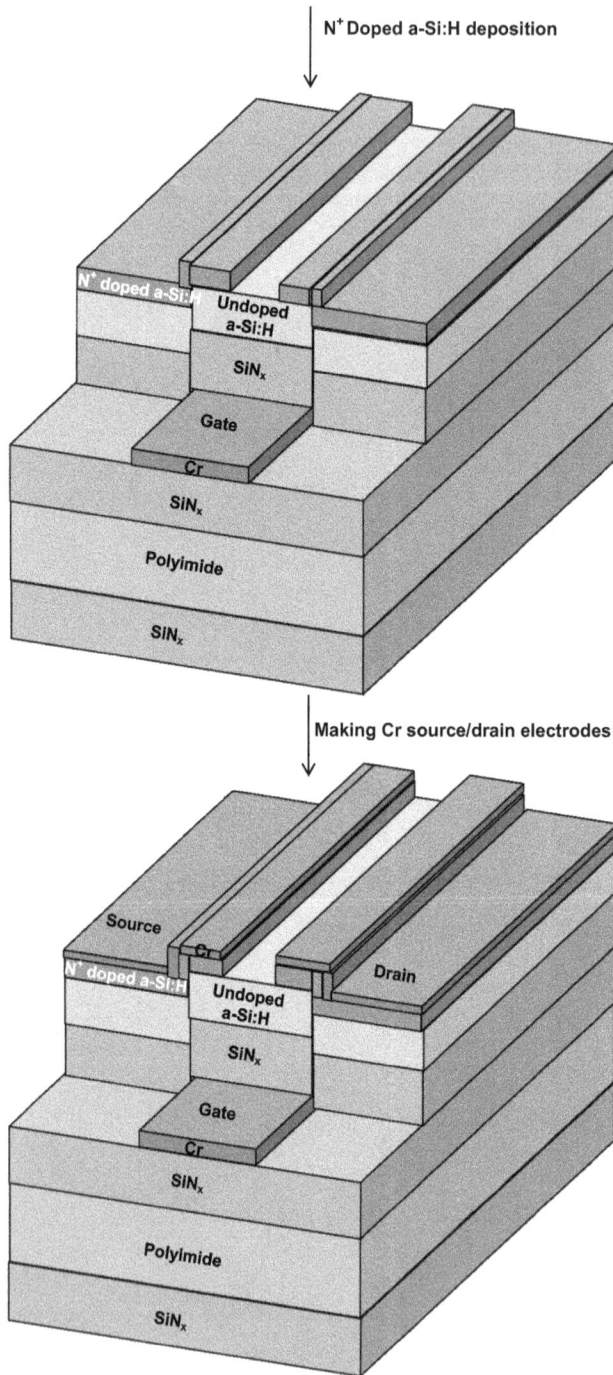

N⁺ Doped a-Si:H deposition

N⁺ doped a-Si:H
Undoped
a-Si:H
SiN$_x$
Gate
Cr
SiN$_x$
Polyimide
SiN$_x$

Making Cr source/drain electrodes

Source
Cr
N⁺ doped a-Si:H
Undoped
a-Si:H
Drain
SiN$_x$
Gate
Cr
SiN$_x$
Polyimide
SiN$_x$

Figure 1.5. (Continued.)

in it. Then a window is dry etched into SiN_x for accessing the gate contact pad. The fabricated TFTs have a channel length $L = 15$ μm and channel width $W = 210$ μm (Gleskova *et al* 2001).

1.3.5 TFT parameters

Parameters of TFTs fabricated on 51 μm thick polyimide foil are: off current of the TFT is 10^{-12} A and the on–off current ratio is $>10^7$. The threshold voltage of the TFT is 3.5 V and the subthreshold slope is 0.5 V/decade (Gleskova *et al* 1998). The carrier mobility in the linear region of operation is around 0.5 $cm^2\,V^{-1}\,s^{-1}$. The parameters tested were defined as: the off current (smallest I_{DS} at $V_{DS} = 10$ V), on current (I_{DS} at $V_{DS} = 10$ V and $V_{GS} = 20$ V) and leakage current (source–gate current at $V_{DS} = 10$ V and $V_{GS} = 20$ V). The off current, on–off current ratio, threshold voltage and linear region mobility for TFTs fabricated on 25 μm thick polyimide foil are the same as those for TFTs on 51 μm foil (Gleskova and Wagner 1999).

1.3.6 Effects of bending on TFTs

Gleskova and Wagner (1999) examined the effects of bending a-Si TFTs on 25 μm polyimide foil by different radii beginning from radius of curvature $R = 4$ mm and ending at $R = 0.5$ mm. For each radius, the TFT is stressed for 1 min and let go. Its characteristics are then measured. It was found that a-Si TFTs showed no noticeable change in electrical characteristics when bent up to a radius of curvature $R = 0.5$ mm nor did they succumb to any catastrophic TFT failure.

Gleskova *et al* (2000) bent a-Si TFTs on 25 μm Kapton foil over mandrels of different radii. The TFTs experience compressive strain when facing inward (negative strain) and tensile strain when they face outward (positive strain). Up to ~2% of compressive strain and ~ 0.5% of tensile strain, the electrical characteristics of TFTs, namely, off current, on current, source–gate leakage current, threshold voltage and mobility of carriers remain unaltered. Tensile strains >0.5% cause mechanical failure of TFTs associated with creation of cracks in TFT layers running perpendicular to the direction of bending.

1.4 Effects of uniaxial and biaxial strain on TFTs

1.4.1 Effects of cylindrical deformation (uniaxial strain) on TFTs

Cylindrical deformation produces one-dimensional strain. Gleskova *et al* (2002) studied the effects of uniaxial compressive or tensile strain on a-Si TFTs on polyimide substrates. Tensile strain increases the field-effect mobility of electrons while compressive strain decreases the same. These effects are reversible and disappear on withdrawal of strain. The dependence of mobility μ on strain ε is expressed as

$$\mu = \mu_0(1 + 26 \times \varepsilon) \tag{1.1}$$

where μ_0 is the linear mobility before application of strain. Tensile strain is taken as positive and compressive strain as negative.

1.4.2 Effects of spherical deformation of TFT islands (biaxial strain) on TFTs

Spherical deformation must be distinguished from the cylindrical case. When the deformation is spherical, the shape of the resultant structure determines the strain and this strain is not reducible by substrate thinning.

Hsu *et al* (2002, 2004) investigated the effects of mechanical strain on TFTs fabricated in an island structure on a 50 μm thick polyimide foil substrate which was permanently deformed into the shape of a spherical dome after fabrication. If the brittle TFT islands of limited size are fabricated on a compliant substrate and the structure is curved to spherical shape such that plastic deformation (permanent change in shape/size without fracturing by applying a stress beyond the elastic limit) is restricted only to substrate regions separating the islands, the TFTs can be interconnected after deformation into circuits. After deformation, the strain in TFT islands is biaxial and dependent on the geometry of the island. It can be either tensile or compressive. Accordingly, the carrier mobility in TFT increases or decreases, equation (1.1).

1.5 TFTs on stainless steel foil substrates

1.5.1 Advantages of stainless steel substrates

While looking for viable flexible alternatives to glass substrates, plastics were a natural first choice but their inability to withstand high temperatures led researchers to think about metal foils, such as stainless steel foils, as possible substitutes. Stainless steel substrates offer ruggedness combined with flexibility. Besides lower costs, their higher temperature processing capability, long-term stability and shielding against electromagnetic interference are distinct advantages.

Grade 430 stainless steel is a better match to silicon regarding coefficient of thermal expansion (CTE) α, which is 2.6×10^{-6} per °C for silicon, 20×10^{-6} per °C for Kapton and 6×10^{-6} per °C for grade 430 stainless steel. Grade 430 is a straight chromium, ferritic steel. It is non hardenable and possesses good corrosion resistance. It resists nitric acid attack and shows heat and oxidation resistance up to 816 °C.

1.5.2 TFTs on thin stainless steel substrates

Theiss and Wagner (1996) used 200 μm thick, one side polished, 430 grade steel of RMS surface roughness (root mean square average of the deviations in profile height with respect to the mean line) = 0.1 μm. Because the substrate is electrically conducting, it is coated with an insulating layer of a-SiN$_x$:H (figure 1.6). A three-chamber 13.56 MHz RF PECVD system is used for depositing a-Si:H, N$^+$-doped a-Si:H and a-SiN$_x$:H films. The deposition pressure is 500 mTorr. However, the process engineer has the liberty to use much higher deposition temperatures than for polymeric substrates, e.g. 250 °C for a-Si:H, 260 °C for N$^+$-doped a-Si:H and 310 °C for a-SiN$_x$:H films. Different chambers are used for the three films.

Cr is thermally evaporated for metallization of contact pads. Dry etching with CF$_4$/O$_2$ gas mixture completes the TFT structure by exposing the gate pads. The

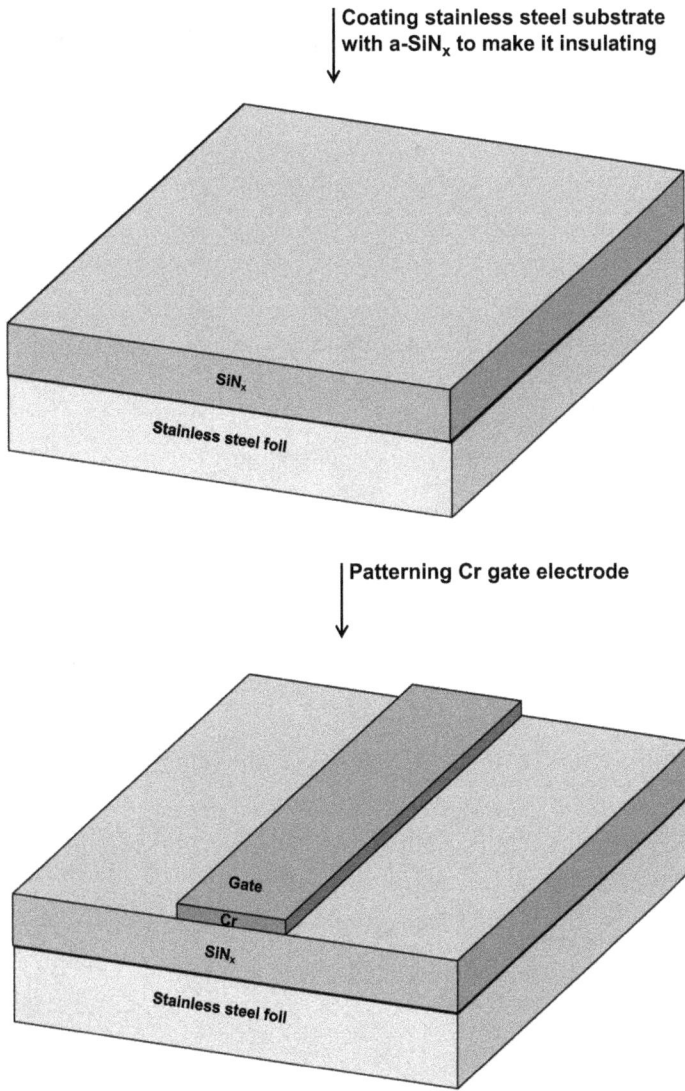

Figure 1.6. a-Si TFT on stainless steel foil.

Depositing SiN$_x$ gate insulator film

Depositing undoped a-Si:H layer

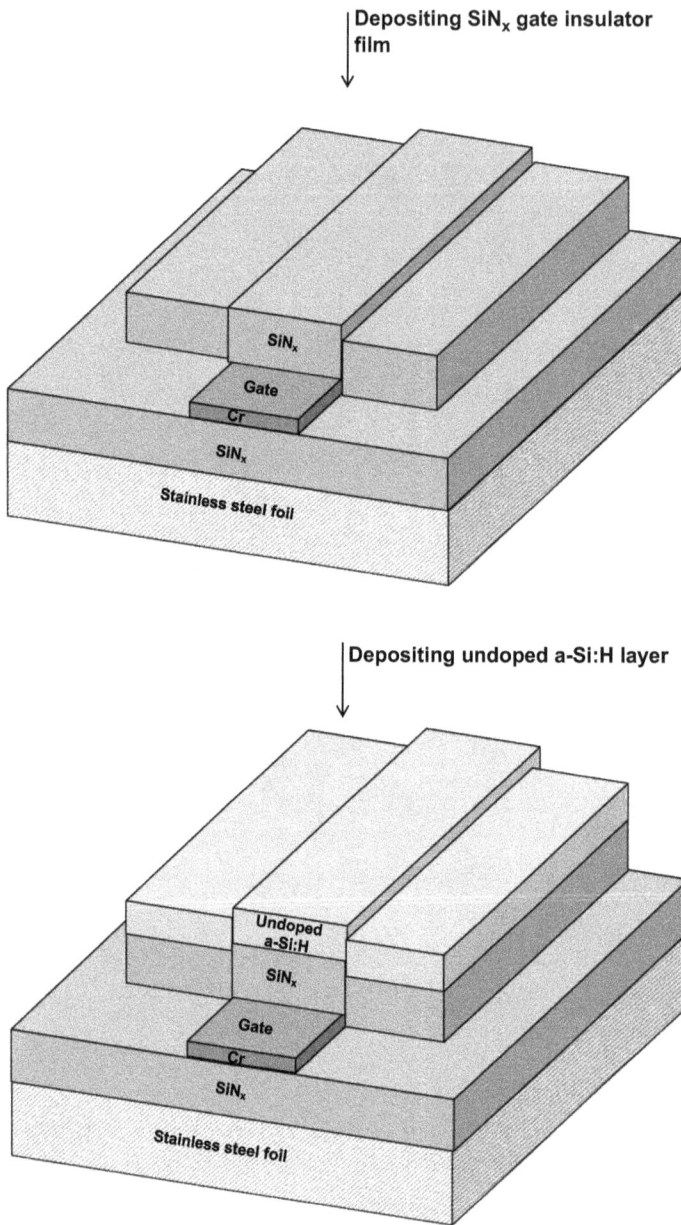

Figure 1.6. (Continued.)

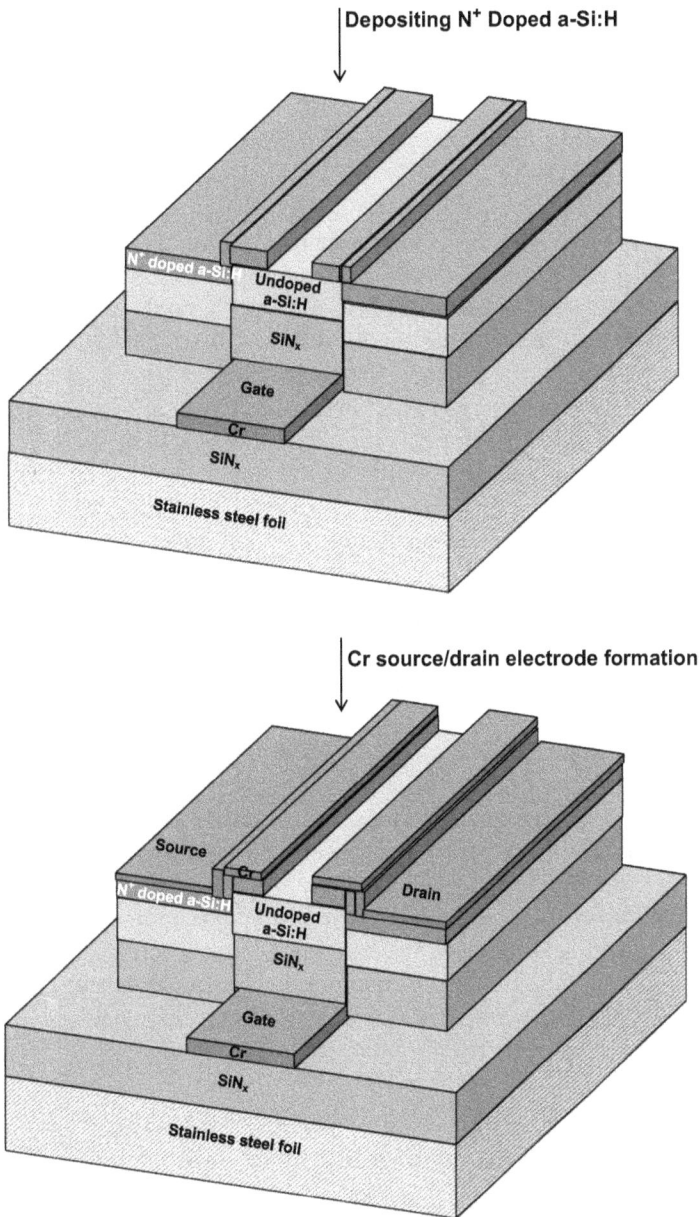

Figure 1.6. (Continued.)

inverted staggered gate TFTs have a channel length $L = 7.5$ µm and channel width $W = 50$ µm. The TFTs show an off current of 10^{-12} A, an on/off current ratio of 10^7, a threshold voltage $= 5$ V, subthreshold slope of 0.5 V/decade, a linear region mobility of 0.5 cm^2 V^{-1} s^{-1} and saturation region mobility of 0.7 cm^2 V^{-1} s^{-1}. The TFTs are mechanically robust, they exhibit a high mechanical strength, tolerance to bending and unbreakability, and they continue functioning after falling from a height of four floors (Theiss and Wagner 1996).

Wu *et al* (1997) describe the integration of OLEDs with a-Si TFT drivers on stainless steel substrates. On a 200 µm thick, one side polished, 430 grade steel foil, 0.95 µm thick a-SiN$_x$ is deposited, then 0.15 µm thick Cr gate electrode covered with 0.32 µm thick a-SiN$_x$. The ensuing three layers are 0.16 µm thick a-Si:H, 0.05 µm thick N$^+$-doped a-Si:H and 0.15 µm thick Cr for source/drain contact. The channel aspect ratio is 776 µm/42 µm = 18.5, the threshold voltage is 3–4 V, the subthreshold slope is 0.5–1 V/decade, the electron mobility is 0.5–0.7 cm^2 V^{-1} s^{-1}, the on/off current ratio exceeds 10^5 and the TFTs drive the OLEDs to ~100 cd m^{-2}.

1.5.3 TFTs on thin stainless steel substrates

Hong *et al* (2006) fabricated TFTs with channel length $L = 5$ µm on stainless steel foil substrates. The threshold voltage is 4.5 V and the mobility is 0.3 cm^2 V^{-1} s^{-1}. They designed a pixel circuit with 2 TFTs. When the AMOLED backplane on 75 µm thick stainless steel foil is operated, a luminance of 500 cd m^{-2} is obtained with a white OLED.

Han *et al* (2007a, 2007b) fabricated TFTs on a 76 µm thick stainless steel foil. After multi-barrier coating of the foil, its RMS surface roughness decreased to 5 nm. The TFT structure is the inverted staggered type. The gate electrode is sputter-deposited aluminum doped with neodymium (Nd) and molybdenum (Mo). The 150 °C-PECVD process is used for depositing a-SiN$_x$, a-Si and phosphorous doped a-Si. The molybdenum source and drain contacts are formed by sputtering. The width of TFT channel is $W = 30$ µm and length is 5 µm; aspect ratio = 30 µm/5 µm = 6. The threshold voltage is 1.0 V. The off current is 10^{-13} A and the field-effect mobility is 0.54 cm^2 V^{-1} s^{-1}. For thermal stabilization, the TFTs are annealed at 230 °C when the field-effect mobility decreases to 0.71 times the original value.

1.6 TFTs on clear plastic (CP) foil substrates

1.6.1 Difficulties with common substrate materials

The high glass transition temperature T_g ~ 350 °C of Kapton E polyimide allows fabrication of TFTs in the processing temperature range 150 °C–250 °C. However, Kapton E lacks optical clarity with a 50 µm thick substrate cutting off at a wavelength of ~ 500 nm, resulting in an orange–brown color (Long *et al* 2006a). Poly(ethylene terephthalate) (PET) with $T_g = 70$ °C–100 °C and poly(ethylene naphthalate) (PEN) with $T_g = 120$ °C are two common optically clear substrates. While the T_g of PET is too low for fabrication of TFTs, the TFTs fabricated on PEN at 130 °C suffer from poor quality in terms of a lower mobility, higher leakage current and drift in characteristics.

1.6.2 TFTs on clear plastic with temperature tolerance limit up to 180 °C

Long *et al* (2004, 2006a) use a clear plastic substrate of thickness 75 μm. It has a T_g value of 326 °C but allows the implementation of process with maximum temperature of 180 °C. This is because its CTE of 50×10^{-6} per °C grossly mismatches with that of a-Si $\sim 3 \times 10^{-6}$ per °C and a-SiN$_x$ $\sim 2.7 \times 10^{-6}$ per °C. The detrimental consequence of this mismatch is that the TFT device layer cracks after deposition if a temperature >180 °C is used. So, this value represents a maximum temperature limit for TFT process on clear plastic.

The inverted, staggered BCE TFT structure is adopted with a gate at the bottom and source/drain contacts at the top. The first step is deposition of an a-SiN$_x$ layer for planarization and passivation of the substrate, and promotion of adhesion of TFT layers to be laid over it (figure 1.7). A 100 nm thick Cr film deposited by sputtering is wet etched to define the gate electrode. Then the process is carried out in a multi-chamber PECVD system. Sequentially a-SiN$_x$ film (300 nm), a-Si film (200 nm) and heavily doped N$^+$a-Si film (30 nm) are deposited at 180 °C without any interruption in the vacuum. The a-SiN$_x$ film acts as the gate dielectric, the undoped a-Si film is the layer in which the active channel is formed and the degeneratively doped N$^+$a-Si film is the layer providing ohmic contact to the source and drain electrodes of TFT. An 80 nm thick Cr film is thermally evaporated and delineated by wet etching for source/drain contacts. The TFT island is defined by dry etching undoped a-Si and the device structure is completed by etching windows into the gate SiN$_x$ film for accessing the pad of the gate contact.

The W/L ratio of the TFT is 80 μm/5 μm = 16. The threshold voltage is 3.4 V and the subthreshold slope is 400 mV/decade. In the linear region of operation, the mobility is 0.73 cm^2 V^{-1} s^{-1}, in the saturation region, it is 0.68 cm^2 V^{-1} s^{-1}. The off current is 10 pA, the on/off current ratio is 10^7 and the gate–source leakage current is <10 pA at V_{GS} = 20 V, which are indicators that the SiN$_x$ film deposited at 180 °C is of premium quality. The TFT–AMOLED integration is done on a clear plastic substrate. The drive current provided by a-Si TFT backplanes on a clear plastic substrate is found to be sufficient to activate AMOLED (Long *et al* 2006a).

1.6.3 TFTs on clear plastic with temperature tolerance limit up to 250 °C–280 °C

In another effort, Long *et al* (2006b) used a proprietary clear plastic with T_g > 315 °C. This plastic is transparent from 400–800 nm; its visible light wavelength extends from 400–750 nm. Another notable feature is that its CTE is $<10 \times 10^{-6}$ per °C. The substrate thickness is 75 μm. It has a 3″× 3″ square shape. Processing temperatures up to 250 °C–280 °C are permitted owing to the closeness of CTE of plastic with the values for CTEs of TFT layers. These temperatures are comparable to those used in TFT fabrication on glass substrates so that the TFT fabricated on clear plastic must be competitive in performance with TFTs on glass. The TFT channel is 40 μm long. The threshold voltage is obtained as 2.5–4.5 V; subthreshold slope = 400 mV/decade. The on–off current ratio is 10^6. Mobility in the linear region is 1.0 cm^2 V^{-1} s^{-1}; that in saturation mode is 0.8 cm^2 V^{-1} s^{-1}. The source–gate leakage current is <4 pA. Threshold voltage stability under gate bias stress and DC characteristics of TFTs on clear

Passivation of both sides of clear
plastic with SiN_x

Buffer SiN_x

Clear plastic

Buffer SiN_x

Evaporating and patterning
Cr/Al/Cr gate electrode

Gate

Cr/Al/Cr

Buffer SiN_x

Clear plastic

Buffer SiN_x

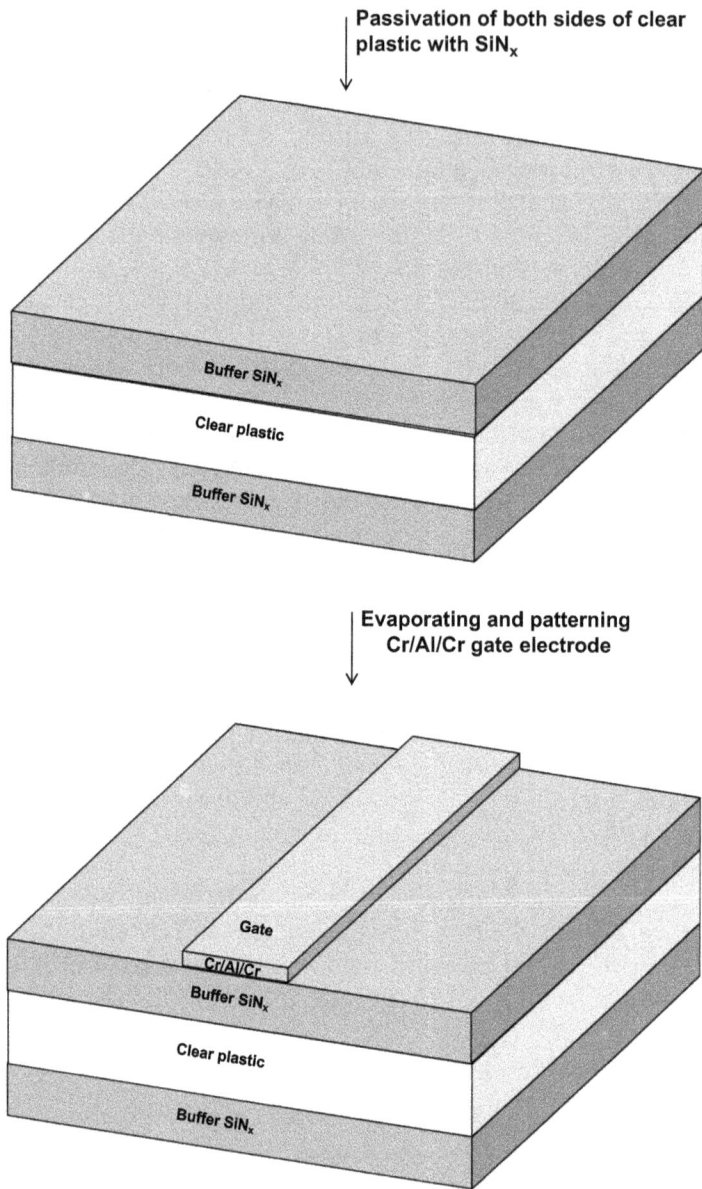

Figure 1.7. a-Si TFT on clear plastic substrate.

SiN$_x$ gate insulator film deposition

SiN$_x$

Gate

Cr/Al/Cr

Buffer SiN$_x$

Clear plastic

Buffer SiN$_x$

Undoped a-Si:H layer deposition

Undoped
a-Si:H

SiN$_x$

Gate

Cr/Al/Cr

Buffer SiN$_x$

Clear plastic

Buffer SiN$_x$

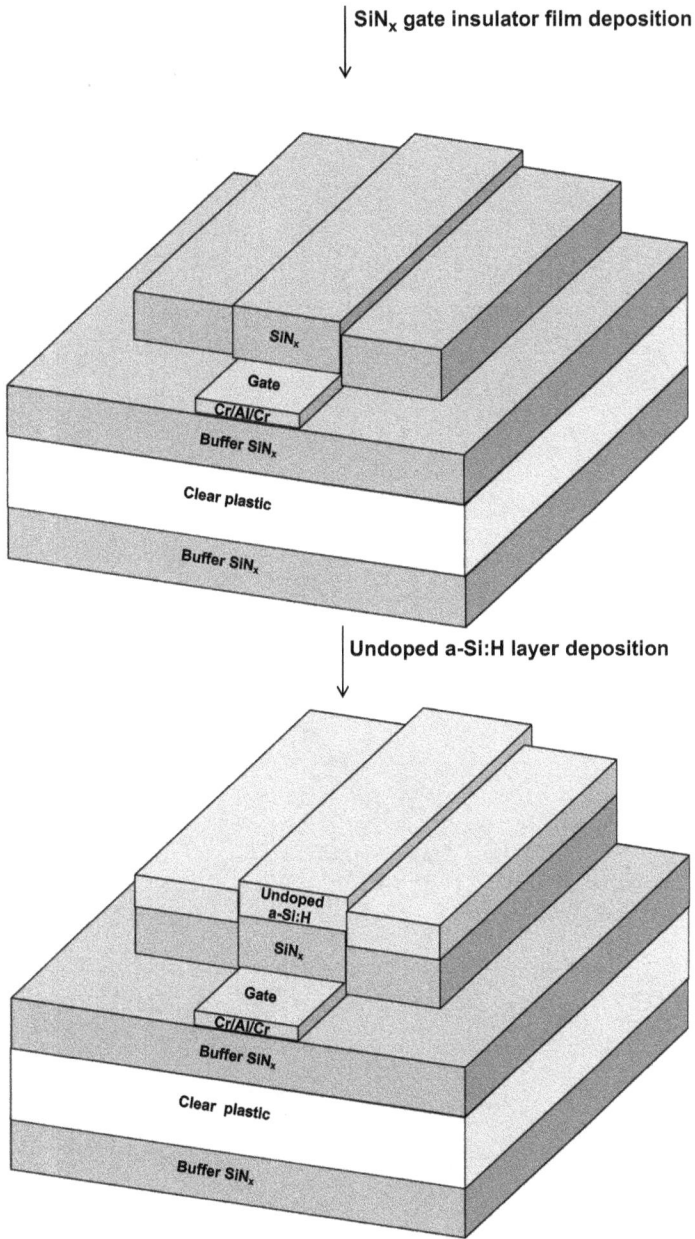

Figure 1.7. (Continued.)

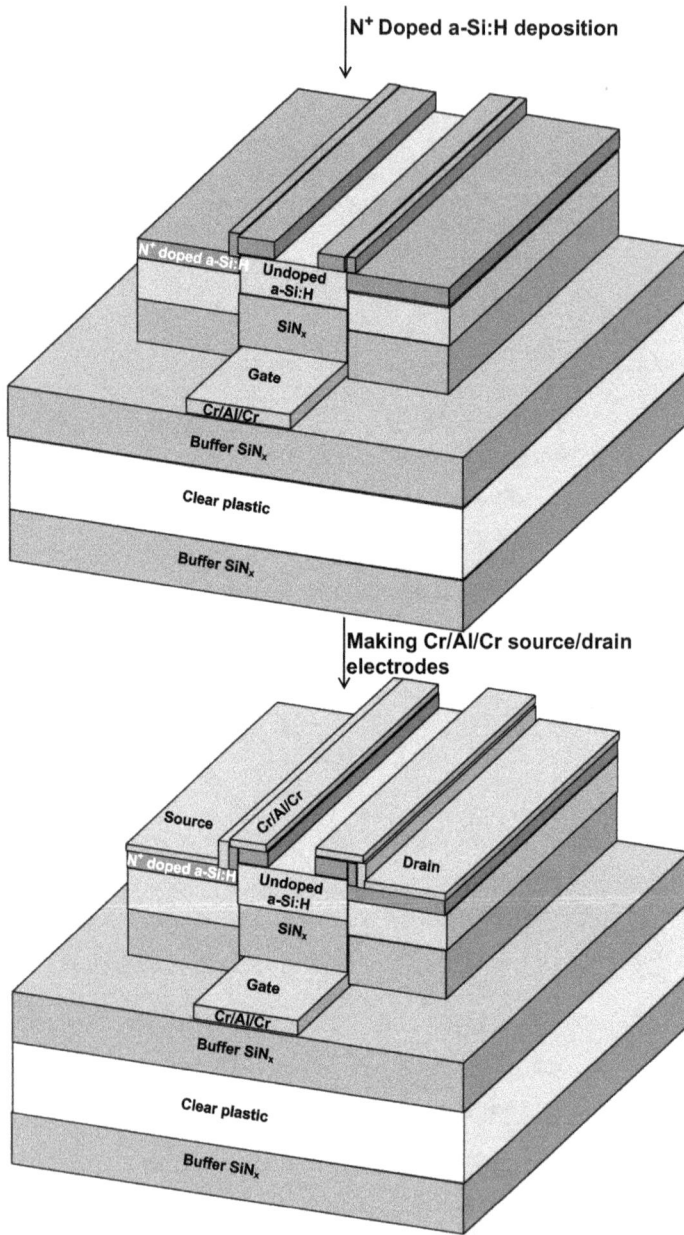

Figure 1.7. (Continued.)

plastic approached those of industry-standard TFTs made on glass substrates at 300 °C–350 °C. Thus, glasslike stability is achieved on clear plastic in accordance with expectation (Long *et al* 2006b).

1.6.4 TFTs on clear plastic by controlling mechanical stresses in substrate passivation and TFT structural layers

Cherenack *et al* (2007) emphasized the necessity of stress design and control during TFT fabrication to obtain functional TFTs on free-standing substrates of clear plastic (CP) (figure 1.8). The relevant process parameter for stress control during PECVD is the RF power aided by deposition temperature. The process is subdivided into two parts: substrate preparation and device fabrication.

Substrate preparation: PECVD processes are carried out on the CP substrate by mounting it in a frame with the substrate surface downwards. This frame allows the substrate to expand and contract to a limited extent. Three layers successively back up the CP substrate in the mount: a Kapton E polyimide foil, a glass substrate and black-body absorber in the form of a graphite plate to provide radiative heating. While the substrate is in the load lock, it is annealed at 200 °C for degassing. Then it is transferred to the PECVD chamber. Here, a 300 nm thick SiN_x layer is deposited at 280 °C on the front side of the substrate. This is the side on which the TFT is to be fabricated. The RF power density of deposition is 20 mW cm^{-2}. The resulting stress in SiN_x film is tensile in nature. Next the substrate is returned to the load lock where

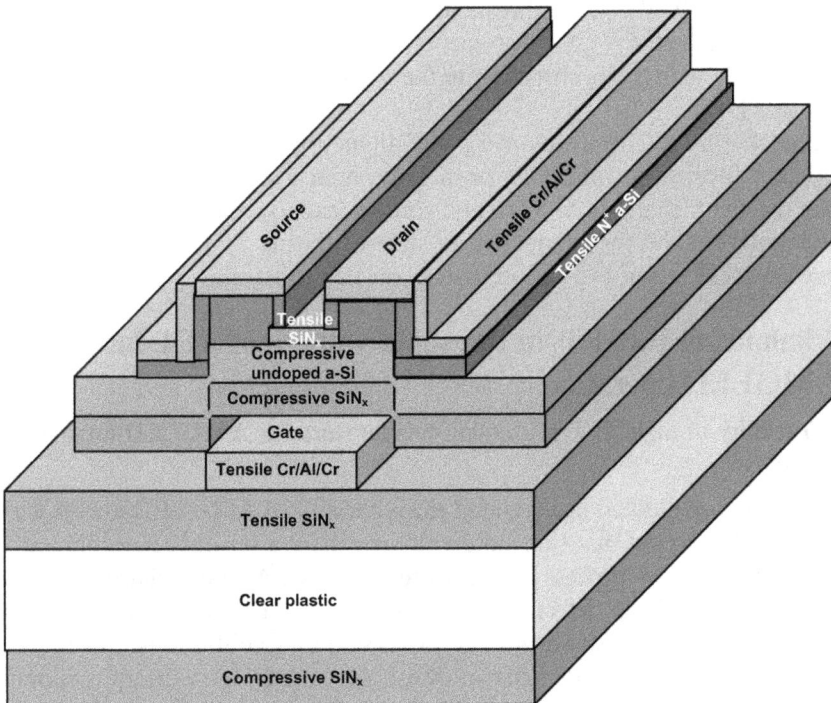

Figure 1.8. TFT fabrication on clear plastic by control of mechanical stresses in the deposited layers.

it is flipped over. So the deposition will now take place on the back side of the substrate. After shifting the substrate again to the PECVD chamber, SiN_x deposition is done at 280 °C on the back side of the substrate but now at a higher RF power density of 90 mW cm^{-2} to build compressive stress in the SiN_x film.

Device fabrication: The channel length of the TFT is $L = 40$ μm and the channel width is $W = 80$ μm so that W/L ratio is 80 μm/40 μm = 2. The gate electrode is made with thermally evaporated trilayers Cr (20 nm)/Al (80 nm)/Cr (20 nm); source/drain contacts too have trilayer composition. The 340 nm thick SiN_x gate dielectric is deposited at 300 °C. The RF power density is 90 mW cm^{-2} for compressive stress. All the a-Si:H layers are deposited at 280 °C. The 300 nm thick a-Si: H active channel layer is deposited at RF power density of 17 mW cm^{-2} for compressive stress. The 55 nm thick N$^+$ a-Si:H layer is deposited at 20 mW cm^{-2} for tensile stress. The 150 nm thick SiN_x layer for back-channel passivation is also deposited at 280 °C at 20 mW cm^{-2} for tensile stress. The fabrication is followed by annealing at 180 °C in air for 30 min. This step is necessary for repairing the damage incurred during plasma processes.

The TFT has a threshold voltage of 3.5 V, subthreshold slope of 500 mV/decade and on–off current ratio $>10^7$. In the linear region of operation, the mobility is 0.95 cm^2 V^{-1} s^{-1}, whereas in the saturation region it is 0.96 cm^2 V^{-1} s^{-1}. Similar results are obtained for TFT fabrication on glass substrates with identical processing parameters confirming that the TFT performance is substrate independent. The TFTs are subjected to gate bias stressing. Cherenack *et al* (2007) found that at a stress field of 1×10^8 V s^{-1}, the threshold voltage shift for TFTs fabricated at 300 °C is only 1.1 V against 4 V, 2 V for earlier-fabricated TFTs at 150 °C and 250 °C respectively. These findings corroborate the influence of process temperature on the stability of TFTs.

Cherenack *et al* (2010) developed a self-aligned process for TFT fabrication on CP allowing fabrication of TFTs on a large area 7 cm × 7 cm CP foil thereby reducing the channel length to 3 μm and source/drain overlap with the gate to 1 μm. This technique enables the production of state-of-the art TFTs on plastic substrates, as proven by realization of ring oscillators on this substrate.

1.7 Minimizing the shift in threshold voltage of TFT for reliable AMOLED operation

1.7.1 Cruciality of high TFT processing temperature for AMOLED luminance stability

Hekmatshoar *et al* (2008a) investigated the reliability of AMOLED arrays with a-Si backplanes. They assert that threshold voltage shift is a serious reliability issue with TFTs fabricated at low process temperatures ~150 °C. As more charges are trapped in the gate SiN_x and the defects are produced in the a-Si film in low temperature processed TFTs, the threshold voltage of these TFTs continuously increases with time. Consequent upon the increase in threshold voltage, less current is supplied by the TFT to the OLED leading to a fall in pixel brightness. By employing higher processing temperatures during gate dielectric (285 °C) formation as well as a-Si deposition (250 °C), the threshold voltage increase is minimized, thereby enhancing the stability of the AMOLED array.

They designed a 2-TFT OLED pixel (figure 1.9). This pixel contains a switching TFT and a driving TFT along with a storage capacitor and an OLED. It also has data, select, power and common ground lines. The TFT process is similar to Cherenack *et al* (2007). Throughout the process, the mechanical stresses in the deposited layers are carefully monitored to prevent cracking of layers and to obtain flat surfaces. The buffer SiN_x layers are deposited on both sides of the CP substrate at 200 mW cm^{-2}. The built-in compressive stresses balance out to make the substrate flat. The Cr–Al–Cr contact trilayers on the top and bottom have low tensile stress Cr layers and low stress Al layer. Gate SiN_x is deposited at 22 mW cm^{-2} and undoped a-Si at 17 mW cm^{-2} resulting in compressive stress. To cancel this compressive stress, the N$^+$doped a-Si is deposited at 17 mW cm^{-2} with tensile stress. Along with the tensile Cr layer, a backplane devoid of cracks is obtained. The backplane also has a 250 nm thick SiN_x passivation layer and a 200 nm thick ITO layer. Both these layers are stress-free.

For experiments, three different gate nitride deposition temperatures are selected: 200 °C, 250 °C, 285 °C. For gate nitride deposited at 200 °C, a-Si deposition is also done at 200 °C while for 250 °C and 285 °C gate nitride films, a-Si is deposited at 250 °C. On pixel stressing by keeping bias voltages on select line and data line constant, the luminance degradation is faster for OLED with TFT processed at lower temperature owing to rapid shift in the threshold voltage of the TFT. Hence the effect of processing temperature is evident.

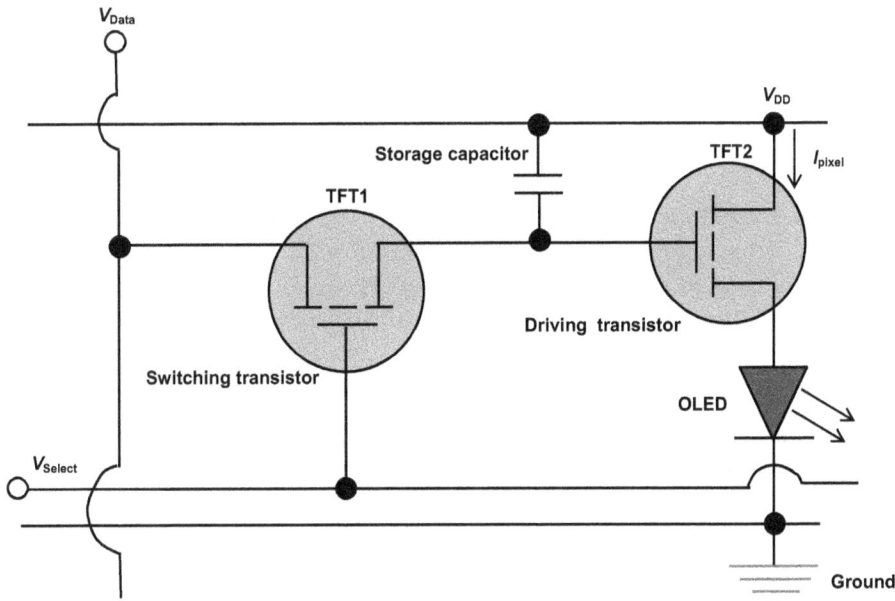

Figure 1.9. Circuit diagram of a 2-TFT OLED pixel.

The driving TFTs have a *W*/*L* ratio = 150 μm/5 μm = 30. The threshold voltage is 2.1 V in the saturation region. The effective mobility is 0.63 cm^2 V^{-1} s^{-1}. At a data voltage of 16.8 V, the luminance intensity is 1000 cd m^{-2} (Hekmatshoar *et al* 2008a).

1.7.2 Conditions to achieve unwavering TFT characteristics

In a further continuation of their work, Hekmatshoar *et al* (2008b) reiterated that the operation of AMOLEDs is more critically dependent on the stability of threshold voltage than that of AMLCDs. The reason is that in the case of AMLCD, the TFT works in a digital switching format with a low duty cycle around 0.1%, which can adapt to variations in threshold voltage. But in AMOLED, the TFT functions in DC mode so that an increase of threshold voltage decreases the drive current which degrades the pixel brightness.

Two TFT structures are contrasted: back-channel etched TFT and back-channel passivated TFT. In the former type of TFT, the top surface of the a-Si channel is naked. It is exposed to the atmosphere. In the later type of TFT, the a-Si channel is sealed *in situ* at the top with SiN$_x$ passivation layer. For this type of TFT, an extra photolithographic step is required in which holes are etched in SiN$_x$ passivation to make contacts with N$^+$ a-Si source and drain regions.

For both etched and passivated TFT structures, Hekmatshoar *et al* (2008b) varied the a-Si PECVD conditions after keeping the gate SiN$_x$ fixed at 285 °C. No major change was observed by changing a-Si deposition temperature in the range 230 °C–280 °C. Dilution of SiH$_4$ with H$_2$ caused slight degradation of passivated TFTs. But when passivated TFTs fabricated with hydrogen dilution was annealed at a temperature of 260 °C, there was a drastic decline in threshold voltage shift ΔV_T. This shift extrapolates to ~ 1.2 V after operating at gate field of 2.5 × 10^5 V cm^{-1} continuously for 10 years. The high stability owes its origin to two causes:

(i) High gate SiN$_x$ deposition temperature decreases charge trapping;
(ii) Hydrogen dilution during a-Si deposition decreases the defect creation in a-Si possibly by removal of weak Si–Si bonds, which break down with the passage of time, leading to temporal instability.

Thus, the conditions to achieve stable TFT characteristics are a combination of selecting a high gate SiN$_x$ deposition temperature, a back-channel passivated TFT structure and a higher degree of hydrogen dilution during a-Si film deposition (Hekmatshoar *et al* 2008b).

1.8 Discussion and conclusions

TFTs are fabricated on plastic substrates. They show bendability up to a radius of 0.5 mm. TFTs on stainless steel substrates show flexing behavior and ruggedness. TFTs on optically clear plastic substrates match the performance of industry-standard TFTs on glass. A low temperature TFT fabrication process is carried out at 150 °C on polyimide. Also, a high temperature process is available at 300 °C on clear plastic and up to 310 °C on stainless steel substrate. A high processing temperature is recommended for stability of threshold voltage of TFT keeping in view the temperature limit of the flexible substrate. All the TFTs described have good, though different, electrical characteristics but favoring usability and catering to the needs of various applications.

Review exercises

1.1 Why can a thin-film transistor be formed on a plastic substrate? On what substrate is a conventional transistor fabricated?

1.2 What is the display using a TFT circuit called? What happens to the cursor on a display in which the TFT is not used?

1.3 What are the relative merits and demerits of etch-stop TFT and back-channel TFT?

1.4 When is a TFT said to be staggered? When is it coplanar?

1.5 What are top gate and bottom gate TFT structures? What is an inverted gate TFT?

1.6 Why does the polyimide substrate need to be covered on both sides with an SiN_x passivation film? Elucidate the functions of the passivating film during device fabrication and during device operation.

1.7 How are the deposition conditions of SiN_x optimized by keeping track of the leakage current? What are the optimum deposition conditions?

1.8 How are the TFTs fabricated using the optimum deposition conditions of SiN_x film? What are the levels of off current and threshold voltage of these TFTs? How do they behave when bent? At what tensile strain value do they fail mechanically?

1.9 What kinds of strains are produced by cylindrical and spherical deformations of a TFT? How does mobility of charge carriers vary with tensile and compressive strains? Write the equation relating mobility with strain.

1.10 Which has a more matching thermal coefficient to silicon: polyimide or stainless steel? Give three reasons for favoring stainless steel over plastic substrates.

1.11 A plastic substrate is insulating, whereas stainless steel is electrically conducting. How is steel made insulating for TFT fabrication? Describe the main process steps of TFT fabrication on steel. Discuss the TFT characteristics.

1.12 Compare Kapton, PET and PEN substrates with reference to their glass transition temperatures and optical clarity.

1.13 Even though the glass transition temperature of an optically clear plastic is 326 °C, why is processing not allowed using a-Si and a-SiN_x at a temperature exceeding 180 °C, whereas processing can be done up to 250 °C–280 °C with another clear plastic having lower T_g of 317 °C?

1.14 Discuss the statement, 'The process parameters for stress control during PECVD of SiN_x are RF power and temperature.' How is this declaration applied to improve TFT performance?

1.15 How does luminance of OLED degrade with a shift in threshold voltage of TFT? What information is obtained from experiments carried out using gate nitride deposition temperatures of 200 °C, 250 °C and 285 °C regarding the impact of processing temperature on stability of threshold voltage of TFT?

1.16 What type of TFT performance is expected if: (i) a high gate nitride deposition temperature is used during TFT fabrication, and (ii) a high degree of hydrogen dilution is used during a-Si film deposition? Give reasons for your answers.

References

Brody T P, Asars J A and Dixon G D 1973 A 6 × 6 inch 20 lines-per-inch liquid-crystal display panel *IEEE Trans. Electron Devices* **20** 995–1001

Cherenack K H, Hekmatshoar B, Sturm J C and Wagner S 2010 Self-aligned amorphous silicon thin-film transistors fabricated on clear plastic at 300 °C *IEEE Trans. Electron Devices* **57** 2381–9

Cherenack K H, Kattamis A Z, Hekmatshoar B, Sturm J C and Wagner S 2007 Amorphous-silicon thin film transistors fabricated at 300 °C on a free-standing foil substrate of clear plastic *IEEE Electron Device Lett.* **28** 1004–6

Correia A P P, Cândido Barquinha P M and Goes J C D P 2016 Thin-film transistors *A Second-Order ΣΔ ADC Using Sputtered IGZO TFTs*, (Springer Briefs in Electrical and Computer Engineering) (Berlin: Springer) ch 2 1–15

Gleskova H and Wagner S 1999 Amorphous silicon thin-film transistors on compliant polyimide foil substrates *IEEE Electron Device Lett.* **20** 473–5

Gleskova H, Wagner S and Suo Z 1998 a-Si:H TFTs made on polyimide foil by PECVD at 150 °C *MRS Online Proc. Library* **508** 73–8

Gleskova H, Wagner S and Suo Z 2000 a-Si:H thin film transistors after very high strain *J. Non Cryst. Solids* **266–9** 1320–4

Gleskova H, Wagner S, Gašparík V and Kováč P 2001 150 °C Amorphous silicon thin-film transistor technology for polyimide substrates *J. Electrochem. Soc.* **148** G370–4

Gleskova H, Wagner S, Soboyejo W and Suo Z 2002 Electrical response of amorphous silicon thin-film transistors under mechanical strain *J. Appl. Phys.* **92** 6224–9

Han C-W, Kim C-D and Chung I-J 2007a Effects of post-annealing on a-Si:H TFT characteristics fabricated on stainless-steel substrate *J. Soc. Inf. Disp.* **15** 439–44

Han C-W, Han M-K, Paek S-H, Kim C-D and Chung I-J 2007b Thermal annealing effect on amorphous silicon thin-film transistors fabricated on a flexible stainless steel substrate *Electrochem. Solid-State Lett.* **10** J65–7

Hekmatshoar B, Cherenack K H, Kattamis A Z, Long K, Wagner S and Sturm J C 2008b Highly stable amorphous-silicon thin-film transistors on clear plastic *Appl. Phys. Lett.* **93** 032103-1–3

Hekmatshoar B, Kattamis A Z, Cherenack K H, Long K, Chen J-Z, Wagner S, Sturm J C, Rajan K and Hack M 2008a Reliability of active-matrix organic light-emitting-diode arrays with amorphous silicon thin-film transistor backplanes on clear plastic *IEEE Electron Device Lett.* **29** 63–6

Hong Y, Heiler G, Kerr R, Kattamis A Z, Cheng I-C and Wagner S 2006 64.3: Amorphous silicon thin-film transistor backplane on stainless steel foil substrates for AMOLEDs *Society for Information Display SID Symp. Digest of Technical Papers* **vol 37** 1862–5

Hsu P I, Bhattacharya R, Gleskova H, Huang M, Xi Z, Suo Z, Wagner S and Sturm J C 2002 Thin-film transistor circuits on large-area spherical surfaces *Appl. Phys. Lett.* **81** 1723–5

Hsu P I, Huang M, Gleskova H, Xi Z, Suo Z, Wagner S and Sturm J C 2004 Effects of mechanical strain on TFTs on spherical domes *IEEE Trans. Electron Devices* **51** 371–7

le Comber P G, Spear W E and Ghaith A 1979 Amorphous-silicon field-effect device and possible application *Electron. Lett.* **15** 179–81

Long K, Gleskova H, Wagner S and Sturm J C 2004 Short channel amorphous-silicon TFTs on high-temperature clear plastic substrates *62nd DRC Conf. Digest Device Research Conf. (21–23 June 2004, Notre Dame, IN, USA)* 89–90

Long K, Kattamis A Z, Cheng I-C, Gleskova H, Wagner S and Sturm J C 2006b Stability of amorphous-silicon TFTs deposited on clear plastic substrates at 250 °C to 280 °C *IEEE Electron Device Lett.* **27** 111–3

Long K, Kattamis A Z, Cheng I-C, Gleskova H, Wagner S, Sturm J C, Stevenson M, Yu G and O' Regan M 2006a Active-matrix amorphous-silicon TFT arrays at 180 °C on clear plastic and glass substrates for organic light-emitting display *IEEE Trans. Electron Devices* **53** 1789–96

Morita S, Ochi M and Kugimiya T 2016 Amorphous oxide semiconductor adopting back-channel-etch type thin-film transistor *Kobelco Technol. Rev.* **34** 52–8

Pappas I, Siskos S and Dimitriadis C A 2009 Active-matrix liquid crystal displays—operation, electronics and analog circuits design *New Developments in Liquid Crystals* ed G V Tkachenko (London: InTech) pp 147–70 https://intechopen.com/books/new-developments-in-liquid-crystals/active-matrix-liquid-crystal-displays-operation-electronics-and-analog-circuits-design (accessed on 2 February 2018)

Theiss S D and Wagner S 1996 Amorphous silicon thin-film transistors on stainless steel foil substrates *IEEE Electron Device Lett.* **17** 578–80

Weimer P K 1962 The TFT—A new thin-film transistor *Proc. Inst. Radio Eng.* **50** 1462–9

Wu C C, Theiss S D, Gu G, Lu M H, Sturm J C, Wagner S and Forrest S R 1997 Integration of organic LED's and amorphous Si TFT's onto flexible and lightweight metal foil substrates *IEEE Electron Device Lett.* **18** 609–12

IOP Publishing

Flexible Electronics, Volume 2
Thin-film transistors
Vinod Kumar Khanna

Chapter 2

PolySi TFT

The superior performance of polySi TFT backplanes over a-Si TFT backplanes encourages development of processes for polySi TFT realization on flexible substrates (Carey *et al* 1997, Theiss *et al* 1998, Kim *et al* 2006, Han and Kim 2006). Thermal constraints of plastic substrates call for revising the high-temperature steps in a polySi process sequence. Plastic films have been invariably coated with multipurpose barrier layer films on both sides before commencing the process. Sputtering is favored instead of PECVD for a-Si deposition to avoid hydrogen incorporation, which might be liberated explosively during laser crystallization, thereby ablating the a-Si film. Process modifications include deposition of aluminum nitride on the substrate to prevent delamination of a-Si during laser crystallization. A buffer oxide layer on the substrate is helpful in preventing ablation of a-Si film by laser. For easy processing, lamination of the plastic film is frequently done over a glass carrier (Lemmi *et al* 2004). The surface-free technology by laser annealing (SUFTLA) process allows all the high-temperature steps to be carried out on an a-Si exfoliation layer on a glass substrate, then shift the exfoliation layer carrying the TFT to a transfer substrate of glass and finally to the plastic substrate. It uses laser action along with water-soluble/insoluble and UV hardenable adhesives (Shimoda and Inoue 1999, Inoue *et al* 2002). Of great importance is the possibility of carrying out furnace annealing of a-Si when using a stainless steel substrate. Uniform transport properties are achieved over a larger area by furnace annealing (Wu *et al* 2000).

2.1 Introduction

Polysilicon TFTs offer higher electron and hole mobilities. The higher mobilities together with the capability of P- and N-channel operation enable integration of matrix switches with driver circuits. Further, the higher mobilities allow downsizing of TFTs to make them capable of carrying on current >100 μA and operation at switching frequency >4 MHz. Increased brightness with lower power consumption

doi:10.1088/2053-2563/ab0d18ch2

<1 mW can be obtained from CMOS circuits using polySi TFTs. Consequently, the displays with polySi backplanes perform significantly better than a-Si TFT backplanes.

High-temperature polysilicon processing (>900 °C) requires costly quartz substrates. Low-temperature polysilicon (LTPS) processing is done below 450 °C using less expensive standard glass. Flexible electronics can avail the benefits of LTPS to realize TFTs on plastic substrates.

2.2 PolySi TFT on PET

Carey *et al* (1997) and Theiss *et al* (1998) chose a 175 μm thick plastic substrate known as polyethylene terephthalate in the form of 4″ diameter PET wafers owing to the low cost, optical transparency and easy availability of this plastic.

2.2.1 Need for process re-optimization

PET has the limitation that the maximum temperature that can be used during processing should not exceed 120 °C. It is reasonable to inquire about the harmful effects of exceeding the temperature limit of 120 °C on a PET substrate. Firstly, a PET substrate shrinks, which may lead to alignment errors in successive photo-lithographic operations. Secondly, its optical transparency deteriorates so that its clarity worsens. Thirdly, PET has a CTE of 18×10^{-6} per °C which is far apart from that of silicon (2.5×10^{-6} per °C). As a result, cracks and delaminations are likely to occur in TFT layers at higher temperatures, which are obviously not wanted.

It is well known that during fabrication of a-Si TFTs, processing temperatures of 225 °C–250 °C are necessary for a-Si deposition and a still higher temperature of ~300 °C is required for SiN_x film. PolySi TFT fabrication requires much higher temperatures than a-Si TFT, in the range of 400 °C–600 °C. Therefore, a PET substrate imposes severe thermal constraints and calls for thorough re-optimization of process steps.

2.2.2 Si crystallization by localized heating

The high-temperature requirement cannot be compromised, but the heating can be confined to a small region of silicon so that the PET substrate remains unaffected. Such localized heating is provided by laser processing, which forms the basis of the 100 °C TFT on plastic process (Carey *et al* 1997 and Theiss *et al* 1998). In this process, the overall temperature is kept ⩽100 °C. However, the temperature in small regions of silicon may momentarily reach the melting point of silicon (1414 °C). But this high temperature is attained for a very short interval of time ~ tens of ns, so short that sufficient heating effect cannot reach the substrate to cause any harm to it. Therefore, this technique is called pulsed laser processing. The pulsed processing is applied in two ways. Firstly, it is used for melting a-Si. The molten a-Si immediately resolidifies to form grains containing crystalline Si, i.e. polySi. This process is known as pulsed laser crystallization. Secondly, the high temperature is used for doping a-Si; this is called pulsed laser doping. Thus the heating effect of the laser pulse is used for a twofold purpose, viz, silicon crystallization and dopant diffusion into silicon without substrate mutilation.

An XeCl excimer laser is used. The wavelength of the laser beam is 308 nm and the laser pulse duration is 35 ns. During the full processing cycle, the total time during which the substrate is exposed to any high temperature is exceedingly small, in the range of a few tens of microseconds. Exposure of the substrate for this short duration is too feeble to cause any damage to it. Thus, excimer laser crystallization (ELC) using pulses of short duration ~ tens of ns efficiently melts the surface silicon layer with minimal effects on the substrate.

2.2.3 TFT fabrication process

The TFT structure is an aluminum top gate self-aligned TFT. Its fabrication involves four photolithographic steps. The self-aligned configuration takes care of any possible misalignment errors arising from shrinkage of the substrate during processing. The steps are (Theiss *et al* 1998 and Carey *et al* 1997) (figure 2.1):

  (i) The starting step is deposition of 0.75 μm thick barrier silicon dioxide film on both sides of the PET wafer (Theiss *et al* 1998). The top SiO_2 layer prevents contamination during processing and acts as a thermal insulation film preventing the heat produced by the laser pulse from penetrating to the PET. The bottom SiO_2 layer protects the substrate from chemical attack during processing and serves as a barrier to moisture for the TFT.

  (ii) Following SiO_2 deposition on both sides of the PET wafer, DC sputtering is used to deposit a 9000 Å thick a-Si film on top of the barrier SiO_2 film. Sputtering is preferred over PECVD for a-Si film formation because it is a room-temperature process. Further, little hydrogen is incorporated ~1–2 at %, which is much lower than hydrogen that inherently accompanies a-Si in PECVD ~10–20 at %. Hydrogen is avoided because it may explosively release during ELC and spoil the process by ablation of the film during laser exposure. The shortcoming of sputtered a-Si film is that it is more likely to delaminate at high laser energy densities, which must be cautiously prevented.

  (iii) For crystallization of the a-Si layer, it is irradiated with laser fluence (energy density) in consecutive steps of typically 180, 250 and 330 mJ cm^{-2}. Multiple laser scanning is done with increasing fluence because lower fluences aid the evolution of any hydrogen in a-Si while higher fluences cause a-Si melting and crystallization. The pulse repetition rate is 25–50 Hz and all the spots on a-Si receive 3–15 pulses at each fluence.

  (iv) Gate oxide is formed by PECVD; thickness = 1000 Å.

  (v) Al gate metal is deposited by sputtering; thickness = 1700 Å.

  (vi) With Al gate as a masking film, gate oxide etching is done over source and drain.

  (vii) Dopant-rich film is deposited over source and drain.

  (viii) By shining a laser beam, the dopant is diffused into a-Si.

Depositing barrier SiO$_2$ on
both sides of PET substrate

Barrier silicon dioxide

PET substrate

Barrier silicon dioxide

Amorphous silicon
deposition by DC sputtering

Amorphous silicon layer

Barrier silicon dioxide

PET substrate

Barrier silicon dioxide

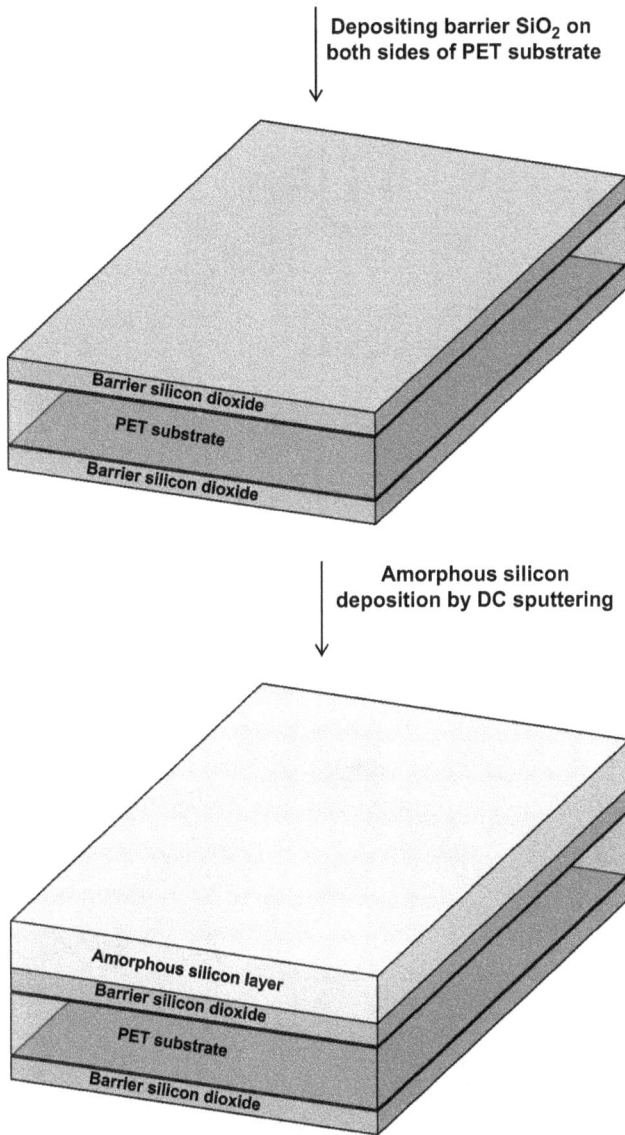

Figure 2.1. Polysilicon TFT fabrication on PET by 100 °C process.

In TFT fabrication by Carey *et al* (1997), doping is done by gas
immersion laser doping (GILD) in which the silicon is melted in an
ambient containing the dopant gas. Doping is performed in the same
apparatus as used for a-Si crystallization except for the incorporation of a
doping environment. This ambient gas is PF$_5$ for n-type doping and BF$_3$
for P-type doping. However, the fluence of the laser beam is less than the

Irradiation with laser

Laser

Amorphous silicon layer
Barrier silicon dioxide
PET substrate
Barrier silicon dioxide

Conversion of amorphous silicon into polysilicon

Polysilicon layer
Barrier silicon dioxide
PET substrate
Barrier silicon dioxide

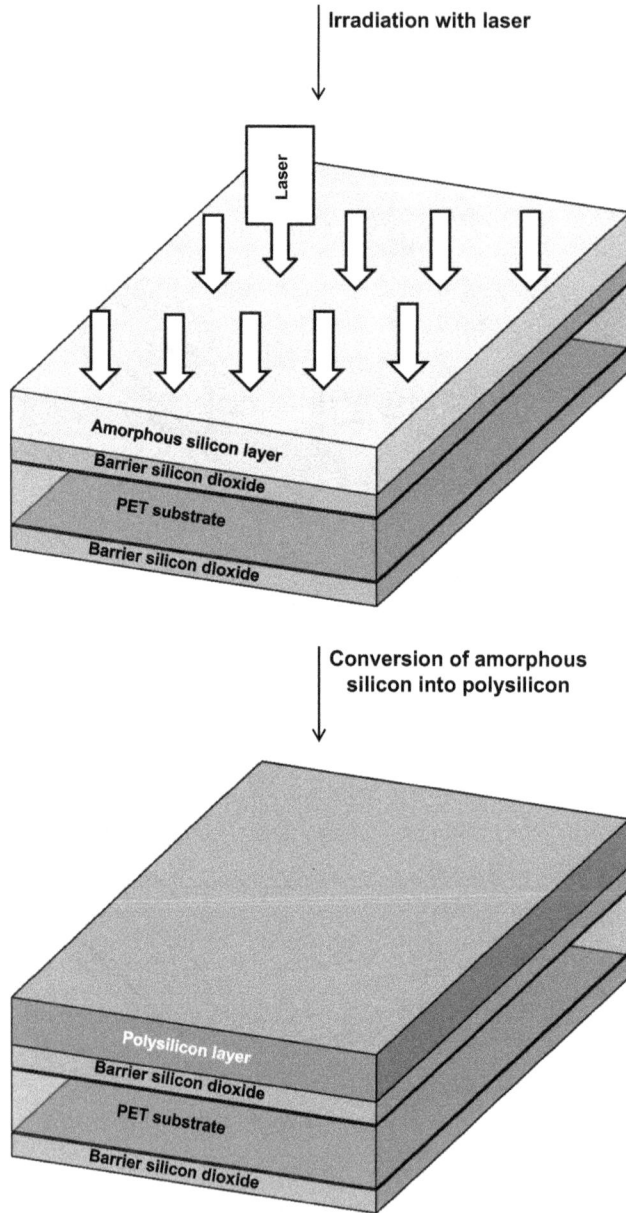

Figure 2.1. (Continued.)

full melting threshold. Each spot receives 50 pulses. The resulting sheet resistance is 500–700 Ω/sq.

(ix) After defining the TFT islands by photomasking, they are separated from each other by dry etching.

Making gate stack

Depositing dopant-rich layers
by PECVD

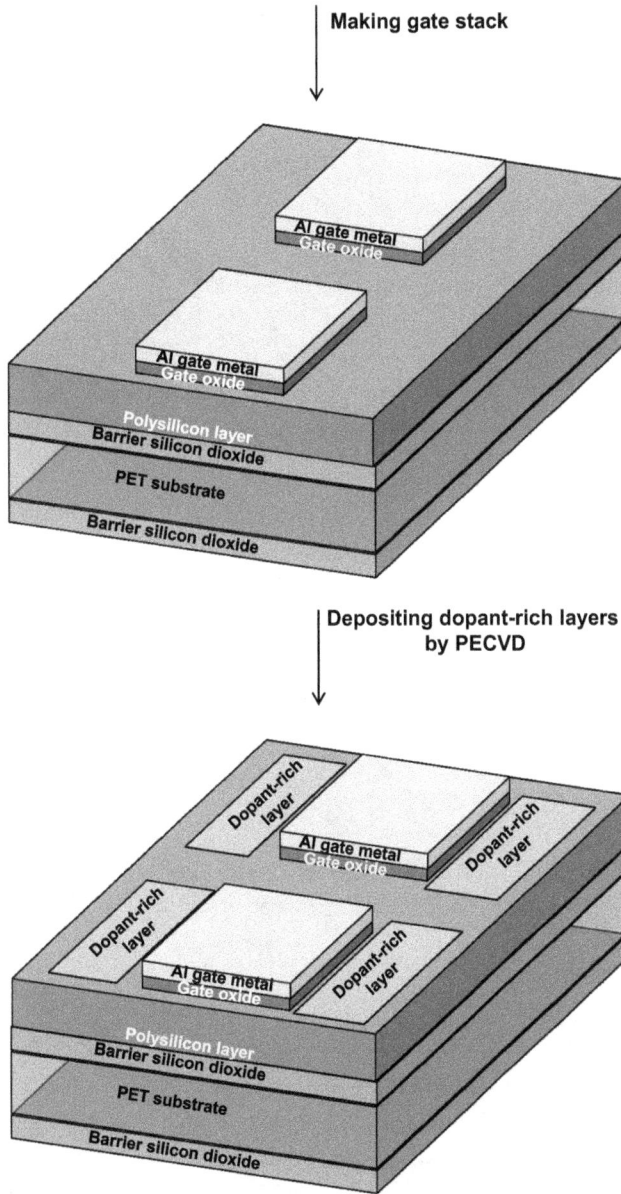

Figure 2.1. (Continued.)

(x) A silicon dioxide layer for contact isolation is deposited by PECVD;
thickness = 4700 Å.

(xi) A contact isolation layer is patterned. Wet etching is done for contact hole
definition.

Dopant activation with laser

After dopant activation

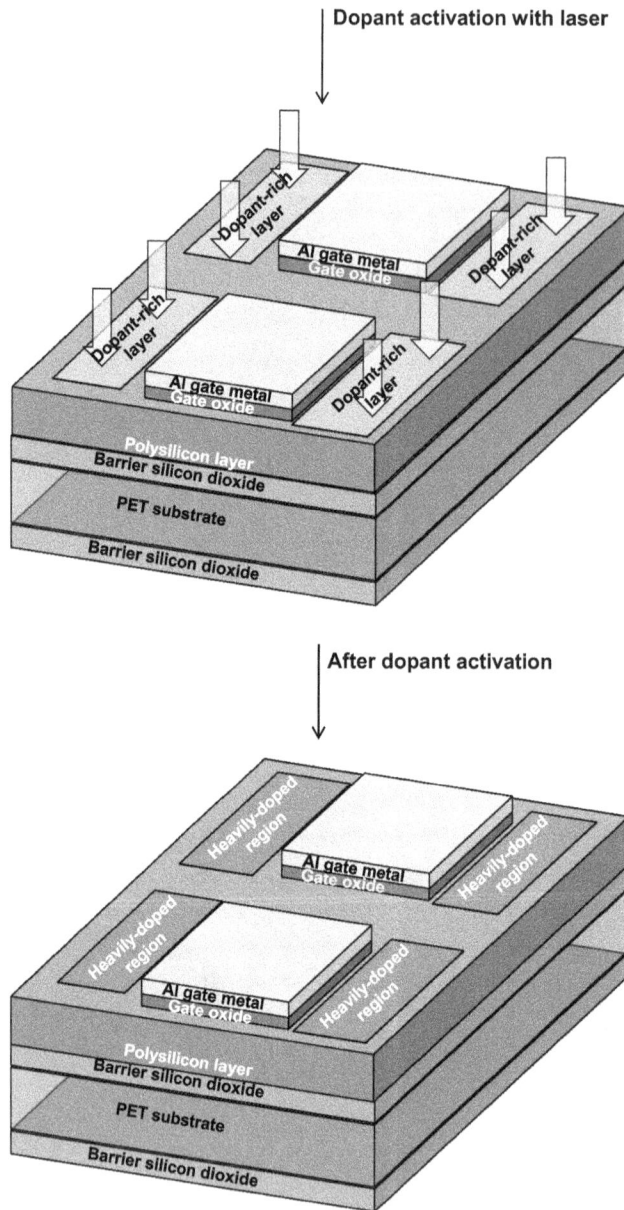

Figure 2.1. (Continued.)

(xii) An aluminum interconnection layer is formed by sputtering; thickness = 0.8 μm.

(xiii) Wet etching follows photolithography. After pattern delineation, the TFT structure is ready.

Dry etching silicon to
separate TFTs

Heavily-doped
region

Al gate metal
Gate oxide

Heavily-doped
region

Heavily-doped
region

Al gate metal
Gate oxide

Heavily-doped
region

Polysilicon layer

Barrier silicon dioxide

PET substrate

Barrier silicon dioxide

Contact isolation SiO$_2$ layer
by PECVD

Heavily-doped
region

Al gate metal
Gate oxide

Heavily-doped
region

Heavily-doped
region

Contact isolation SiO$_2$ layer

Al gate metal
Gate oxide

Heavily-doped
region

Polysilicon layer

Barrier silicon dioxide

PET substrate

Barrier silicon dioxide

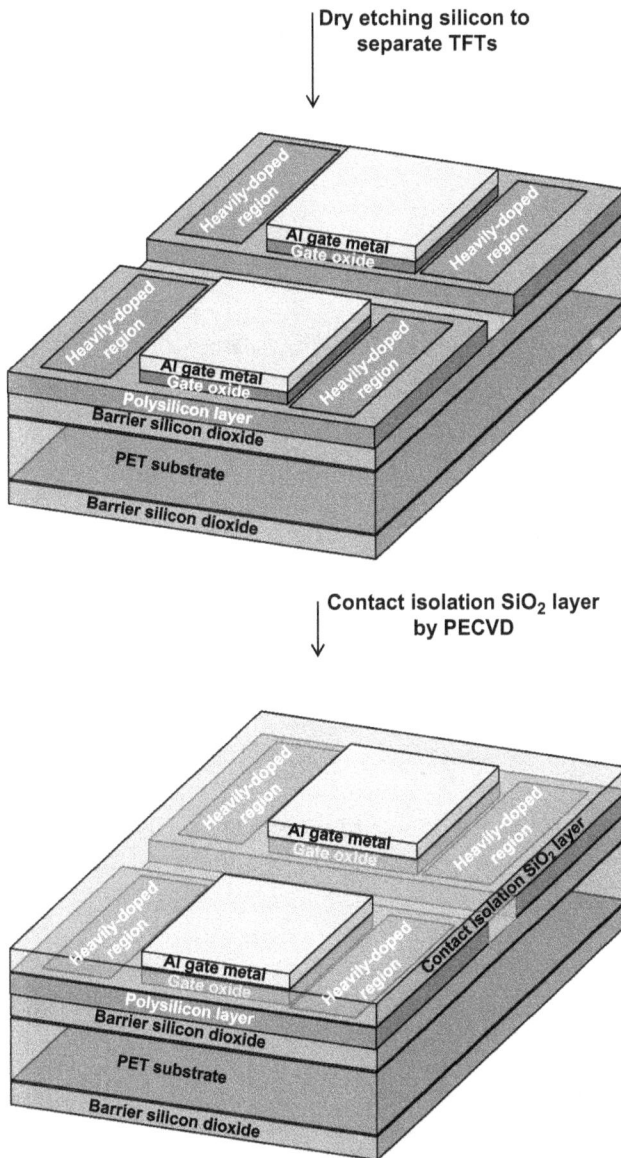

Figure 2.1. (Continued.)

(xiv) Post-fabrication annealing is done at 150 °C, which is higher than the allowed temperature. Annealing is done for 3 h. The substrate can tolerate the high annealing temperature at this stage because all photolithographic steps have been completed and no further photolithographic steps are to be done. This post-fabrication annealing step greatly improves the device characteristics.

Sputtering and patterning Al
interconnect layer

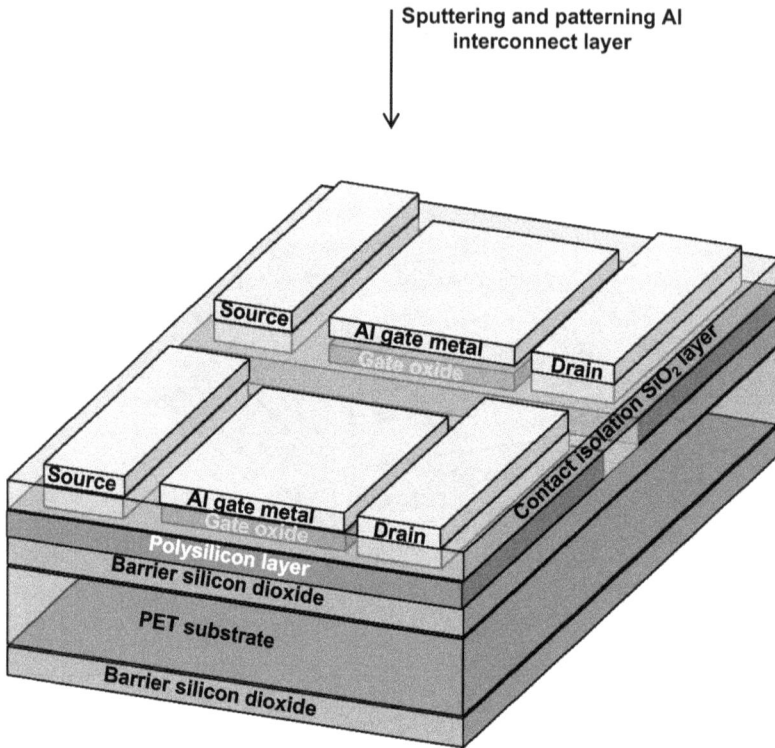

Figure 2.1. (Continued.)

2.2.4 TFT parameters

For the TFT with a W/L ratio of 20 μm/10 μm = 2, the on current is >100 μA, the off current is <100 pA/μm and the on–off current ratio is $>10^6$. The threshold voltage in the linear regime is 8 V. The effective mobility is >7.5 cm^2 V^{-1} s^{-1} (Carey *et al* 1997).

For the TFT with a W/L ratio of 100 μm/50 μm = 2, at $V_{DS} = 10$ V, the on current is 0.5 mA, the off current is <40 pA/μm and the on–off current ratio is 5×10^5. The threshold voltage in linear regime is 18 V with subthreshold slope of 2.9 V/decade. The saturation mobility is >60 cm^2 V^{-1} s^{-1} (Theiss *et al* 1998).

2.3 PolySi TFT on PES

2.3.1 TFT fabrication

Kim *et al* (2006) fabricated TFTs on PES (polyethersulfone) plastic substrates of thickness 200 μm, restricting the temperature to $\leqslant 200$ °C. PES plastic is highly resistant to heat and moisture effects. It has a high transparency ~76%. The PES substrate has an organic coating on both sides for the LTPS process. The sequence of process steps is as follows (figure 2.2):

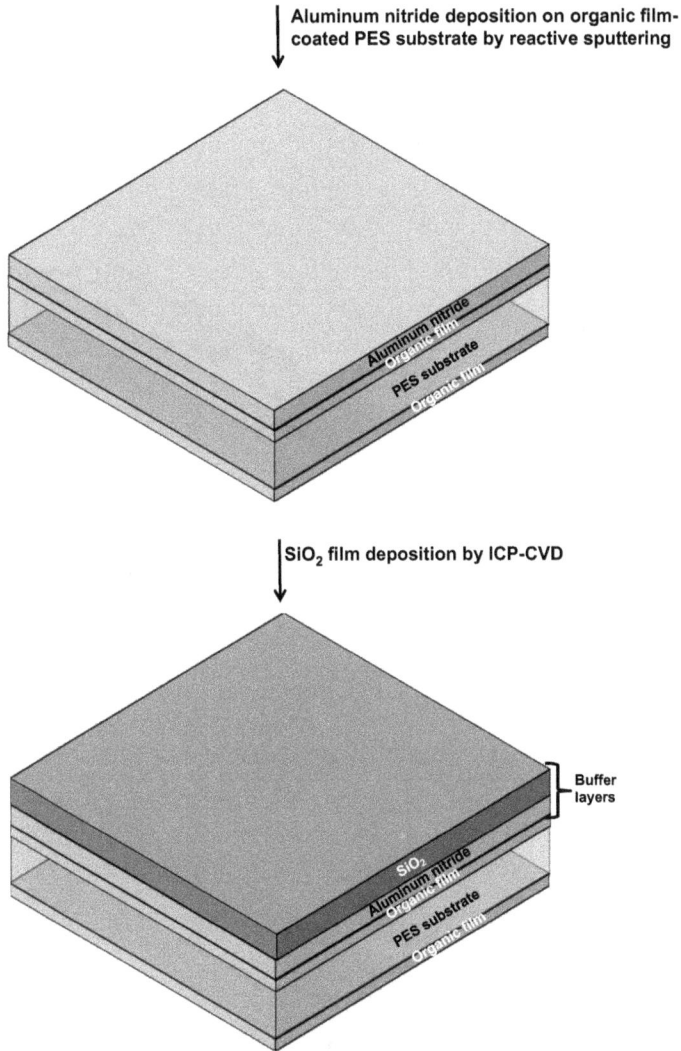

Figure 2.2. Polysilicon TFT fabrication on an organic-film-coated PES substrate.

(i) Over the organic-film-coated PES substrate, reactive sputtering is done in a nitrogen ambient to deposit a 100 nm thick aluminum nitride layer. This layer of high thermal conductivity increases the tolerable limit of laser energy density by absorbing thermal stress, thereby inhibiting delamination of sputtered a-Si film during ELC.

(ii) A low-temperature inductively-coupled plasma CVD (ICP-CVD) technique is used to cover the AlN layer with 200 nm thick SiO_2 film. The deposition temperature is 170 °C.

(iii) RF plasma sputtering is used with Xe inert gas at 200 W power for a-Si deposition. Pressure = 5 mTorr, deposition rate = 0.05 nm s^{-1}. The a-Si

RF sputtering of a-Si

Buffer layers

Amorphous silicon
SiO₂
Aluminum nitride
Organic film
PES substrate
Organic film

Annealing with excimer laser

Buffer layers

Amorphous silicon
SiO₂
Aluminum nitride
Organic film
PES substrate
Organic film

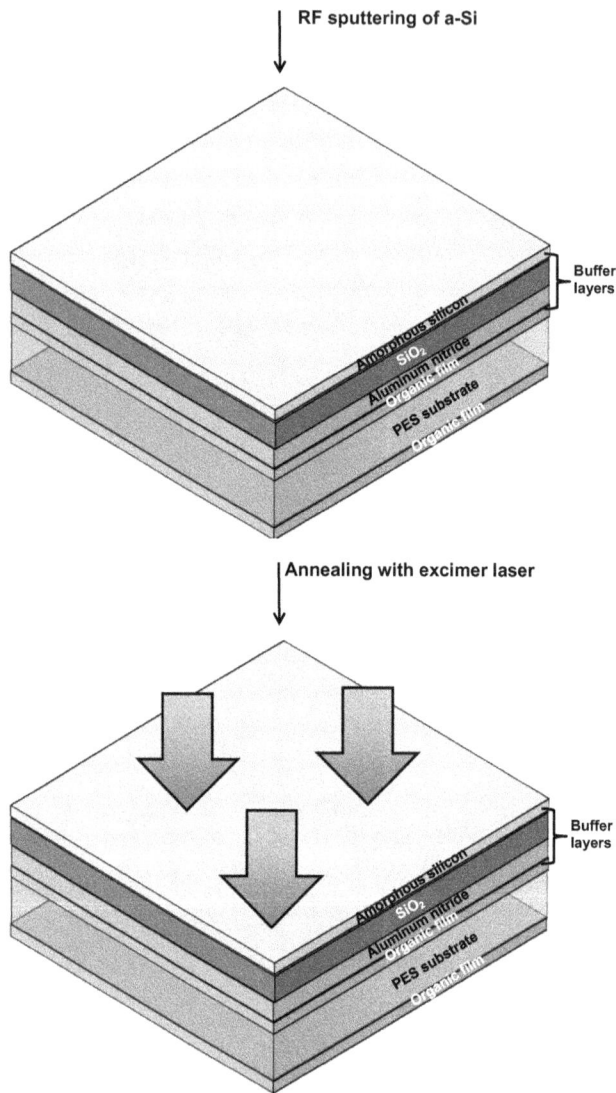

Figure 2.2. (Continued.)

layer thickness is 50 nm. The reason for preferring xenon gas over argon is the prevention of contamination of the sputtered film. Since the mass of Xe gas is three times that of Ar, the Xe sputtering enables the formation of a-Si film with comparatively low impurity content ∼0.39% with Xe as opposed to 1.1% with Ar.

(iv) The a-Si is crystallized with 308 nm XeCl laser.

 (v) After crystallization, the a-Si film is patterned into an island structure by photolithography.

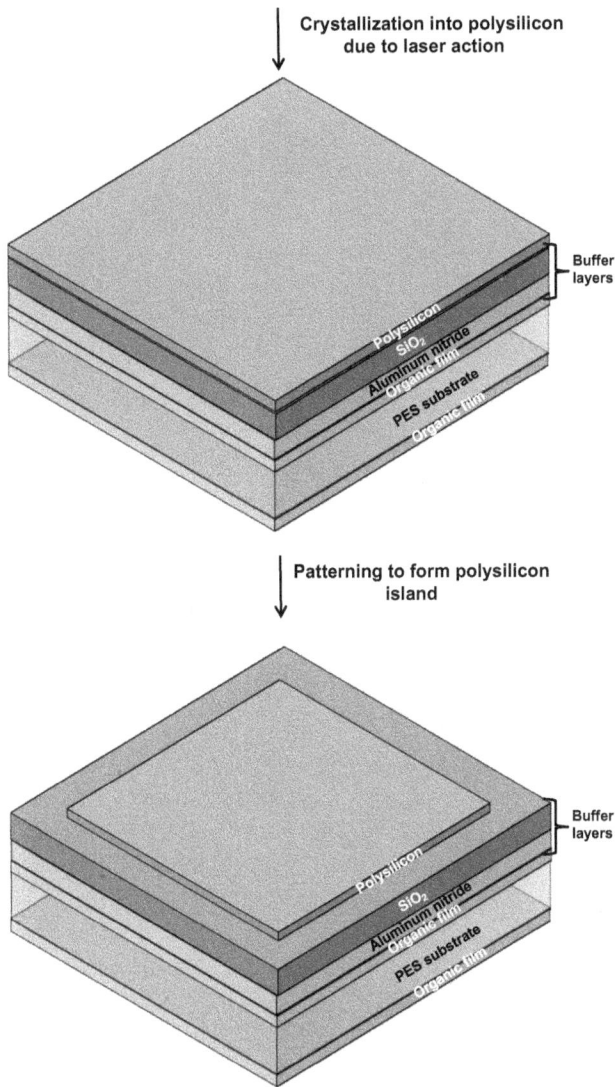

Figure 2.2. (Continued.)

(vi) ICP-CVD is used to deposit SiO_2 for gate oxidation.

(vii) Sputtering is used to deposit Al:Nd for gate metallization.

(viii) SiO_2 and Al:Nd films are patterned by photolithography to form the gate stack.

(ix) Source and drain implantations are done; dose $= 5 \times 10^{15}$ cm^{-2}.

(x) Dopant activation is done with XeCl laser at 150 mJ cm^{-2}.

(xi) Source and drain contacts are formed with aluminum.

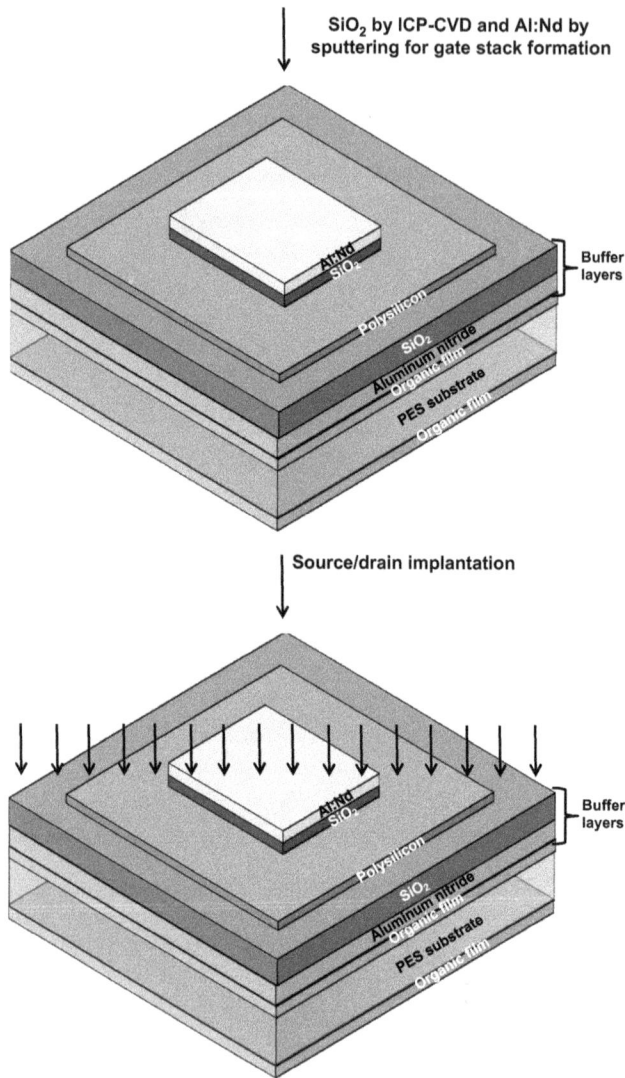

SiO$_2$ by ICP-CVD and Al:Nd by sputtering for gate stack formation

Buffer layers

Al:Nd
SiO$_2$
Polysilicon
SiO$_2$
Aluminum nitride
Organic film
PES substrate
Organic film

Source/drain implantation

Al:Nd
SiO$_2$
Polysilicon
SiO$_2$
Aluminum nitride
Organic film
PES substrate
Organic film

Buffer layers

Figure 2.2. (Continued.)

2.3.2 TFT performance

The field-effect mobility in TFTs is 15 cm^2 V^{-1} s^{-1}. The maximum grain size in polySi film is 1 µm. The TFTs are able to drive pixels of LCDs and OLEDs (Kim *et al* 2006).

2.4 PolySi TFT on PES or PAR

Han and Kim (2006) fabricated polySi PMOS TFTs on polyethersulfone (PES) or polyarylate (PAR) substrates (figure 2.3). Polyarylate is a transparent thermoplastic material (softening when heated and hardening on cooling).

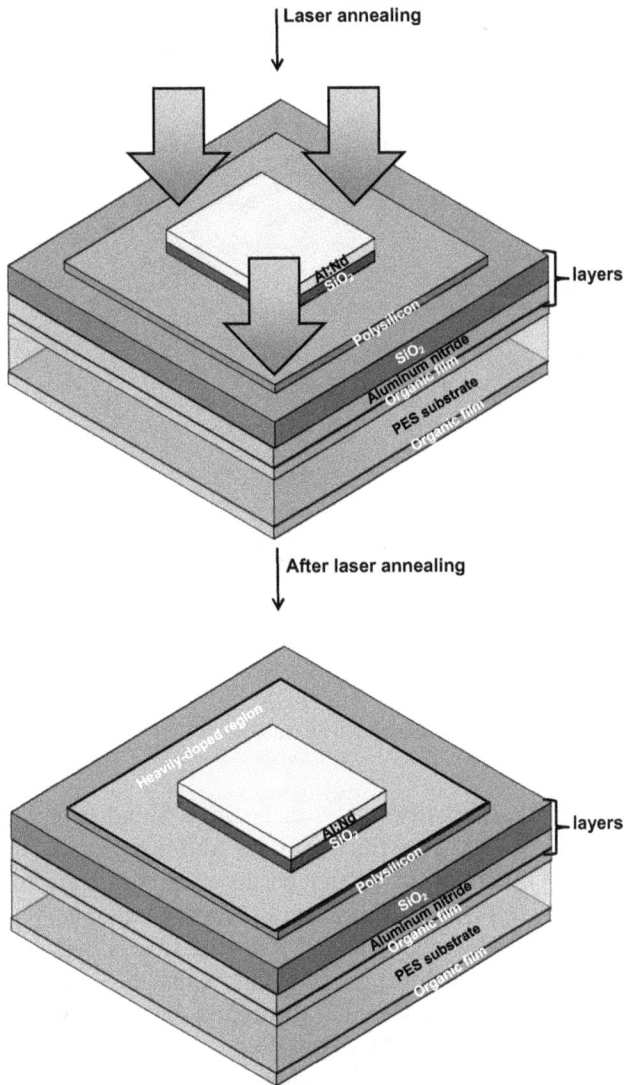

Figure 2.2. (Continued.)

2.4.1 TFT fabrication

For stress minimization, the substrates are vacuum annealed prior to the process: PES at 180 °C and PAR at 250 °C. A 600 nm thick SiO_2 buffer layer is deposited on the substrate followed by a-Si precursor film deposition by RF sputtering in Ar/He mixture at a ratio 2:20 to reduce Ar content in a-Si film. A XeCl laser (wavelength = 308 nm) with a flat top beam profile (area = 45×0.2 mm^2) is used for a-Si crystallization at a pulse duration of 35 ns. The polySi grain size increases from 10–400 nm as the laser energy density rises from 200–289 mJ cm^{-2}. Thickness of the

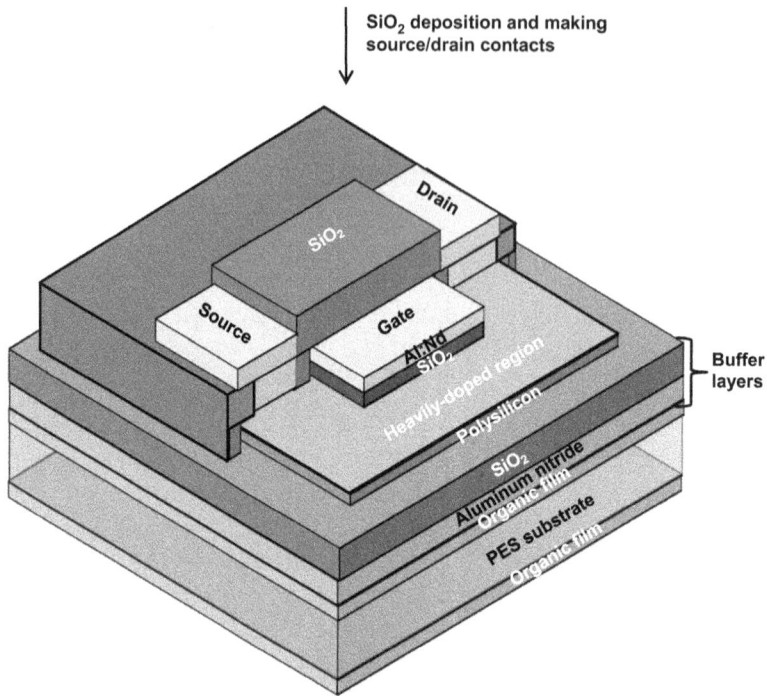

Figure 2.2. (Continued.)

SiO$_2$ buffer layer is critical because the laser beam is found to ablate the a-Si film when the buffer layer is 400 nm thick but no ablation occurs for buffer layer thickness $\geqslant$600 nm. This ablation occurs because the substrate of poor thermal stability incurs a high-temperature rise during the laser pulse.

A 200 nm thick gate dielectric for the PMOS structure is deposited by PECVD. The gate electrode is AlNd of thickness 200 nm deposited by sputtering. Then ion implantation is done for ohmic contact formation between Si and source/drain metal. The acceleration voltage is 10 kV. Ion dose is 1×10^{15} cm^{-2} for phosphorous, for boron, it is 8.2×10^{15} cm^{-2}. The sheet resistance decreases rapidly with increasing laser energy density. For P, it decreases from 11.5 to 2.5 kΩ/sq in the energy density range 150–213 mJ cm^{-2}, for B, it decreases from 4.5 to 0.5 kΩ/sq in the energy density range 150–267 mJ cm^{-2}. After ion activation by laser, interlayer dielectric is deposited; film thickness is 400 nm. Contact holes are defined. Finally, 300 nm thick AlNd is deposited for source and drain metallization. Source and drain electrodes are patterned.

2.4.2 TFT parameters

The PMOS TFTs have a threshold voltage = -1.5 V. The current on–off ratio is 10^5 and the field-effect mobility is 63.6 cm^2 V^{-1} s^{-1} (Han and Kim 2006).

SiO$_2$ buffer layer deposition on PES/PAR
substate

SiO$_2$ buffer layer
PES/PAR substrate

Amorphous silicon film deposition by RF
sputtering

Amorphous silicon
SiO$_2$ buffer layer
PES/PAR substrate

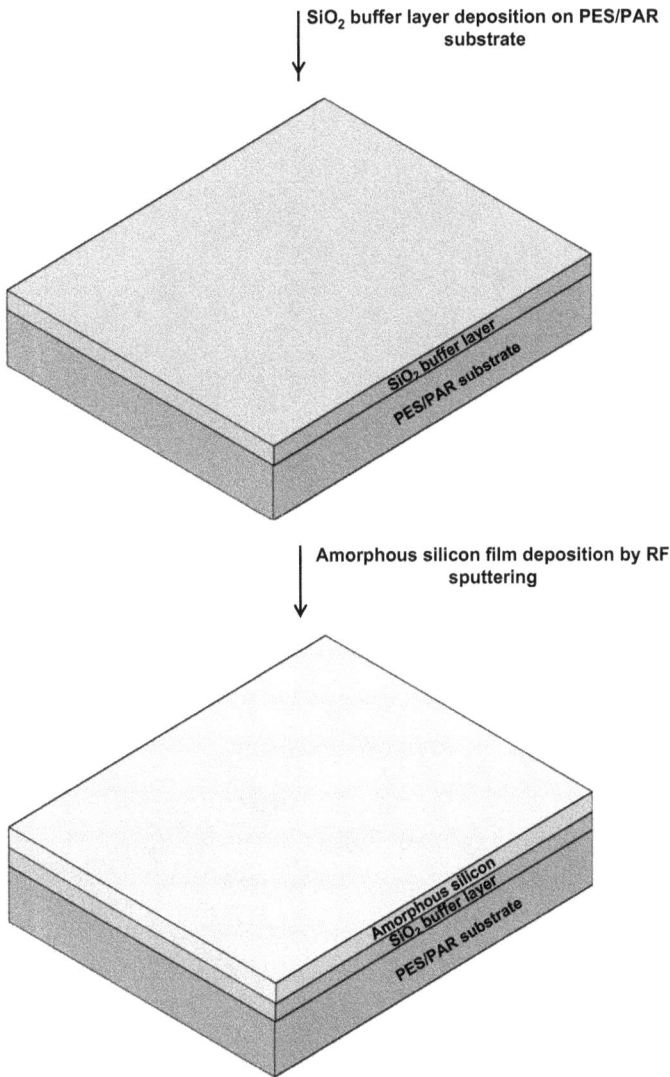

Figure 2.3. Polysilicon TFT on PES/PAR substrate.

2.5 PolySi TFT on plastic film by laminating on glass carrier

Lemmi *et al* (2004) report a technique which obviates practical difficulties in processing thin, flexible plastic substrates using standard automated manufacturing equipment. Here, the process begins by laminating the plastic film over a carrier glass wafer (figure 2.4). The convenience derived by this lamination step is that the plastic film can be handled easily. Various processing steps can be easily executed in the same way as for glass wafers using the same standard equipment. However, the

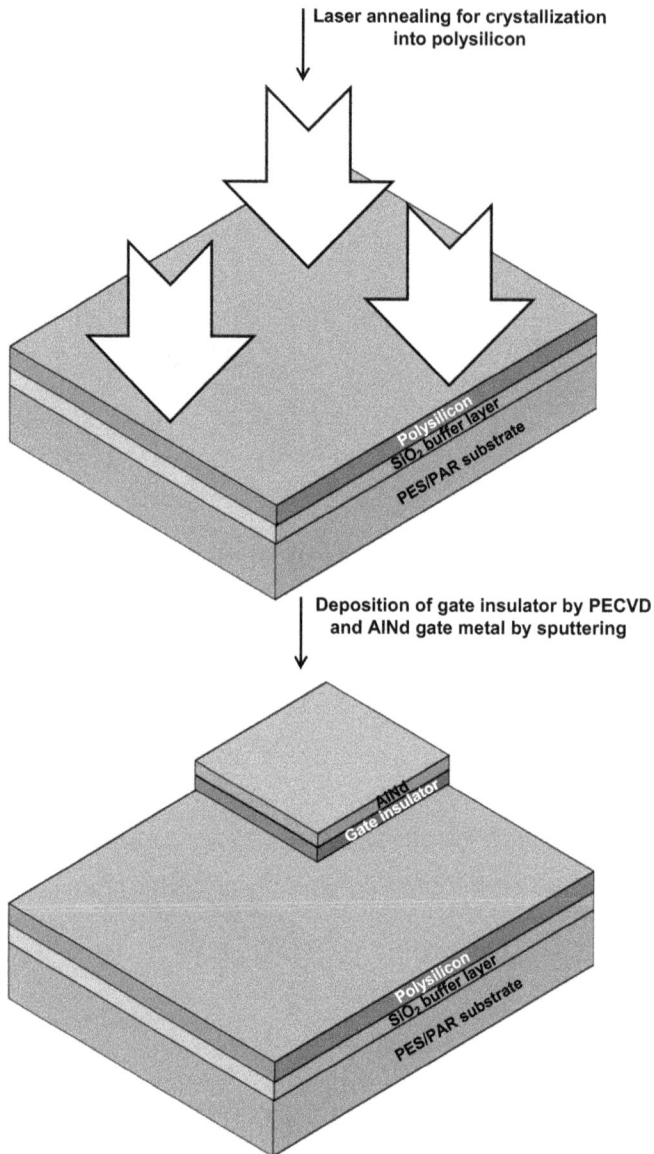

Figure 2.3. (Continued.)

process has to be carried out a temperature <105 °C because the adhesion between plastic and glass may be weakened at higher temperatures. After the process has been completed, the plastic film is delaminated and removed from carrier glass wafer. In case the laminated film is able to sustain higher temperatures such as for polyimide, the TFT is annealed at 320 °C in forming gas. This step serves to reduce leakage current.

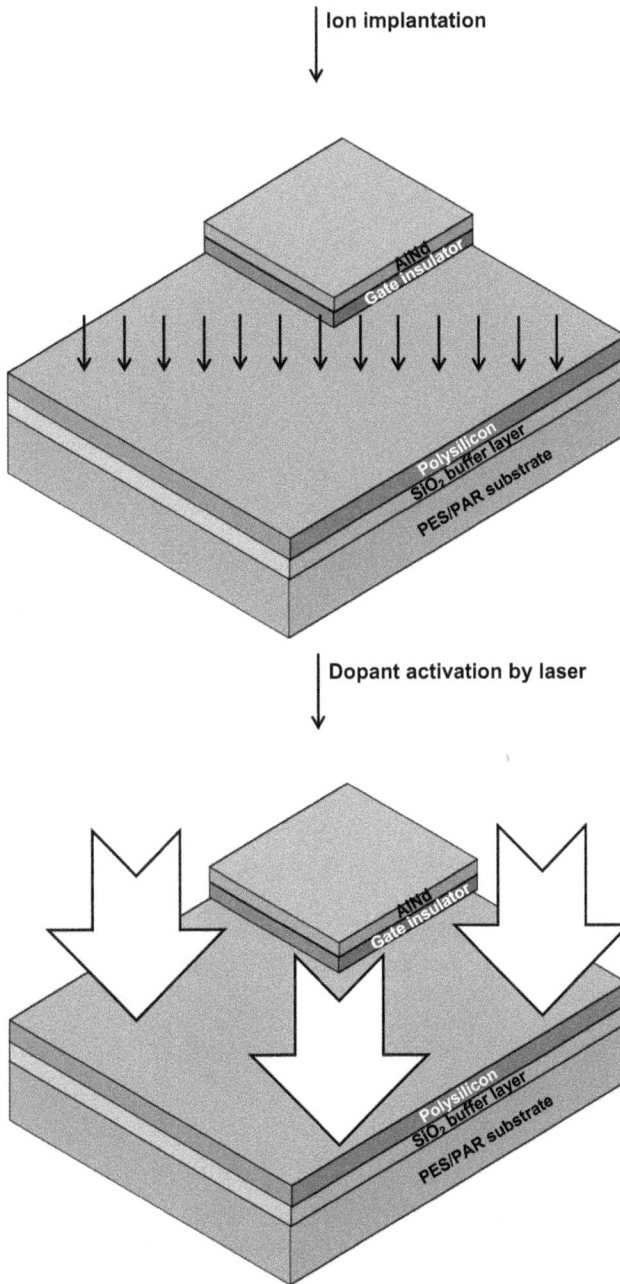

Figure 2.3. (Continued.)

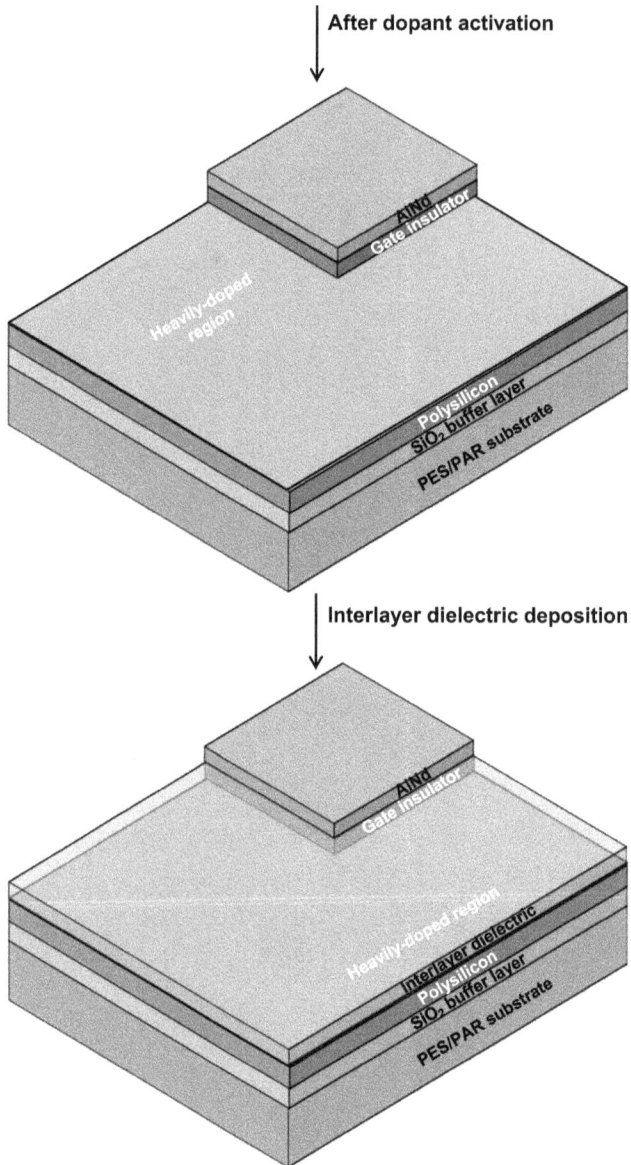

Figure 2.3. (Continued.)

2.5.1 TFT fabrication

The fabrication process begins with barrier layer deposition on both sides of the wafer followed by its lamination on a glass wafer. Then a quarter wavelength Bragg reflector is deposited using PECVD SiO$_2$ and SiN$_x$ films to maximize reflectivity towards XeCl laser (Lemmi *et al* 2004). Amorphous Si is deposited by sputtering. Crystallization is done with an excimer laser keeping the fluence below the full-melt

Making contact holes and depositing AlNd
as source/drain metal

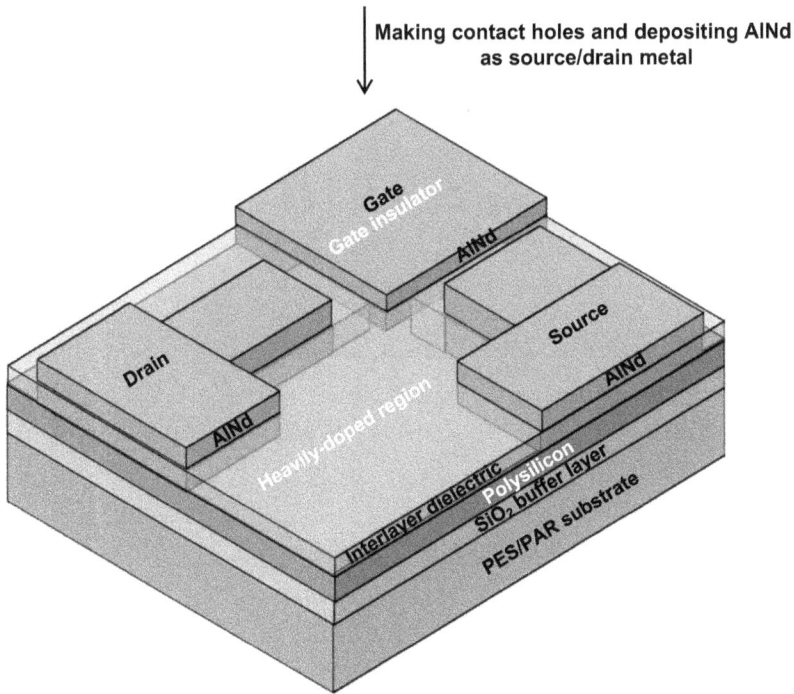

Figure 2.3. (Continued.)

threshold. By this fluence adjustment, a uniform polySi grain size is obtained across the wafer. Channel thickness is 500 Å. The CVD gate oxide is 1000 Å thick. The gate electrode is defined and the gate oxide is etched to expose the source/drain areas. Boron ions are implanted; energy 10 keV and dose 2×10^5 atoms cm^{-2}. Dopants are activated with an excimer laser. Al film protects the channel region from laser effects. The quarter wavelength reflector protects the plastic film between the TFTs. Depositions of interlayer dielectrics, metal interconnections and passivation films complete the process.

2.5.2 TFT parameters

The P-channel TFT has a threshold voltage of -4.5 V and a subthreshold slope of 600 mV/decade. Its leakage current is 2 pA. The field-effect mobility is 65 cm^2 V^{-1} s^{-1} (Lemmi *et al* 2004).

2.6 Low-temperature <425 °C polySi TFT by SUFTLA

Shimoda and Inoue (1999), and Inoue *et al* (2002) developed a technology by which polySi TFT is fabricated on a conventional hard substrate such as glass, and post-fabrication transferred to another substrate, e.g. plastic without any effect on device performance characteristics. By this arrangement, the advantages of processing on

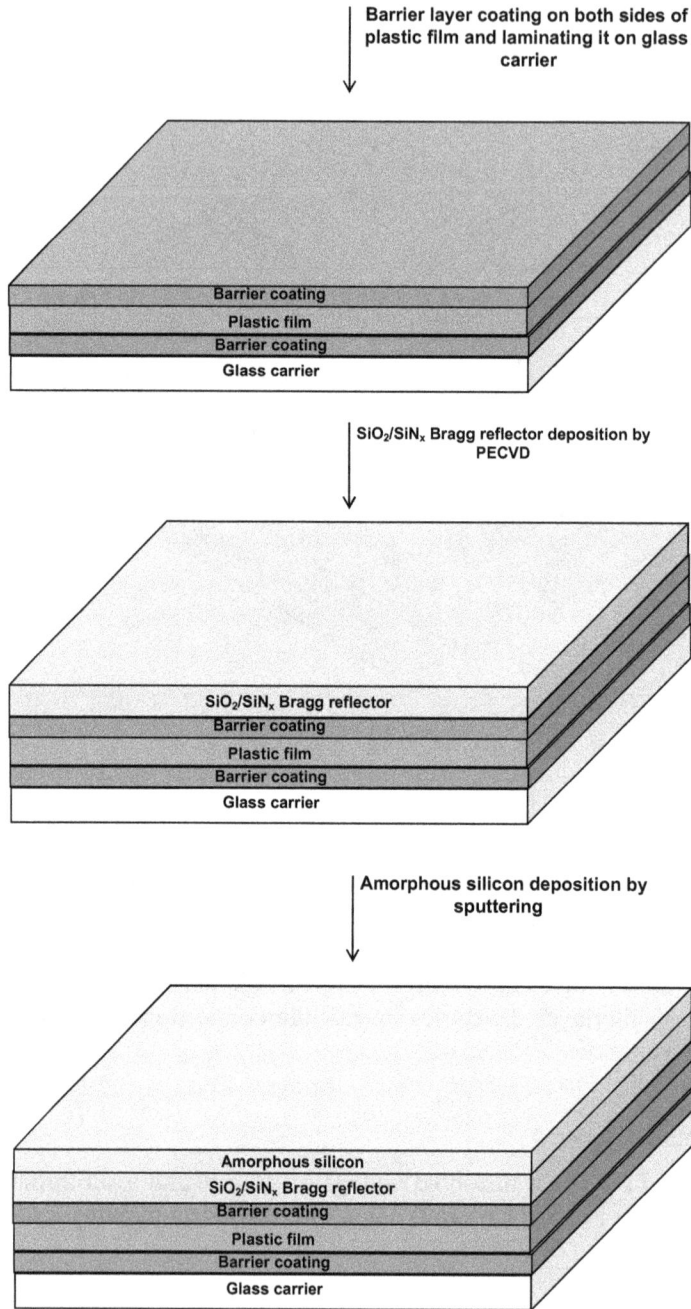

Figure 2.4. Polysilicon TFT fabrication by laminating plastic film on a glass carrier.

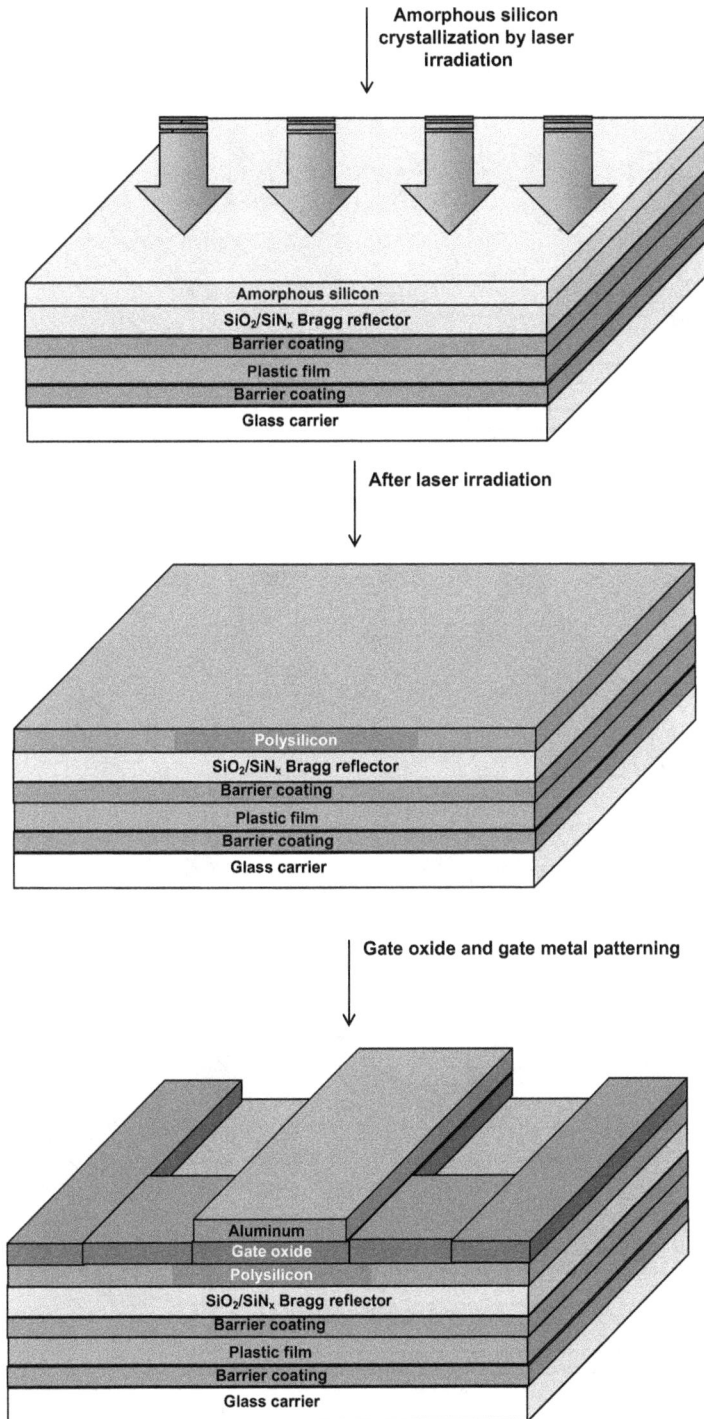

Figure 2.4. (Continued.)

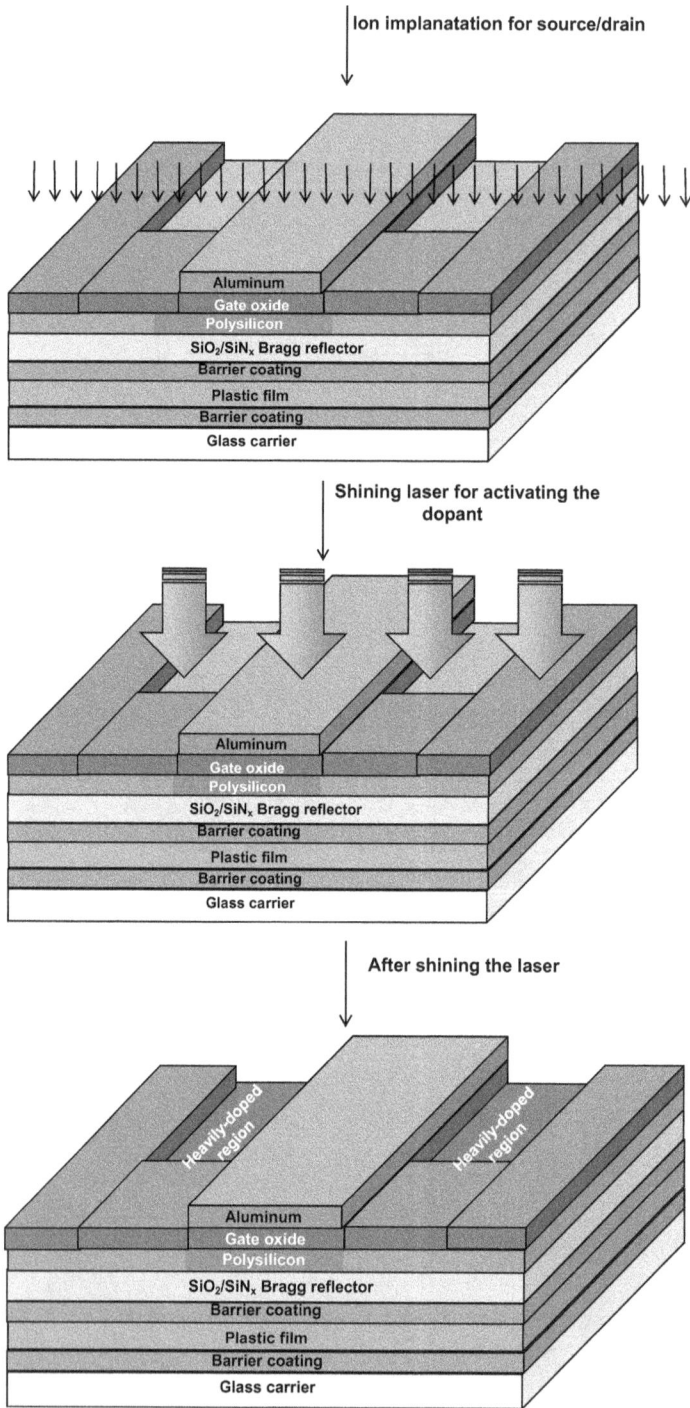

Figure 2.4. (Continued.)

Interlayer dielectric deposition and
source/drain metallization

Source | Drain

Mo/Al | Mo/Al

Gate

Aluminum
Gate oxide
Polysilicon
SiO₂/SiNₓ Bragg reflector
Barrier coating
Plastic film
Barrier coating
Glass carrier

Delamination of plastic film from
glass carrier

Source | Drain

Mo/Al | Mo/Al

Gate

Aluminum
Gate oxide
Polysilicon
SiO₂/SiNₓ Bragg reflector
Barrier coating
Plastic film
Barrier coating

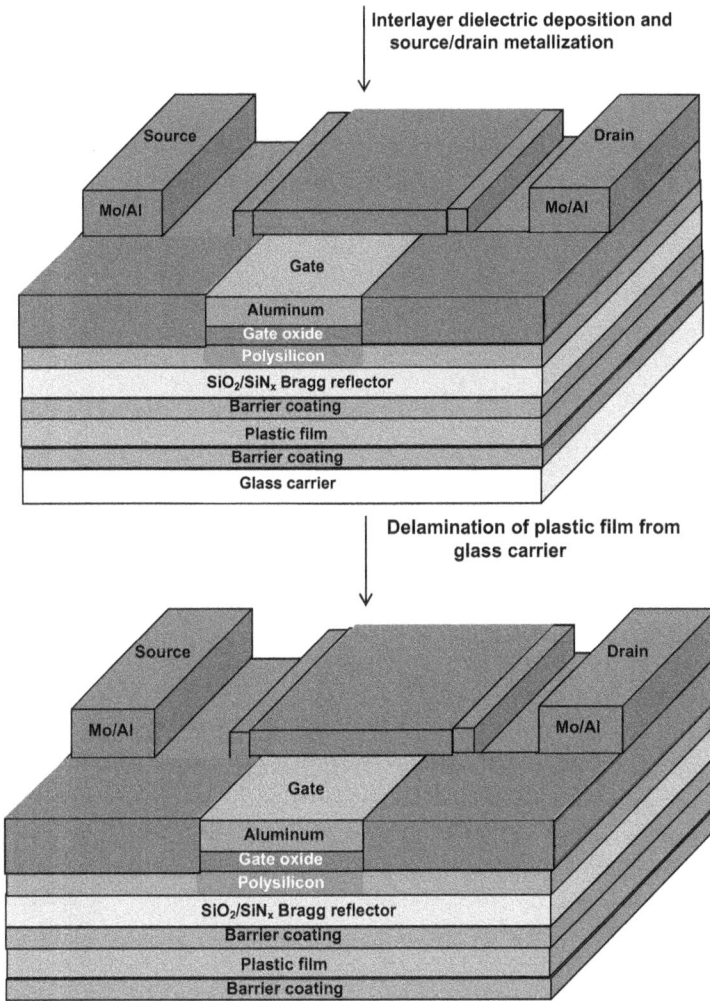

Figure 2.4. (Continued.)

conventional glass substrates can be extended to plastic substrates. The arrangement
has manifold advantages.

Firstly, as the deposition temperatures for active channel layer and gate dielectric
of TFT are drastically lowered for plastic substrates, the qualities of these films are
degraded so that the full capability of polysilicon remains unutilized. By making the
TFT on glass, the full capability of polysilicon possible with high-quality films
deposited at high processing temperatures is realized.

Secondly, the expansion of the plastic film poses problems in alignment of
successive masking layers during photolithography. If the processing is done on a
rigid glass substrate, these errors will be eliminated, greatly improving the manufacturing yield.

Thirdly, processing on plastic substrates is altogether different from that on glass substrates in terms of chemicals used and thermal treatments. Therefore, many processes need to be thought out again for successfully carrying the plastic substrate through the various process stages. Obviously, non-requirement of changing processes or facilities is a great boon. Thus, the possibility of transferring the fabricated TFT from the glass substrate to plastic substrate brings enormous opportunities.

2.6.1 TFT fabrication

TFT fabrication steps are as follows (Shimoda and Inoue 1999, and Inoue *et al* 2002) (figure 2.5):

(i) The main idea of SUFTLA is that the TFT is fabricated on an exfoliation layer on the glass substrate. After completion of the process, this layer is peeled away or shed off by laser irradiation to facilitate TFT transfer to the plastic substrate. The exfoliation layer is a 100 nm thick a-Si film. The film is deposited by low-pressure chemical vapor deposition (LPCVD) in Si_2H_6 at 425 °C.

(ii) Buffer oxide film of 100 nm thickness is deposited on the a-Si exfoliation layer by electron cyclotron resonance CVD (ECR-CVD) in vacuum at 100 °C.

(iii) LPCVD facility is used for amorphous silicon film formation over the buffer oxide. This a-Si film is the layer in which the active channel of the device is formed. The deposition parameters for a-Si are taken to be the same as in step (i). The thickness of a-Si film is 50 nm.

(iv) The a-Si is crystallized by irradiation with XeCl excimer laser. The laser has a wavelength of 308 nm. Its beam size is 150 mm × 0.4 mm. Energy density of 380 mJ cm^{-2} is used.

(v) ECR-CVD method is used for gate oxide formation. Oxide thickness is 120 nm.

(vi) TaN gate electrode is formed.

(vii) Source and drain regions are formed by ion implantation. Phosphorous ions are implanted for N-channel TFT. Energy = 80 keV, dose = 7×10^{15} cm^{-2}. Boron ions are implanted for P-channel TFT. Energy = 80 keV, dose = 5×10^{15} cm^{-2}.

(viii) Dopants are activated at 300 °C for 3 h in N_2.

(ix) The remaining process includes deposition of SiO_2 as interlayer insulator, ITO for pixel electrodes, Al interconnections after contact hole opening, and SiO_2 protection layer.

2.6.2 SUFTLA process

SUFTLA process steps are (Shimoda and Inoue 1999, and Inoue *et al* 2002):

First transfer step: The glass substrate carrying processed TFT (hereafter called the original substrate) is glued to a transfer substrate of glass by a temporary

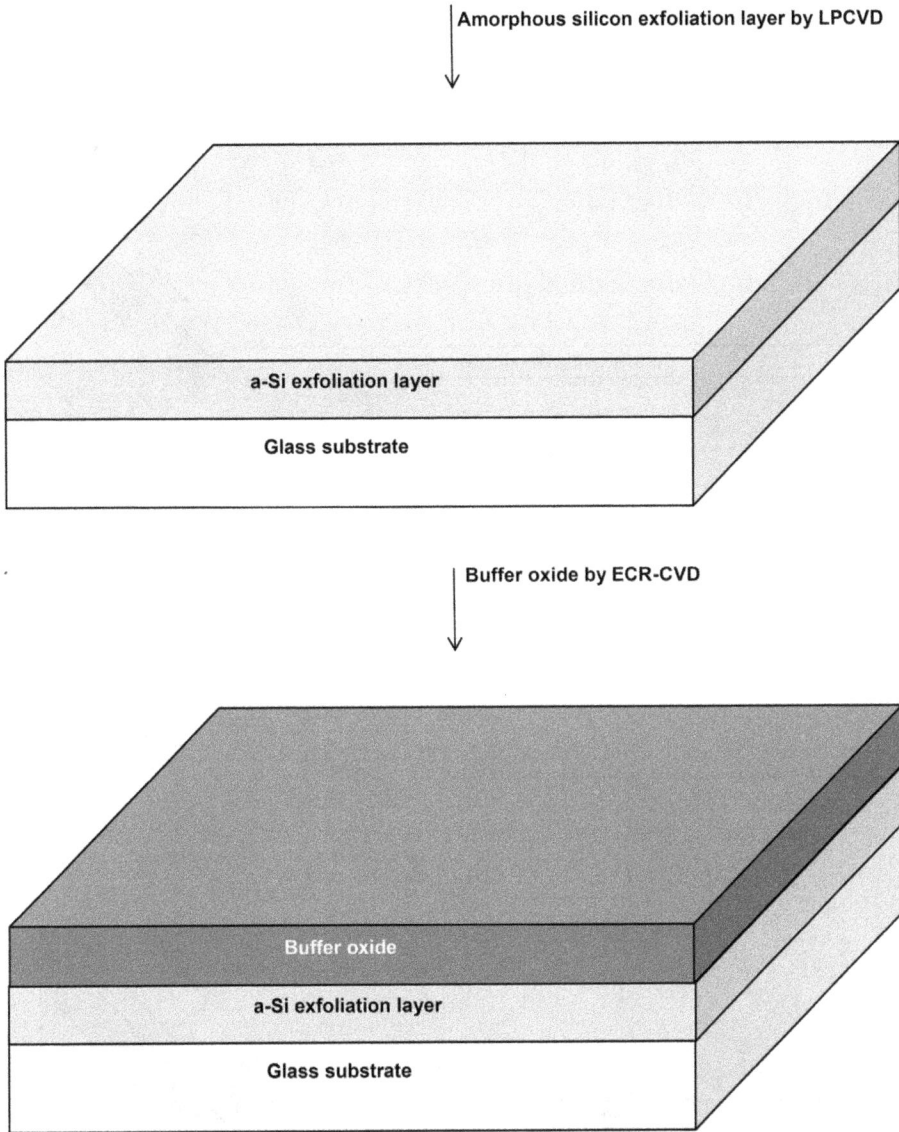

Amorphous silicon exfoliation layer by LPCVD

a-Si exfoliation layer

Glass substrate

Buffer oxide by ECR-CVD

Buffer oxide

a-Si exfoliation layer

Glass substrate

Figure 2.5. SUFTLA process of polysilicon TFT fabrication.

water-soluble adhesive, which is hardened by UV exposure. The TFT side of the original substrate is in contact with the transfer substrate while the back side of the original substrate is exposed. When the XeCl laser beam strikes the back side of the original substrate, the a-Si exfoliation layer containing 5% hydrogen melts releasing the hydrogen and weakening the adhesion of this layer with the original

Amorphous silicon active channel layer by LPCVD

a-Si active channel layer

Buffer oxide

a-Si exfoliation layer

Glass substrate

XeCl excimer laser irradiation

a-Si active channel layer

Buffer oxide

a-Si exfoliation layer

Glass substrate

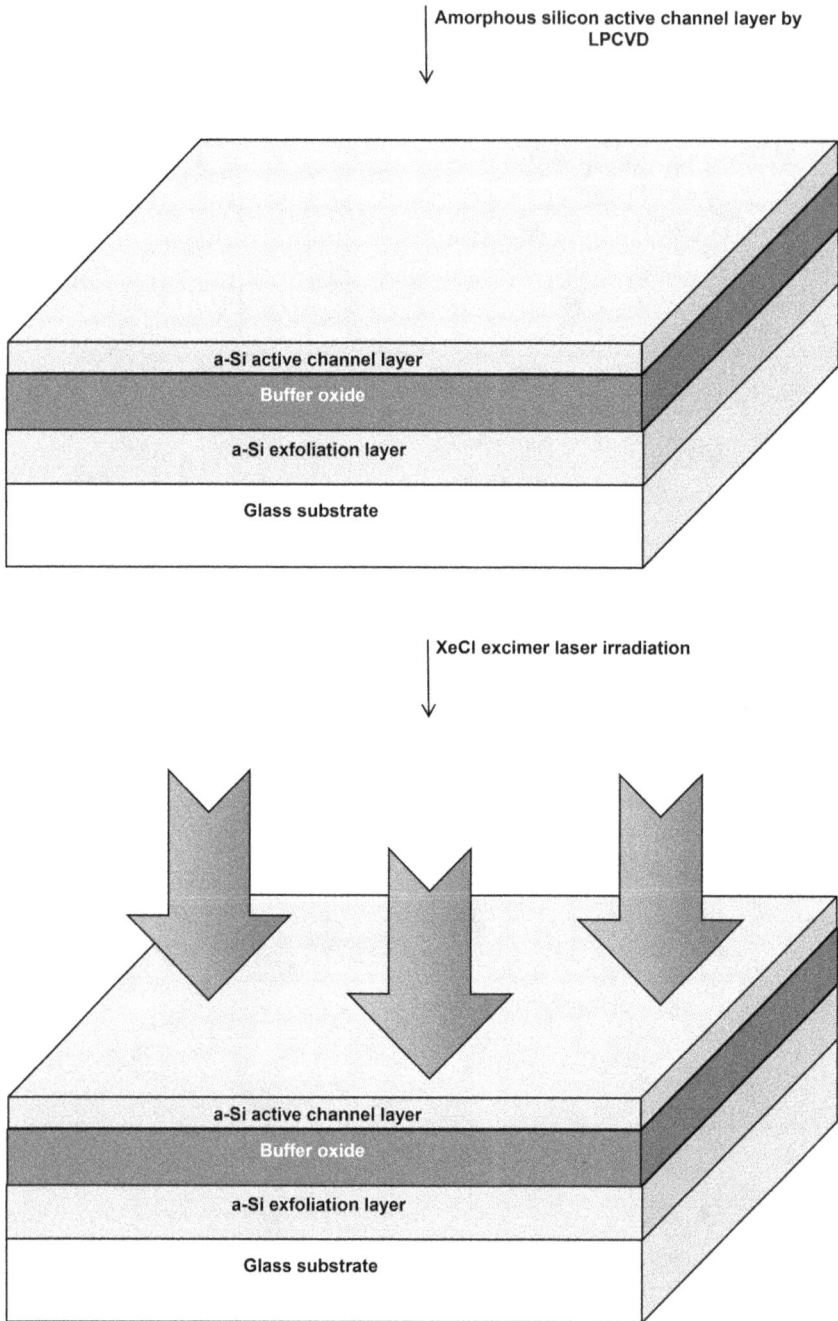

Figure 2.5. (Continued.)

After laser irradiation

Polysilicon active channel layer

Buffer oxide

a-Si exfoliation layer

Glass substrate

Gate oxide by ECR-CVD and TaN
gate electrode

TaN gate electrode

Gate oxide

Polysilicon active channel layer

Buffer oxide

a-Si exfoliation layer

Glass substrate

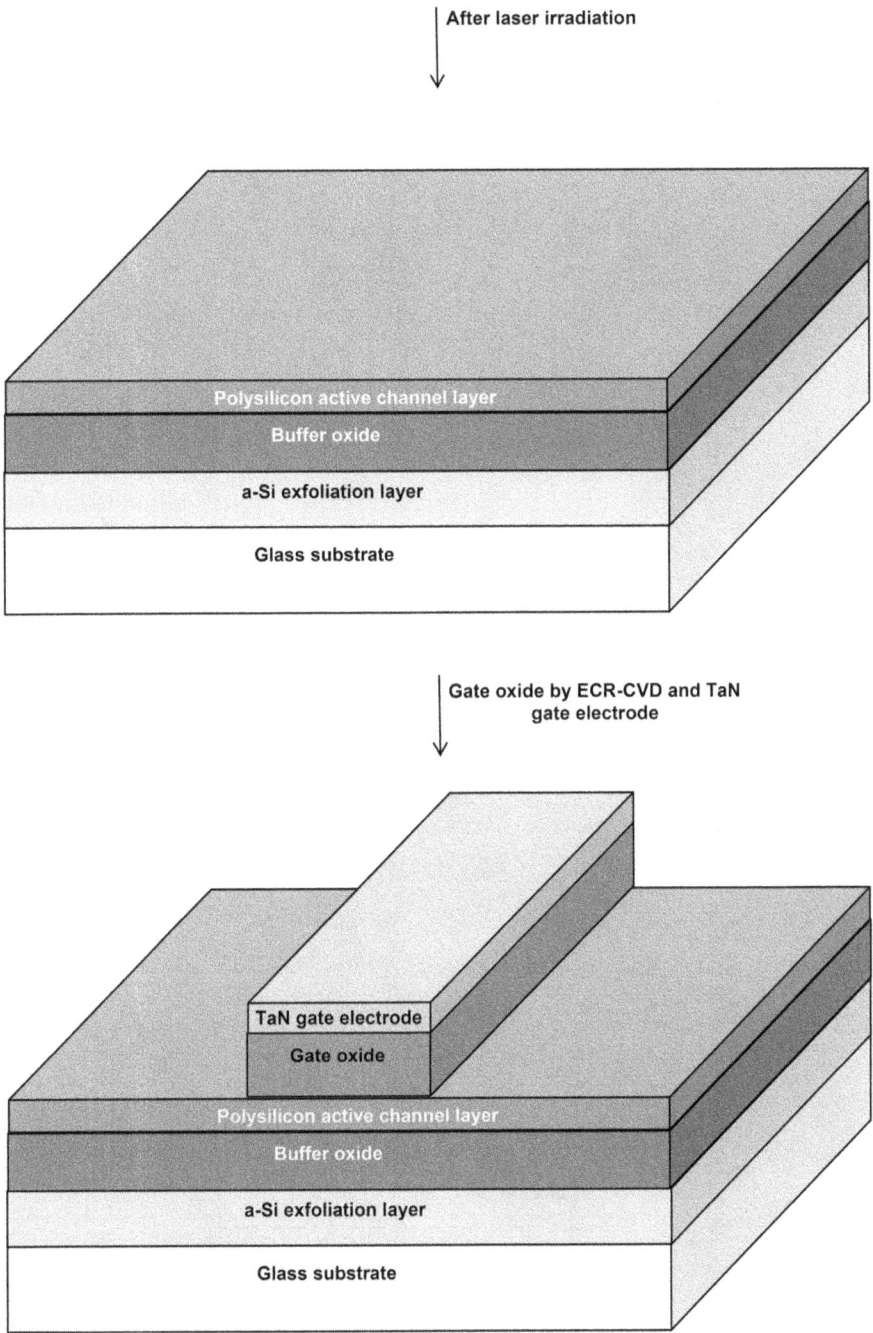

Figure 2.5. (Continued.)

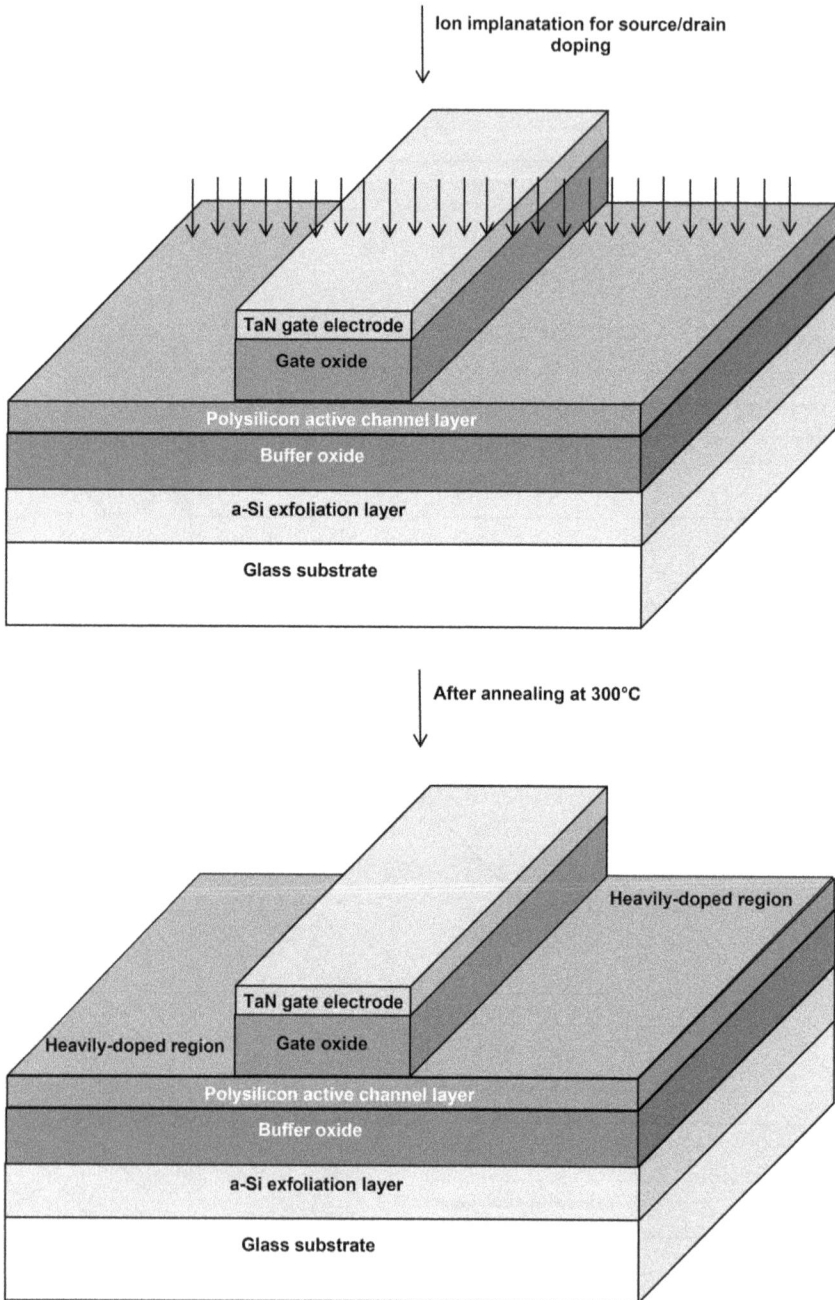

Figure 2.5. (Continued.)

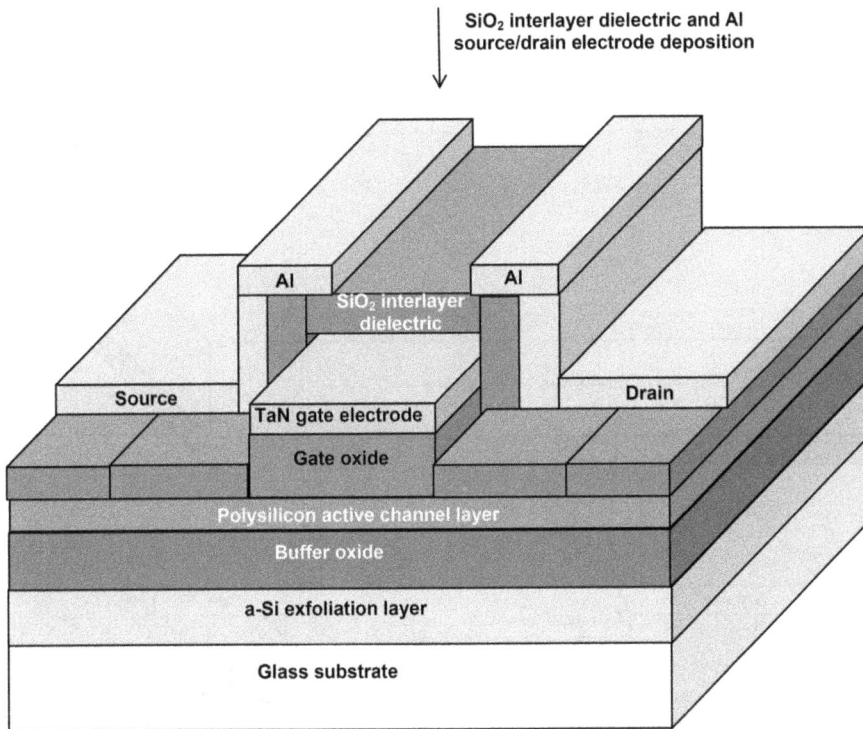

Figure 2.5. (Continued.)

substrate. Hence, the original substrate is set free. The TFT remains glued to the transfer substrate by temporary water-soluble adhesive. The question arises as to why the laser beam does not cause any damage to the TFTs? The reason is that for a laser wavelength of 308 nm, the absorption coefficient of silicon is 0.18 nm^{-1} so that 5.6 nm Si thickness completely absorbs the laser beam. Hence, it cannot penetrate beyond this depth to pose any danger to the TFT.

Second transfer step: The TFT glued to the transfer substrate by temporary water-soluble adhesive is stuck to 400 μm thick plastic film by a permanent adhesive which is water-insoluble and stiffens in UV radiation. The TFT side is in contact with the transfer substrate. The (transfer substrate + TFT + plastic film) combination is immersed in water where the water-soluble adhesive joining the transfer glass substrate and TFT is dissolved. The plastic film joined to the TFT with water-insoluble adhesive is separated and shed away.

TFT backplanes for LCDs with integrated drivers are transferred from the quartz–glass substrate to the plastic substrate by SUFTLA, and the TFT character-istics are found to be same after transfer as before transfer. An all-plastic substrate TFT-LCD is fabricated. Its display area is 0.7 in measured diagonally. The pixel count is 428×238 (Inoue *et al* 2002).

Fixing first transfer substrate with water-soluble adhesive

First transfer substrate

Water-soluble adhesive

Al   Al

SiO₂ interlayer dielectric

Source   Drain

TaN gate electrode

Gate oxide

Polysilicon active channel layer

Buffer oxide

a-Si exfoliation layer

Glass substrate (Original substrate)

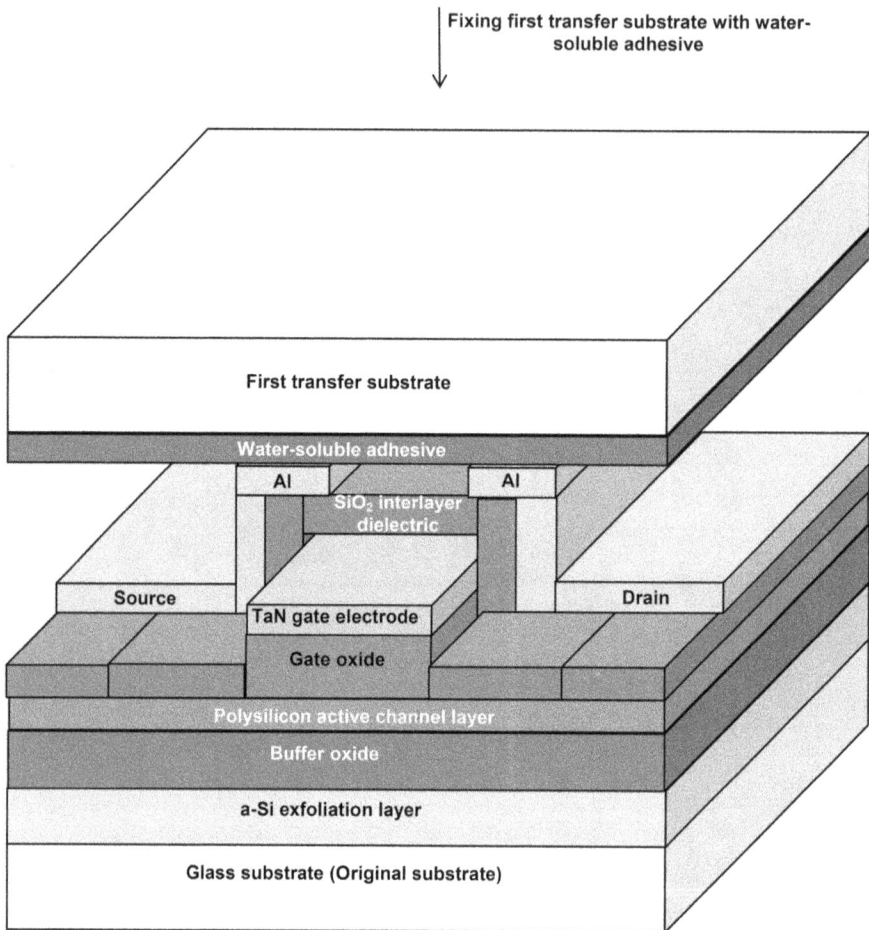

Figure 2.5. (Continued.)

2.7 TFTs on stainless steel foil

2.7.1 PolySi TFT on thin stainless steel foil

Serikawa and Omata (1999) used 100 μm thick stainless steel foils as substrates. Glow discharge sputtering is used for active Si and gate SiO_2 deposition; the latter is done with SiO_2 target in oxygen atmosphere. KrF excimer laser is used for crystallization. The on–off current ratio of polySi TFT is 10^6 and the mobility is $106 \ cm^2 \ V^{-1} \ s^{-1}$.

Serikawa and Omata 2000 reported N-channel and P-channel polySi TFTs on stainless steel foils. Process temperatures are below 200 °C. Mobility in the N-channel TFT is $106 \ cm^2 \ V^{-1} \ s^{-1}$; that in the P-channel TFT is $66 \ cm^2 \ V^{-1} \ s^{-1}$. The off current is 10^{-10} A with an on–off current ratio $= 1 \times 10^6$.

Shining laser beam on the back side of
original glass substrate causing melting of a-
Si exfoliation layer accompanied by hydrogen
liberation to reduce a-Si exfoliation
layer/substrate adhesion

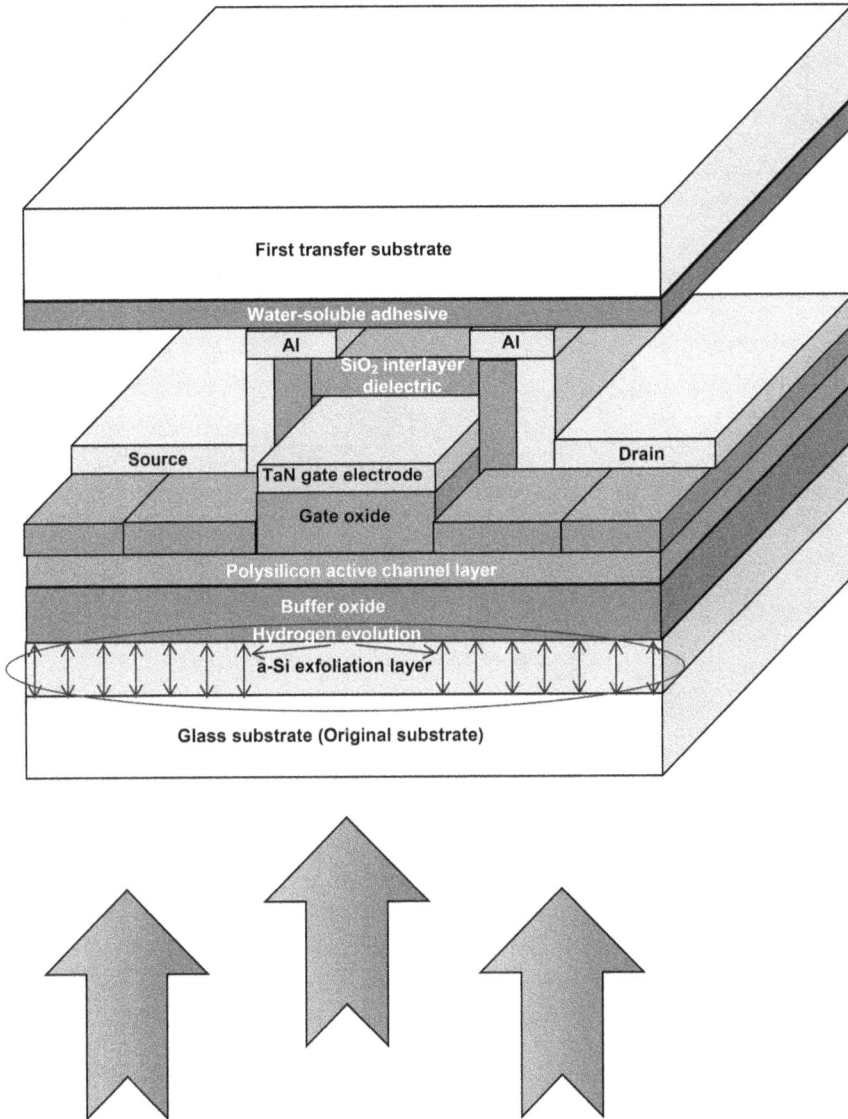

First transfer substrate

Water-soluble adhesive

Al

Al

SiO₂ interlayer
dielectric

Source

Drain

TaN gate electrode

Gate oxide

Polysilicon active channel layer

Buffer oxide

Hydrogen evolution

a-Si exfoliation layer

Glass substrate (Original substrate)

Figure 2.5. (Continued.)

Detachment of original glass substrate

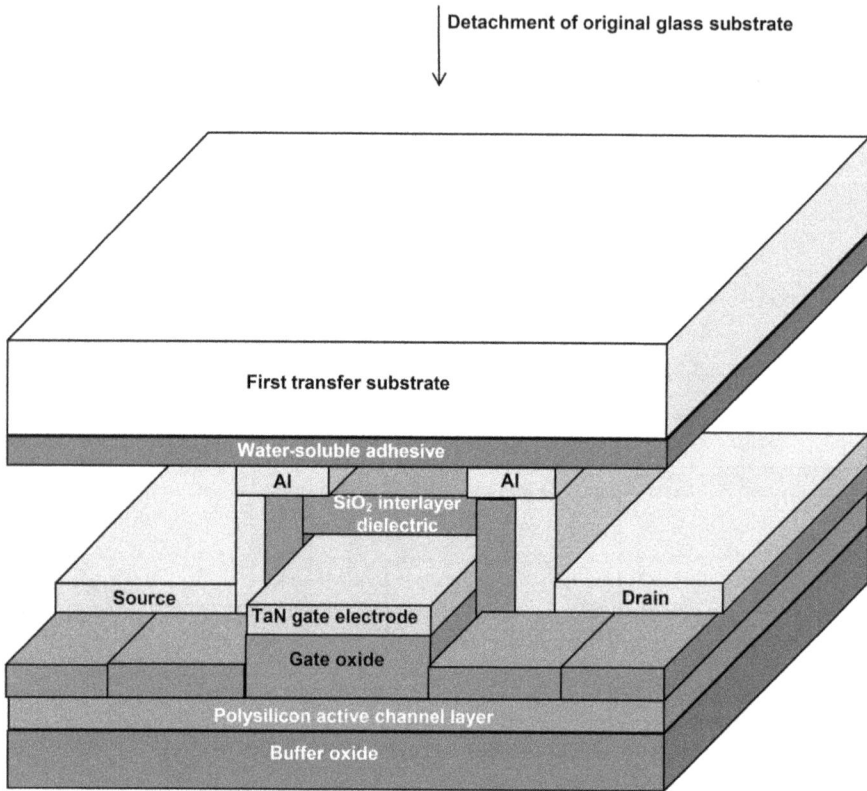

Figure 2.5. (Continued.)

2.7.2 PolySi TFT on thick stainless steel foil

Wu *et al* (2000) fabricated Al top-gated polySi TFTs with W/L ratio $= 180$ μm/ 45 μm $= 4$ on 304 stainless steel substrates with composition: iron 72 wt%, chromium 18 wt% and nickel 10 wt% (figure 2.6). They preferred furnace annealing of a-Si instead of laser induced crystallization. Furnace annealing of a-Si provides uniform transport properties over large areas due to its isothermal nature. It can be done in the case of a stainless steel substrate because it can be exposed to much higher temperatures than plastics.

The substrate thickness is 200 μm. Substrate cleaning with acetone and methanol is followed by application of spin-on-glass (210 nm) on both sides and then PECVD SiO_2 (270 nm) at 250 °C. The substrate temperature is ramped up from 450 °C to 810 °C at 5 °C per min for annealing SiO_2. Hydrogenated precursor a-Si film (160 nm) deposited at 150 °C is subjected to crystallization in a furnace at 650 °C for 1 h after exposing to hydrogen discharge for 1 h for inducing seeding.

After a-Si crystallization, N^+μc-Si (75 nm) is deposited at 350 °C to later serve as source/drain. By reactive ion etching (RIE), TFT islands are formed in the polySi layer. After μc-Si patterning by RIE, PECVD gate oxide (200 nm) is formed at 250 °C.

Flipping and fixing to plastic film with
permanent non-water
soluble adhesive

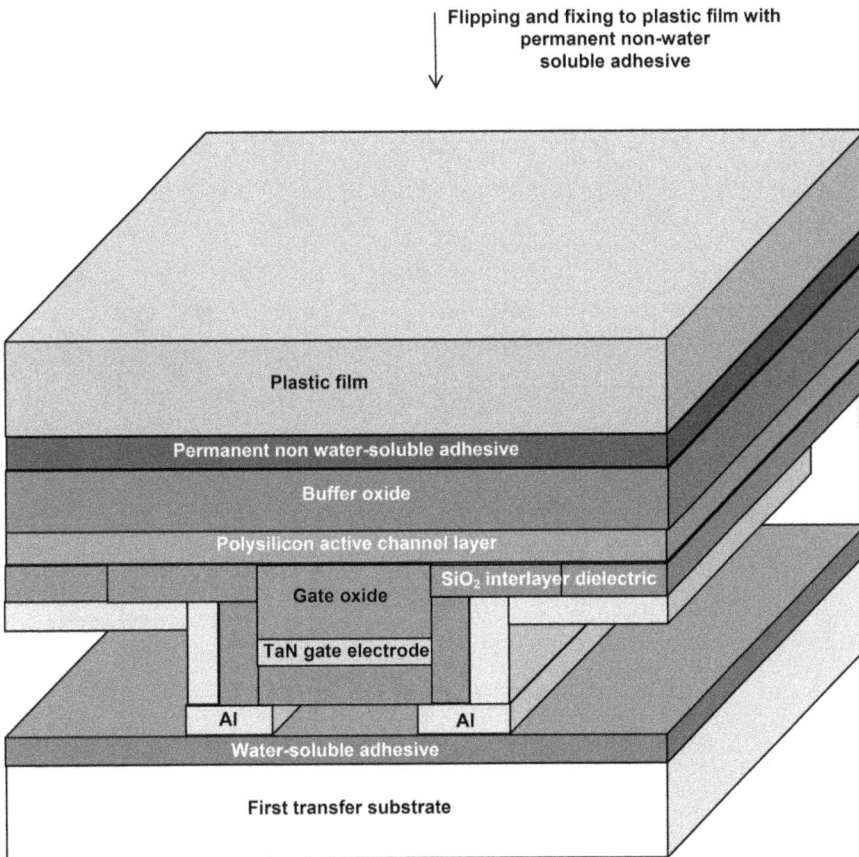

Figure 2.5. (Continued.)

Wet etching is done to open source/drain contact windows. Thermal evaporation of Al (200 nm) is followed by wet etching for defining gate, source and drain contacts. Annealing is done at 250 °C in (85 vol% nitrogen + 15 vol% hydrogen) for 15 min.

With $V_{DS} = 10$ V, the threshold voltage is 7.2 V, the subthreshold slope is 400 mV/decade and the off current is 10^{-9} A. The field-effect mobility of electrons is 64 cm^2 V^{-1} s^{-1} in both linear and saturation modes. (Wu *et al* 2000).

2.7.3 PolySi TFT circuit on stainless steel

Afentakis *et al* (2006) discuss the stainless steel substrate preparation, and examine the crystallization approaches such as solid-phase and laser methods to obtain polySi as well as techniques of thermal growth and CVD for gate dielectric formation, to develop an optimized TFT process.

Based on this process, they fabricate CMOS circuits. The CMOS inverter block in a ring oscillator circuit has a delay of 1.2 ns. The maximum frequency of shift register with static architecture is 1.25 MHz while that of shift register with dynamic architecture is 1.45 MHz.

Putting in water to dissolve the water-soluble adhesive and shed the first transfer substrate

Plastic film

Permanent non water-soluble adhesive

Buffer oxide

Polysilicon active channel layer

SiO₂ interlayer dielectric

Gate oxide

TaN gate electrode

Al   Al

First transfer substrate

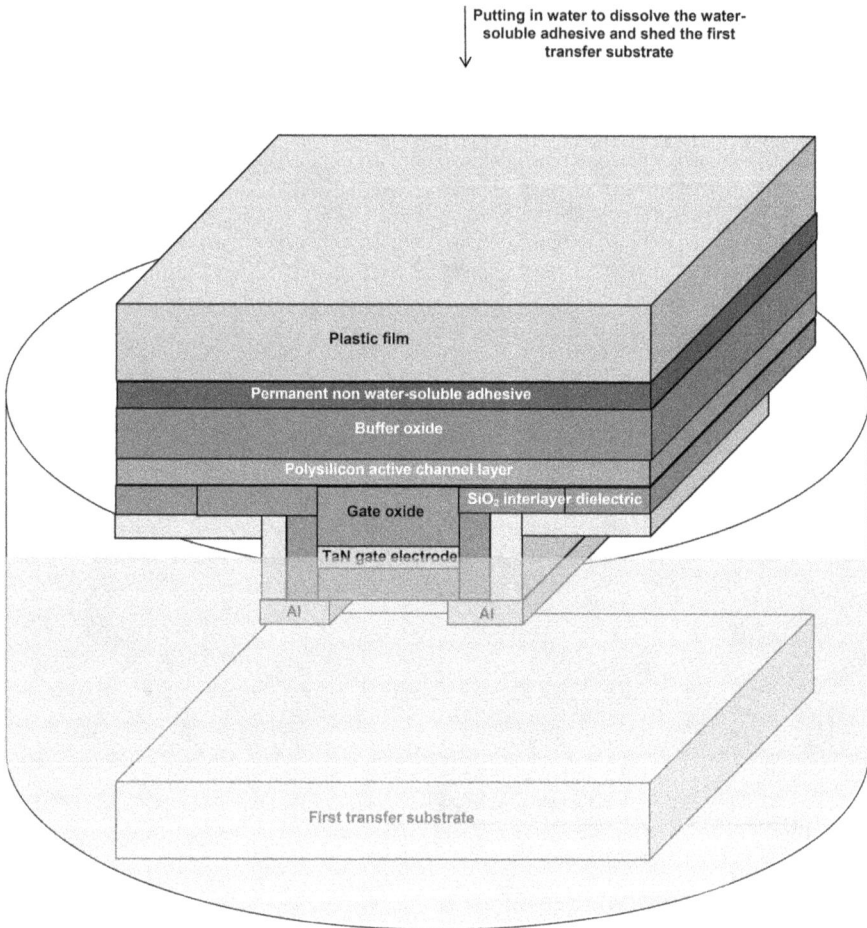

Figure 2.5. (Continued.)

The on–off current ratio of polySi TFT is $\sim 10^7$ and the field-effect mobility is 200 cm^2 V^{-1} s^{-1}. The stainless steel foils used in this study have thicknesses of 125 μm and 500 μm (Afentakis *et al* 2006).

2.8 Discussion and conclusions

The high-temperature process for polysilicon TFT fabrication on a glass substrate is modified for implementation on plastic substrates with crystallization of a-Si by localized heating through irradiation with laser pulse. Suitable barrier layers and buffer layers are necessary for plastic substrates. These layers play a multiplicity of roles from protecting the substrate to prevention of ablation of the a-Si film. But stainless steel substrates are more rugged and allow crystallization of a-Si by annealing in furnace.

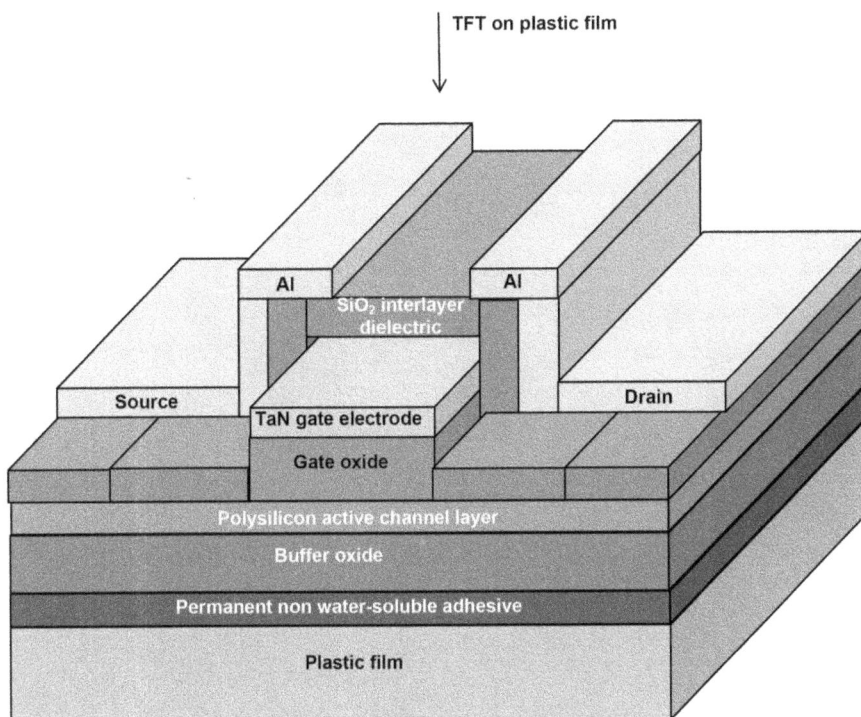

Figure 2.5. (Continued.)

Review exercises

2.1 Which TFT backplanes perform better: a-Si TFT or polySi TFT? In what way? Why?

2.2 At what temperatures is high-temperature polysilicon processing done? What is the temperature used in low-temperature polysilicon processing?

2.3 Give three detrimental effects of exceeding the temperature above 120 °C for a PET substrate.

2.4 Why is it necessary to revise the process steps when trying to fabricate polysilicon TFTs on PET? Discuss the process revision in the context of thermal restrictions.

2.5 How does pulsed laser processing permit the crystallization of amorphous silicon to polysilicon without any adverse effects on the PET substrate? Apart from crystallization, what additional process is performed by the laser pulse?

2.6 What is the laser source used for silicon crystallization? What is its wavelength? What is the duration of the laser pulse?

2.7 What is the temperature limit of the plastic substrate process? What is the melting point of silicon?

2.8 Why is it essential to deposit a silicon dioxide film on both sides of the PET substrate as the first step of processing?

Applying spin-on-glass on both sides
of stainless steel substrate

Spin-on-glass
Stainless steel substrate
Spin-on-glass

SiO$_2$ by PECVD on both sides and its
annealing

PECVD SiO$_2$
Spin-on-glass
Stainless steel substrate
Spin-on-glass
PECVD SiO$_2$

Insulation and
passivation
layers

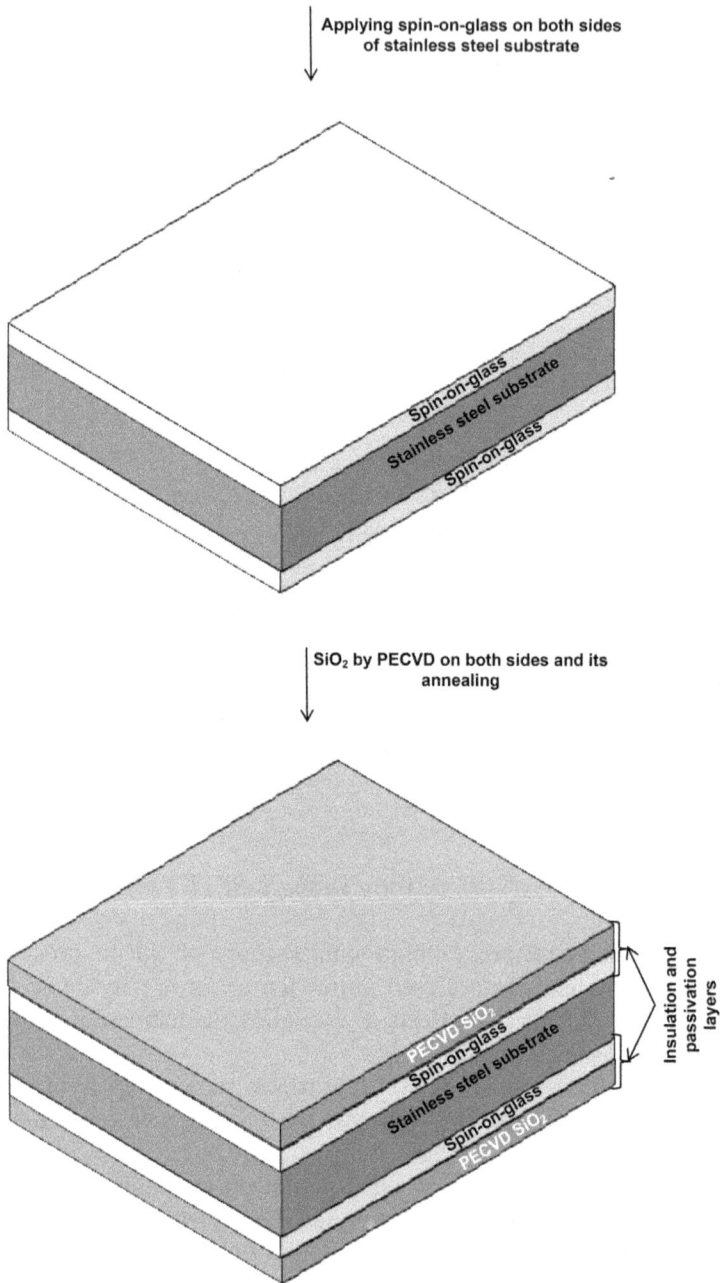

Figure 2.6. Polysilicon TFT fabrication on a stainless steel substrate.

Hydrogenated a-Si deposition

Hydrogenated a-Si
PECVD SiO₂
Spin-on-glass
Stainless steel substrate
Spin-on-glass
PECVD SiO₂

Crystallization of a-Si at 650°C

Polysilicon
PECVD SiO₂
Spin-on-glass
Stainless steel substrate
Spin-on-glass
PECVD SiO₂

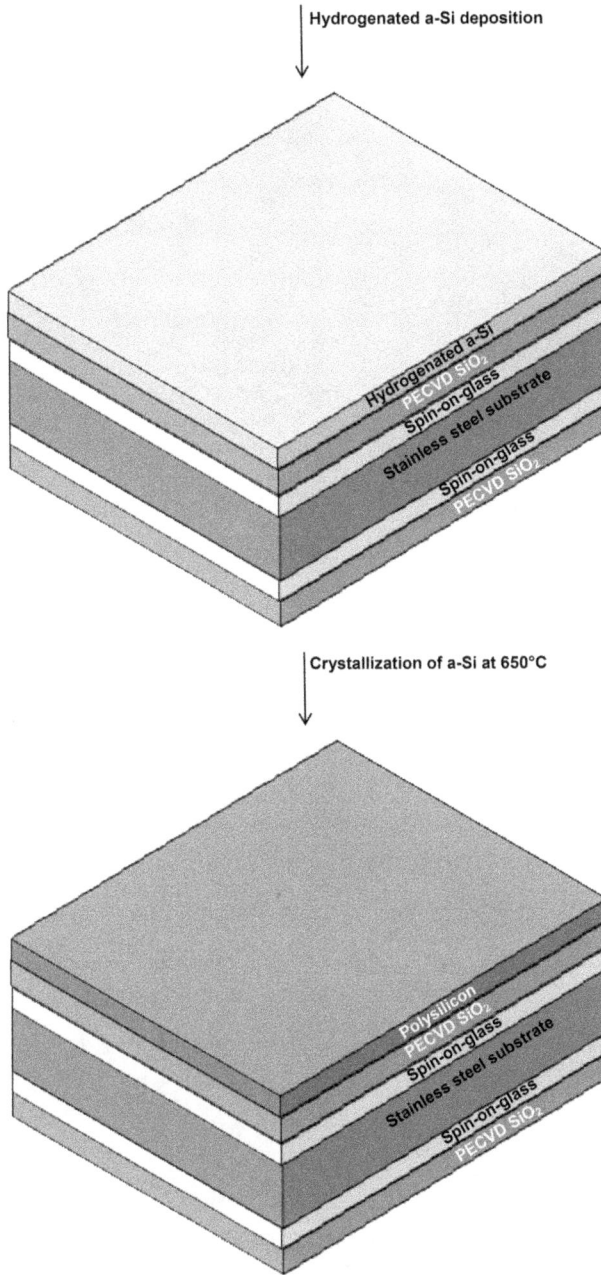

Figure 2.6. (Continued.)

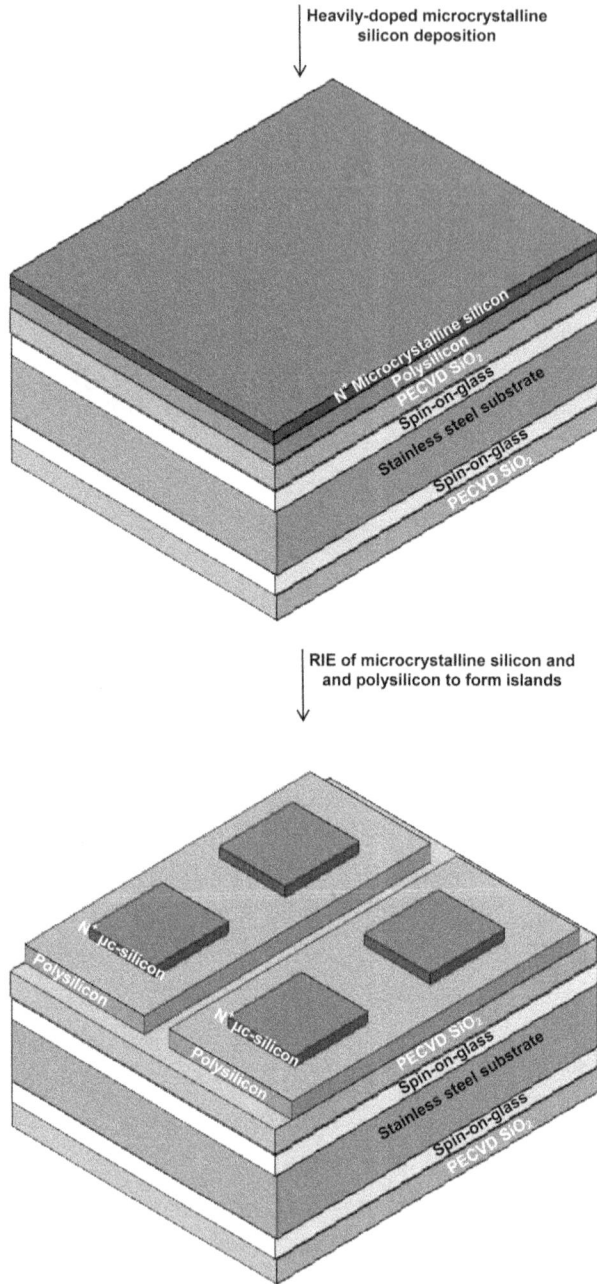

Heavily-doped microcrystalline
silicon deposition

N+ Microcrystalline silicon
Polysilicon
PECVD SiO₂
Spin-on-glass
Stainless steel substrate
Spin-on-glass
PECVD SiO₂

RIE of microcrystalline silicon and
and polysilicon to form islands

N+ μc-silicon
Polysilicon
N+ μc-silicon
Polysilicon
PECVD SiO₂
Spin-on-glass
Stainless steel substrate
Spin-on-glass
PECVD SiO₂

Figure 2.6. (Continued.)

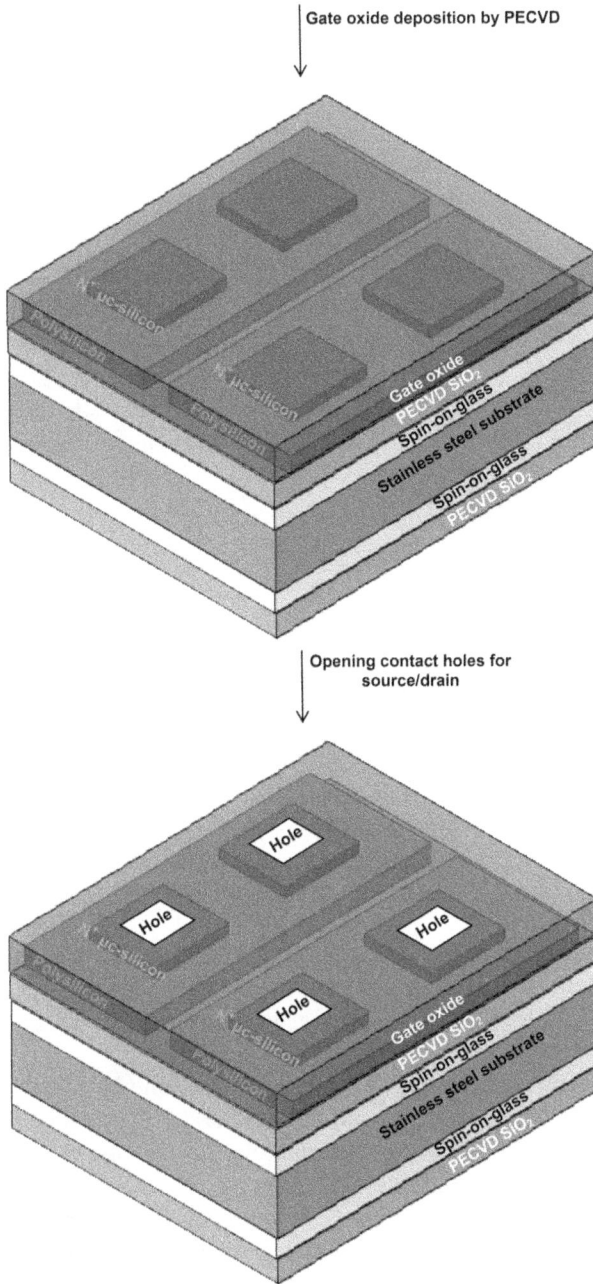

Figure 2.6. (Continued.)

Source/drain and gate
contact formation

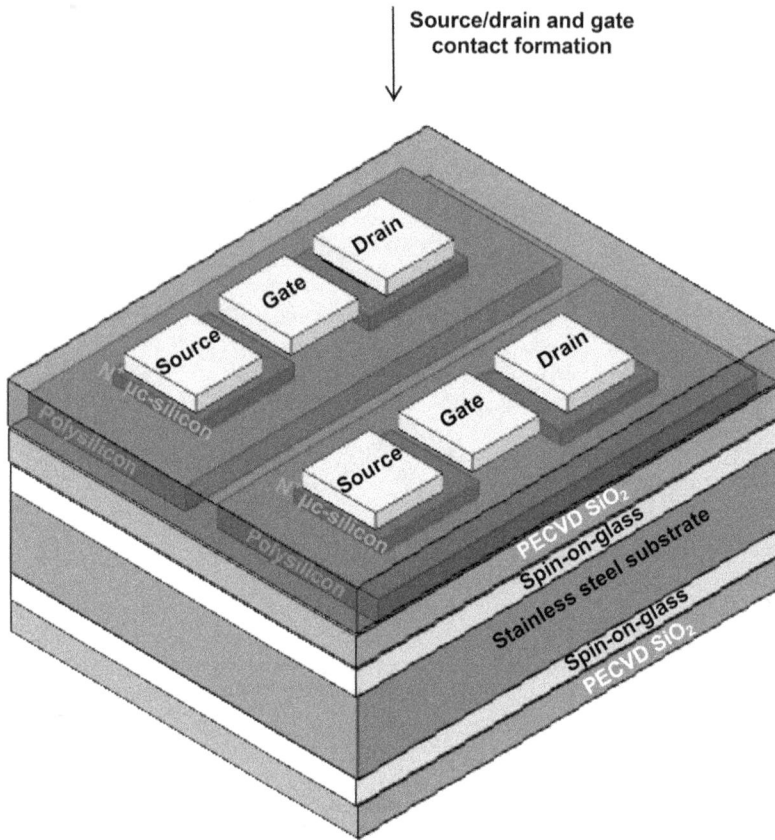

Figure 2.6. (Continued.)

2.9 Why is DC sputtering preferred over PECVD for deposition of the amorphous silicon film?

2.10 What will happen during excimer laser crystallization if a high percentage of hydrogen is incorporated in the a-Si film?

2.11 What is gas immersion laser doping? Name one N-type dopant source and one P-type dopant source used in GILD.

2.12 What is it necessary to deposit an aluminum nitride layer over the organic-film-coated PES substrate?

2.13 Which gas gives a-Si film with lower impurity content: Xe or Ar? Why?

2.14 Why is an SiO_2 buffer layer deposited on the PES substrate before deposition of amorphous silicon film? Discuss the critical role of buffer layer thickness.

2.15 Why is a plastic film laminated over a glass carrier to carry out the fabrication steps?

2.16 What notable advantages accrue if the polysilicon TFT process is completed on glass and the completed TFT is transferred to a plastic substrate, post-processing?

2.17 What is the exfoliation layer used in the SUFTLA process? How is the TFT fabricated on this exfoliation layer? How is the exfoliation layer separated from the original glass substrate? How is this layer transferred to a plastic film?

2.18 Describe a process of polySi TFT fabrication on stainless steel substrates based on furnace annealing of a-Si for crystallization.

References

Afentakis T, Hatalis M, Voutsas A T and Hartzell J 2006 Design and fabrication of high-performance polycrystalline silicon thin-film transistor circuits on flexible steel foils *IEEE Trans. Electron Devices* **53** 815–22

Carey P G, Smith P M, Wickboldt P, Thompson M O and Sigmon T W 1997 Polysilicon TFT fabrication on plastic substrates *Int. Display Research Conf. (Toronto, September 13–16, 1997)* pp 4 https://www.osti.gov/servlets/purl/641353 (accessed on 12 May 2018)

Han J-I and Kim Y-H 2006 Low temperature polysilicon thin-film transistors on flexible substrates *Proc. of the 9th Asian Symp. on Information Display, ASID '06 (India Habitat Centre, New Delhi 8–12 Oct)* pp 308–11

Inoue S, Utsunomiya S, Saeki T and Shimoda T 2002 Surface-free technology by laser annealing (SUFTLA) and its application to poly-Si TFT-LCDs on plastic film with integrated drivers *IEEE Trans. Electron Devices* **49** 1353–60

Kim D Y, Kwon J-Y, Jung J S, Park K B, Cho H S, Lim H, Kim J M, Yin H, Zhang X and Noguchi T 2006 Ultra-low temperature poly-Si thin film transistor for plastic substrate *J. Korean Phys. Soc.* **48** S61–3

Lemmi F, Chung W, Lin S, Smith P M, Sasagawa T, Drews B C, Hua A, Stern J R and Chen J Y 2004 High-Performance TFTs fabricated on plastic substrates *IEEE Electron Device Lett.* **25** 486–8

Serikawa T and Omata F 1999 High-mobility poly-Si TFTs fabricated on flexible stainless-steel substrates *IEEE Electron Device Lett.* **20** 574–6

Serikawa T and Omata F 2000 High-mobility poly-Si thin film transistors fabricated on stainless-steel foils by low-temperature processes using sputter-depositions *Japan. J. Appl. Phys.* **39** L393–5

Shimoda T and Inoue S 1999 Surface free technology by laser annealing (SUFTLA) *Int. Electron Devices Meeting, IEDM '99. Technical Digest (Washington, DC, 5–8 Dec. 1999)* pp 289–92

Theiss S D, Carey P G, Smith P M, Wickboldt P, Sigmon T W, Tung Y J and King T-J 1998 Polysilicon thin-film transistors fabricated at 100 °C on a flexible plastic substrate *Int. Electron Devices Meeting, 1998. IEDM '98. Technical Digest (San Francisco, CA, 6–9 Dec. 1998)* pp 257–60

Wu M, Chen Y, Pangal K, Sturm J C and Wagner S 2000 High-performance polysilicon thin film transistors on steel substrates *J. Non-Cryst. Solids* **266–9** 1284–8

IOP Publishing

Flexible Electronics, Volume 2
Thin-film transistors
Vinod Kumar Khanna

Chapter 3

Single-crystal Si TFT

The application of nanoelectronics to macroelectronics is exemplified by the silicon nanomembrane concept. These membranes are made by etching trenches in the template layer of SOI wafer either by dry or wet etching, and then removing the buried oxide by undercutting to release the membranes (Zhang *et al* 2012). Membrane release is preceded by completion of all high-temperature/corrosive chemical-based steps of the process. Three types of transfer printing processes are described. One process works by direct flip transfer and another through elastomeric stamp. But in both these processes, an adhesive is applied on the acceptor substrate. The third process, called dry transfer printing, needs no adhesive on the acceptor substrate. Instead, the adhesion of nanomembranes is kinetically controlled. For single-crystal TFT fabrication, the standard self-aligned gate process is modified. In the modified gate-after-source/drain process, the source and drain can be heavily doped at high temperatures but the self-alignment advantage is lost. Based on the silicon nanomembrane method, the fabrication of a microwave TFT is described (Seo *et al* 2011). Also explained is a strain-engineered TFT using a Si/SiGe/Si trilayer structure. In both cases, the silicon nanomembranes are transfer printed on PET substrates (Zhou *et al* 2013).

3.1 Introduction

Amorphous and polycrystalline silicon TFTs on flexible substrates suffer from low charge carrier mobilities, albeit to different extents, and are, therefore, unable to deliver the full capabilities of crystalline silicon on plastics. A convenient delivery vehicle for crystalline silicon must be found. One such vehicle is a silicon nanomembrane. Here, we speak of a nanomembrane, not in the usual parlance of nanoelectronics but in the realm of macroelectronics and bendability. This circumstance may be visualized as nanomaterials coming to the rescue of large-area electronics.

3.2 Transferrable single-crystal silicon nanomembranes (NMs)

These are extremely thin free-standing membranes or diaphragms made of single-crystal or monocrystalline silicon, the widely pervasive semiconductor device material, supreme in many respects, displaying excellent physical and chemical properties and high carrier mobilities. NMs have thicknesses in the range 1–100 nm. On these membranes, the high-temperature processing steps such as thermal diffusion, annealing etc, and treatment with corrosive chemicals are performed when they are mounted on the host substrate called the donor substrate. This substrate is usually silicon itself.

The low-temperature processing steps such as PECVD, metallization, etc, as well as less harsh chemical treatments are carried out after transferring the nano-membrane to a receiving substrate known as the acceptor substrate, generally of inferior physical/chemical stability to the donor one. The acceptor substrate is a flexible substrate such as a plastic substrate, which is neither liberal towards high temperatures nor resistant to aggressive chemicals.

Thus the idea of NMs is threefold:

  (i) To complexly decouple the high-temperature/chemically corrosive processing from low-temperature/less corrosive processing.
  (ii) To utilize the superior electrical properties of single-crystal silicon.
  (iii) To derive the benefits of small thickness of NM rendering it easily bendable and conformable to irregular surfaces so that when mounted on a plastic substrate, the (NM + plastic substrate) combination fulfills the criteria of a flexible electronics component.

3.3 SOI wafer process for Si NMs production, doping and transfer

3.3.1 Basic questions

Two questions immediately come to mind: How do we fabricate Si NMs? How do we transfer them from a donor to an acceptor substrate? To answer these questions, we take the SOI (silicon on insulator) wafers as the starting material (Zhang *et al* 2012). These wafers are widely known to readers familiar with MEMS and CMOS processes. They are trilayer wafers (figure 3.1), the three layers are called the device or template layer, the buried oxide (BOX) and the handle layer. The template layer is made of premium quality silicon and typically has a thickness in the nanoscale range. The buried oxide is a thermally grown oxide with a thickness of several hundred nm while the handle layer is bulk silicon serving as a mechanical support for easy handling and is, therefore, several hundred microns thick. Notwithstanding, the acrimonious reality that they are expensive, the SOI wafers are available as manufacturing solutions for semiconductor device fabrication.

3.3.2 Formation of Si nanoribbons

We have to consider what pattern or design is to be drawn on the top template layer of silicon? A common geometry consists of long, narrow bars or strips of silicon known as nanowires or nanoribbons. Of course, any other geometry can also be

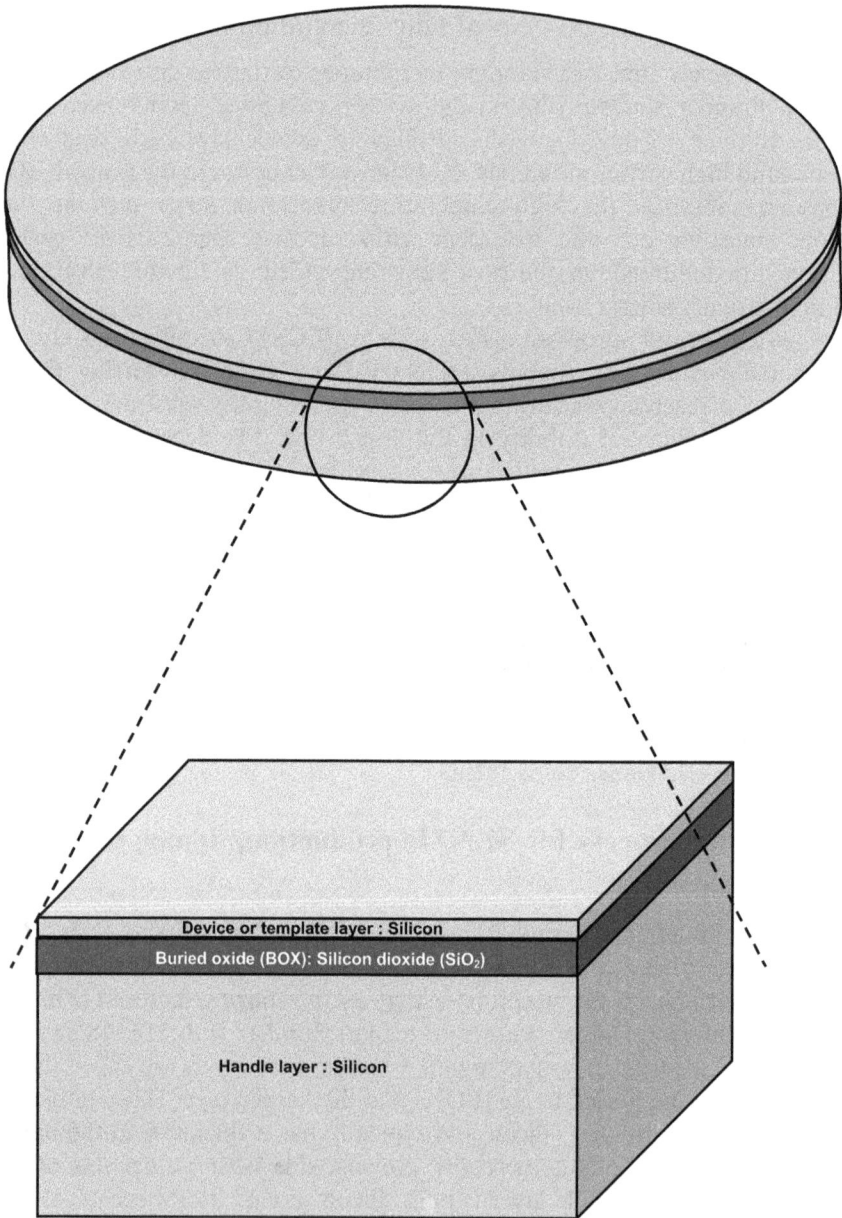

Device or template layer : Silicon

Buried oxide (BOX): Silicon dioxide (SiO$_2$)

Handle layer : Silicon

Figure 3.1. SOI wafer structure.

drawn. These bars are defined by photolithography in the photoresist. Through this photoresist layer, silicon is etched from areas left uncovered by photoresist by reactive ion etching in SF$_6$ plasma (Menard *et al* 2005) (figure 3.2). Thus, trenches with vertical sidewalls are dug in the Si template layer and the etching is stopped when we reach the underlying BOX layer.

Taking SOI wafer

Device layer (thickness =100 nm)
BOX layer (thickness =150 nm)
Handle layer (thickness =400 µm)

Coating with photoresist

Photoresist
Device layer (thickness =100 nm)
BOX layer (thickness =150 nm)
Handle layer (thickness =400 µm)

Photolithography for
defining nanoribbons

Photoresist
Device layer (thickness =100 nm)
BOX layer (thickness =150 nm)
Handle layer (thickness =400 µm)

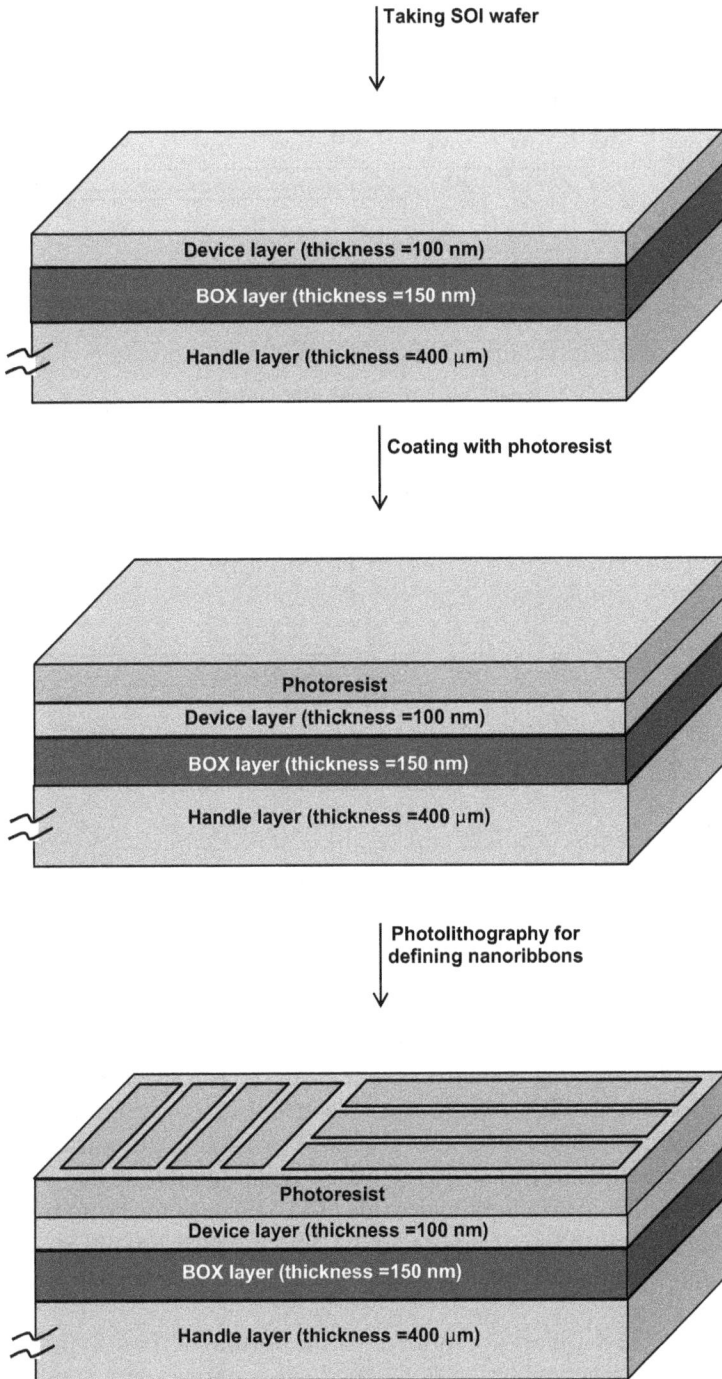

Figure 3.2. Nanoribbon fabrication by making a pattern in photoresist on SOI wafer and dry etching of silicon device layer.

Dry etching of silicon

Photoresist
Silicon nanoribbon
BOX layer (thickness =150 nm)
Handle layer (thickness =400 μm)

BOX etching in HF

Photoresist
Silicon nanoribbon
Handle layer (thickness =400 μm)

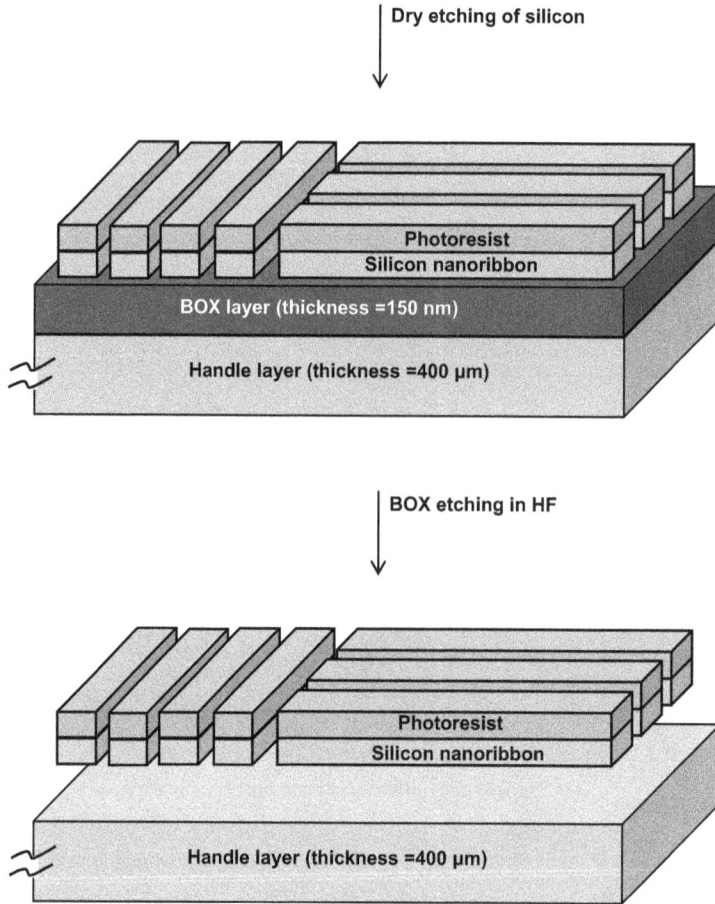

Figure 3.2. (Continued.)

Another way of etching trenches into the Si template layer of SOI wafer (figure 3.3) consists of first etching the native oxide from the top surface of this layer, then immediately loading the SOI wafer into an electron-beam evaporation unit to coat it sequentially with Al (20 nm) and gold (100 nm), photolithographically defining the nanowire pattern on the metal, etching gold and aluminum followed by wet etching silicon in tetramethyl ammonium hydroxide (TMAH) until the bottom BOX layer is reached (Menard *et al* 2004). It takes about 3.5 min to etch 100 nm Si. The etch rate is temperature dependent. It also depends on TMAH concentration and the type of silicon, N- or P-type.

In both cases, the Si pillars are formed. At the bottom of these Si pillars as well as at the bases of the trenches lies the BOX layer. To separate these Si pillars from the BOX layer, the silicon dioxide is etched in concentrated HF (49%). This acid attacks the oxide at the trench bases and then penetrates towards the bottoms of the pillars. This undercutting releases the pillars from the BOX layer. After oxide etching, the

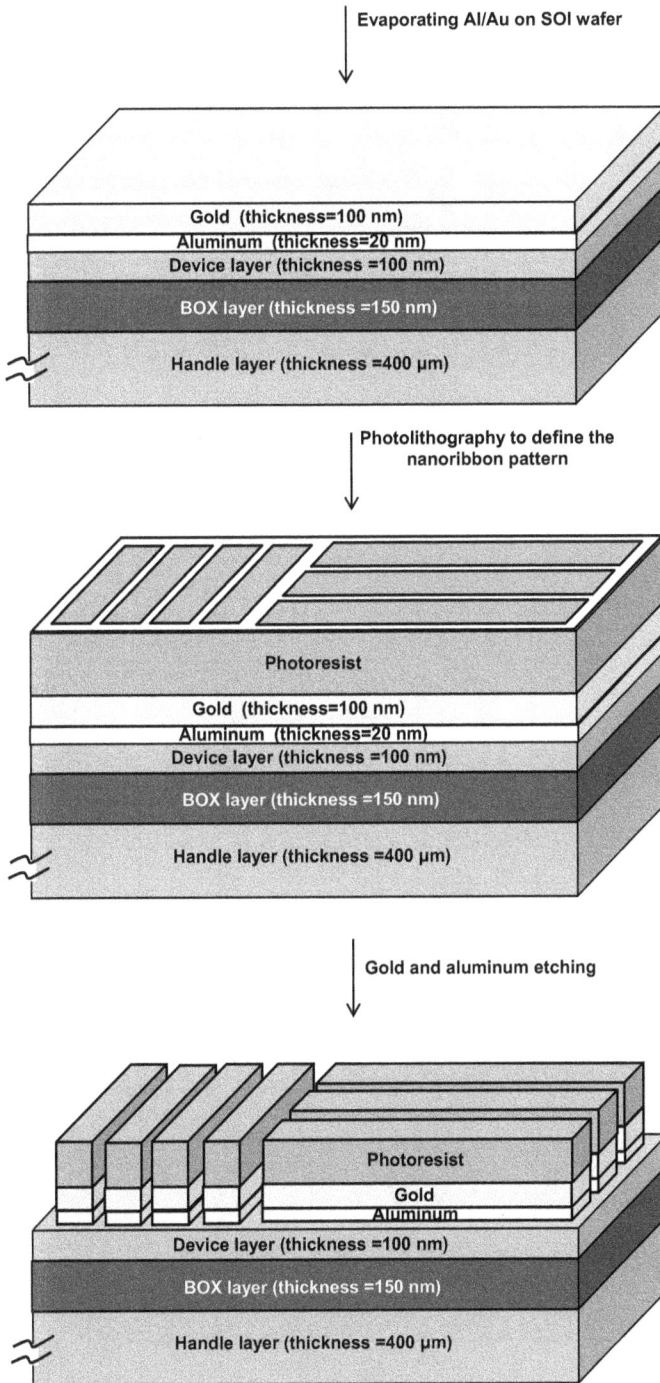

Evaporating Al/Au on SOI wafer

Gold (thickness=100 nm)
Aluminum (thickness=20 nm)
Device layer (thickness =100 nm)
BOX layer (thickness =150 nm)
Handle layer (thickness =400 μm)

Photolithography to define the nanoribbon pattern

Photoresist
Gold (thickness=100 nm)
Aluminum (thickness=20 nm)
Device layer (thickness =100 nm)
BOX layer (thickness =150 nm)
Handle layer (thickness =400 μm)

Gold and aluminum etching

Photoresist
Gold
Aluminum
Device layer (thickness =100 nm)
BOX layer (thickness =150 nm)
Handle layer (thickness =400 μm)

Figure 3.3. Nanomembrane fabrication by using Al/Au bilayer patterns as masks against silicon wet etching in TMAH and oxide etching in HF.

Photoresist removal

Gold
Aluminum
Device layer (thickness =100 nm)
BOX layer (thickness =150 nm)
Handle layer (thickness =400 µm)

Silicon wet etching in TMAH

Gold
Aluminum
Silicon nanoribbon
BOX layer (thickness =150 nm)
Handle layer (thickness =400 µm)

BOX etching in HF

Gold
Aluminum
Silicon nanoribbon
Handle layer (thickness =400 µm)

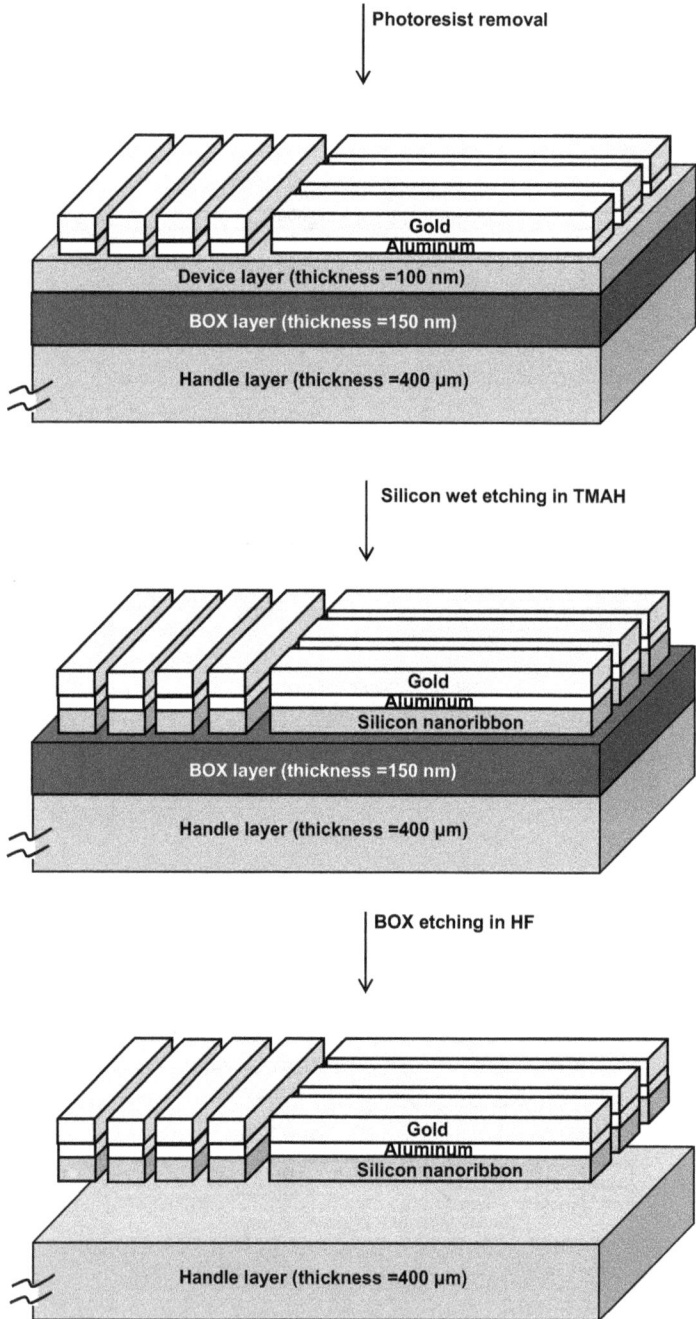

Figure 3.3. (Continued.)

photoresist and Al/Au protective layers are removed. In some cases, the Al/Au layers are retained for integration into the device structure that is fabricated over the pillars. These Si pillars, typically 1 μm broad, 100 μm long and <100 nm thick are the nanoribbons. The Si nanoribbons constitute the nanomembranes.

3.3.3 Pre-release doping of nanomembranes

Prior to their release from SOI wafer, the doping areas in the nanomembrane are defined by photolithography. The doping profile in these areas is properly tailored.

One method involves photolithographically defining and chemically etching windows for diffusion through a PECVD grown silicon dioxide and then performing thermal diffusion, e.g. for phosphorous doping using a solid source (Ahn *et al* 2006). But this process is acceptable only for larger geometries because the lateral spreading of dopant may create problems in case of long diffusion times. Moreover, the doping concentration may fall rapidly with depth leading to unduly large sheet resistance values in a region away from the silicon surface.

Keeping such possibilities in view, ion implantation is the preferred method for introduction of dopant into Si NM. Post-implantation annealing at 950 °C provides a uniform doping concentration from the top to the bottom of the nanomembrane. Such a uniform doping concentration is highly desirable because when the nano-membrane is released from the donor substrate for transference to acceptor substrate, it is conveyed either straight up or upside down depending on the method used. Therefore, such a doping profile will make the NM behavior in a device independent of the transfer method used. In addition, when the doped region serves as the source or drain of a TFT, the doping concentration should be high so that the doped region does not contribute any parasitic resistance in series.

3.3.4 Transfer printing of nanomembranes

Transfer printing of nanomembranes is done by any one of the following three methods:

(i) By direct flip transfer using an adhesive layer on the acceptor substrate: When the SOI wafer is dipped in HF to release the NMs formed in the top silicon layer, the Si NMs settle down over the handling substrate and become weakly bonded with it. This state is referred to as in-place bonding. A glue such as SU-8 photoresist is applied on the acceptor substrate and the SU-8 covered acceptor substrate is contacted with the donor substrate carrying the nanomembranes on its upper surface (figure 3.4) with the SU-8 layer and the nanomembranes facing each other. When the two substrates are separated, the nanomembranes are detached from the donor substrate and are stuck with the SU-8 photoresist on the acceptor substrate due to the stronger bonding force of SU-8 for Si NMs compared to their rather loose in-place bonding. However, the surface, which was formerly the upper free surface of the nanomembranes unbound to the donor substrate now becomes the lower bound surface on the acceptor substrate. Also, the previous lower surface of nanomembranes on the donor substrate becomes

Coating the acceptor flexible
substrate with SU-8 in preparation for
nanoribbon transfer upon it

SU-8 photoresist

ITO

Acceptor plastic substrate

Flipping the acceptor plastic
substrate and bringing it down
over the donor silicon substrate
carrying the nanoribbons

Acceptor plastic substrate

ITO

SU-8 photoresist

Photoresist

Silicon nanoribbon

Handle layer (thickness =400 μm)

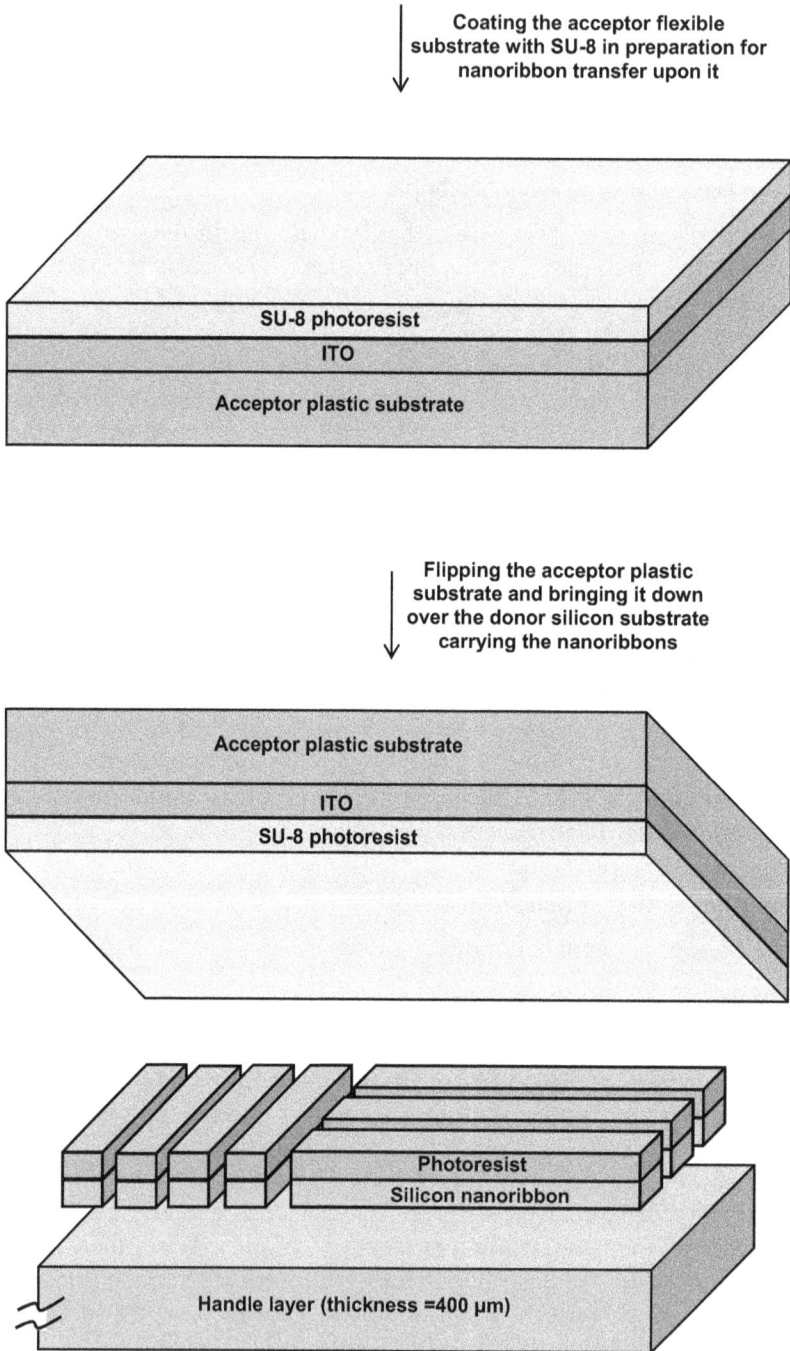

Figure 3.4. Printing of silicon nanoribbons by direct flip transfer.

Contacting the acceptor plastic
substrate with donor silicon
substrate

Acceptor plastic substrate

ITO

SU-8 photoresist

Photoresist

Silicon nanoribbon

Handle layer (thickness =400 μm)

Separating the acceptor and donor
substrates and flipping the acceptor
substrate

Silicon nanoribbon

Photoresist

SU-8 photoresist

ITO

Acceptor plastic substrate

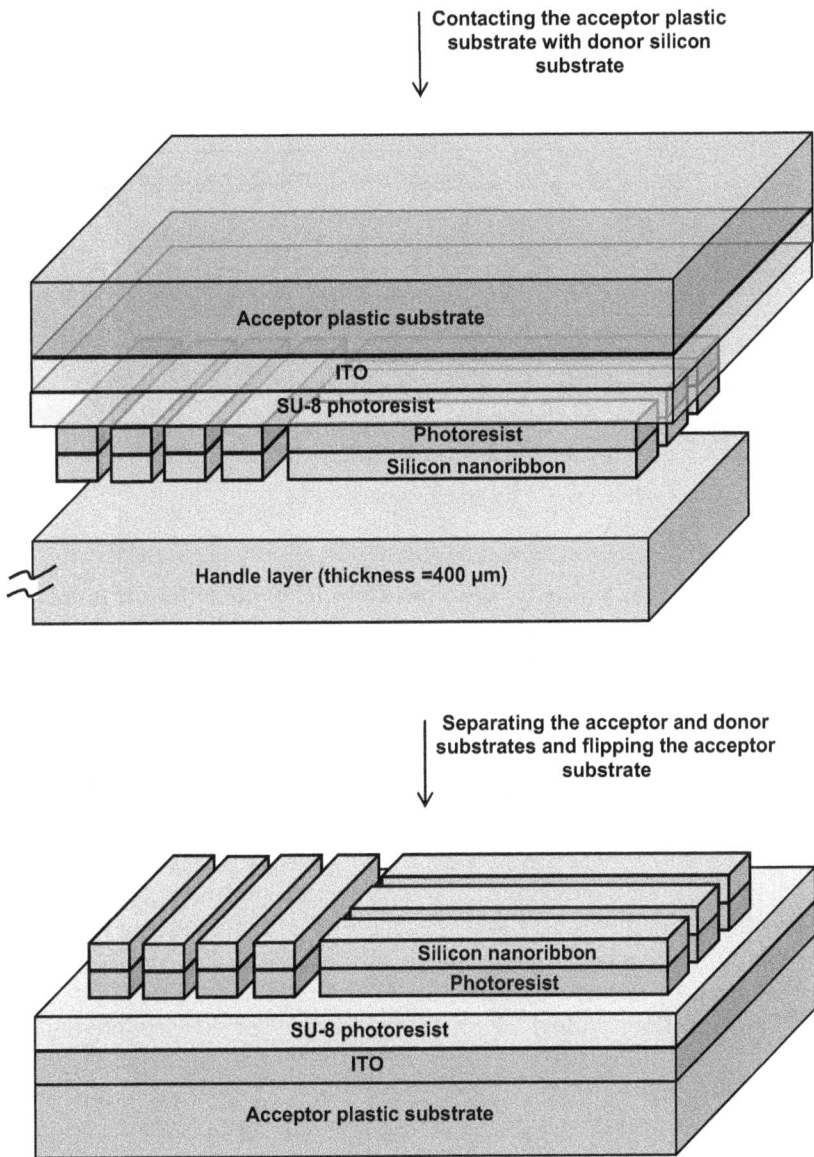

Figure 3.4. (Continued.)

the upper surface on the acceptor substrate. After the Si NM transfer is completed, the SU-8 photoresist is hardened by UV exposure so that the NMs are firmly fixed in their places on the acceptor substrate.

(ii) Stamp-assisted transfer using an adhesive layer on the acceptor substrate: In this method also, the surface of the acceptor substrate is coated with glue e.g. SU-8 photoresist (figure 3.5). An elastomeric stamp made is used. The stamp is made of a viscoelastic polymer, i.e. a material showing the

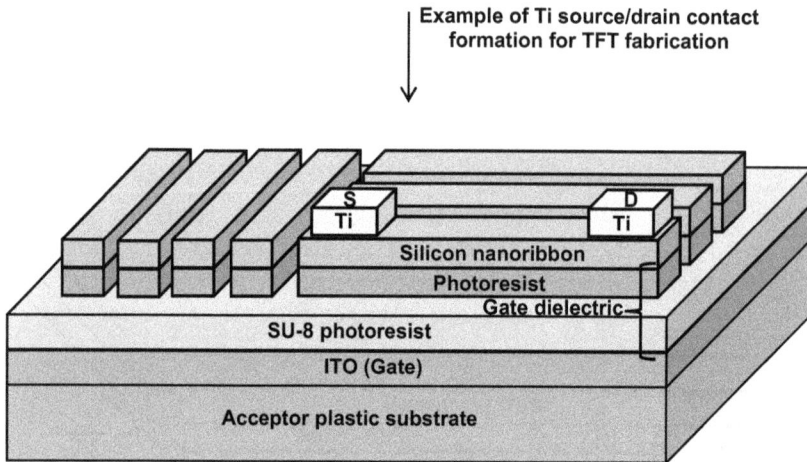

Example of Ti source/drain contact
formation for TFT fabrication

Figure 3.4. (Continued.)

properties of viscosity and elasticity. Polydimethylsiloxane (PDMS), $(C_2H_6OSi)_n$, is one such material. The transfer printing involves two steps. In the first step, the stamp is pressed face-to-face against the surface of the donor substrate on which the nanomembranes have been made. After contacting conformally, the stamp is withdrawn. The withdrawn stamp picks up the nanomembranes along with itself. The stamp with the nanomembranes attached to it is said to be in inked state. In the second step, the inked stamp is brought in contact with the SU-8 covered surface of the acceptor substrate. On pulling the stamp away from the acceptor substrate, the nanomembranes cling to this substrate. Notice that in this method, the upper and lower surfaces of NMs on the acceptor substrate remain the same as in donor substrate. As in (i), the NM transfer is followed by fixation by exposing SU-8 to UV radiation.

(iii) Dry transfer printing: Here no adhesive layer is used on the acceptor substrate. The peeling velocity from the donor substrate to elastomeric stamp is high $\sim$10 cm s^{-1} or faster while that from elastomeric stamp to acceptor substrate is slow $\sim$1 mm s^{-1} or smaller. The remaining processes are the same as (ii). Since, dry transfer printing does not use any adhesive layer, the question arises how do the nanomembranes stick to the stamp or with the acceptor substrate? Because of the viscoelastic properties of the elastomer, the adhesion of the nanomembranes is kinetically controlled, i.e. it is sensitive to the rate of pulling away the stamp (Meitl *et al* 2006). When the stamp is laminated against and thereafter pulled away very fast from the donor substrate, the adhesive forces are sufficiently strong to grab and attach the nanomembranes to the surface of the stamp in preference to that of the donor substrate, thereby lifting them away from that substrate. Likewise, when the stamp is laminated against and then pulled away very

Bringing down the PDMS stamp
over the donor silicon substrate
carrying the nanoribbons

PDMS stamp

Photoresist
Silicon nanoribbon

Handle layer (thickness =400 µm)

Contacting the PDMS stamp with the
photoresist layer on silicon
nanoribbons on the donor silicon
substrate

PDMS stamp

Photoresist
Silicon nanoribbon

Handle layer (thickness =400 µm)

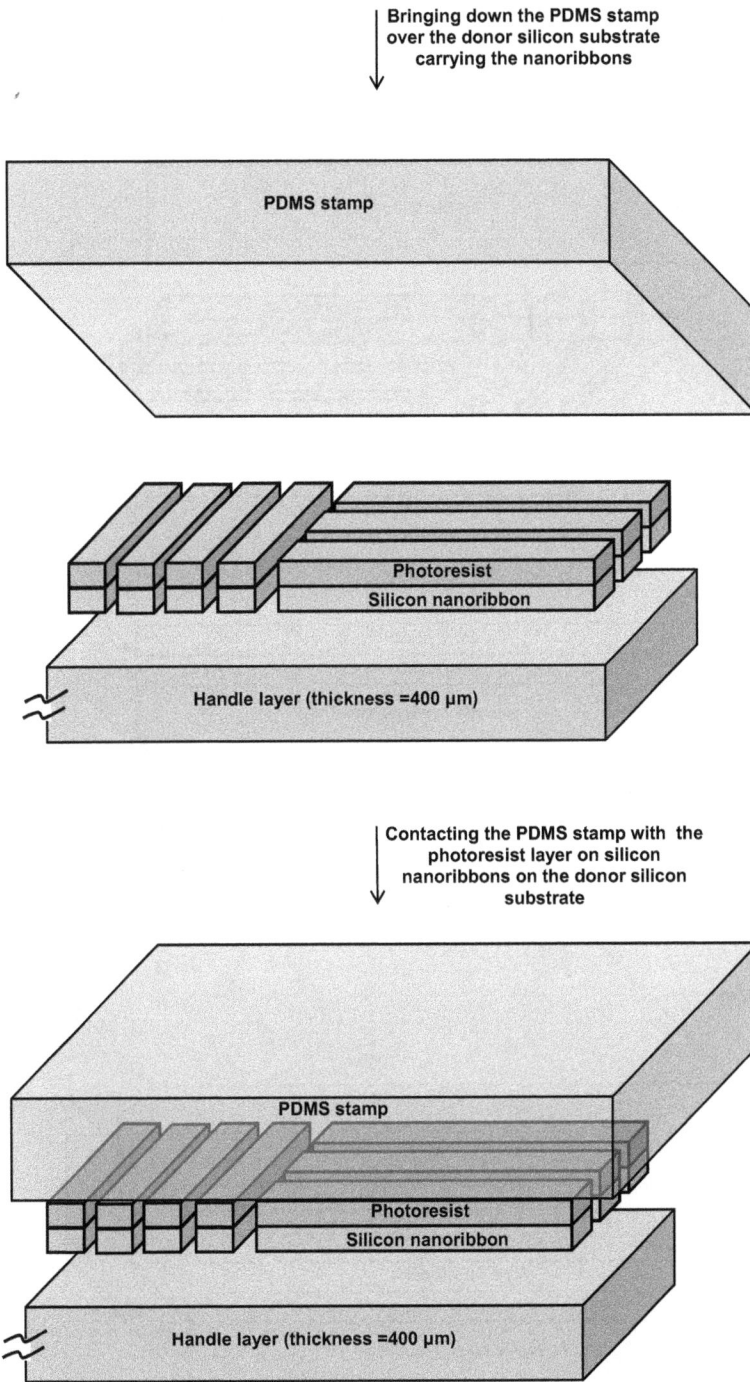

Figure 3.5. Stamp-assisted transfer of silicon nanoribbons.

Taking away the PDMS stamp with
attached silicon nanoribbons from
the donor silicon substrate and
lowering it over the SU-8 coated
ITO/acceptor plastic substrate

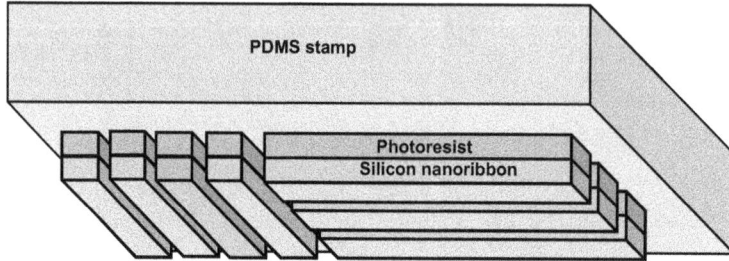

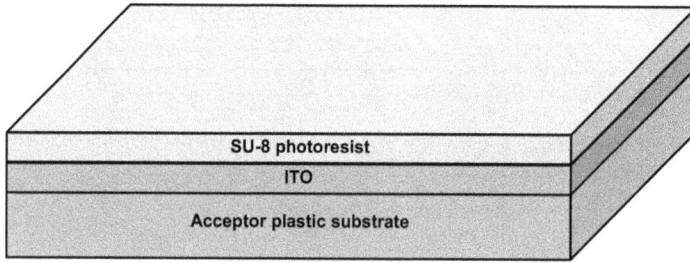

PDMS stamp

Photoresist

Silicon nanoribbon

SU-8 photoresist

ITO

Acceptor plastic substrate

Contacting the PDMS stamp with the
SU-8 coated ITO/acceptor plastic
substrate

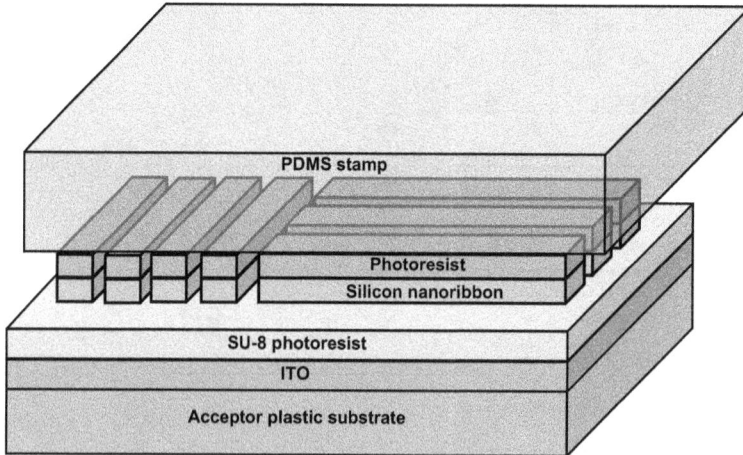

PDMS stamp

Photoresist

Silicon nanoribbon

SU-8 photoresist

ITO

Acceptor plastic substrate

Figure 3.5. (Continued.)

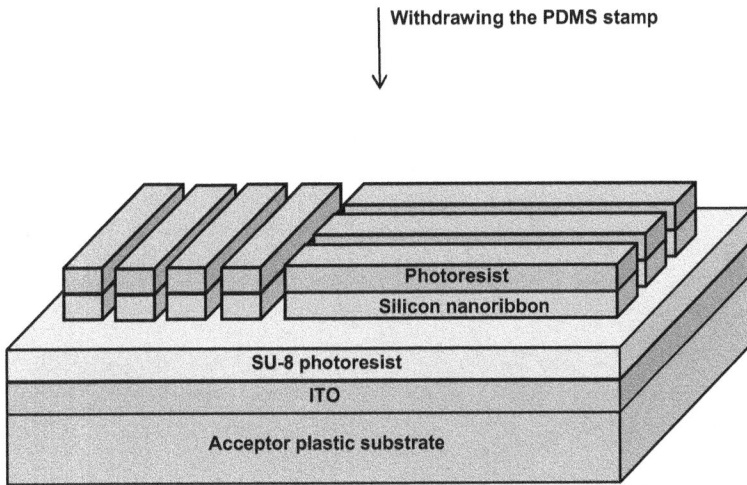

Figure 3.5. (Continued.)

slowly from the acceptor substrate, the adhesive forces are adequate to snatch and stick the nanomembranes to the surface of the acceptor substrate in preference to that of the stamp, thereby causing the nano-membranes to be retained by the acceptor substrate. Since there is no adhesive layer, the adhesion forces are dominated by Van der Waals interactions.

All the transfer printing methods preserve the spatial order, positions and orientations of the nanomembranes. Registration of Si NMs is of high fidelity. In method (i), the top and bottom surfaces of nanomembranes are interchanged but in methods (ii) and (iii), they remain same. In methods (i) and (ii), a glue is applied to acceptor substrate but method (iii) is glue-free. In all the methods, the process yield is critically dependent on controlling the adhesion between the nanomembranes, the donor/acceptor surfaces and the stamp where applicable.

3.4 Microwave TFT fabrication using Si NMs

3.4.1 Process steps on SOI wafer

Seo *et al* (2011) started with SOI wafer with 200–300 nm template layer of P-type doped Si (figure 3.6). The source and drain areas are heavily phosphorous doped by ion implantation. Annealing is done in a furnace at 850 °C. The Si template layer is patterned into strips or membranes. The NMs are released by wet etching of the BOX layer. Upon immersion in HF, the acid removes the buried oxide by under-cutting. The membranes are transferred by direct flip method on SU-8 coated polyethylene terephthalate (PET) substrate whose softening temperature is 170 °C (Yuan *et al* 2009). Since direct flip transfer method is used here, the phosphorous doping profile is tailored in such a manner that the bottom surface of the silicon

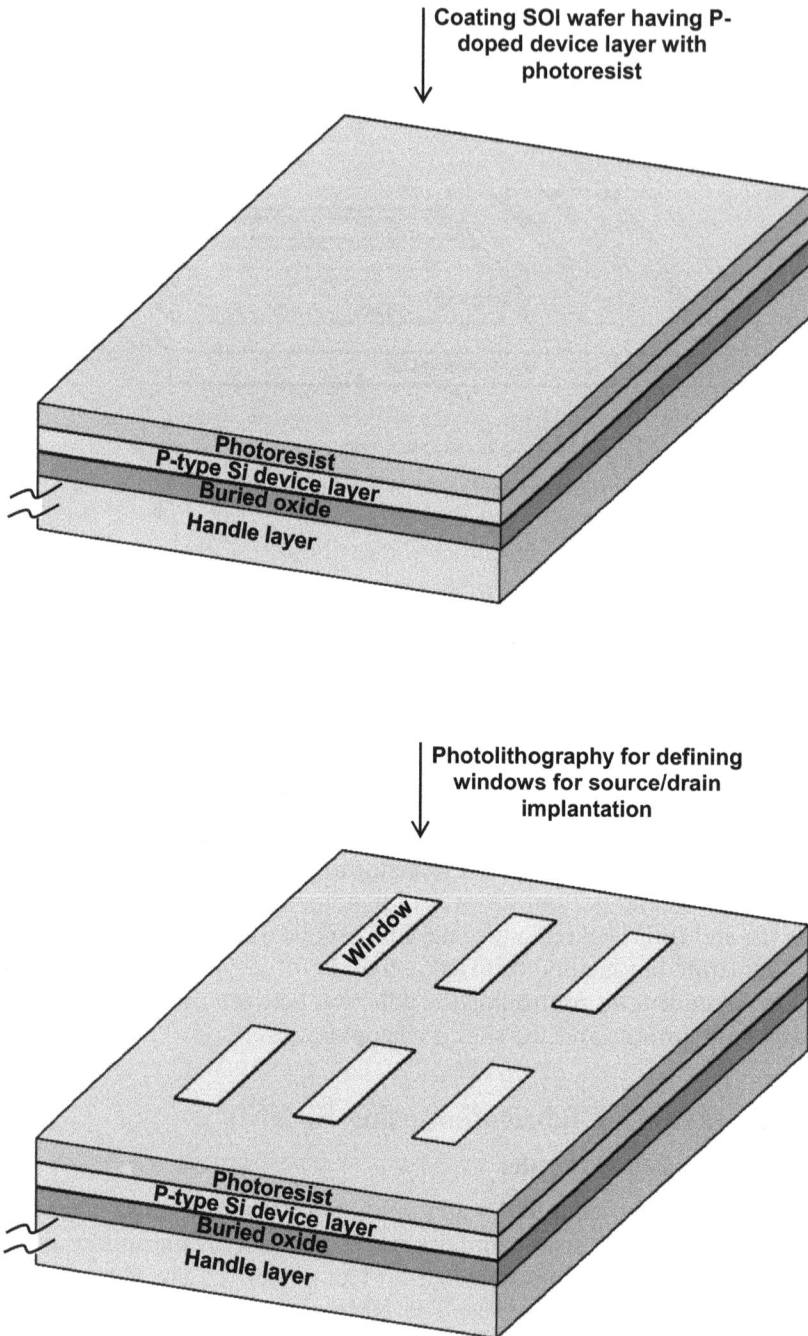

Coating SOI wafer having P-doped device layer with photoresist

Photoresist
P-type Si device layer
Buried oxide
Handle layer

Photolithography for defining windows for source/drain implantation

Window

Photoresist
P-type Si device layer
Buried oxide
Handle layer

Figure 3.6. Microwave TFT fabrication using the silicon nanomembranes technique.

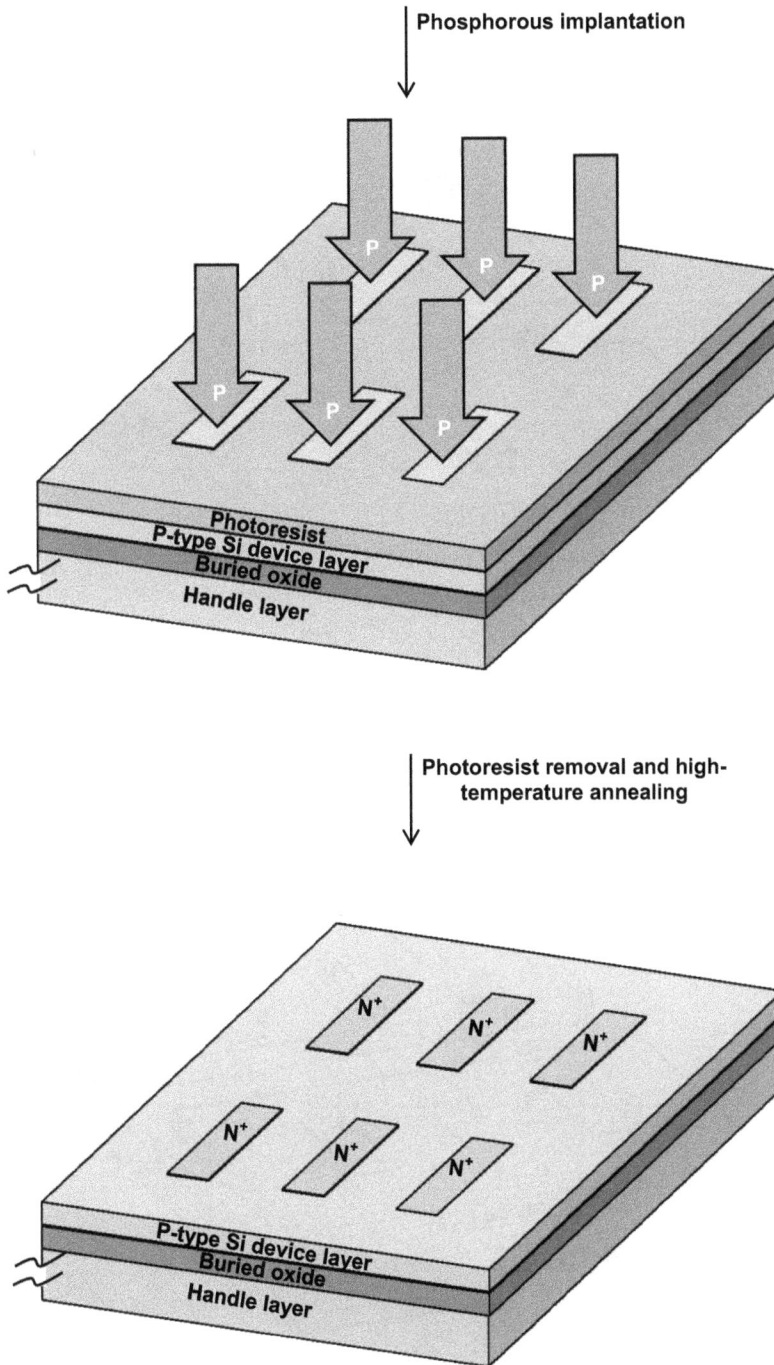

Figure 3.6. (Continued.)

Patterning into strips by photolithography and silicon etching

N⁺ N⁺ N⁺

N⁺ N⁺ N⁺

P-type Si device layer
Buried oxide
Handle layer

BOX etching

N⁺ N⁺ N⁺

N⁺ N⁺ N⁺

P-type Si device layer
Handle layer

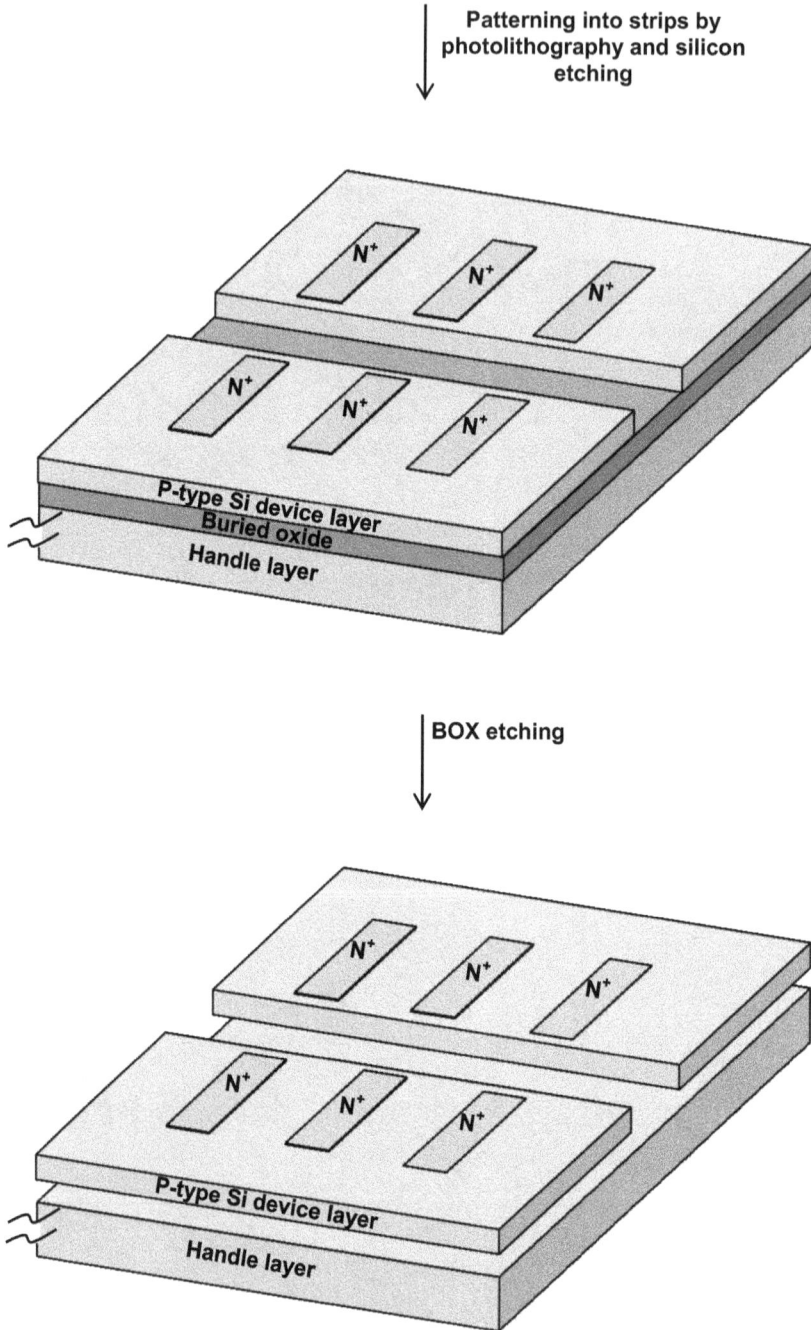

Figure 3.6. (Continued.)

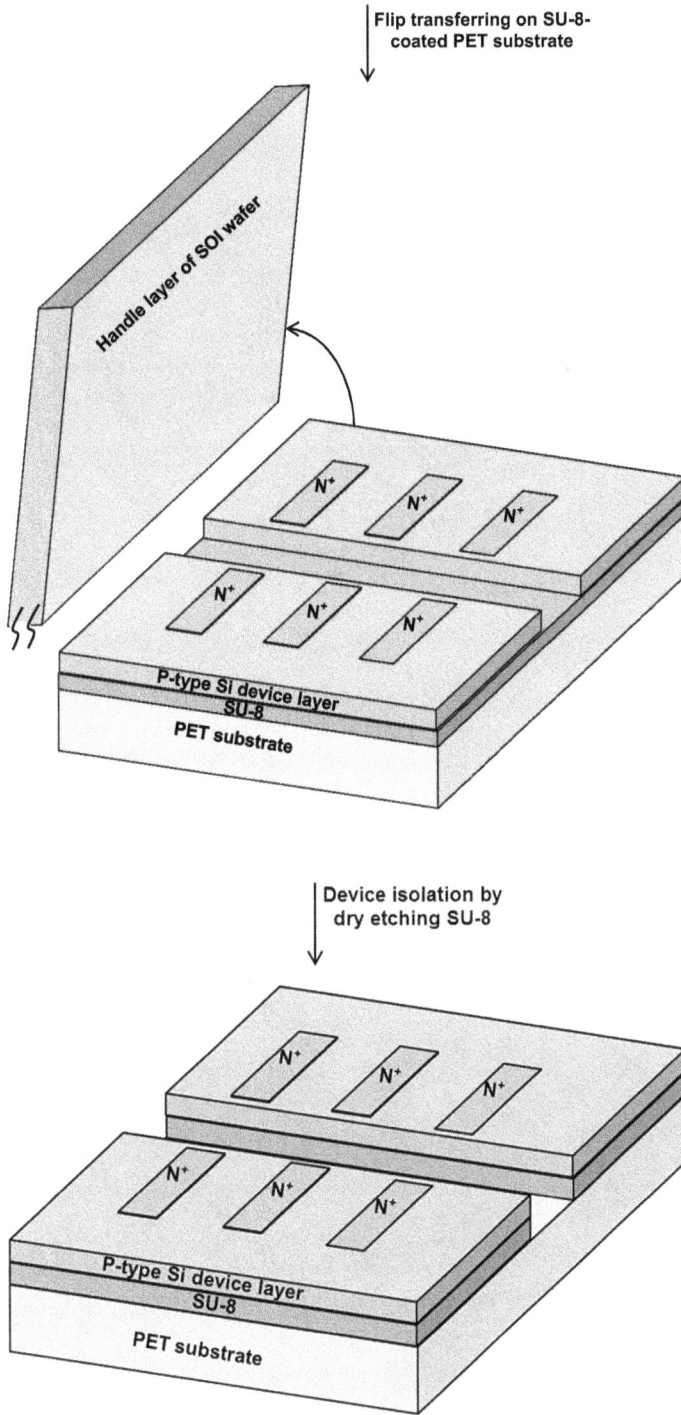

Flip transferring on SU-8-coated PET substrate

Handle layer of SOI wafer

N+ N+ N+

N+ N+ N+

P-type Si device layer
SU-8
PET substrate

Device isolation by dry etching SU-8

N+ N+ N+

N+ N+ N+

P-type Si device layer
SU-8
PET substrate

Figure 3.6. (Continued.)

Gate stack formation

Evaporated SiO

Gate

N⁺

N⁺

N⁺

N⁺

N⁺

N⁺

Gate

P-type Si device layer

SU-8

PET substrate

Deposition of source/drain metal layers

Evaporated SiO

Gate

Source N⁺

Drain N⁺

Source N⁺

Source

Drain

Source N⁺

Gate

P-type Si device layer

SU-8

PET substrate

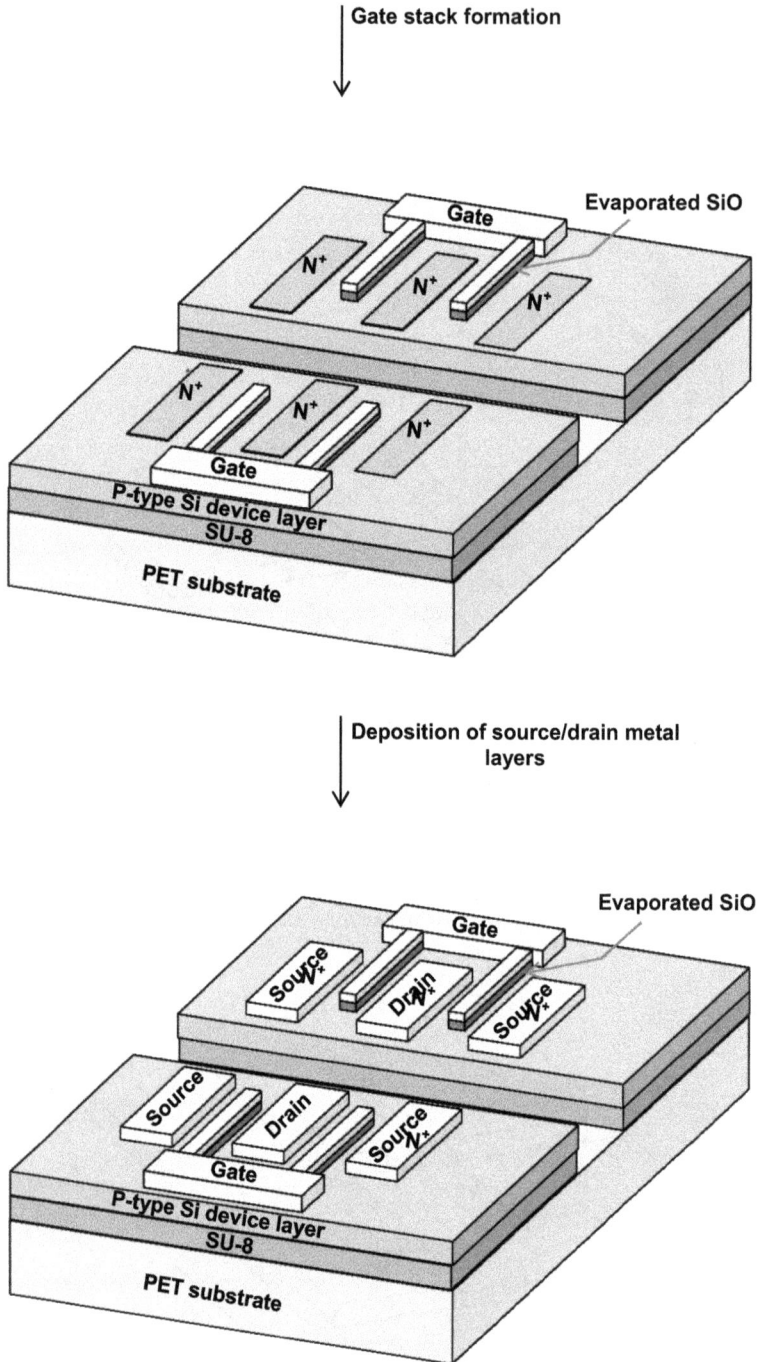

Figure 3.6. (Continued.)

nanomembranes has a high N^+ doping concentration. This tailoring is decided from the consideration that the bottom surface of NM becomes the top surface after transfer. In the case of stamp-assisted transfer, the opposite is true with the top NM surface having a high N^+ doping concentration instead of the bottom surface. The choice of doping profile, therefore, depends on the method of NM transfer, whether direct flip or stamp-aided.

The transferred NMs are permanently fixed on the PET substrate by hardening the SU-8 photoresist by UV exposure. After this stage, the remaining process steps to be carried out on NMs mounted on PET substrate are done at lower temperatures.

3.4.2 Process steps after transferring to PET substrate

By dry etching of SU-8 in plasma, the devices are isolated. The silicon oxide (SiO) gate stack is formed by evaporation. Then gate, source and drain regions are metallized. Since source and drain regions have already been defined and are heavily doped in the nanomembrane itself, and the gate region is formed afterwards when the nanomembrane is fixed to the PET substrate, this method of FET fabrication is known as a gate-after-source/drain process distinct from the standard self-aligned FET fabrication (Seo *et al* 2011). In this process, the high-temperature requirement for source/drain formation is not disturbed. This helps to reduce the series resistances associated with these contacts. However, the advantage of self-alignment between the gate and source/drain regions is sacrificed.

3.4.3 TFT parameters

The unit-gate-length transconductances and effective mobilities for three TFTs having different critical dimensions are 146.8 $\mu S\ mm^{-1}$, 230 $cm^2\ V^{-1}\ s^{-1}$; 1777.5 $\mu S\ mm^{-1}$, 331 $cm^2\ V^{-1}\ s^{-1}$ and 9315 $\mu S\ mm^{-1}$, 423 $cm^2\ V^{-1}\ s^{-1}$. RF circuits are realized with cut-off frequency f_T of TFTs = 3.8 GHz and the maximum oscillation frequency f_{max} = 12 GHz (Seo *et al* 2011). These values are an improvement over f_T values of TFTs = 2.04 GHz and f_{max} = 7.8 GHz, reported in an earlier publication (Yuan *et al* 2007).

RF characterization by Qin *et al* (2013) showed that the Si NM TFTs showed slight but monotonic performance improvement under uniaxial convex mechanical bending, which was less than that for thinned silicon bulk MOSFETs under similar tensile strains. The latter devices are those obtained by back side wafer grinding.

3.5 TFTs on strained Si/SiGe/Si NMs

Besides dimensional downscaling, device speed can be increased by strain engineering. Zhou *et al* (2013) utilize the nanomembrane approach to fabricate strained silicon RF transistors on flexible PET substrates.

3.5.1 Ion implantation, annealing and reduction of device layer thickness

They begin with SOI wafer having 200 nm thick lightly P-doped Si (001) template layer; BOX layer is 145 nm thick (figure 3.7). After phosphorous implantation at 20 keV energy and 2×10^{15} cm^{-2} dose, annealing is done at 950 °C in N_2 done for 30 min for dopant activation. The Si template thickness is reduced to 48 nm by treatment in RIE machine (SF$_6$/O$_2$, 40 s) and dry oxidation at 1050 °C, 10 min Cleaning is done in Piranha, water rinsing; $H_2O + NH_4OH + H_2O_2$, water rinsing; and 10% HF, water rinsing.

3.5.2 Growth of SiGe alloy and Si films

Over the 48 nm thick Si template of SOI wafer, undoped Si$_{0.795}$Ge$_{0.205}$ alloy film and then an undoped Si film (46 nm) are grown by molecular beam epitaxy (MBE) to realize the trilayer structure Si (46 nm)/Si$_{0.795}$Ge$_{0.205}$ (80 nm)/Si (48 nm). The topmost undoped Si film is of approximately the same thickness as the template layer over which the Si$_{0.795}$Ge$_{0.205}$ layer is grown, so that the (Si/Si$_{0.795}$Ge$_{0.205}$/Si) trilayer structure is symmetrical in composition. Owing to the lattice mismatch between Si and Si$_{0.795}$Ge$_{0.205}$, the Si$_{0.795}$Ge$_{0.205}$ layer in the trilayer structure experiences compressive strain.

3.5.3 Separation and release of the strips

After photolithography, active layers are patterned into strips (40 μm width) separated by gaps of 10 μm by reactive ion etching. The BOX layer is etched in 4:1 dilute HF (49%) releasing the strips.

3.5.4 Strain distribution in the trilayer after release of strips

The distribution of strain in the trilayer Si/Si$_{0.795}$Ge$_{0.205}$/Si completely changes after release (Yuan *et al* 2008). As long as the Si/Si$_{0.795}$Ge$_{0.205}$/Si was part of the SOI wafer, the strain in the trilayer was sustained by the rigid silicon substrate. But as soon as the (Si/Si$_{0.795}$Ge$_{0.205}$/Si) NM is released by etching away the BOX layer, all the three layers are straightaway spontaneously elastically strained. The compressive strain in Si$_{0.795}$Ge$_{0.205}$ is shared with the covering Si layers. As a consequence, tensile strain is produced in the two silicon layers while the compressive strain in the Si$_{0.795}$Ge$_{0.205}$ layer decreases. Since the two Si layers in this structure have the same thickness, equal tensile strains are created in the silicon layers. Due to equal strain sharing, the structure remains flat and is not curled up or down. The forces in the free-standing NM are equipoised.

Figure 3.8 displays the Si and SiGe lattices in relaxed condition. In figure 3.9, the strains in the three layers are indicated after setting the trilayer free from the SOI wafer.

The amount of strain transferred by the Si$_{0.795}$Ge$_{0.205}$ layer to Si layers depends on the ratio of thicknesses of Si$_{0.795}$Ge$_{0.205}$ layer and Si layer, and the Ge content of

Taking SOI wafer after phosphorous implantation, annealing and thinning down device layer from 200 nm to 48 nm by dry oxidation

P-Si (48 nm)

N⁺ N⁺ N⁺

N⁺ N⁺ N⁺

Buried oxide
Handle layer

MBE growth to form trilayer : Si(48 nm) /Si$_{0.795}$Ge$_{0.205}$/Si (46 nm)

Undoped-Si (46 nm)

Si$_{0.795}$Ge$_{0.205}$

N⁺ N⁺ N⁺

P-Si (48 nm)

N⁺ N⁺ N⁺

Buried oxide
Handle layer

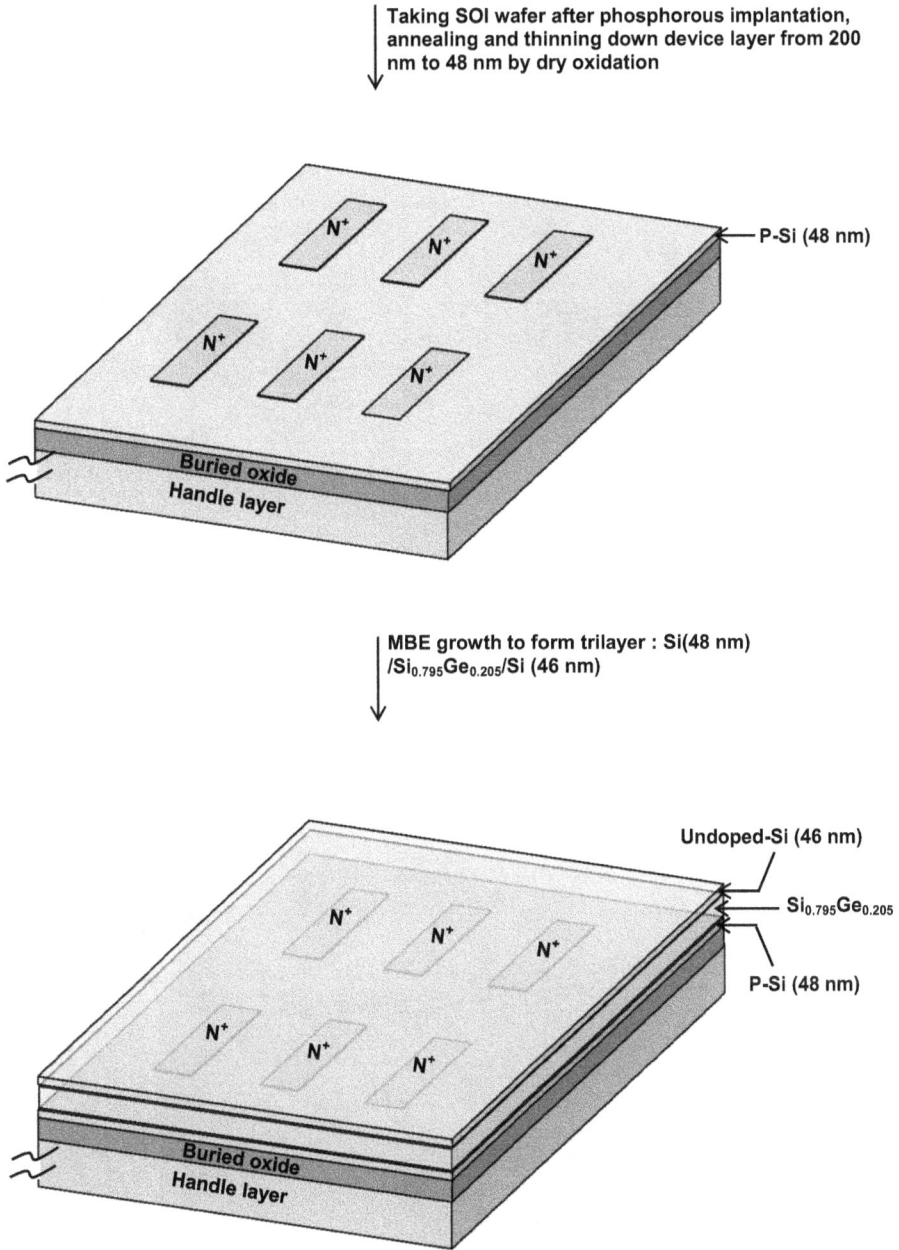

Figure 3.7. TFT on strained Si/SiGe/Si nanomembrane.

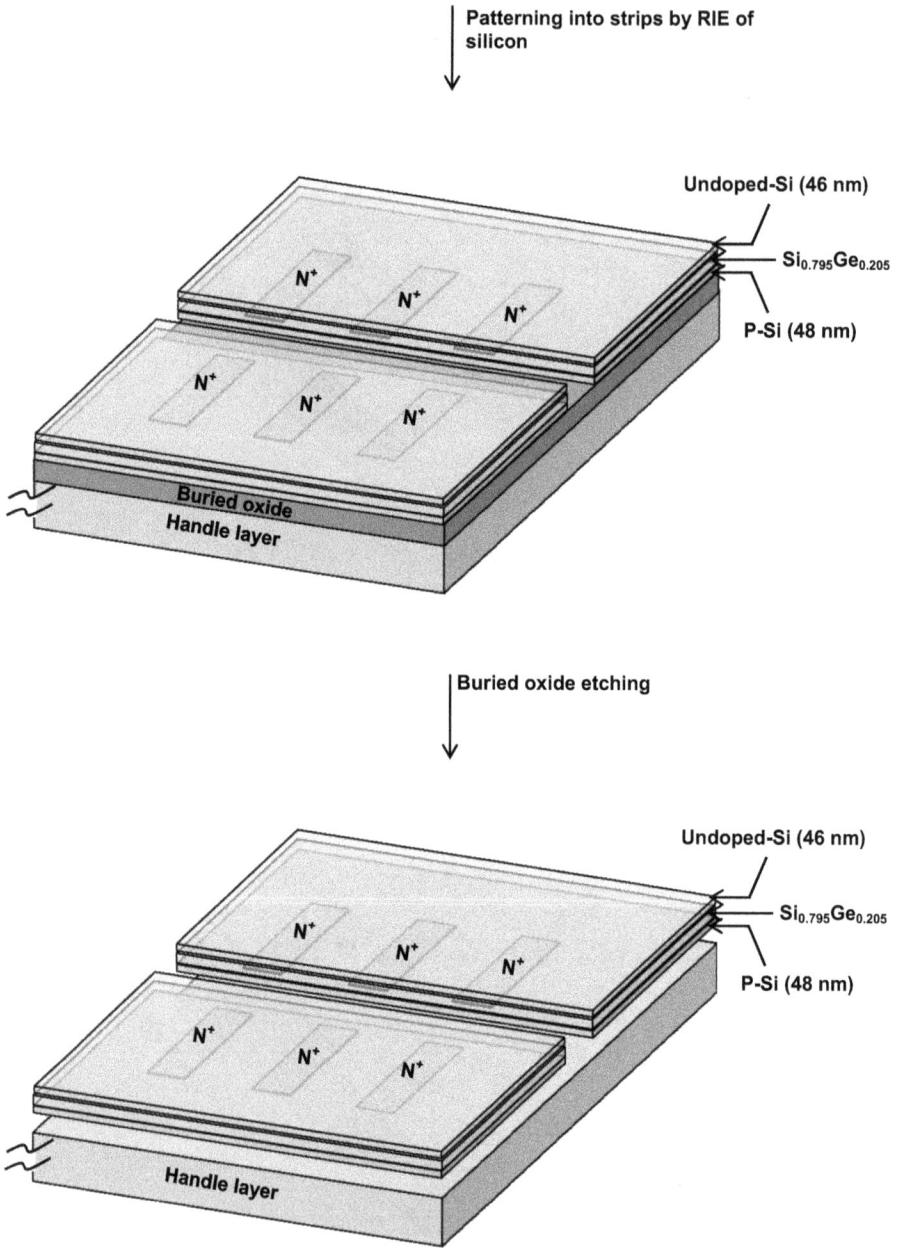

Figure 3.7. (Continued.)

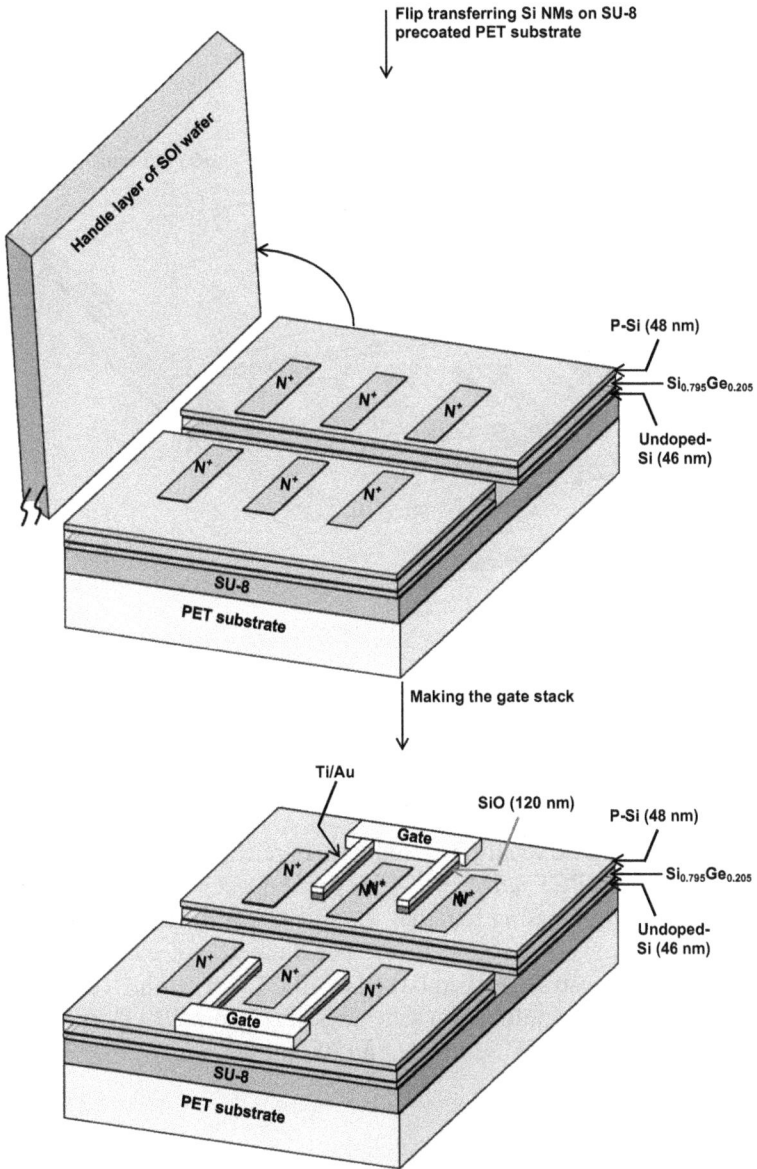

Figure 3.7. (Continued.)

the $Si_{0.795}Ge_{0.205}$ layer. An increase in thickness of the $Si_{0.795}Ge_{0.205}$ layer augments the strain sharing. However, the thickness must be kept below the kinetic critical thickness at which dislocations appear.

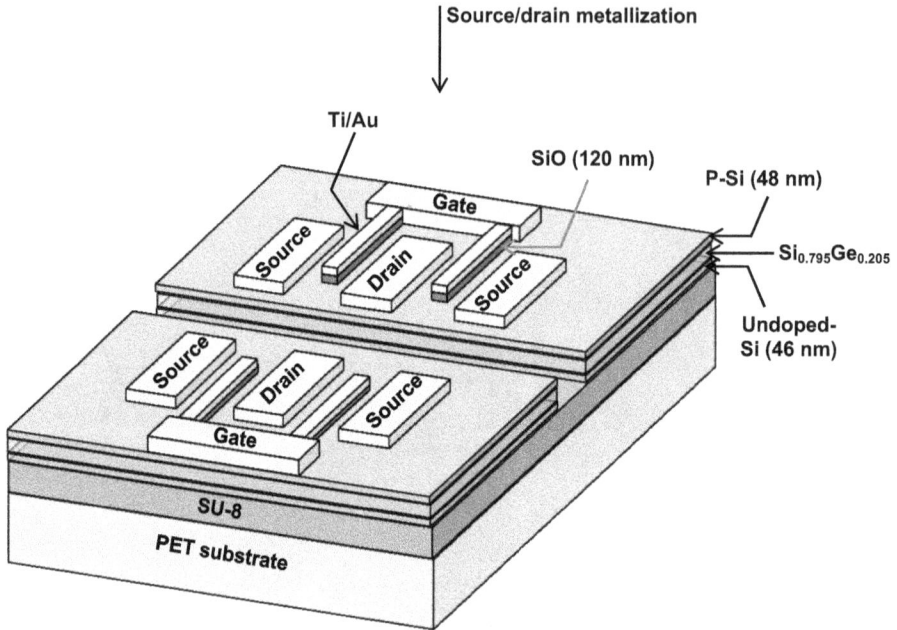

Figure 3.7. (Continued.)

3.5.5 Transfer of strips to PET substrate

The free-standing elastically relaxed strips are transferred upside down to the SU-8 layer spin-coated PET substrate. Adhesion between the SU-8 layer and the strips is secured by gently pressing together the silicon substrate on which the strips came to rest after release with the SU-8 layer. The back side of the PET substrate is exposed to UV and the substrate is then baked at 105 °C for 5 min. The gate dielectric is SiO (120 nm), the gate metallization scheme is: Ti (20 nm)/Au (150 nm) while the source/drain metallization is: Ti (20 nm)/Au (300 nm).

3.5.6 TFT parameters

The peak transconductance value of the strained-channel TFT is 386 µS; that of unstrained-channel TFT is 262 µS. The 47.3% increase in the peak transconductance of strained-channel TFT with respect to the unstrained channel device originates from the difference in their carrier mobilities. The cut-off frequency (f_T) of the strained-channel TFT is 5.1 GHz and its maximum oscillation frequency (f_{max}) is 15.1 GHz. The corresponding values of unstrained-channel TFT of similar dimensions are 3.3 GHz and 10.3 GHz respectively. The increase in speed (54.5% for f_T and 46.6% for f_{max}) arises from the 47.3% mobility enhancement (Zhou *et al* 2013).

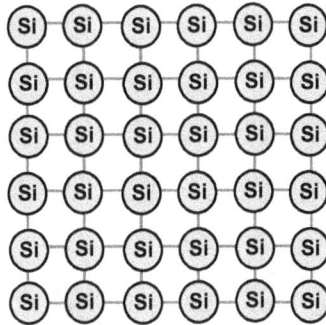

(a)

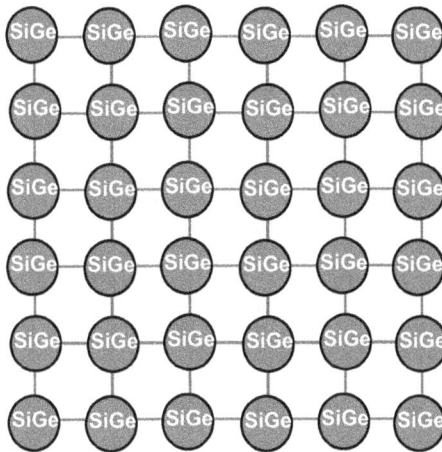

(b)

Figure 3.8. Lattices in relaxed state: (a) silicon and (b) silicon–germanium.

3.6 Discussion and conclusions

The idea of TFT fabrication using a silicon nanomembrane can be looked upon as a nanotechnology-enabled macroelectronics concept. The nanomembrane has all the properties of single-crystal silicon. It can also be flexed because of its small thickness. By mounting it on a plastic substrate, it receives mechanical support while retaining its bending capability. The concept is based on well-proven silicon-on-insulator technology ensuring its practicability.

Review exercises

3.1 What is a silicon nanomembrane? State the three objectives of making and using silicon nanomembranes for electronic device fabrication.

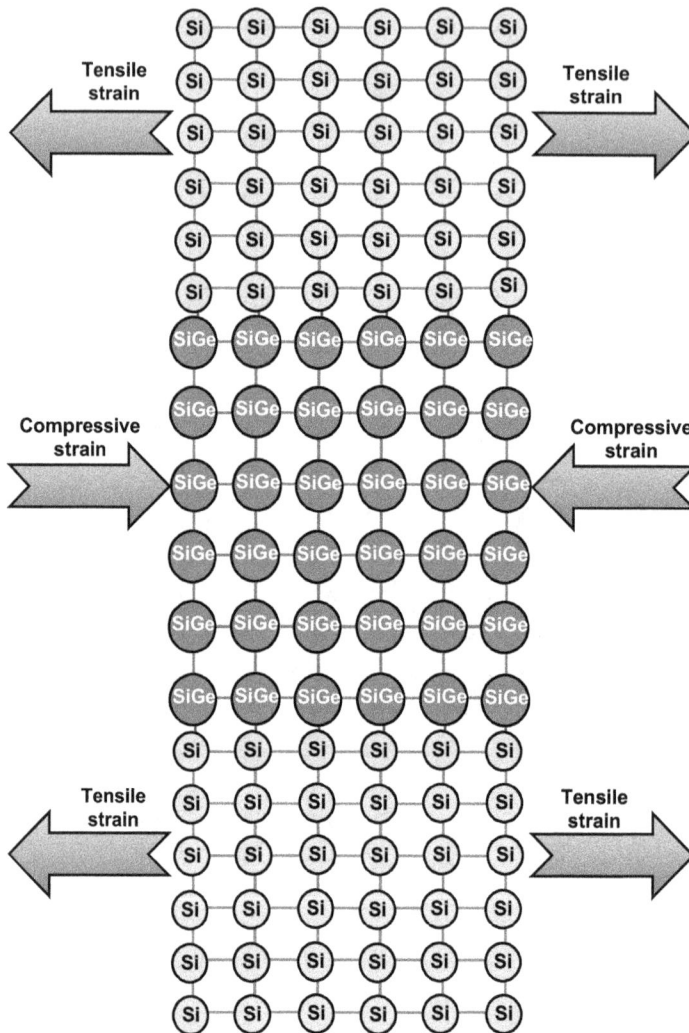

Figure 3.9. Strain distribution in the trilayer Si/SiGe/Si after release from SOI wafer.

3.2 What is a silicon-on-insulator (SOI) wafer? What are the three layers in this wafer called? What is the reason for deciding such a three-layer arrangement? What are the typical thicknesses of the three layers?

3.3 Describe two methods of etching trenches in the silicon template layer of SOI wafer to make nanoribbons. How are the silicon pillars tied at their bases to the BOX layer, released from its captivity?

3.4 For what type of nanomembrane geometries is doping by thermal diffusion used? When is ion implantation used?

3.5 Name two methods of transfer printing the nanomembranes from a donor substrate to an acceptor one in which the acceptor substrate is coated with an adhesive.

3.6 Name: (i) a material used for making stamp for transfer printing nano-membranes, (ii) a material used as an adhesive on the acceptor substrate.

3.7 Can transfer printing be done without applying any adhesive on the acceptor substrate? What is the method used for such printing called? Explain how it is done?

3.8 How is the adhesion of nanomembranes controlled in dry transfer printing? How does the peeling speed from the donor substrate to the stamp differ from that from the stamp to the acceptor substrate?

3.9 How is the gate-after-source/drain process carried out? Does this process provide self-alignment?

3.10 A SiGe layer is epitaxially grown over the silicon template layer of an SOI wafer and the SiGe layer itself is covered with an epitaxial silicon layer. What kind of strain is developed in the SiGe layer? When this Si/SiGe/Si trilayer is released from the buried oxide as a nanomembrane, how does the strain in SiGe layer and the top and bottom silicon layers change?

3.11 How does the carrier mobility in the strained-channel TFT differ from that in the unstrained channel TFT?

3.12 Which has a higher transconductance: Strained-channel TFT or unstrained channel TFT? Why?

3.13 Describe the process for realizing a strain-engineered TFT on a flexible substrate.

References

Ahn J-H, Kim H-S, Lee K J, Zhu Z, Menard E, Nuzzo R G and Rogers J A 2006 High-speed mechanically flexible single-crystal silicon thin-film transistors on plastic substrates *IEEE Electron Device Lett.* **27** 460–2

Meitl M A, Zhu Z-T, Kumar V, Lee K J, Feng X, Huang Y Y, Adesida I, Nuzzo R G and Rogers J A 2006 Transfer printing by kinetic control of adhesion to an elastomeric stamp *Nat. Mater.* **5** 33–8

Menard E, Lee K J, Khang D-Y, Nuzzo R G and Rogers J A 2004 A printable form of silicon for high performance thin film transistors on plastic substrates *Appl. Phys. Lett.* **84** 5398–400

Menard E, Nuzzo R G and Rogers J A 2005 Bendable single crystal silicon thin film transistors formed by printing on plastic substrates *Appl. Phys. Lett.* **86** 093507-1–3

Qin G, Seo J-H, Zhang Y, Zhou H, Zhou W, Wang Y, Ma J and Ma Z 2013 RF Characterization of gigahertz flexible silicon thin-film transistor on plastic substrates under bending conditions *IEEE Electron Device Lett.* **34** 262–4

Seo J-H, Yuan H-C, Sun L, Zhou W and Ma Z 2011 Transferrable single-crystal silicon nanomembranes and their application to flexible microwave systems *J. Inf. Disp.* **12** 109–13

Yuan H-C, Celler G K and Ma Z 2007 7.8-GHz Flexible thin-film transistors on a low-temperature plastic substrate *J. Appl. Phys.* **102** 034501-1–4

Yuan H-C, Kelly (Roberts) M M, Savage D E, Lagally M G, Celler G K and Ma Z 2008 Thermally processed high-mobility MOS thin-film transistors on transferable single-crystal elastically strain-sharing Si/SiGe/Si nanomembranes *IEEE Trans. Electron Devices* **55** 810–5

Yuan H-C, Qin G, Celler G K and Ma Z 2009 Bendable high-frequency microwave switches formed with single-crystal silicon nanomembranes on plastic substrates *Appl. Phys. Lett.* **95** 043109-1–3

Zhang K, Seo J-H, Zhou W and Ma Z 2012 Fast flexible electronics using transferrable silicon nanomembranes *J. Phys. D: Appl. Phys.* **45** 143001

Zhou H, Seo J-H, Paskiewicz D M, Zhu Y, Celler G K, Voyles P M, Zhou W, Lagally M G and Ma Z 2013 Fast flexible electronics with strained silicon nanomembranes *Sci. Rep.* **3** 1291

IOP Publishing

Flexible Electronics, Volume 2
Thin-film transistors
Vinod Kumar Khanna

Chapter 4

Metal-oxide TFT

Bestowed with the special feature of optical transparency and possessing a higher mobility than a-Si, oxide semiconductors have carved a niche for themselves in TFT fabrication on plastic substrates. Indium gallium zinc oxide TFT on a PEN substrate is able to work up to a bending radius of 1 mm at a strain level of 0.7% (Tripathi *et al* 2015, Heremans *et al* 2016). An IGZO TFT fabricated on both sides of paper utilizes paper in the dual role of a substrate and gate dielectric. In this TFT, the gate electrode is made of IZO film (Fortunato *et al* 2008). The essential high-temperature annealing step in sol–gel IGZO film preparation for TFT fabrication on polyarylate film is avoided by deep UV photo-annealing at low temperature (Kim *et al* 2012). An IGZO TFT is fabricated on a polyimide film using block copolymer for moisture barrier and gate dielectric functions. Upon bending, the TFT characteristics remain stable up to 4 mm radius (strain = 1.25%), behaving differently when bent along the channel direction and perpendicularly (Kumaresan *et al* 2016). Looking at the scarcity and toxicity of indium and gallium, an indium-free and gallium-free TFT is made on a transparent PET substrate using nickel-doped ZnO as the semiconductor. This transparent TFT on a see-through substrate is useful for transparent flexible electronics (Huang *et al* 2015). High-speed circuits are realized with ZnO TFTs on polyimide substrates using a plasma-enhanced atomic layer deposition process for ZnO film (Zhao *et al* 2010).

4.1 Introduction

Oxide semiconductors are optically transparent. Also, the mobility in these semi-conductors is higher than in hydrogenated amorphous silicon. Additionally, by making oxide semiconductor TFTs on plastic substrates, tolerance towards bending is conferred. These oxide semiconductor TFTs are endowed with a useful combination of properties making them highly favorable for flexible electronics.

4.2 IGZO TFT with ESL on PEN substrate

Tripathi *et al* (2015) and Heremans *et al* (2016) described an IGZO TFT on a planarized PEN substrate of thickness 25 μm with an etch stop layer (ESL) over the semiconducting IGZO layer for the backplane of the display. An 8b transponder chip bendable up to a radius of 2 mm is reported.

Figure 4.1 shows the cross-section of the TFT with ESL. The structure of the ESL TFT starts with a bottom PEN substrate. Upon this substrate is a silicon nitride SiN_x moisture barrier layer inhibiting the penetration of water vapor through the substrate in humid environments. Over the barrier layer lies the gate metal. The gate metal is covered by the gate dielectric. Now comes the IGZO semiconductor. The IGZO layer is coated with an etch stop layer. Then the holes are drilled in the ESL for contacting the source and drain; and source/drain metallization is done. Finally, the device is encapsulated by covering its top surface with an SU-8 layer. Thus the device configuration is essentially a bottom-gate structure with source and drain contacts from the top surface. During the device fabrication, the maximum temperature used in any process step is restricted below 200 °C. The sequence of steps involved in TFT fabrication is given in the following subsections (figure 4.2).

4.2.1 Coating the PEN substrate with a moisture barrier layer

The planarized PEN substrate is passivated against moisture with an SiN_x layer. The thickness of this layer is 150 nm.

4.2.2 Deposition of gate metal

The gate metal is molybdenum–chromium (Mo–Cr). It is deposited by sputtering. The metal pattern is defined photolithographically.

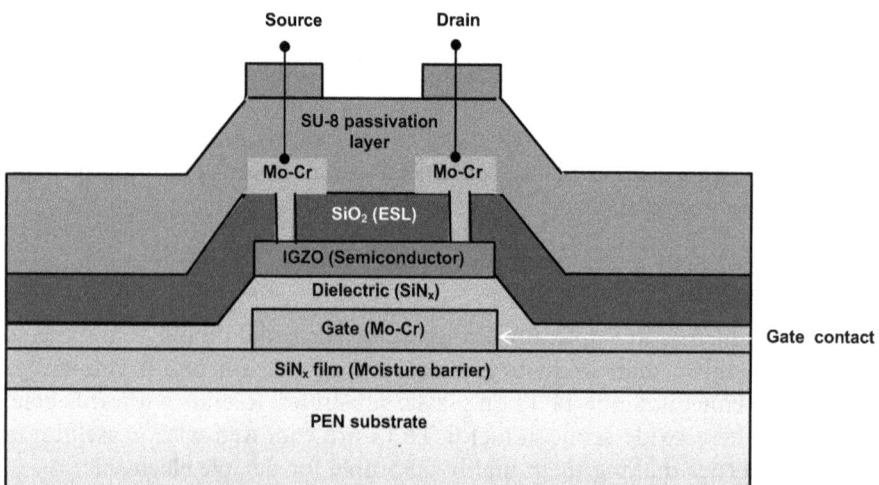

Figure 4.1. Schematic cross-section of TFT with IGZO semiconducting layer.

Moisture barrier layer (SiN$_x$) deposition
on PEN substrate

SiN$_x$ film (Moisture barrier)

PEN substrate

Gate metal (Mo-Cr) sputtering

Gate (Mo-Cr)

SiN$_x$ film (Moisture barrier)

PEN substrate

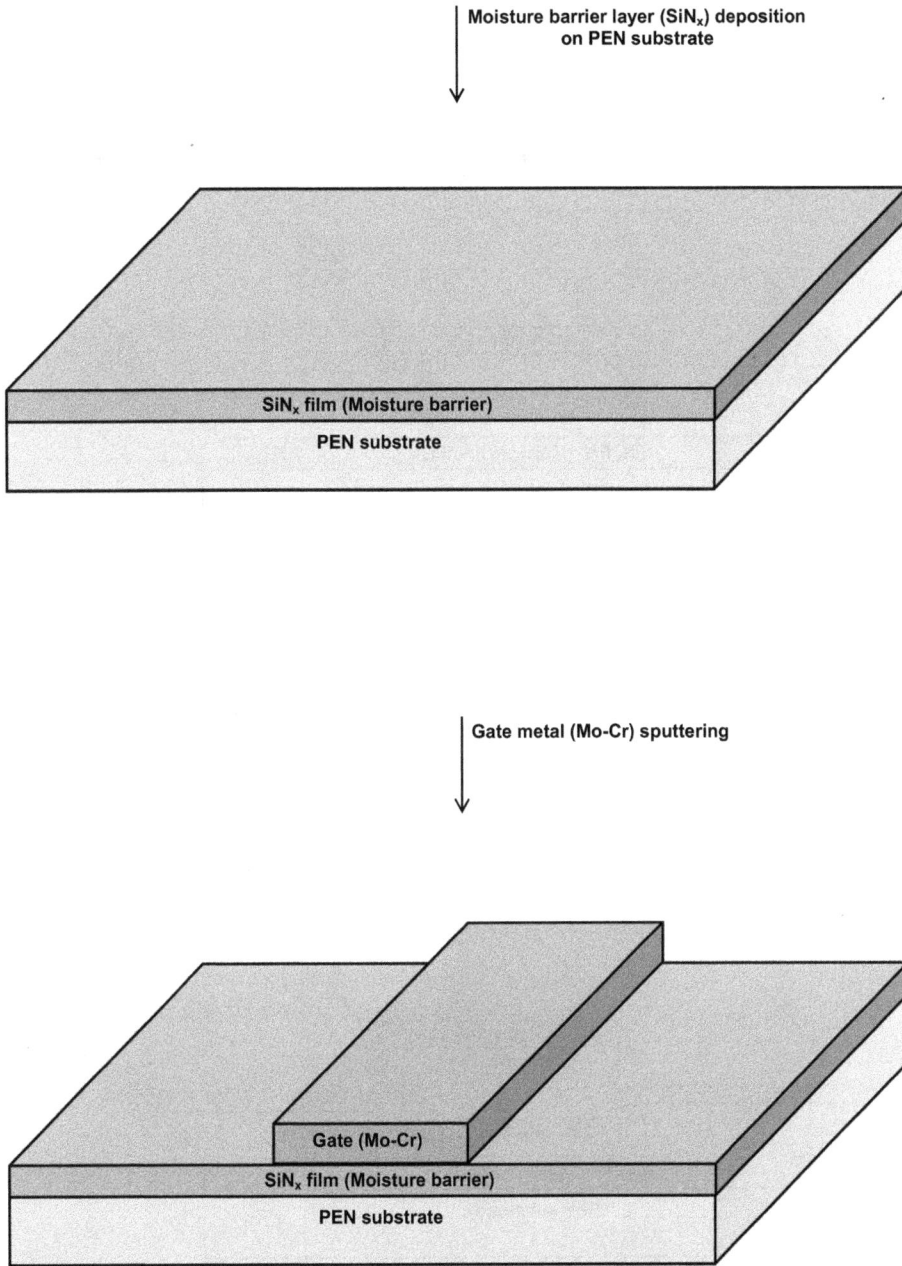

Figure 4.2. TFT fabricated with IGZO as semiconducting layer and PECVD SiO$_2$ as etch stop layer.

4.2.3 Gate dielectric deposition

The gate dielectric is SiN$_x$. The deposition method used is PECVD. The SiN$_x$ film is deposited at 180 °C. SiN$_x$ film thickness is 200 nm. This film has a breakdown field of 6×10^6 V cm^{-1}.

Gate dielectric (SiN$_x$) by PECVD

Dielectric (SiN$_x$)

Gate (Mo-Cr)

SiN$_x$ film (Moisture barrier)

PEN substrate

Semiconductor (IGZO) sputtering

IGZO (Semiconductor)

Dielectric (SiN$_x$)

Gate (Mo-Cr)

SiN$_x$ film (Moisture barrier)

PEN substrate

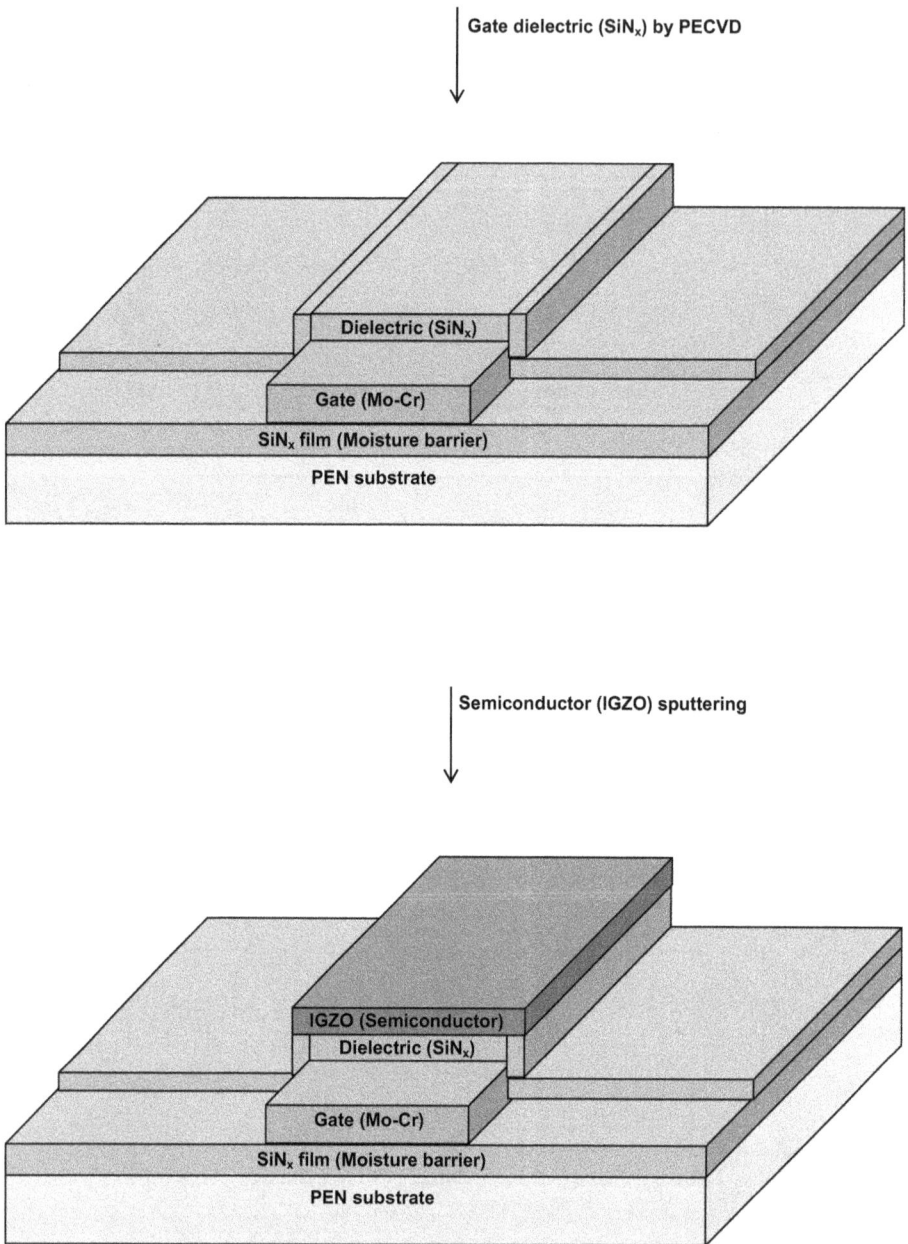

Figure 4.2. (Continued.)

4.2.4 IGZO semiconductor deposition

Before IGZO deposition, annealing is done in air at 200 °C for 1 h for desorption of any loosely bound impurities. DC sputtering method is applied. The target used for sputtering has the composition In:Ga:Zn in the atomic ratio 1:1:1. The sputtering is

Etch stop layer (ESL) SiO₂ by PECVD

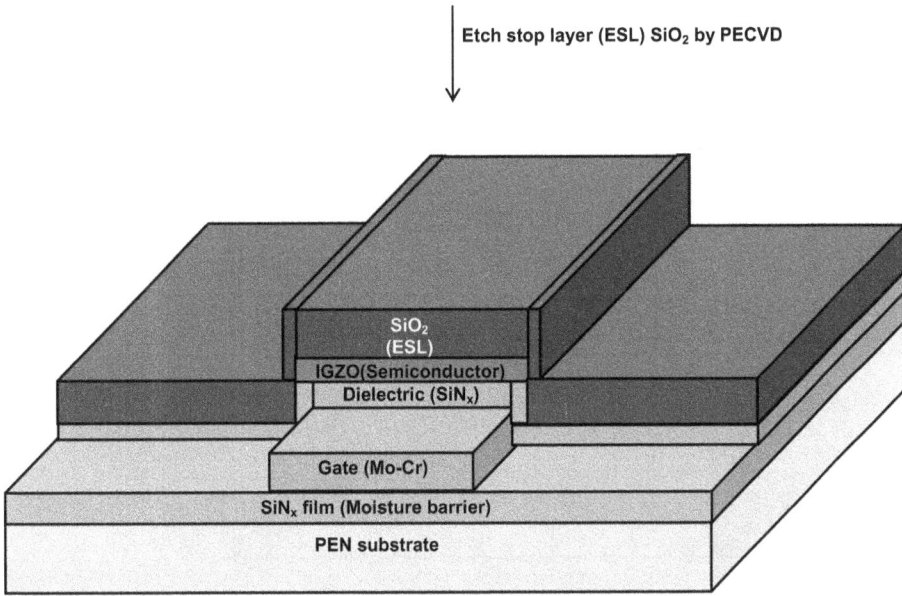

SiO₂ (ESL)
IGZO(Semiconductor)
Dielectric (SiNₓ)
Gate (Mo-Cr)
SiNₓ film (Moisture barrier)
PEN substrate

Drilling holes in SiO₂ for source/drain contacts by dry etching

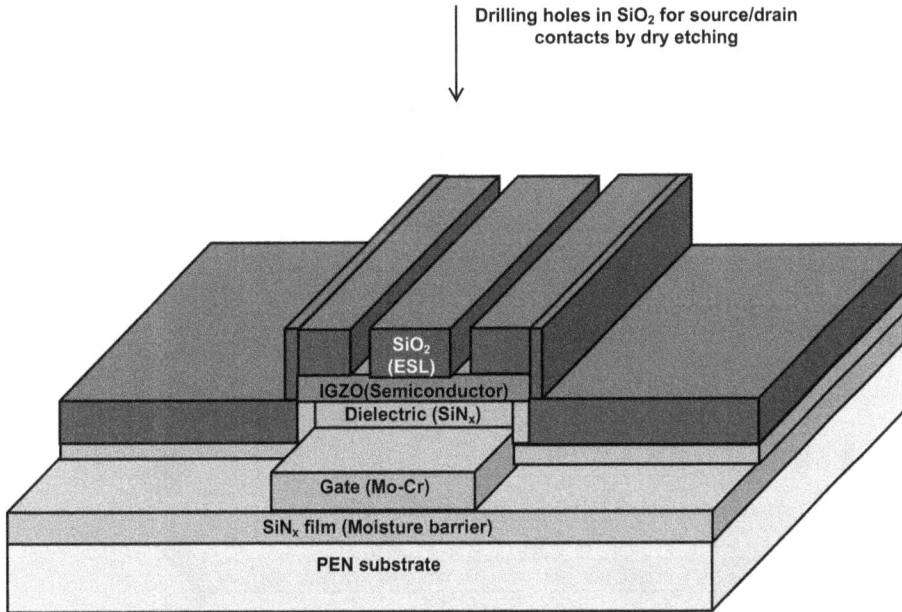

SiO₂ (ESL)
IGZO(Semiconductor)
Dielectric (SiNₓ)
Gate (Mo-Cr)
SiNₓ film (Moisture barrier)
PEN substrate

Figure 4.2. (Continued.)

Mo-Cr sputtering

Mo-Cr

Mo-Cr

SiO$_2$ (ESL)

IGZO (Semiconductor)

Dielectric (SiN$_x$)

Gate (Mo-Cr)

SiN$_x$ film (Moisture barrier)

PEN substrate

SU-8 coating and taking out electrical connections

Source   Drain

SU-8 passivation layer

Mo-Cr

Mo-Cr

SiO$_2$ (ESL)

IGZO (Semiconductor)

Dielectric (SiN$_x$)

Gate (Mo-Cr)

SiN$_x$ film (Moisture barrier)

PEN substrate

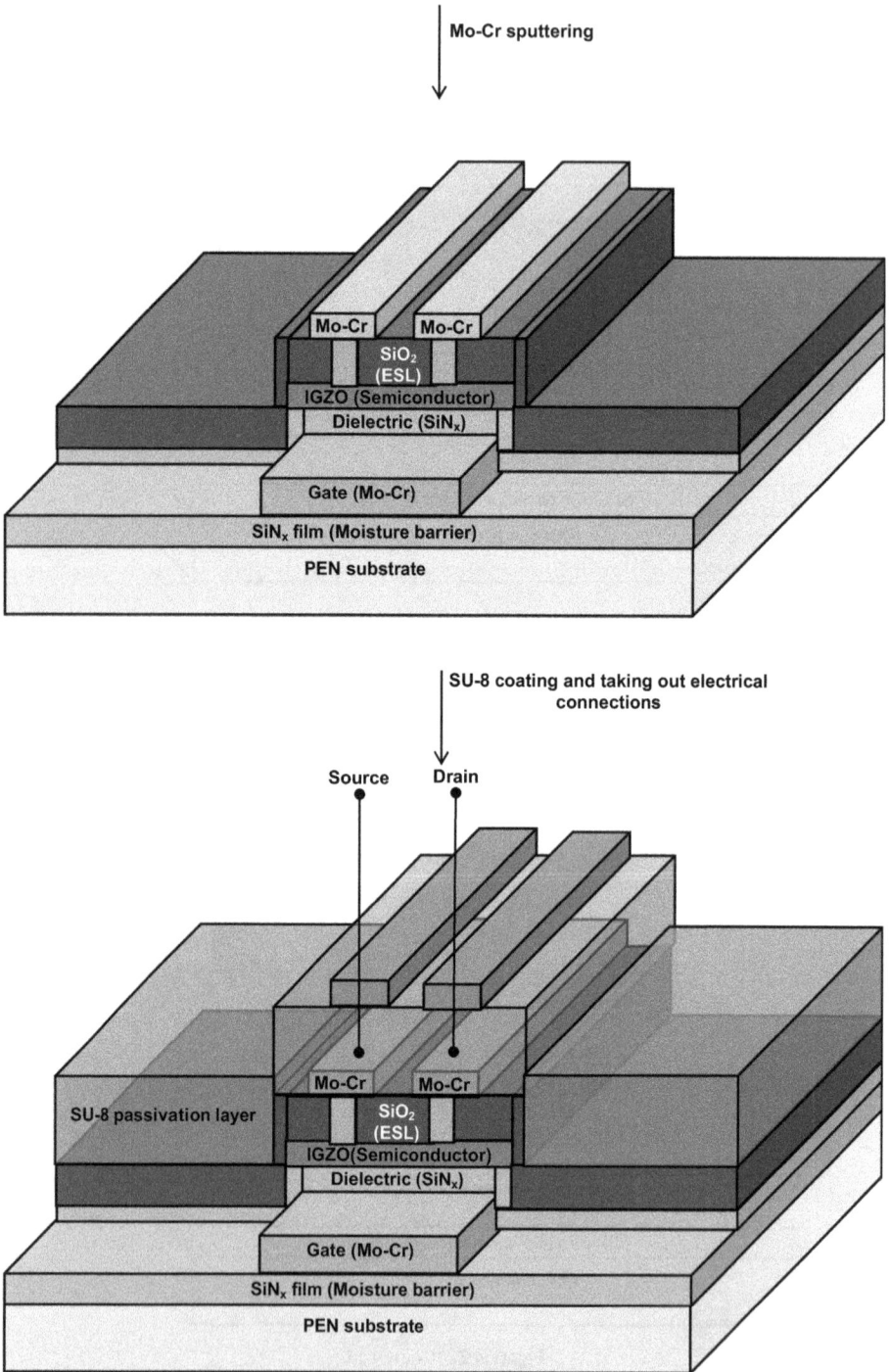

Figure 4.2. (Continued.)

carried out under oxygen flow of 6%, which is optimized in order to achieve satisfactory performance of TFT fabricated at a low temperature. The thickness of the sputtered IGZO film is 15 nm.

4.2.5 Etch stop layer (ESL)

After the semiconductor film has been patterned, PECVD is applied to deposit silicon dioxide ESL film over this semiconductor film. The temperature during the PECVD process is 200 °C and the SiO_2 film thickness is 100 nm.

4.2.6 Making contact holes

Dry plasma etching is used to make the contact holes.

4.2.7 Source/drain metallization and interconnections

Sputtering is done using an Mo–Cr target.

4.2.8 Encapsulation with SU-8

The SU-8 film has a thickness 2 μm. Annealing is done in air at 180 °C for 3 h.

4.2.9 Effect of bending on the device and circuit

The threshold voltage of the TFT is 3.1 V and the subthreshold slope is 330 mV/decade. The carrier mobility is 13 cm^2 V^{-1} s^{-1}. The on–off ratio is 10^8.

To determine the failure limit, the TFT is mounted on cylinders of different radii, thereby inducing mechanical strains of varying degrees in them. At each radius, the TFT is bent for 10 s, then flattened and re-measured. A bending radius up to 1 mm is unable to cause any damage to the TFT. The corresponding strain is 0.7%. Irreversible failure of TFT occurs at a radius of 0.5 mm at which the strain is 1.4%. The failure is observed in the form of an abnormally high gate leakage current and off current of the TFT. Cracks produced in the dielectric film are responsible for this failure.

In general, the TFT maintains a gate leakage current and off current below 1 pA up to a bending radius of 2 mm. There is no degradation in its current–voltage characteristics up to this radius value. However, with respect to the flat condition, the threshold voltage shifts marginally negatively with maximum shift <0.3 V at 2 mm radius, irrespective of whether the TFT is bent parallel to the channel or orthogonally.

The mobility μ is weakly dependent on strain ε. When strained along the channel, the strain dependence of mobility is expressed by the linear relationship

$$\frac{\mu(R)}{\mu_0} = 1 + 2.1\,\varepsilon \tag{4.1}$$

where μ_0 is the mobility in flat state of TFT and $\mu(R)$ is the mobility when TFT is bent. R is the bending radius.

For straining perpendicular to the channel, we have

$$\frac{\mu(R)}{\mu_0} = 1 + 2.2\,\varepsilon \qquad (4.2)$$

The constant of proportionality shown as 2.1 or 2.2, varies between 0.5 to 3 for both the cases, parallel and perpendicular.

Gate-bias stress at $\pm 1 \times 10^6$ V cm^{-1} shows a threshold voltage shift $< \pm 0.8$ V after 3800 s, both for flat and bent TFTs.

A working 8b transponder chip with TFT integration level ~300 is demonstrated at a bending radius of 2 mm. Its rate of data transfer is 31.6 kb s^{-1}. This is the speed of the 19 stage ring oscillator inside the chip. The code is produced in ~253 µs (Tripathi *et al* 2015).

4.3 IGZO TFT with cellulose fiber-based paper as substrate cum gate dielectric

Fortunato *et al* (2008) used the paper in an 'interstrate' structure (figure 4.3). The word 'interstrate' is used to mean both sides in analogy to 'substrate' which refers to an underlying layer. The uniqueness of the so-called interstrate structure is that the device is built on both sides or surfaces of the paper. On the bottom surface of the paper, the gate electrode is deposited so that we get the structure: Gate electrode/ paper. On the top surface of the paper, the semiconductor layer is deposited. Upon the semiconductor layer, source and drain electrodes are formed giving the structure: paper/semiconductor/source–drain contacts. Thus combining the two structures we get the total structure as: gate electrode/paper/semiconductor/source–drain contacts. The paper (cellulose) of thickness ~75 µm acts as the gate dielectric.

4.3.1 IZO gate electrode deposition

Two types of paper designated as cellulose type A and cellulose type B are used as substrates. The conductor used is IZO composed of In_2O_3 and ZnO in the ratio 5:2 mol%. The deposition technique is RF sputtering. The thickness of IZO film is 160 nm.

4.3.2 IGZO semiconductor layer deposition

IGZO is made of In_2O_3, Ga_2O_3 and ZnO in the ratio 2:1:1 mol%. RF magnetron sputtering is used to deposit a 40 nm thick film of this semiconductor at room temperature.

4.3.3 Source/drain contacts

Aluminum metal is used. It is deposited by electron-beam evaporation. The thickness of the aluminum film is 180 nm.

IZO gate electrode deposition on the backside
of paper substrate by RF sputtering

Cellulose-fiber based paper substrate

IZO gate electrode

IGZO semiconductor layer deposition on the
front side of paper substrate by RF sputtering

IGZO semiconductor layer

Cellulose-fiber based paper substrate

IZO gate electrode

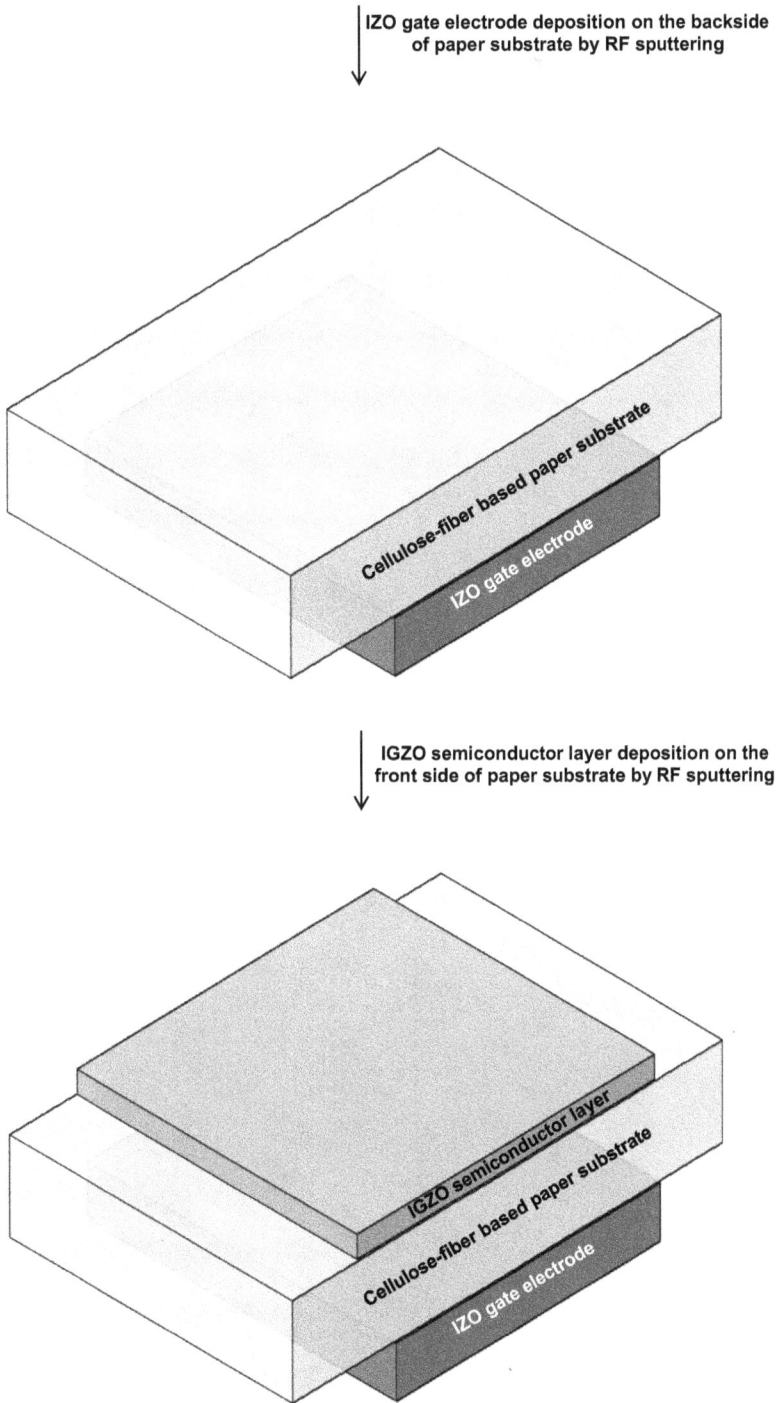

Figure 4.3. TFT with IZO gate electrode and IGZO semiconductor film, fabricated using both sides of the paper substrate.

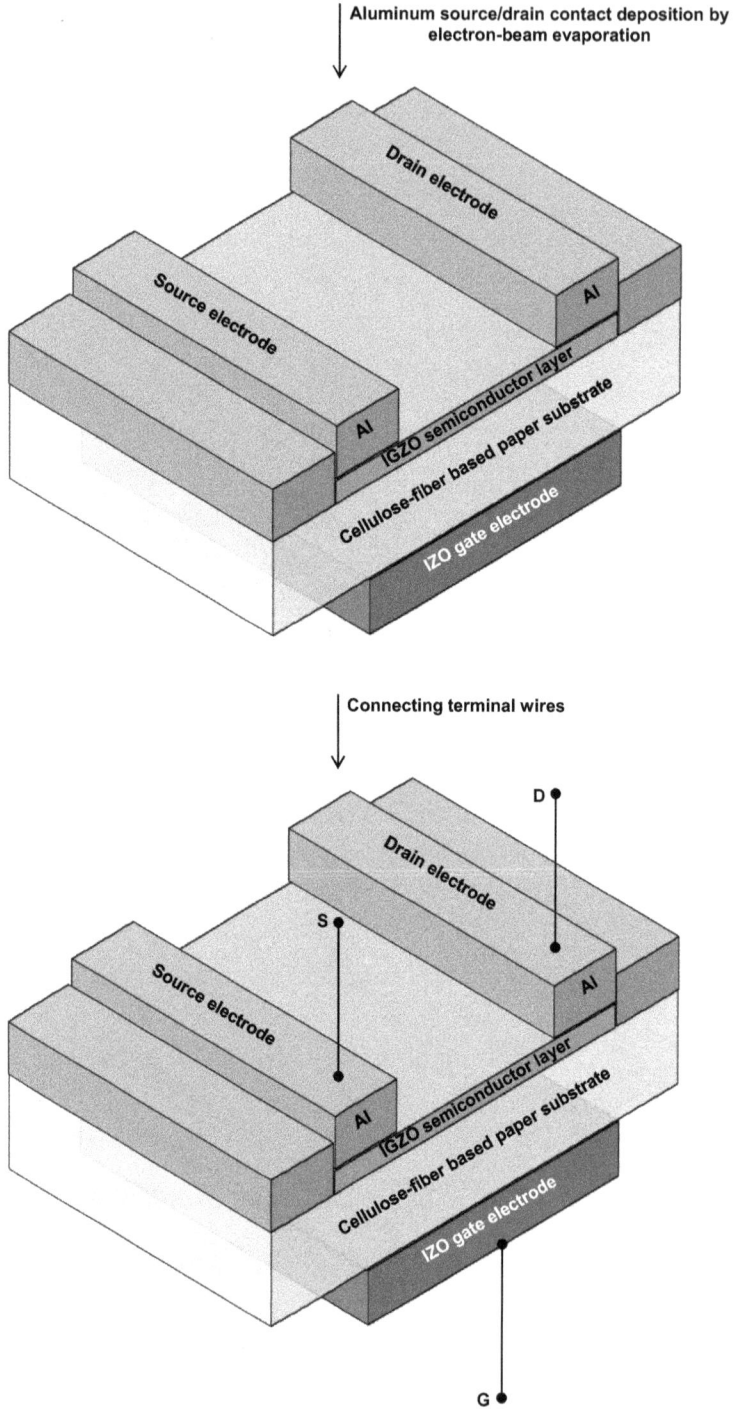

Figure 4.3. (Continued.)

4.3.4 TFT parameters on paper types A and B

On paper A, the threshold voltage is −0.6 V with subthreshold slope of 80 mV/ decade. The current on/off ratio is 5.9×10^3. The saturation mode mobility is 30 cm^2 V^{-1} s^{-1}. On paper B, the values of these parameters are +1.9 V, 80 mV/ decade, 2.9×10^4, and 34 cm^2 V^{-1} s^{-1} respectively (Fortunato *et al* 2008).

4.4 IGZO TFT fabrication process by sol–gel route

Kim *et al* (2012) overcame the difficulty faced in metal-oxide formation by the sol–gel method (figure 4.4). The problem here is the requirement of a high-temperature annealing step for film densification and stabilization of characteristics. The high temperature cannot be used with a polymer substrate. The bottleneck arising from temperature limitation is resolved by photochemically activating the sol–gel film by exposure to deep UV at low temperature. A low-pressure mercury lamp is used for irradiation. The exposure is done in an inert ambient to prevent formation of ozone which is highly reactive.

4.4.1 Gate electrode

The substrate is a polyarylate (PAR) film of thickness 200 μm. Titanium (3 nm)/gold (80 nm) or molybdenum (100 nm) film is patterned to make the gate electrode. The metal films are deposited by thermal evaporation.

4.4.2 Gate dielectric

Al_2O_3 film is formed by atomic layer deposition over the substrate with patterned gate electrode. The deposition temperature is 100 °C. The thickness of the Al_2O_3 film is 35 nm. Trimethylaluminum, $(CH_3)_3Al$, existing as a dimer $C_6H_{18}Al_2$, is used.

4.4.3 Preparation of IGZO precursor solution

The solvent used is 2-ME (2-methoxyethanol), $C_3H_8O_2$. Into this solvent, the following three compounds are dissolved: indium (III) nitrate hydrate, $In(NO_3)_3 . xH_2O$; gallium nitrate hydrate, $Ga(NO_3)_3 . xH_2O$; and zinc acetate dihydrate, $Zn(CH_3COO)_2 . 2H_2O$. The solution contains 0.085 M $In(NO_3)_3 . xH_2O$; 0.0125 M $Ga(NO_3)_3 . xH_2O$; and 0.0275 M $Zn(CH_3COO)_2 . 2H_2O$.

4.4.4 Semiconductor channel layer

On the gate dielectric layer formed in the previous step, the IGZO precursor solution is spin coated. Densification of IGZO is accomplished by deep UV photo-annealing. A high-density deep UV treatment system is used. The low-pressure Hg lamp has two emission peaks, one peak at 253.7 nm (0.9) and the other peak at 184.9 nm (0.1). The distance between the Hg lamp and film is ~1–5 cm and the power density is 25–28 mW cm^{-2}. Nitrogen gas is constantly flowed during annealing. The annealing is done for 1.5–2 h.

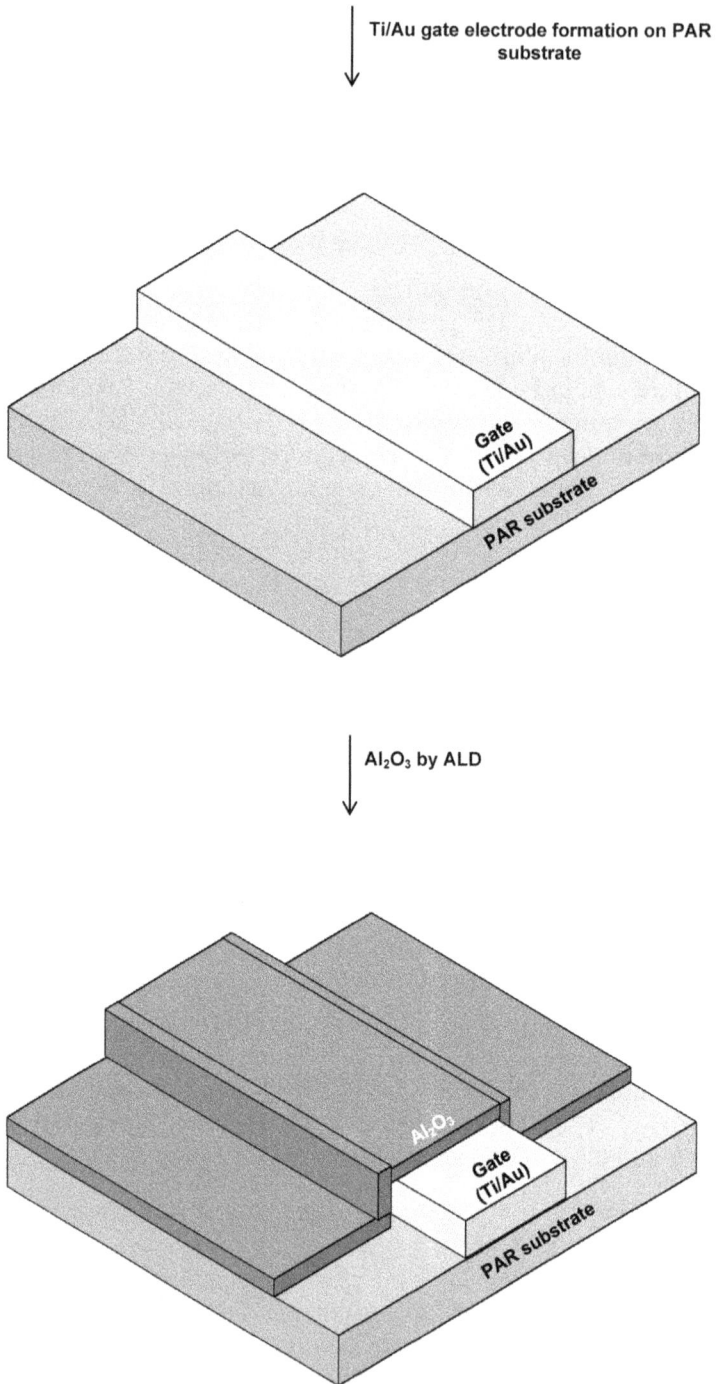

Ti/Au gate electrode formation on PAR substrate

Gate (Ti/Au)

PAR substrate

Al_2O_3 by ALD

Al_2O_3

Gate (Ti/Au)

PAR substrate

Figure 4.4. IGZO TFT fabrication by sol–gel technique.

Spin coating IGZO

IGZO

Al₂O₃

Gate
(Ti/Au)

PAR substrate

Patterning IGZO

IGZO

Al₂O₃

Gate
(Ti/Au)

PAR substrate

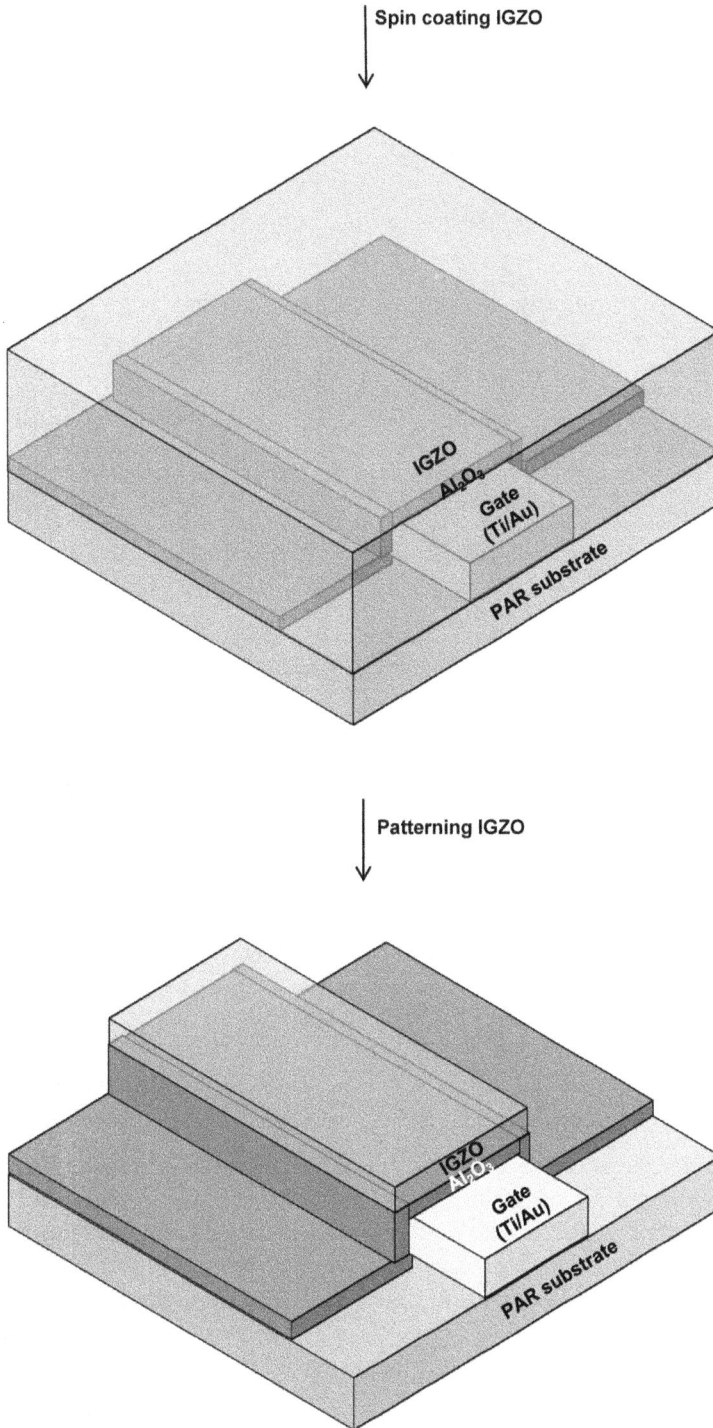

Figure 4.4. (Continued.)

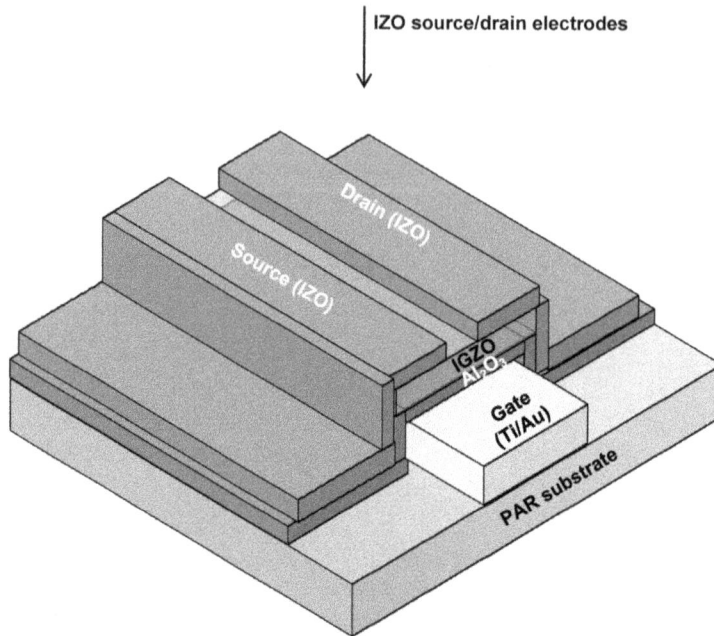

Figure 4.4. (Continued.)

4.4.5 Via holes

These holes are formed after patterning of the photo-annealed semiconductor channel layer by wet etching.

4.4.6 Source/drain electrodes

These electrodes are made of indium zinc oxide (IZO). Composition of the IZO precursor solution is: 0.05 M $In(NO_3)_3.xH_2O$ and 0.05 M of $Zn(CH_3COO)_2.2H_2O$. The solvent is 2-ME. The IZO film thickness is 100 nm. Lift-off process is used for creating the electrode patterns.

4.4.7 Electrical characteristics of TFT

The threshold voltage of the TFT is 2.7 V. The subthreshold swing (SS) is 95.8 mV/ decade. The current on–off ratio is 10^8. The PAR substrate is yellowed by deep UV but this yellow coloration is restricted only to the surface and does not move deeper inside the device. For the 49 devices fabricated by Kim *et al* 2012 on a polyarylate (PAR) film as substrate using an alumina gate insulator, the field-effect mobilities are spread around a value of 3.77 cm^2 V^{-1} s^{-1}; the maximum mobility is 7 cm^2 V^{-1} s^{-1}. A seven stage ring oscillator fabricated on PAR substrate shows an oscillation frequency >340 kHz. The propagation delay per stage is 210 ns (Kim *et al* 2012).

4.5 IGZO TFT with organic gate dielectric/moisture barrier layers

Kumaresan *et al* (2016) reported a low-temperature TFT fabrication process on a polyimide substrate using a thermally stable organic film as the gate insulator and the same film also as a water vapor blocker (figure 4.5). During sputtering of IGZO, the argon–oxygen ratio is controlled to achieve optimal transfer characteristics of the TFT. The TFT has a bottom-gated structure. The ratio, channel width: channel length is 2000 μm:100 μm = 20.

4.5.1 Attachment of PI film to rigid glass substrate

The PI film is 100 μm thick. To keep it flat during processing and to prevent any cracks developing in the film, it is fixed on a rigid substrate of glass using PDMS as an adhesive. The PI is fixed to PDMS/glass and annealed at 120 °C under 2.25×10^{-6} Torr pressure.

4.5.2 Organic barrier layer deposition

The material used is a commercial block copolymer SA7. It consists of the monomers: glycidyl methacrylate, $C_7H_{10}O_3$; methacrylic acid, $H_2C = C(CH_3)$ COOH or $C_4H_6O_2$; isobornyl acrylate, $C_{13}H_{20}O_2$; 2-hydroxyethyl acrylate, $CH_2 = CHCOOCH_2CH_2OH$; and styrene, $C_6H_5CHCH_2$ or C_8H_8. It has a dielectric constant of 2.5. The copolymer SA7 is spin-coated on the PI substrate and hardened by curing at 120 °C for 2 h. The thickness of the SA7 layer is 2 μm.

4.5.3 Organic gate electrode deposition

The organic material used is poly(3, 4-ethylenedioxythiophene):poly(styrenesulfonate) (PEDOT:PSS) = $(C_6H_4O_2S)_n$: $(C_8H_7SO_3^-)_n$. It is spun over the barrier layer. The spin-coated layer is treated with methanol, CH_3OH, to increase its conductivity. Its thickness is <50 nm.

4.5.4 Organic gate dielectric deposition

Upon the PEDOT:PSS layer, the SA7 copolymer layer is formed by spin coating and curing at 120 °C for 2 h. The thickness of the SA7 layer is 1.2 μm.

4.5.5 IGZO active layer deposition

DC sputtering technique is used. The sputtering target is composed of three oxides, namely indium oxide, In_2O_3; gallium oxide, Ga_2O_3; and zinc oxide, ZnO. The target composition is: Indium oxide: gallium oxide: zinc oxide = 1:1:1 atomic ratio. The chamber base pressure is 2.25×10^{-6} Torr. Keeping the argon flow rate at 100 sccm and maintaining the chamber pressure at 5 mTorr, the oxygen flow rate is changed from 1.01 sccm to 19 sccm. The sputtering is done at seven different oxygen partial pressures: 0%, 1%, 2%, 3%, 4%, 8%, 16%. The IGZO layer has N-type polarity, its thickness is 50 nm.

Fixing PI film on glass substrate
with PDMS

PI film

PDMS

Rigid glass substrate

Spin coating organic barrier
layer (SA7)

Organic barrier layer (SA7)

PI film

PDMS

Rigid glass substrate

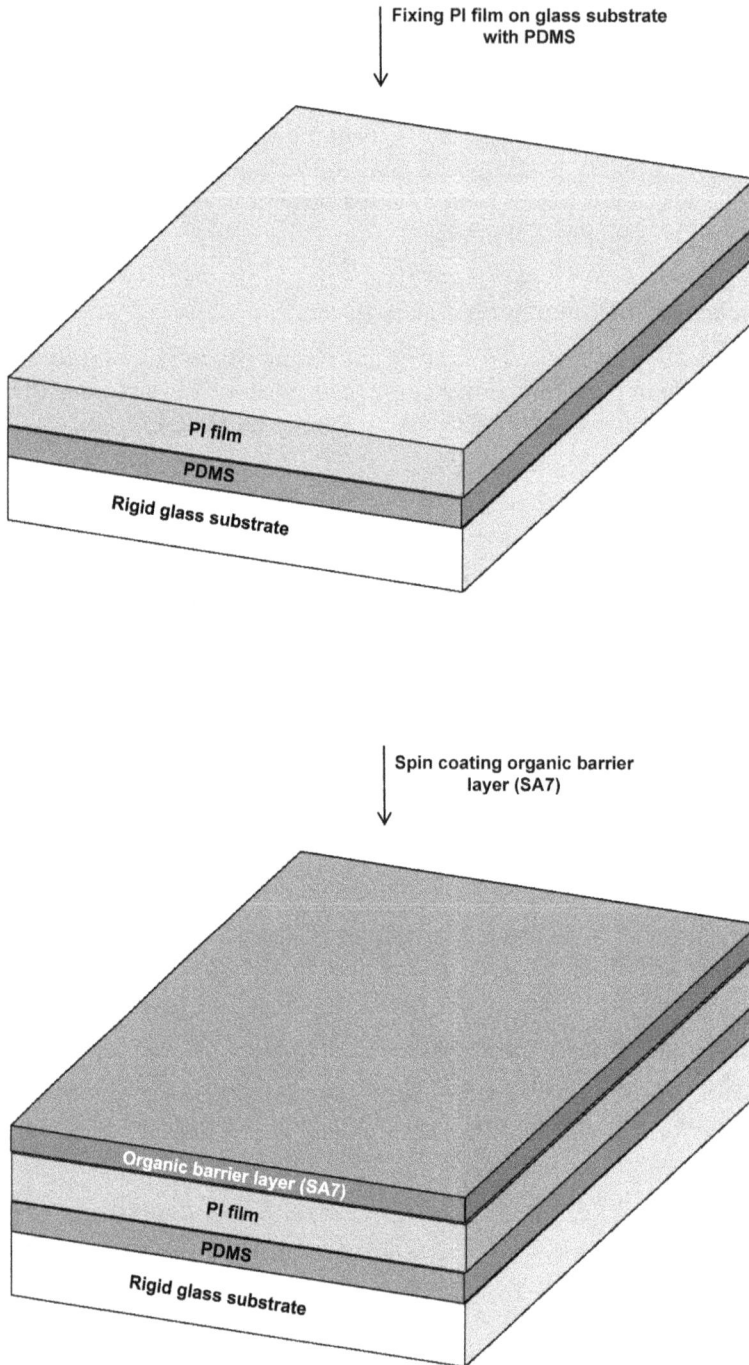

Figure 4.5. Bottom-gated IGZO TFT on PI film with organic barrier layer.

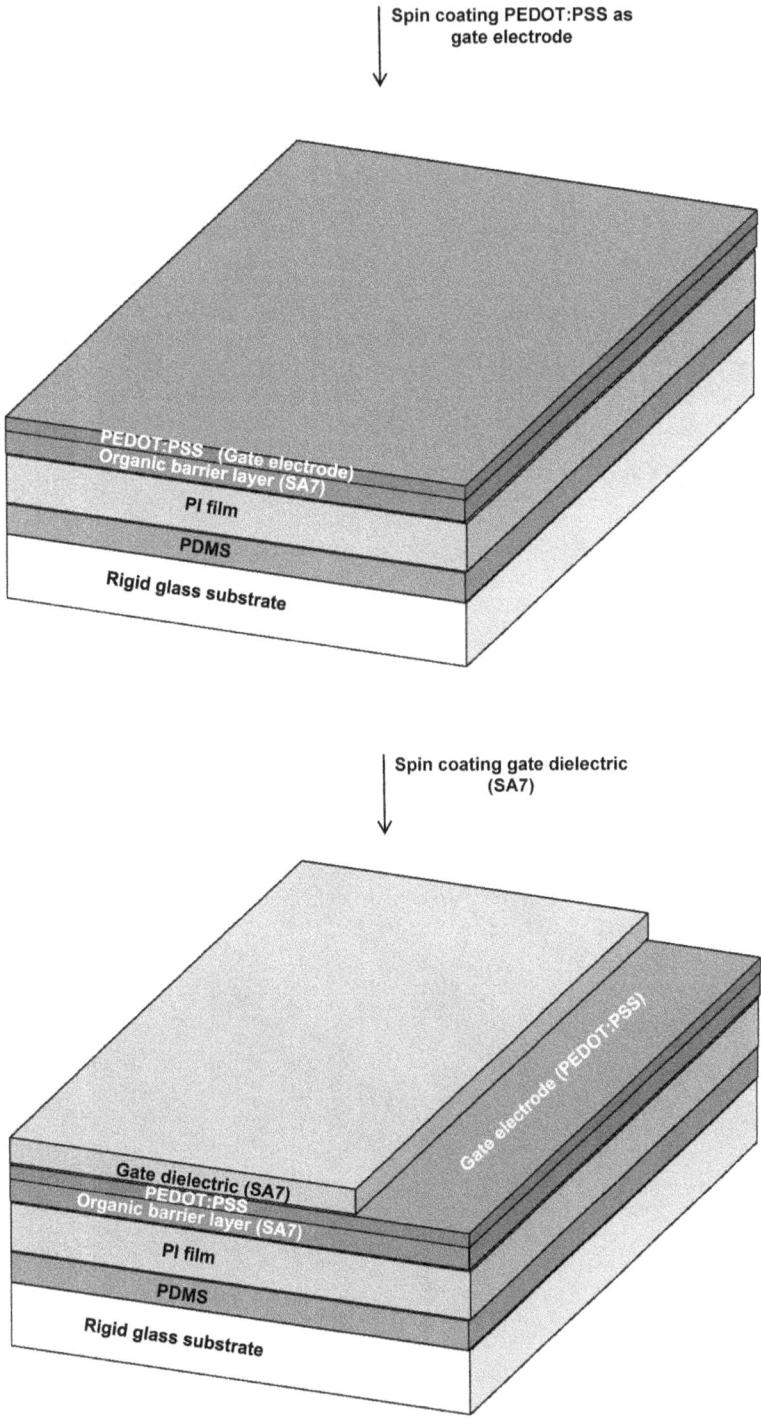

Figure 4.5. (Continued.)

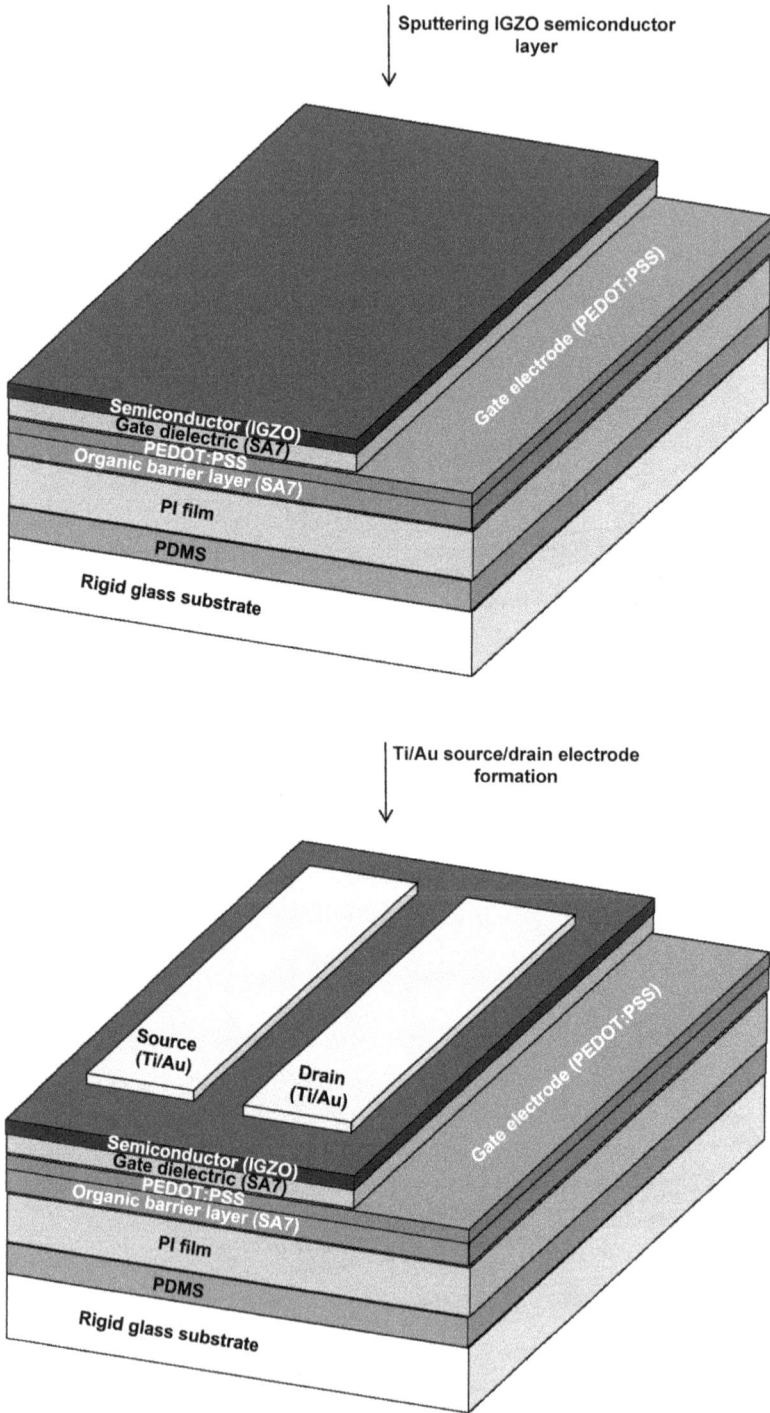

Figure 4.5. (Continued.)

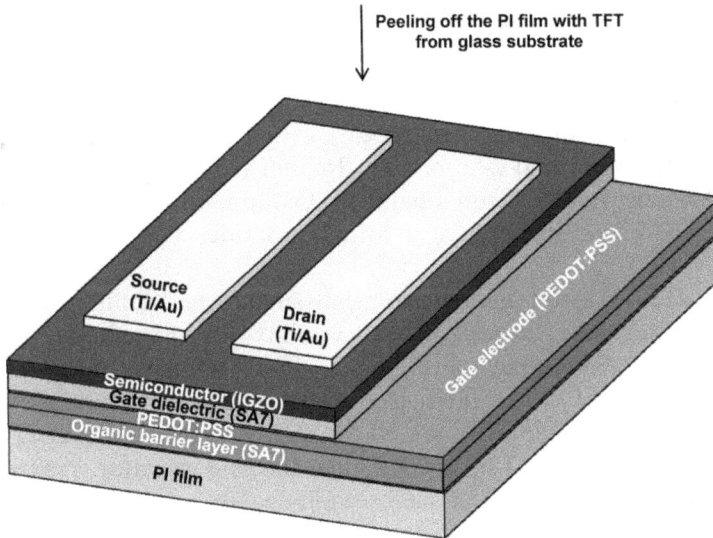

Figure 4.5. (Continued.)

4.5.6 Source/drain electrode deposition

Titanium (5 nm)/gold (50 nm) electrodes are deposited through a shadow mask.

4.5.7 Detachment of TFT from the PDMS/glass substrate

The function of the PDMS/glass supporting substrate is fulfilled, and it is now unnecessary. The whole structure is cured at 120 °C. Then by applying a small external force, the TFT is easily peeled off at the junction between the PI and the PDMS/glass substrate.

4.5.8 Electrical and bending characteristics

The best TFT performance is achieved for the case in which argon: oxygen flow rate is 100 sccm:1 sccm during IGZO layer deposition. This TFT has a threshold voltage = 6.4 V and on–off current ratio = 4.5×10^5. The carrier mobility is 15.64 cm^2 V^{-1} s^{-1}.

Bending experiments involve bending the TFTs along the directions of channel length and channel width. The bending radii are decreased from 20 mm to 1.5 mm. After bending, the stress is released before carrying out measurement. For bending along the channel length direction, the TFT characteristics remain undisturbed up to 1.5 mm radius; the related strain is 3.33%. Along the channel width direction, the TFT behaves differently. Its transfer characteristics remain unaffected up to 3 mm radius. However, the on-current decreases on further bending beyond this point. Measurements on TFTs in the bent condition indicate good behavioral stability up to 4 mm radius, strain = 1.25% (Kumaresan *et al* 2016).

4.6 Transparent Ni-doped ZnO TFT

Indium content in the Earth's crust is ~0.000 016%. Gallium is also a scarce metal present in around 0.0019% of the Earth's crust. Although not found in the Earth's crust in pure metallic form, gallium appears with aluminum in bauxite ore, and is produced in aluminum and zinc mining. Indium is mildly toxic upon ingestion. Further, it is moderately toxic on inhalation. Gallium is irritating to skin and eyes. When exposed to the atmosphere it forms an oxide. This oxide is toxic when inhaled.

Huang *et al* (2015) assert that the two metals, indium and gallium, are rare as well as toxic. Due to these reasons, their suitability for manufacturing devices is questionable. Therefore, they developed a TFT, which is both indium- and gallium-free (figure 4.6). This TFT has a bottom-gate structure. A transparent PET substrate is used. Ni-doped zinc oxide (NZO), NiO 3 wt% + ZnO 97 wt%, is the active channel layer. PECVD silicon dioxide is the gate insulator. Transparent ITO film is used for gate electrode and source/drain contacts.

4.6.1 Deposition of bottom ITO film as gate electrode

RF magnetron sputtering is done in pure argon. The ITO film thickness is 150 nm.

4.6.2 SiO$_2$ gate insulator deposition

This process is done by PECVD. The deposition temperature is kept at ~80 °C because of the low glass transition temperature of PET substrate (76 °C). The silicon dioxide film is 150 nm thick.

4.6.3 NZO channel layer deposition

RF magnetron sputtering is done in pure argon at room temperature immediately after the gate insulator deposition. This is essential to ensure good SiO$_2$–NZO interface properties. For optimization of deposition parameters, four sputtering pressures are tried: 0.4, 0.8, 1.2 and 1.6 Pa.

4.6.4 ITO source/drain contact deposition

This is done by RF magnetron sputtering in pure argon. The ITO film is 150 nm thick.

4.6.5 Optimized TFT characteristics

Amongst the four sputtering pressures tried, the 1.6 Pa pressure gave the optimal results. The 1.6 Pa TFT has a positive threshold voltage of 2.36 V. Its subthreshold swing (SS) is 89 mV/decade. Mobility in saturation mode is 172 cm^2 V^{-1} s^{-1}. The ratio I_{on}/I_{off} for drain current is 10^8.

This TFT fabricated on a transparent substrate using transparent materials is very useful for completely transparent displays (Huang *et al* 2015).

ITO gate electrode by RF sputtering

SiO$_2$ gate insulator by PECVD

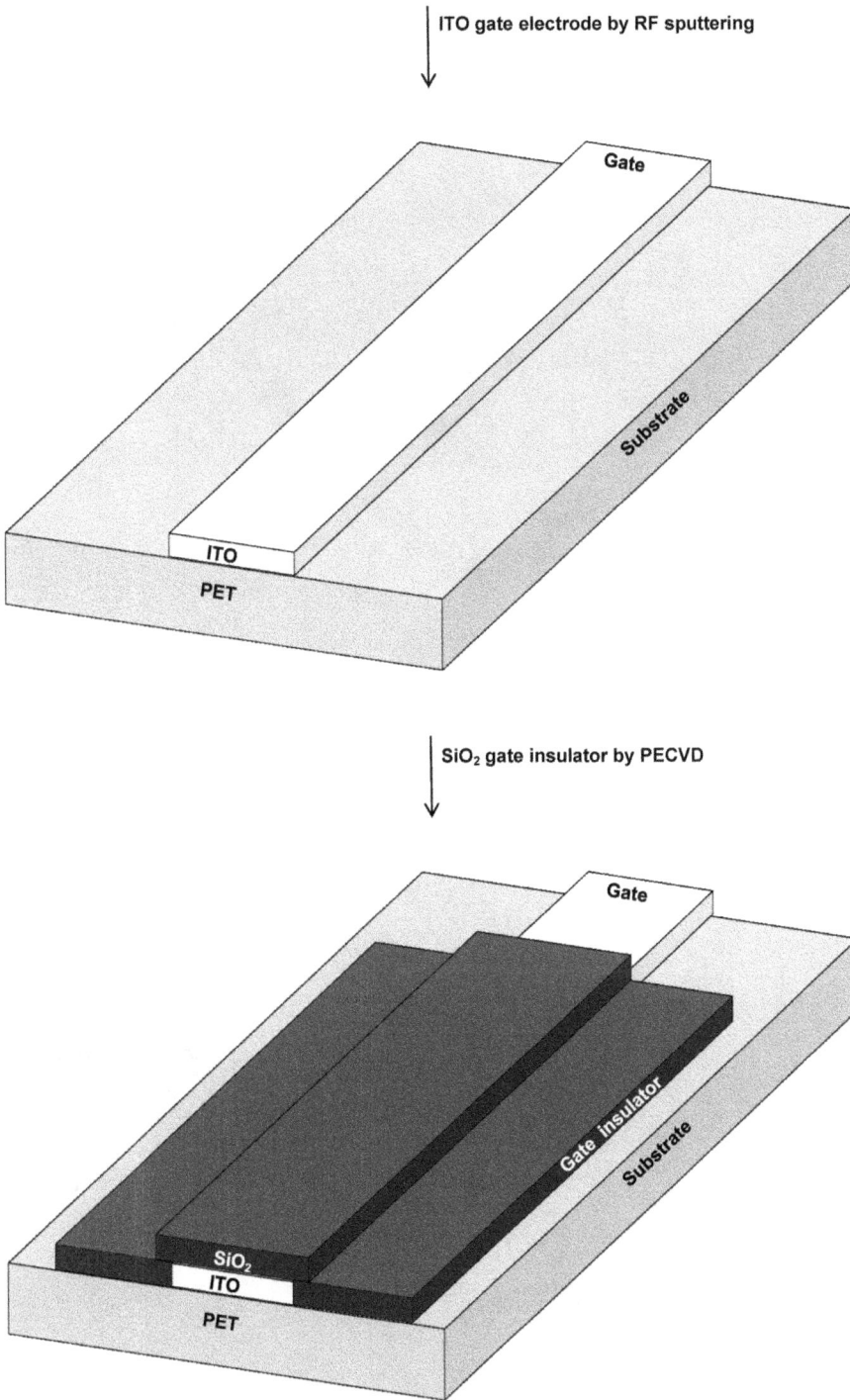

Figure 4.6. Nickel-doped ZnO TFT.

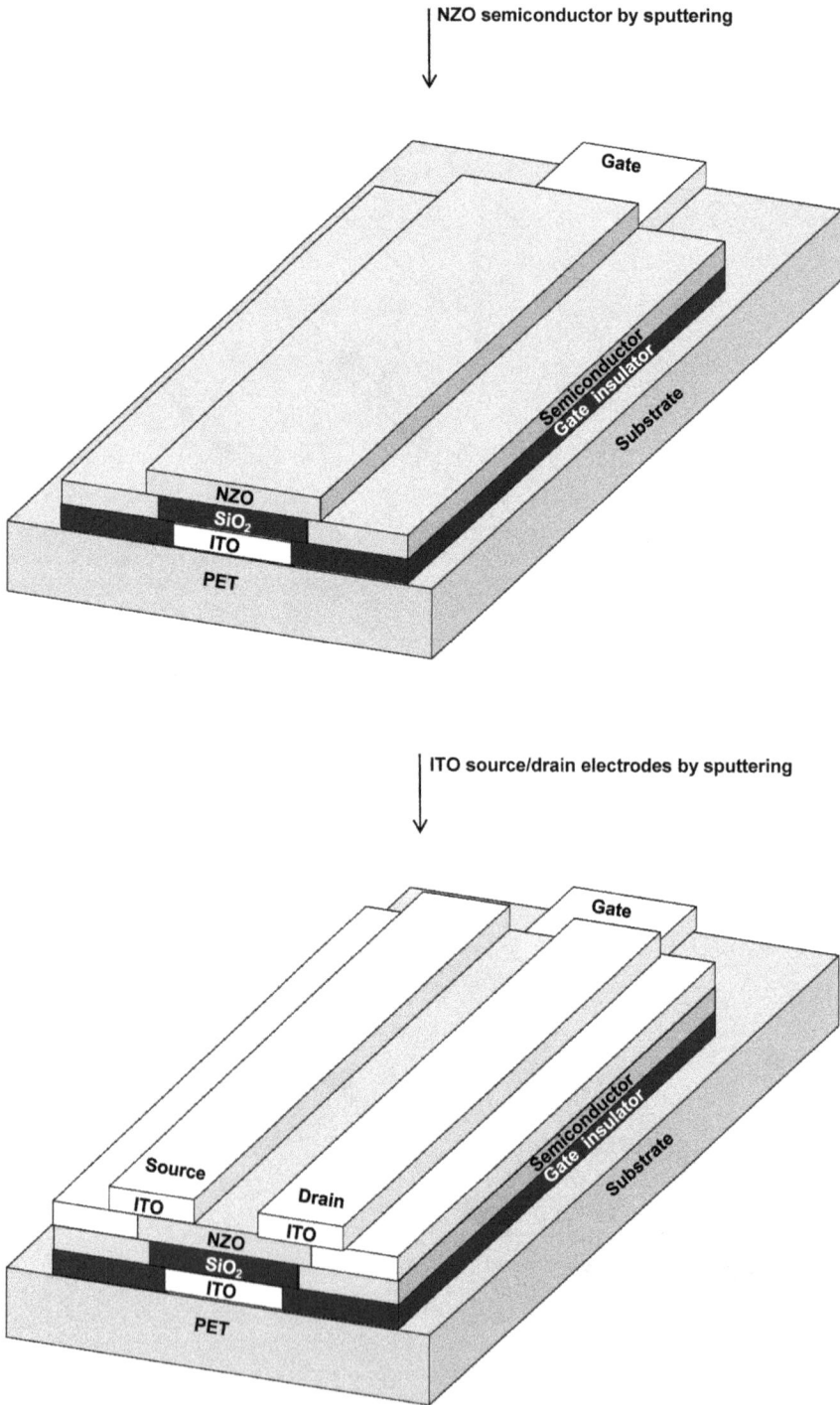

NZO semiconductor by sputtering

Gate

Semiconductor
Gate insulator
Substrate

NZO
SiO$_2$
ITO
PET

ITO source/drain electrodes by sputtering

Gate

Semiconductor
Gate insulator
Substrate

Source
ITO
Drain
ITO
NZO
SiO$_2$
ITO
PET

Figure 4.6. (Continued.)

4.7 TFT with PEALD ZnO layer

Zhao *et al* (2009, 2010) reported TFTs fabricated on polyimide substrates (glass transition temperature ~354 °C; coefficient of thermal expansion ~16 ppm/°C) with the ZnO layer deposited by PEALD at 200 °C. PEALD is able to provide high-quality, uniform thickness ZnO films on PI surface of RMS roughness ~30 nm. The staggered, bottom-gate TFT has a channel length of 20 μm and channel width of 50 μm with channel width/channel length ratio = 50 μm/20 μm = 2.5 (figure 4.7).

4.7.1 Laminating the PI substrate on a glass carrier

The PI substrate is 125 μm thick. Before lamination, it is prebaked at 200 °C in a vacuum oven for 24 h. For carrying out the processing operations on the PI easily, it is pasted on a glass carrier with silicone gel.

4.7.2 Gate metal sputtering and patterning

Ion beam sputtering technique is used for metal deposition. The gate metal is chromium. Chromium film thickness is 100 nm. The pattern of gate metal is defined by wet etching chemistry.

4.7.3 Al_2O_3 and ZnO films by PEALD

For both the films, the same deposition temperature (200 °C) is used. First Al_2O_3 film is deposited, this is the gate dielectric. The reactants are trimethylaluminum (TMA), $Al_2(CH_3)_6$, an organo aluminum compound; and carbon dioxide, CO_2 gas. Al_2O_3 film thickness is 50 nm. Next, ZnO film is deposited, this is the semiconductor layer. The reactants are diethylzinc (DEZn), $(C_2H_5)_2Zn$, an organozinc compound; and nitrous oxide, N_2O. ZnO film thickness is 30 nm. The inclusion of plasma in the ALD process makes this process very adaptable. Consequently, deposition can be carried out at a relatively low temperature. Moreover, a uniform conformal coating is obtained. (See volume 1, section 13.7.8 for details of the PEALD process.)

4.7.4 Patterning ZnO and Al_2O_3 films

Both oxides are selectively removed by wet etching, first ZnO and then Al_2O_3. The etchant used for ZnO is dilute hydrochloric acid, HCl while that for Al_2O_3 is hot phosphoric acid, H_3PO_4 at 80 °C.

4.7.5 Patterning and sputtering of titanium source/drain electrodes

Lift-off photolithography is done. The pattern is defined in photoresist. Then titanium is sputtered. On removal of the photoresist, the metal is carried away from regions where it was deposited over the photoresist. It is retained in all areas where there was no underlying photoresist.

Sticking Kapton substrate on glass carrier
with silicone gel

Kapton

Silicone gel

Glass

Chromium gate electrode deposition by
sputtering

Gate

Cr

Kapton

Silicone gel

Glass

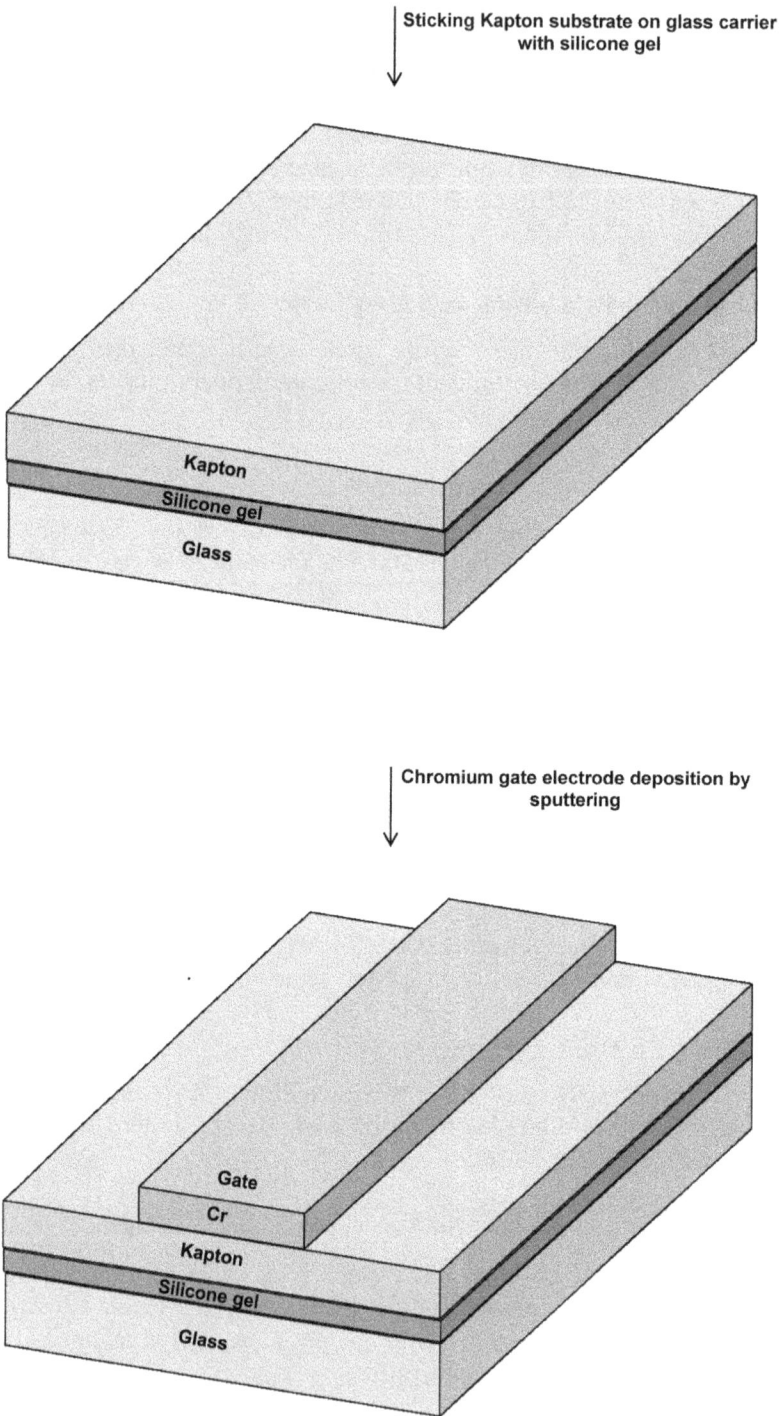

Figure 4.7. TFT fabrication using PEALD ZnO layer; step of passivation with Al_2O_3 not shown.

Al$_2$O$_3$ gate dielectric by PEALD

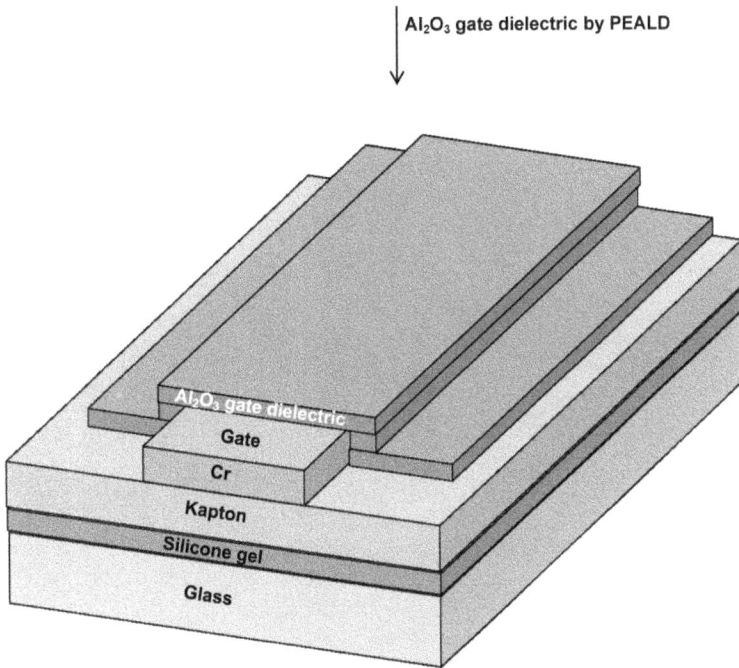

ZnO semiconductor film by PEALD

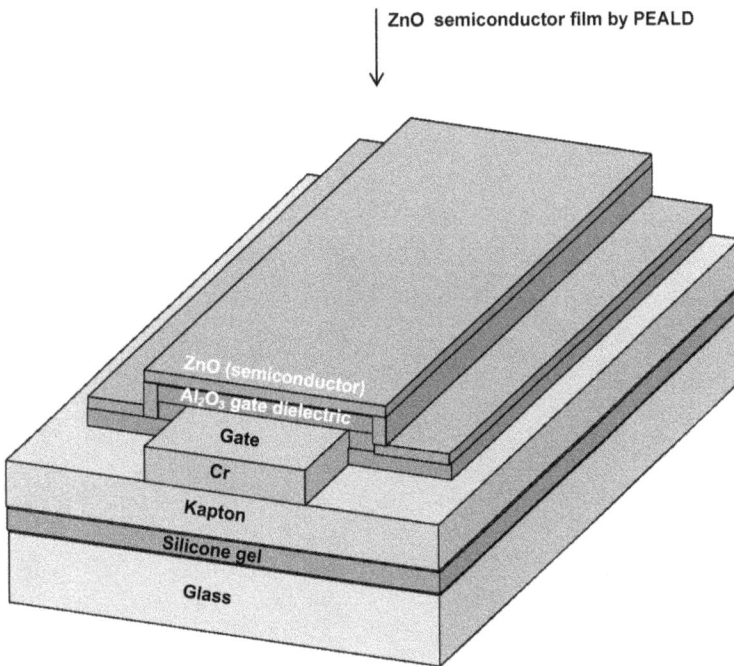

Figure 4.7. (Continued.)

Titanium sputtering for source/drain electrodes

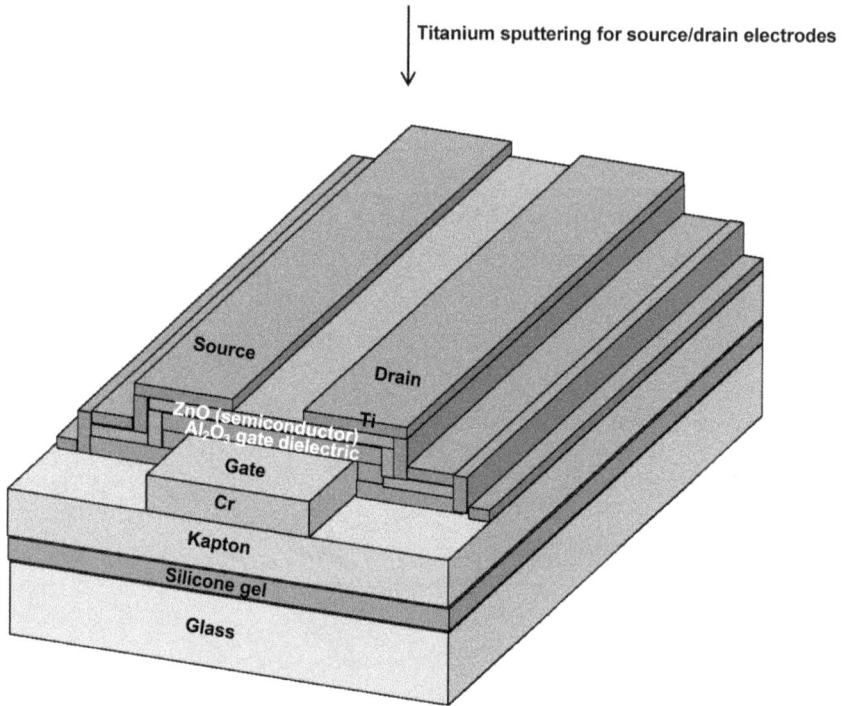

Taking away Kapton film from glass carrier

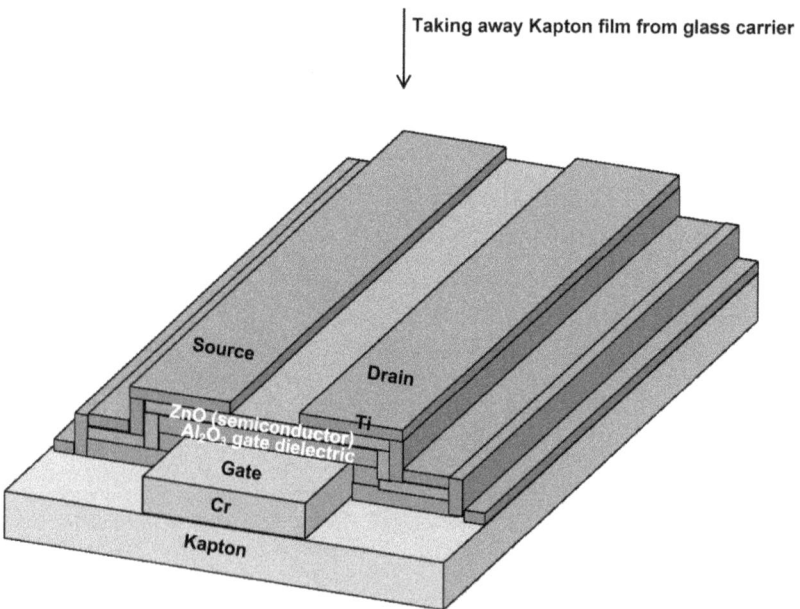

Figure 4.7. (Continued.)

4.7.6 Passivation

The passivation layer is Al_2O_3. ALD process is used. The reactants are trimethylaluminum (TMA) and water. The reaction temperature is 200 °C. Al_2O_3 film thickness is 30 nm. The passivation layer is defined by wet etching.

4.7.7 Delamination of PI

The PI substrate is separated from the glass carrier. The device and circuit characteristics after delamination are the same as before delamination.

4.7.8 Typical electrical characteristics

The threshold voltage of the TFT is 2 V with subthreshold swing of 0.35 V/decade. The ratio I_{on}/I_{off} is $>10^7$. Field-effect mobility is 20 cm^2 V^{-1} s^{-1} in the linear region. For checking the bias stress stability, the device is continuously biased with gate voltage = 3 V and drain voltage = 3 V for 40 000 s. The threshold voltage shift is <50 mV. When the 15 stage ring oscillators fabricated with ZnO TFTs are operated with V_{DD} = 18 V, the oscillation frequency is >2 MHz. For each stage, the propagation delay is <20 ns. The high speed of ZnO circuits validates the achievability of excellent device performance on flexible substrates (Zhao *et al* 2010).

4.8 Discussion and conclusions

TFTs with an IGZO semiconductor layer are fabricated on both transparent and opaque plastic substrates. These TFTs exhibit satisfactory stability of electrical characteristics under bending conditions. The TFTs behave differently on bending along the channel direction and in the direction orthogonal to the channel. Avoidance of rare and toxic elements, indium and gallium, has led to TFT fabrication using ZnO semiconductor. These TFTs are applied to fast ZnO circuits.

Review exercises

4.1 Name one special property of oxide semiconductors, which makes them very useful for TFT fabrication.

4.2 Draw the cross-sectional diagram of an IGZO TFT fabricated with an etch stop layer on a PEN substrate. List the following for the TFT: (i) the substrate used, (ii) the barrier layer to humidity effects, (iii) the gate metal, (iv) the gate dielectric, (v) the semiconductor film, (vi) the etch stop layer, (vii) source and drain metals and (viii) the encapsulation.

4.3 How is the performance of TFT assessed with respect to bending? Up to what bending radius does the TFT remain undamaged? What is the strain at this radius? At what bending radius is the TFT irreversibly damaged? What is the strain at which failure occurs?

4.4 Write the equation relating the mobility with strain for the TFT with ESL, both when the straining is done along the channel and perpendicular to it.

4.5 What is meant by the 'interstrate' structure in fabrication of TFT? How is this kind of structure utilized to build an IGZO TFT? Draw a cross-sectional diagram of the TFT and show the different layers of the TFT.

4.6 What is the gate electrode made of in IGZO TFT on paper substrate? What process is used to deposit the gate electrode?

4.7 What is the problem faced in TFT fabrication when IGZO film is deposited by the sol–gel route? How is the problem solved?

4.8 What is meant by photo-annealing? How is it applied for densification of sol–gel deposited IGZO film for TFT semiconductor layer?

4.9 How is the precursor solution prepared for IGZO film deposition by the sol–gel route? How is this solution applied on the gate dielectric of the TFT? How is the annealing done without damaging the PAR film?

4.10 In the IGZO TFT with organic gate dielectric/moisture barrier layers, name the following: (i) material used for gate dielectric/moisture barrier layer and (ii) material used for gate electrode.

4.11 What is the difference in electrical behavior observed when the IGZO TFT with organic gate dielectric/moisture barrier layers is bent along the directions of channel length and channel width?

4.12 Argue in favor of making TFTs which are free of indium and gallium.

4.13 What process is used to deposit the ITO gate/source/drain electrodes and the nickel-doped ZnO semiconductor of the TFT?

4.14 What functions are performed by the aluminum oxide and zinc oxide in TFT with PEALD ZnO layer?

4.15 What is the advantage of PEALD over ALD?

4.16 What are the etchants used for zinc oxide and aluminum oxide?

References

Fortunato E, Correia N, Barquinha P, Pereira L, Gonçalves G and Martins R 2008 High-performance flexible hybrid field-effect transistors based on cellulose fiber paper *IEEE Electron Device Lett.* **29** 988–90

Heremans P, Papadopoulos N, de Meux A D J, Nag M, Steudel S, Rockelè M, Gelinck G and Tripathi A 2016 Flexible metal-oxide thin film transistor circuits for RFID and health patches *IEEE Int. Electron Devices Meeting (IEDM) IEDM (San Francisco CA, 3–7 December 2016)* pp 151–4

Huang L, Han D, Zhang Y, Shi P, Yu W, Cui G, Cong Y, Dong J, Zhang S, Zhang X and Wang Y 2015 High mobility transparent flexible nickel-doped zinc oxide thin-film transistors with small subthreshold swing *Electron. Lett.* **51** 1595–6

Kim Y-H, Heo J-S, Kim T-H, Park S, Yoon M-H, Kim J, Oh M S, Yi G-R, Noh Y-Y and Park S K 2012 Flexible metal-oxide devices made by room-temperature photochemical activation of sol–gel films *Nature* **489** 128–33

Kumaresan Y, Pak Y, Lim N, kim Y, Park M-J, Yoon S-M, Youn H-M, Lee H, Lee B H and Jung G Y 2016 Highly bendable In-Ga-ZnO thin film transistors by using a thermally stable organic dielectric layer *Sci. Rep.* **6** 37764

Tripathi A K, Myny K, Hou B, Wezenberg K and Gelinck G H 2015 Electrical characterization of flexible InGaZnO transistors and 8-b transponder chip down to a bending radius of 2 mm *IEEE Trans. Electron Devices* **62** 4063–8

Zhao D A, Mourey D A and Jackson T N 2009 Flexible plastic substrate ZnO thin film transistor circuits *Device Research Conf., 2009. DRC 2009 (University Park, PA, 22–24 June 2009)* pp 177–8

Zhao D, Moure D A and Jackson T N 2010 Fast flexible plastic substrate ZnO circuits *IEEE Electron Device Lett.* **31** 323–5

IOP Publishing

Flexible Electronics, Volume 2
Thin-film transistors
Vinod Kumar Khanna

Chapter 5

Small organic molecule TFT

Pentacene TFTs fabricated on PEN substrate with polyimide as the gate dielectric remain functional up to a bending radius of 4.6 mm or strain of 1.4%, failing beyond this limit by buckling of gold metallization films (Kato *et al* 2004, Seitan *et al* 2005). A complementary circuit is realized using P-channel pentacene TFT and N-channel $F_{16}CuPc$ TFT by adjusting the aspect ratios of these TFTs taking into consideration the higher mobility in pentacene TFT and lower mobility in $F_{16}CuPc$ TFT to obtain similar drive currents (Klauk *et al* 2005). During comparison of P-channel TIPS-pentacene TFT, P-channel PTAA TFT and N-channel perylene diimide TFT, TIPS-pentacene exhibits the highest mobility followed by PTAA TFT and then perylene diimide TFT but the mobility of perylene diimide TFT is higher by four orders of magnitude in the measured voltage range, finally climbing to the PTAA mobility value at high gate voltages (Castro-Carranza *et al* 2012). Temporal decline of mobility of organic semiconductor TFTs through oxidation of conjugated molecules under the influence of atmospheric oxygen, ozone and water vapor is avoidable by choosing a semiconductor of higher ionization potential such as DNTT. Because this material is not easily oxidizable in air, TFTs fabricated by using DNTT showed better speed and stability than pentacene TFTs, both at the beginning and after aging by exposure to yellow light and 50% RH over an eight-month period. Inverters using DNTT TFTs maintained correct logic operation over this period (Zschieschang *et al* 2010). A digital library exclusively based on P-channel DNTT TFTs was developed owing to the comparatively poor performance of N-channel devices (Elsobky *et al* 2017).

5.1 Introduction

The low temperatures required, and reduced thermal budget in processing organic semiconductors as opposed to inorganic semiconductors, makes them compatible with flexible polymeric substrates. The structure of small organic molecules has a high crystallinity. Consequently, they exhibit higher mobility than polymeric

doi:10.1088/2053-2563/ab0d18ch5

materials. The carrier mobility in small organic molecules is equivalent to that in amorphous silicon. Most small organic molecules are vacuum-deposited due to their poor solubility. Some solution-processed options have also been proposed. On the whole, the small organic molecules are more difficult to process than polymers. As compared to polymers, processability of small organic molecules is less facile.

5.2 Pentacene TFT on PEN substrate

Kato *et al* (2004) made high-quality organic TFTs using a pentacene semiconductor on a PEN substrate with a solution-processed polymeric gate dielectric (figure 5.1). The main features of their TFTs and process flow are briefly presented in the subsections below.

5.2.1 Base film (substrate)

This is polyethylenenaphthalate or poly(ethylene 2,6-naphthalate), abbreviated as PEN.

5.2.2 Gate dielectric

This is made with a commercial polyimide precursor which can be cured at a low temperature ~180 °C compared with high curing temperatures ~300 °C, commonly used with polyimide.

5.2.3 Semiconductor channel layer

This is pentacene, a P-type semiconductor.

5.2.4 Process

The gate electrodes, Cr (5 nm)/Au (150 nm), are thermally evaporated through a shadow mask on the 125 μm thick PEN base film (Kato *et al* 2004).

Over the PEN base film with the patterned gate electrodes, a high-purity polyimide precursor is deposited by spin coating. Spinning speeds of 1500, 3000 and 6000 RPM yielded polyimide films of post-curing thickness 1900 nm, 990 nm and 540 nm respectively. The solvent in the polyimide precursor film is vaporized by keeping it in a clean oven at 90 °C for 10 min. In the same oven, the film is heated at 180 °C in N_2 for 1 h followed by natural cooling to obtain a stable and reliable dielectric film.

When the oven temperature has fallen below 100 °C, the film is transferred from the oven to the vacuum system. In the vacuum system, pentacene is sublimed through a shadow mask to form a channel layer of thickness 50 nm. During pentacene sublimation, the background pressure is 2.25×10^{-7} Torr. The substrate is kept at ambient temperature. About 1 nm of pentacene is deposited per min.

Next come the source/drain electrodes which are made of 60 nm thick gold. They are formed by evaporation through a shadow mask. For the 990 nm thick gate dielectric TFT, the channel aspect ratio is 1900 μm/100 μm = 19, whereas for the 540 nm thick gate dielectric TFT, this ratio is 1600 μm/50 μm = 32.

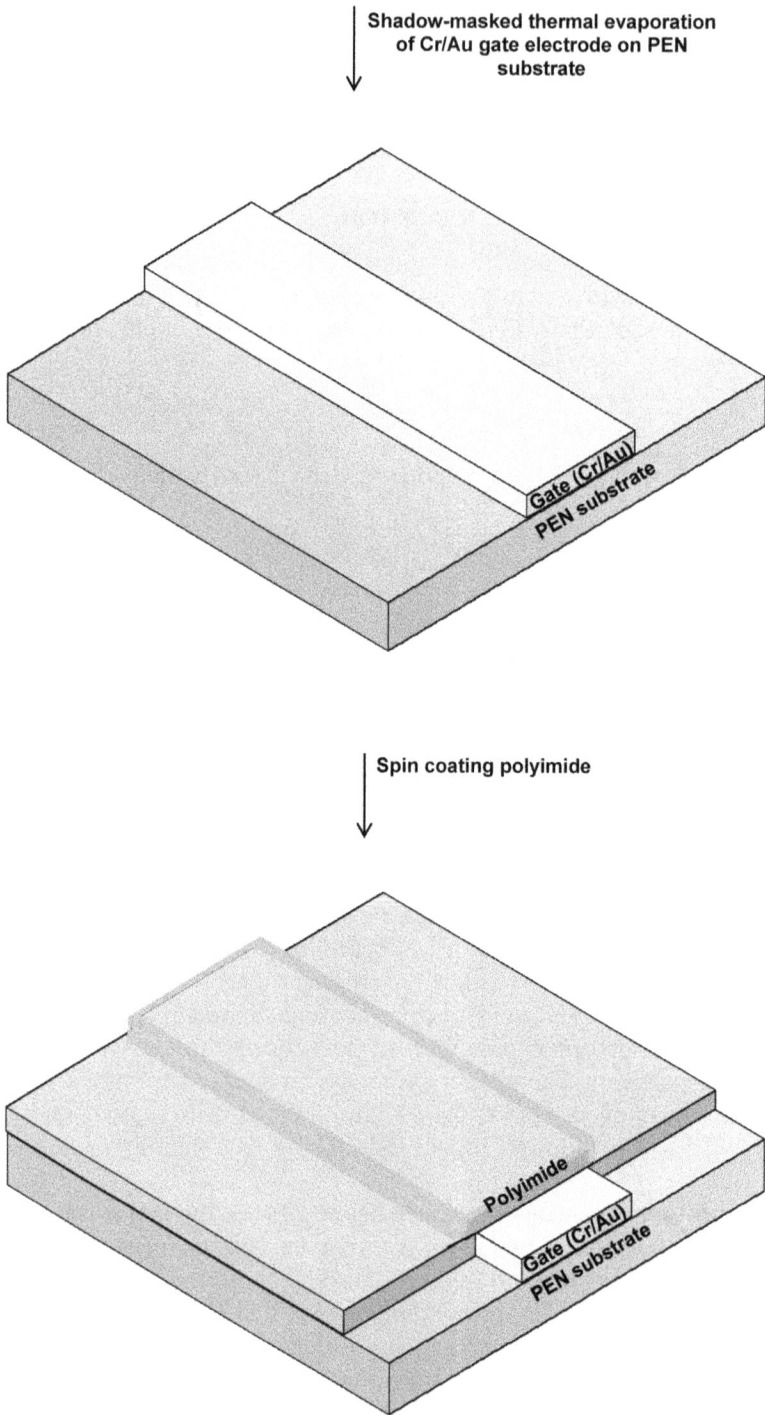

Shadow-masked thermal evaporation
of Cr/Au gate electrode on PEN
substrate

Gate (Cr/Au)

PEN substrate

Spin coating polyimide

Polyimide

Gate (Cr/Au)

PEN substrate

Figure 5.1. TFT fabricated with pentacene semiconducting film using PEN as the substrate.

Pentacene sublimation

Evaporation of gold source/drain
electrodes through shadow mask

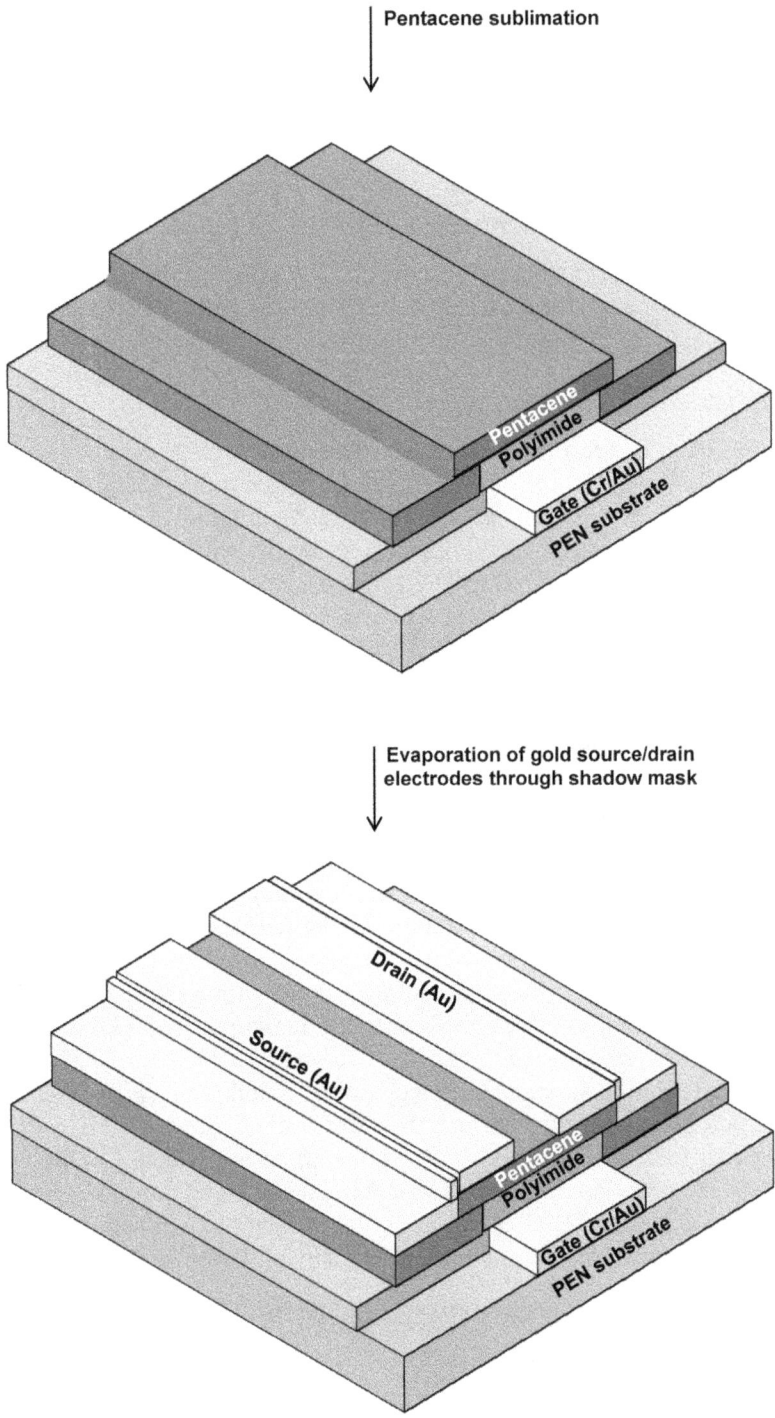

Figure 5.1. (Continued.)

5.2.5 Measurements

The RMS surface roughness of the PEN base film is 1 nm while that of the polyimide gate dielectric film is only 0.2 nm. For TFTs on a 990 nm gate dielectric, the mobility is 0.3 cm^2 V^{-1} s^{-1} while the current on–off ratio is 10^6. The mobility for a 540 nm gate dielectric TFT is 1 cm^2 V^{-1} s^{-1} (Kato *et al* 2004).

5.3 Bending effects on pentacene TFT

Sekitani *et al* (2005) fabricated pentacene TFT on a PEN substrate with polyimide as the gate dielectric and studied its electrical behavior under compressive and tensile strains.

5.3.1 TFT fabrication and characteristics

A Cr (5 nm)/Au (100 nm) gate electrode is deposited by evaporation on a PEN substrate of thickness 125 μm. For the gate insulator, polyimide precursor is spin coated and cured at 180 °C. The thickness of the PI film is 900 nm. Next, the pentacene semiconductor film is evaporated in the vacuum coating unit. The pentacene film thickness is 50 nm. A shadow mask is used to deposit source/drain electrodes in the vacuum evaporator. These electrodes are 60 nm thick.

The TFT has a W/L ratio = 1000 μm/100 μm = 10. The mobility is 0.3 cm^2 V^{-1} s^{-1}. The TFT has a current on–off ratio of 10^5.

5.3.2 Bending experiments

The gate voltage V_{GS} is swept from 20 V to −40 V keeping the drain–source voltage constant at $V_{DS} = -40$ V. The bending radius is varied from 17 mm (flattening of the base film) to 4.6 mm. It is found that, under compressive strain, the saturation drain–source current $I_{DS(sat)}$ increases with increasing the compressive strain. The change in $I_{DS(sat)}$ is = +11% at $R = 4.6$ mm. Under tensile strain, $I_{DS(sat)}$ decreases on increasing the tensile strain. The change in $I_{DS(sat)}$ is = −26% at $R = 4.6$ mm which corresponds to a strain of 1.5%. These changes under compressive and tensile strains are reproducible as well as reversible. Since the functionality of TFTs is maintained up to $R = 4.6$ mm, it is evident that they will work satisfactorily when wrapped around a bar of radius ~4.6 mm. The mechanical flexibility of TFT is ascribed to the soft polyimide gate insulator film and the substrate.

For bending radii <4.6 mm or strain >1.5%, the changes in $I_{DS(sat)}$ are no longer irreversible. The harder components of this chip are the metal films. The irreversibility of characteristics is due to buckling of gold films taking place at strain levels between 1%–2%. Thus, the gold electrodes limit the flexibility of the device (Sekitani *et al* 2005).

5.4 Pentacene and $F_{16}CuPc$ TFTs on PEN substrate for organic complementary circuit

Klauk *et al* (2005) fabricated organic complementary circuit and ring oscillator using TFTs of opposite polarities on a PEN substrate. The TFTs are made in bottom-gate, bottom contact configuration (figure 5.2). The P-type semiconductor is

Making aluminum gate
electrodes on PEN substrate

PVP gate dielectric film
deposition

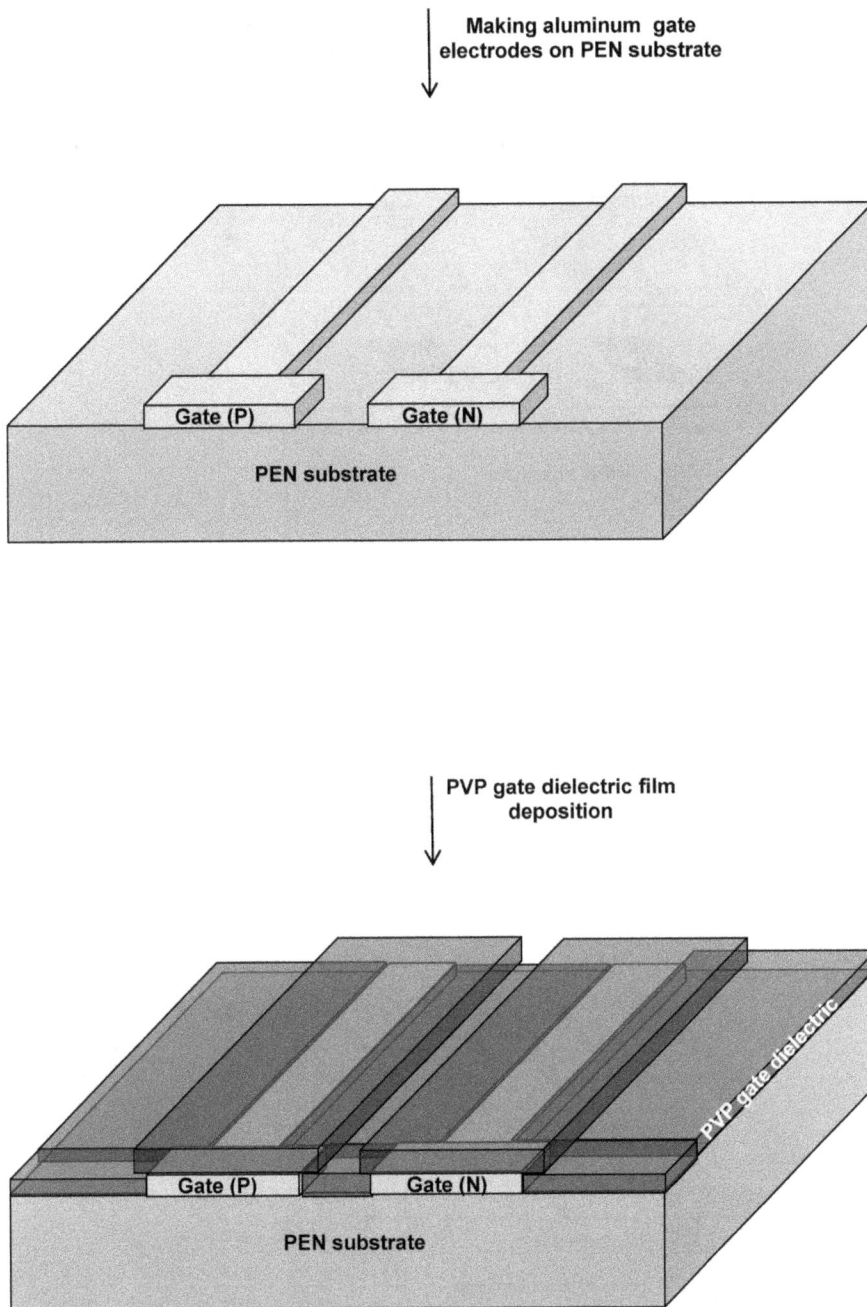

Figure 5.2. Pentacene (P-channel) and $F_{16}CuPc$ (N-channel) TFTs on PEN substrate.

Via creation through PVP film for
gate contact

Au source/drain contact fornation and
via filling for gate electrode

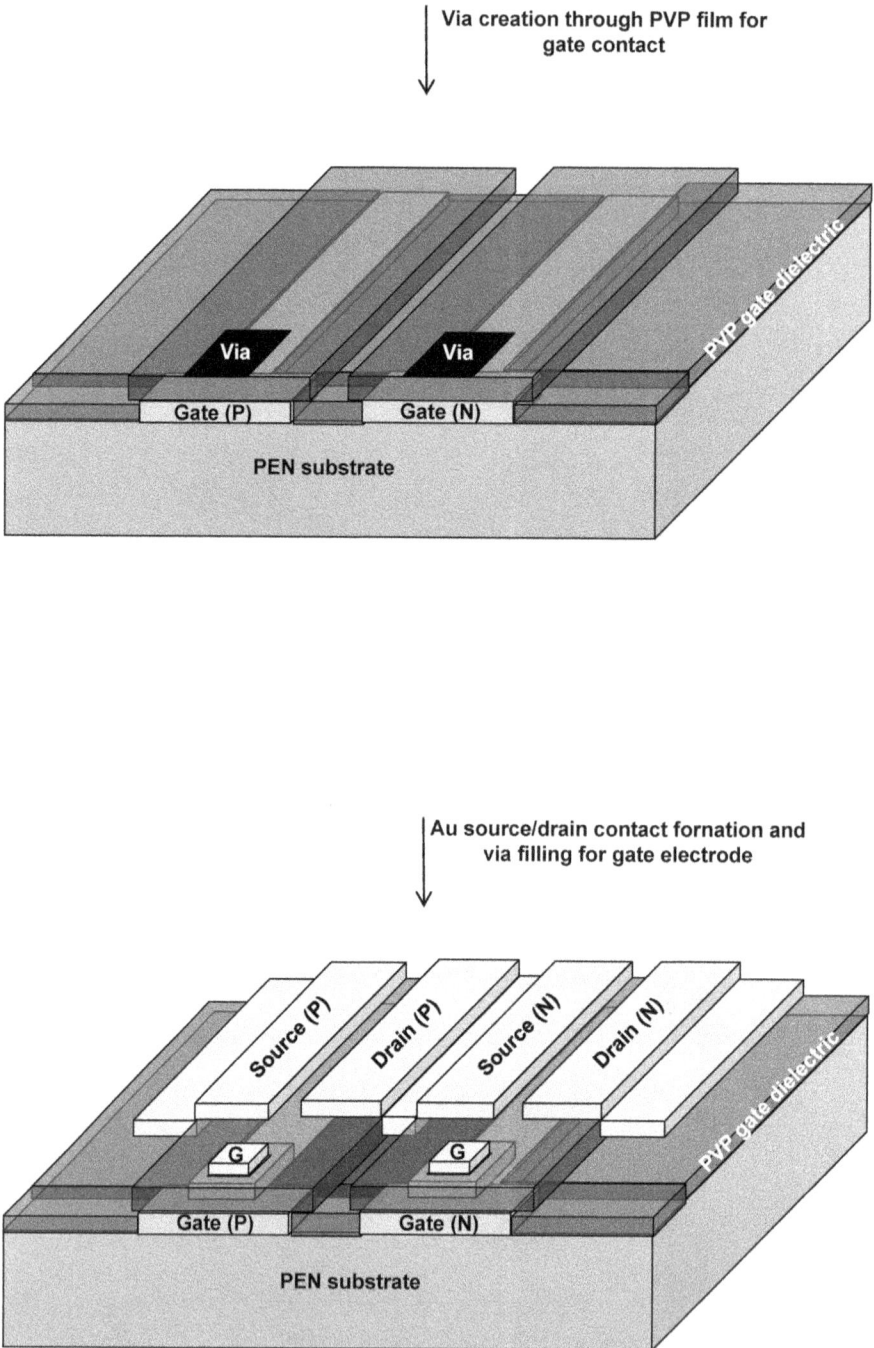

Figure 5.2. (Continued.)

Pentacene deposition/patterning
for P-channel TFT

Source (P)
Pentacene
Drain (P)
Source (N)
Drain (N)
PVP gate dielectric
G
G
Gate (P)
Gate (N)
PEN substrate

F$_{16}$CuPc film deposition/patterning for
N-channel TFT

Source (P)
Pentacene
Drain (P)
Source (N)
F$_{16}$CuPc
Drain (N)
PVP gate dielectric
G
G
Gate (P)
Gate (N)
PEN substrate

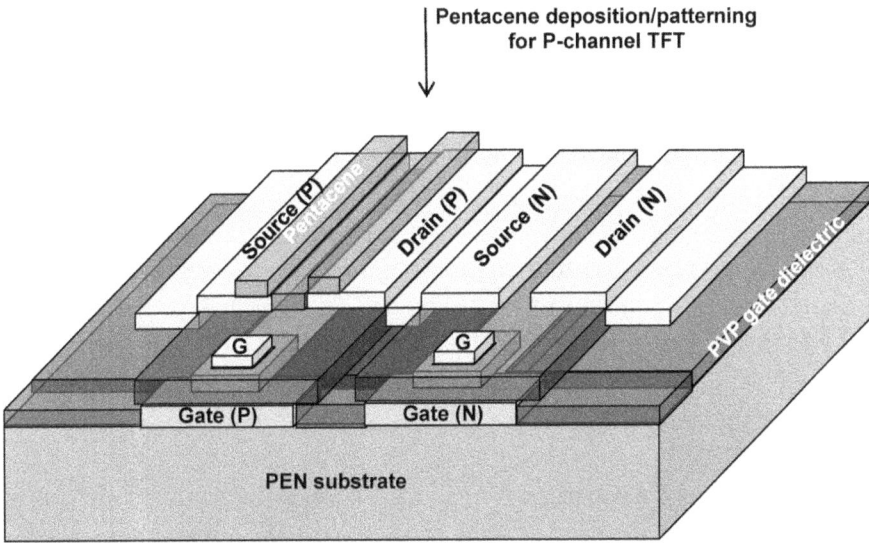

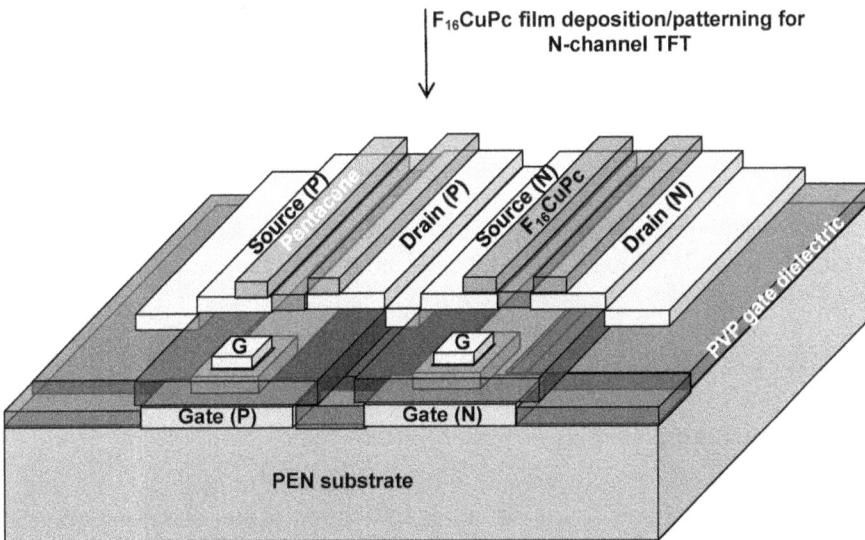

Figure 5.2. (Continued.)

pentacene and the N-type semiconductor is copper (II) hexadecafluorophthalocyanine ($F_{16}CuPc$). Both the semiconductors are deposited in vacuum while the gate dielectric polyvinylphenol (PVP), $[CH_2CH(C_6H_4OH)]_n$ is solution-processed.

5.4.1 Substrate preshrinking

The PEN substrate is 125 µm thick. The substrate is subjected to a prefabrication shrinking step at 200 °C. This step ensures that it maintains its dimensional stability throughout the processing so that no errors are introduced by size variations.

5.4.2 Definition of gate electrode and first interconnect level

Aluminum is evaporated. The aluminum pattern is defined by photolithography and etched in sodium hydroxide, NaOH.

5.4.3 Gate dielectric deposition

Polyvinylphenol is spin coated and cross-linked by baking at 200 °C. Its thickness is 50 nm.

5.4.4 Creation of vias between first and second interconnect levels

Vias are defined in the hardened PVP film by photolithography. They are dug out by dry etching in oxygen plasma.

5.4.5 Definition of source/drain contacts and second interconnect level

For this role, a high work function metal such as gold is chosen. The work function of gold is 5.1 eV. Additionally, gold offers the advantage that it is not degraded by oxidation on exposure to atmosphere.

It is worth noting that gold is better suited for making contact with the P-type semiconductor. For the N-type semiconductor, a lower work function material like aluminum (work function = 4.08 eV) is more suitable, but since the low work function metal tends to oxidize in air, gold is used as the contact metal for both P-type and N-type organic semiconductors.

The gold film is formed by thermal evaporation and the pattern is delineated in a dilute mixture of potassium iodide (KI) and iodine (I_2).

5.4.6 Deposition and patterning of pentacene film

Pentacene film is deposited by evaporation in vacuum. Its thickness is 30 nm. The pentacene pattern is formed using a water-soluble polyvinyl alcohol-based photoresist. Dry etching in oxygen plasma carves out the pentacene pattern. During the dry etching of pentacene, the region where $F_{16}CuPc$ is to be deposited is opened up. The etching period is chosen to be just sufficient to remove pentacene without harming the gate dielectric film.

5.4.7 Deposition of $F_{16}CuPc$ film

Like pentacene, the $F_{16}CuPc$ film is also deposited by vacuum evaporation. Its thickness is also 30 nm.

5.4.8 P-Channel and N-channel TFTs

A low supply voltage of ~8 V is required for operation of TFTs. For the P-channel pentacene TFT, the subthreshold swing SS is 600 mV/decade, the gate current is <10 pA at a gate–source voltage of −8 V. The current on–off ratio I_{on}/I_{off} is 10^4 and the mobility μ is 0.1 cm^2 V^{-1} s^{-1}. Performance of the N-channel $F_{16}CuPc$ TFT is inferior to the pentacene TFT. For the $F_{16}CuPc$ TFT, SS is 1400 mV/decade, the ratio I_{on}/I_{off} is 10^3 and μ is 0.002 cm^2 V^{-1} s^{-1}.

5.4.9 Complementary circuit

In the complementary circuit, both TFTs have the same channel length (2 μm). The aspect ratios of the P-channel pentacene TFT and N-channel $F_{16}CuPc$ TFT are adjusted in such a way that these TFTs supply the same drive currents. As the P-channel TFT is superior to the N-channel TFT, its channel width is taken as 20 μm. The channel width of the N-channel TFT is 400 μm. The small signal gain of the inverter is six at a supply voltage of 8 V. The area of the five stage ring oscillator is <0.5 mm^2. The dynamic performance of the circuit is limited by the poor mobility in the N-type semiconductor (Klauk *et al* 2005).

5.5 N-type small-molecule perylene diimide TFT

5.5.1 Need of N-type TFTs

A large number of organic semiconductors are P-type materials. To realize complementary circuits, both P-type and N-type TFTs are needed. Castro-Carranza *et al* (2012) fabricated TFTs using perylene diimide (figure 5.3), which shows N-type semiconducting properties. For comparison, TFTs of P-type small-molecule 6,13-bis (triisopropylsilylethynyl)pentacene (TIPS-pentacene), empirical formula: $C_{44}H_{54}Si_2$ and molecular weight = 639.07; and P-type polymer poly (triarylamine) (PTAA), linear formula $[C_6H_4N(C_6H_2(CH_3)_3)C_6H_4]_n$ are also fabricated.

5.5.2 Fabrication of TFTs

The TFT has a staggered, bottom contact configuration. A PEN substrate is used. The plasma vapor deposition technique is used to deposit gold for source/drain electrodes. The thickness of the gold film is 30 nm. The electrode pattern is defined by photolithography. Solutions of perylene diimide, TIPS-pentacene and PTAA are deposited by spin coating. They are cured at 115 °C to dry up the solvents. CYTOP insulating films are spin coated over the semiconductor layers to form the gate dielectrics. The gate electrode is printed with an inkjet printer using silver ink.

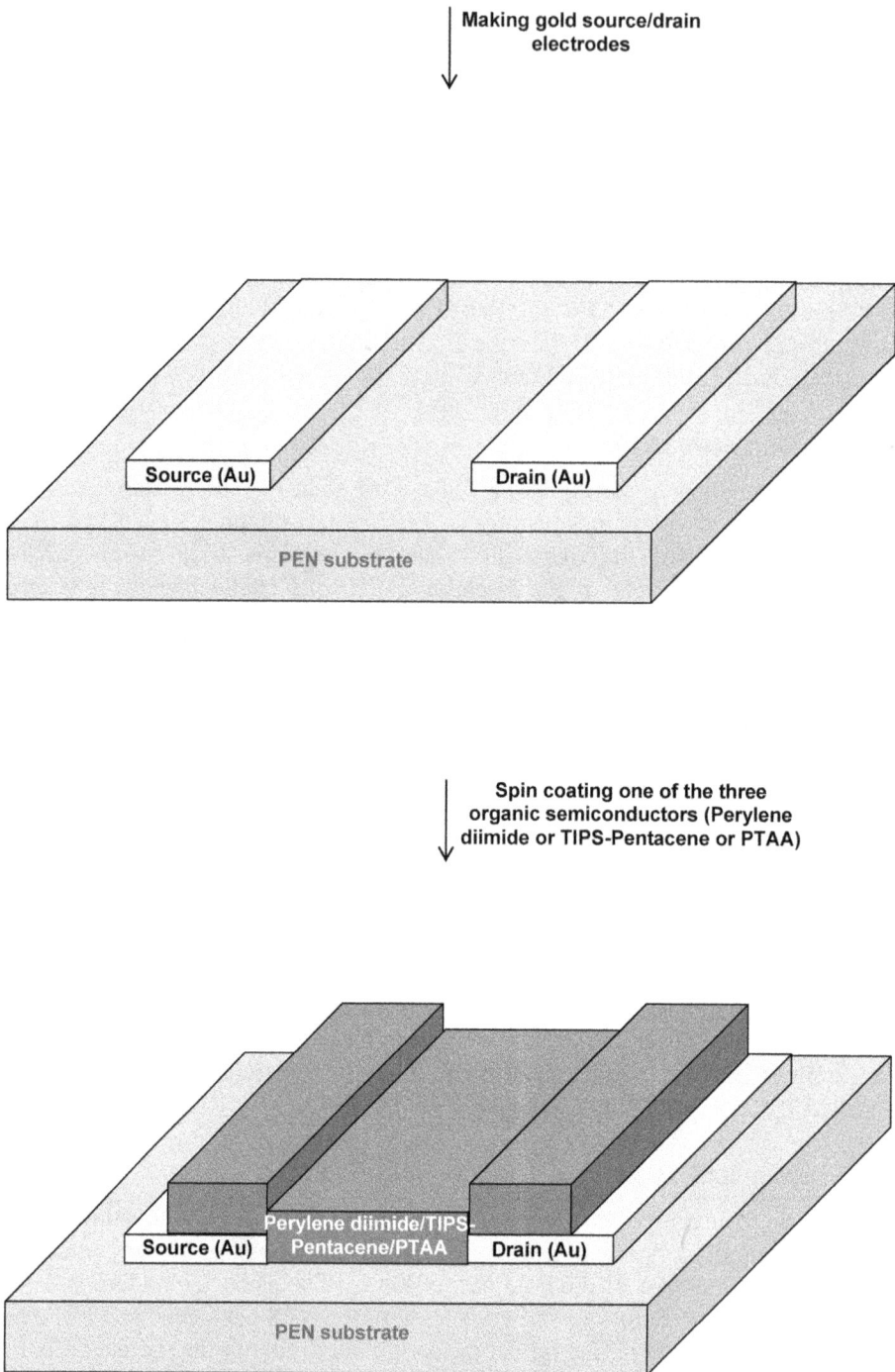

Figure 5.3. Fabricating N-channel perylene diimide TFT or P-channel TIPS-pentacene and PTAA TFTs on PEN substrate.

Spin coating CYTOP gate insulator film

CYTOP

Perylene diimide/TIPS-Pentacene/PTAA

Source (Au)

Drain (Au)

PEN substrate

Inkjet printing silver ink

Gate (Ag)

CYTOP

Perylene diimide/TIPS-Pentacene/PTAA

Source (Au)

Drain (Au)

PEN substrate

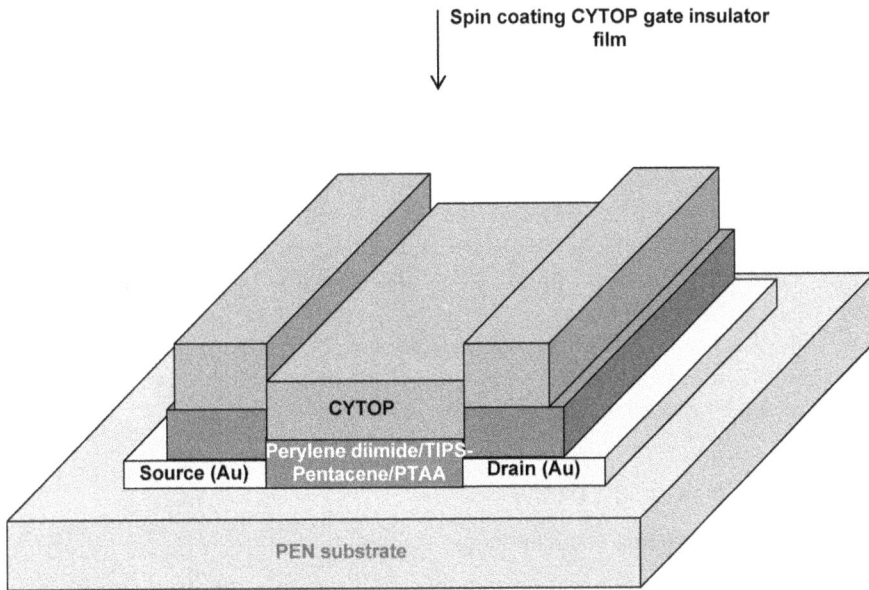

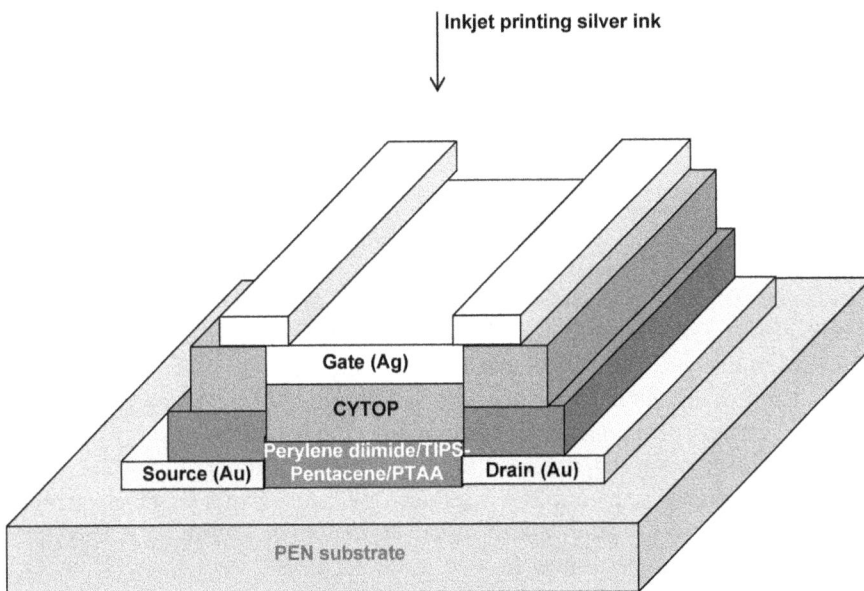

Figure 5.3. (Continued.)

5.5.3 Comparison of TIPS-pentacene TFT, PTAA TFT and perylene diimide TFT

For the devices of different W/L ratios tested, mobility $\mu_{FET,LV}$ = mobility at $|V_{GS} - V_{FB}| = 1$ V (V_{FB} is the flat-band voltage) ranges between 1.2×10^{-5} cm^2 V^{-1} s^{-1} to 3.6×10^{-5} cm^2 V^{-1} s^{-1} for the perylene diimide TFT, 4.4×10^{-3} to 5×10^{-3} cm^2 V^{-1} s^{-1} for the PTAA TFT and 0.49 to 0.58 cm^2 V^{-1} s^{-1} for the TIPS-pentacene TFT. Further, the mobility $\mu_{FET,HV}$ = mobility at $|V_{GS}| = 60$ V range is 0.035–0.108 cm^2 V^{-1} s^{-1} for the perylene diimide TFT, 0.041–0.048 cm^2 V^{-1} s^{-1} for the PTAA TFT and 2.051–2.868 cm^2 V^{-1} s^{-1} for the TIPS-pentacene TFT. Thus, the highest mobility is obtained for TIPS-pentacene TFT, then comes PTAA and finally, in the last position is perylene diimide TFT. But the mobility $\mu_{FET,HV}$ for perylene diimide TFT is comparable with that for PTAA. The mobility of perylene diimide TFT increases by four orders of magnitude in the measured range of voltages. Ultimately it reaches the PTAA mobility value at high gate voltages (Castro-Carranza *et al* 2012).

5.6 DNTT TFTs and circuits

5.6.1 Proneness of organic semiconductor to oxidation

The performance of OTFTs suffer from temporal degradation because of oxidation of conjugated molecules under the joint action of oxygen, moisture and ozone in the atmosphere. This material oxidation alters the energies of molecular orbitals. As a result, the charge transport in the material is affected. It is observed as a decline of carrier mobility with time. A way to avoid this atmospheric effect is to use a material with a high ionization potential, which is difficult to oxidize. Derivatives of the benzothienobenzothiophene (BTBT) family have an ionization potential between 5.4 and 5.7 eV. One member of this family is dinaphtho[2,3-b:2′,3′-f]thieno[3,2-b] thiophene (DNTT), empirical formula $C_{22}H_{12}S_2$ and molecular weight 340.46. Zschieschang *et al* (2010) reported TFTs, inverters and ring oscillators fabricated using DNTT, all operating at low voltages.

5.6.2 Synthesis of DNTT

This is done by a three-step synthetic procedure established by Yamamoto and Takimiya (2007). The three steps in this procedure are:
   (i) Selective functionalization of 2-naphthaldehyde, linear formula $C_{10}H_7CHO$, molecular weight 156.18 with methylthio, CH_3S substituents by ortho-directing lithiation (reaction with lithium; ortho refers to substituents occupying two neighboring positions on the aromatic ring).
   (ii) McMurry coupling reaction (a reductive coupling reaction consisting of two steps in which coupling is induced by transfer of a single electron to the carbonyl groups from alkali metal, after which 1, 2-diol is deoxygenated with low-valent titanium to give the alkene).
   (iii) A ring closure reaction using excess iodine.

The HOMO energy level of DNTT is −5.4 eV as opposed to −5.0 eV for pentacene. The optical bandgap of DNTT is 3.0 eV as compared to 1.8 eV for pentacene.

5.6.3 The substrate

This is a PEN film of thickness 125 μm (figure 5.4).

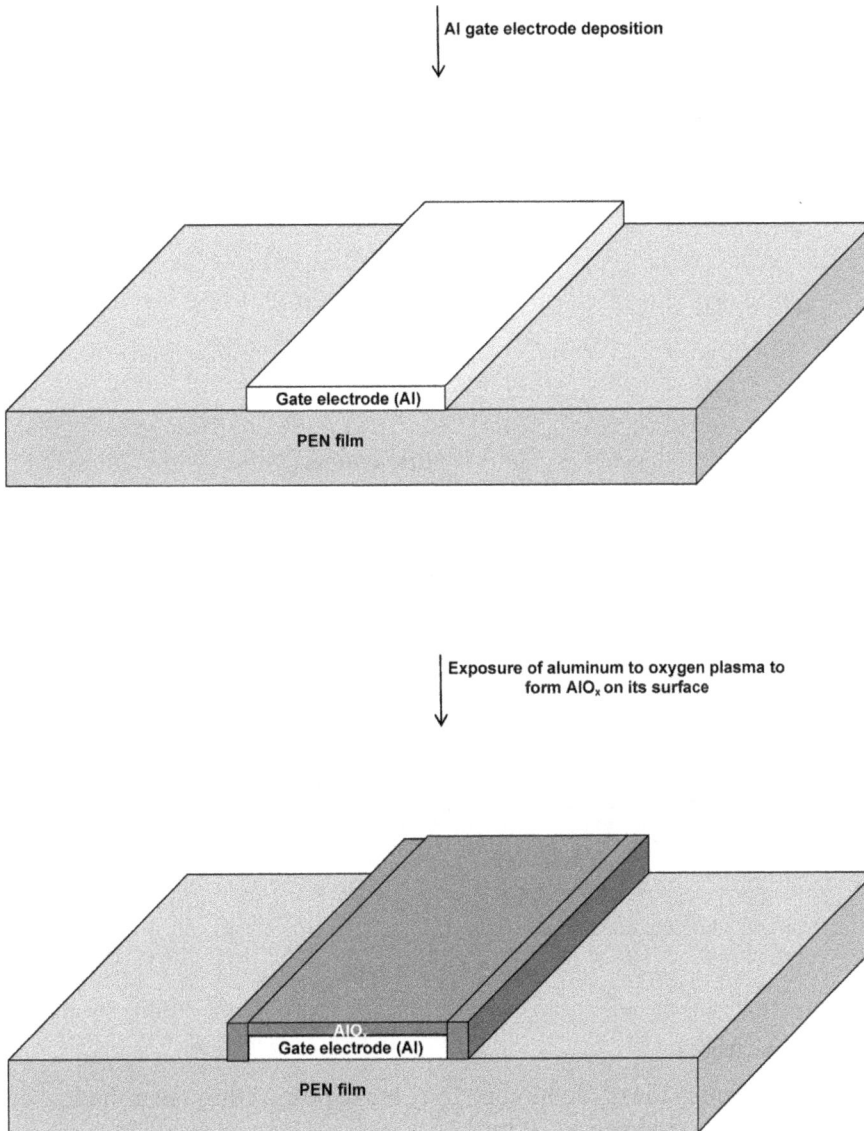

Al gate electrode deposition

Gate electrode (Al)

PEN film

Exposure of aluminum to oxygen plasma to form AlO$_x$ on its surface

AlO$_x$

Gate electrode (Al)

PEN film

Figure 5.4. TFT using DNTT as semiconductor material.

SAM formation

SAM
AlO$_x$
Gate electrode (Al)
PEN film

DNTT sublimation

DNTT film
SAM
AlO$_x$
Gate electrode (Al)
PEN film

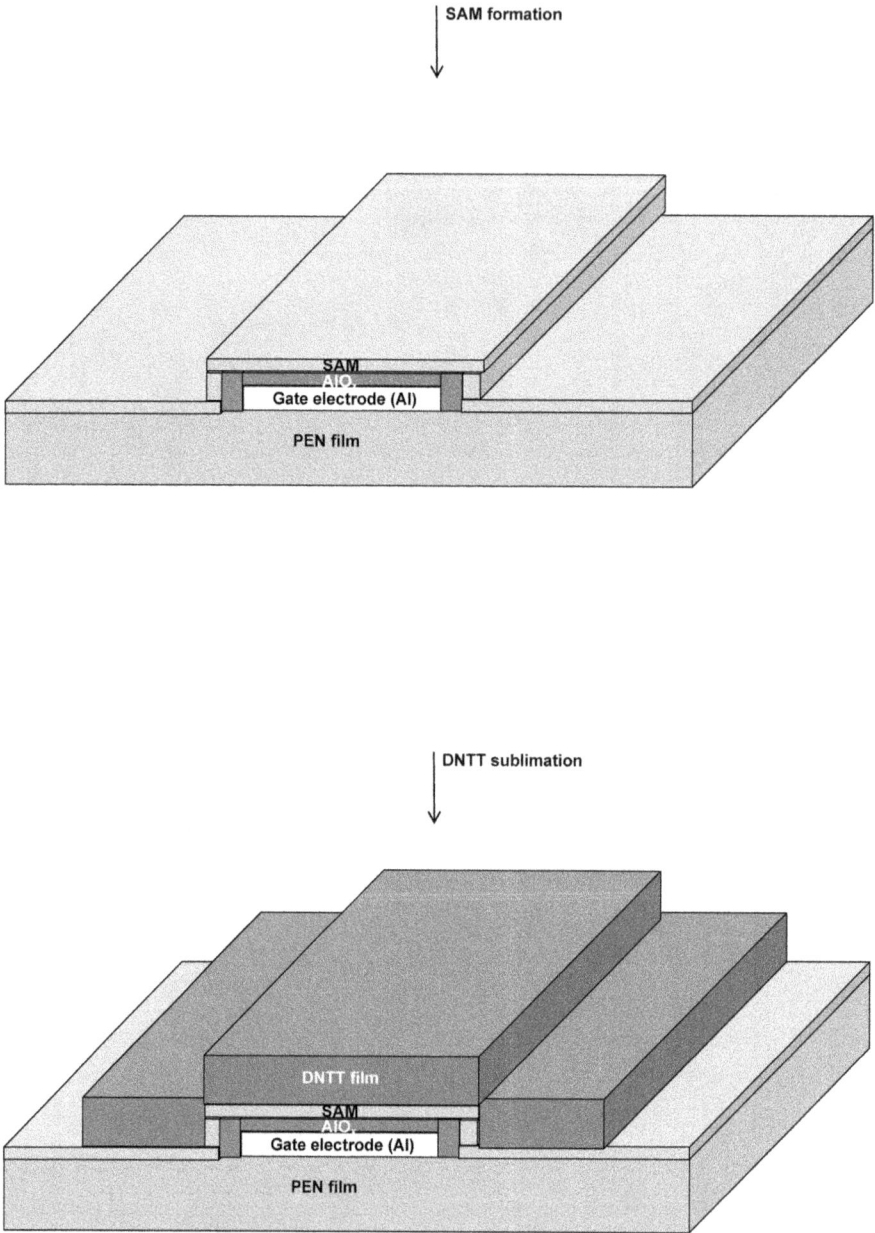

Figure 5.4. (Continued.)

5.6.4 Gate electrodes

Gate electrodes are made of aluminum. They are deposited through a shadow mask. The aluminum film thickness is 20 nm.

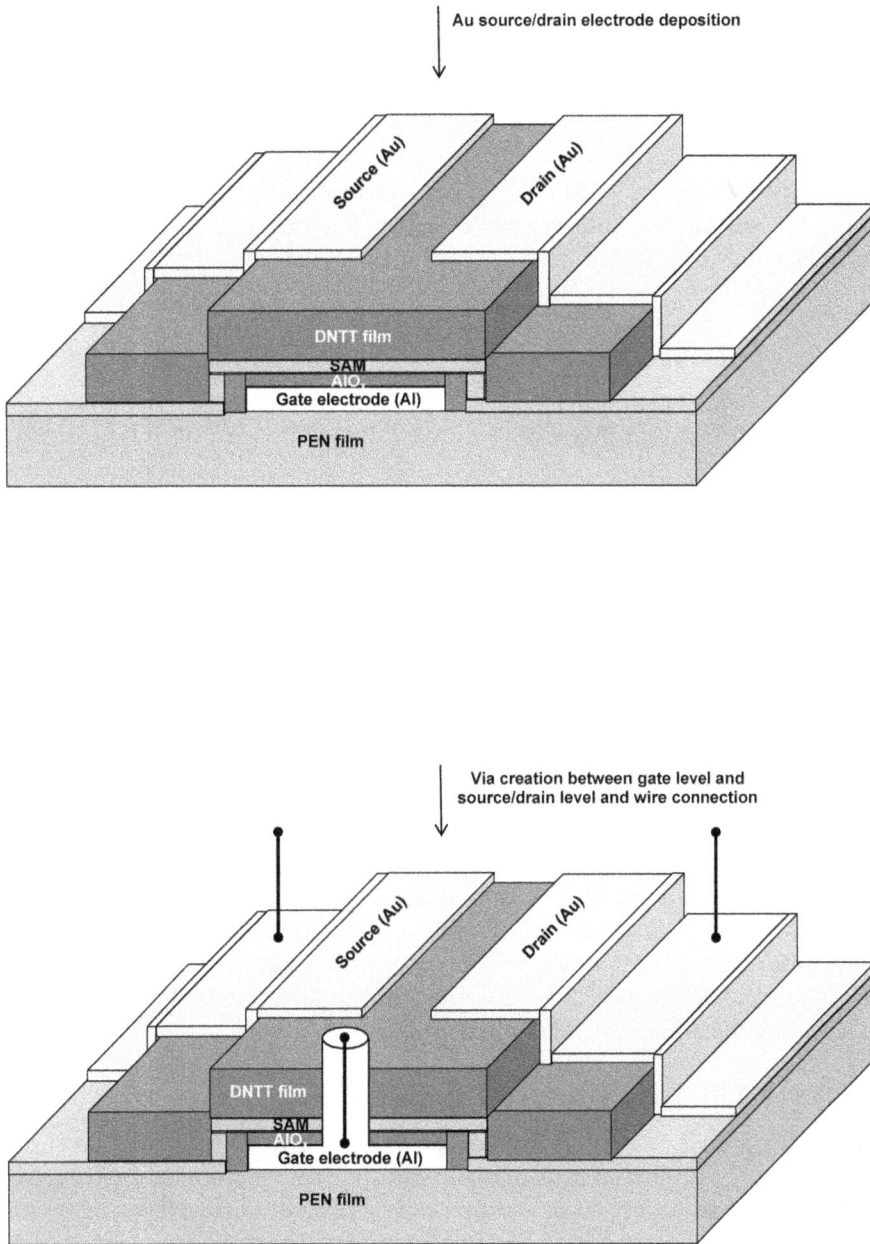

Figure 5.4. (Continued.)

5.6.5 SAM/AlO$_x$ gate dielectric

Exposure of the aluminum gate electrode for a short time interval to an oxygen plasma produces an AlO$_x$ film. The AlO$_x$ film is 3.6 nm thick. Over this AlO$_x$ a self-assembled monolayer is formed by dipping it in a solution of n-tetradecylphosphonic acid, linear formula $CH_3(CH_2)_{13}P(O)(OH)_2$, molecular weight 278.37 in 2-propanol, $(CH_3)_2CHOH$, molecular weight 60.10. The SAM has a thickness of 1.7 nm. Therefore, the total thickness of the gate dielectric = 3.6 + 1.7 = 5.3 nm.

5.6.6 DNTT film

This film is formed by sublimation of the synthesized DNTT in a vacuum evaporating system. The sublimation area is defined by shadow masking. During sublimation, the substrate temperature is 60 °C. This is the maximum temperature of TFT fabrication process. The thickness of the DNTT film is 30 nm.

5.6.7 Source/drain contacts

They are made gold. They are deposited through a shadow mask over defined areas. The thickness of the gold film is 30 nm.

5.6.8 Vias between gate level and source/drain level

The vias are made by a process based on the surface selectivity of self-assembling phosphonic acid, $H_2O_3P^+$, molecular formula 80.987 g mol^{-1} molecules during absorption.

5.6.9 Stability of TFT parameters on exposure to air and light

For the as-fabricated DNTT TFT, the on–off current ratio I_{on}/I_{off} is 10^6, the subthreshold swing SS is 100 mV/decade, and the mobility μ is 0.6 cm^2 V^{-1} s^{-1}. The DNTT TFT is exposed to yellow light in the laboratory and a relative humidity of 50% for a period of eight months (246 days). For the eight months exposed TFT, the ratio I_{on}/I_{off} is 10^6, SS is 150 mV/decade, and μ is 0.3 cm^2 V^{-1} s^{-1} showing that the SS of TFT increases by 50 mV/decade while μ is halved. This stability of parameters is better than that of pentacene TFTs whose mobility decreases by more than an order of magnitude over a much shorter span of barely three months. The higher ionization potential of DNTT film (5.4 eV) than pentacene (5.0 eV) accounts for the improved stability of DNTT TFT parameters.

5.6.10 DNTT TFT-based inverters and ring oscillators

Logic circuits are fabricated on PEN substrates using DNTT TFTs. The electrical characteristics of the inverter measured in as-fabricated condition and after aging for eight months are recorded. After eight months, the inverters displayed correct logic functioning. However, the variation of mobility with time causes a displacement of transfer curves of the inverter towards more positive voltages. Moreover, the maximum small signal gain falls from 2.3 to 1.9.

A five stage ring oscillator is made with drive TFTs having channel aspect ratio of 100 μm/10 μm. The ring oscillator operates with a low supply voltage of 2.2 V due to the high capacitance of the gate insulator. The initial value of stage delay of a DNTT ring oscillator is 37 μs at −3 V and rises to 100 μs after eight months. For the sake of comparison, pentacene TFTs have an initial delay of 500 μs at −3 V which worsens by many orders of magnitude over eight months. DNTT TFTs are much faster than pentacene TFTs both initially and after aging by storage over several months. Thus DNTT TFTs exhibit relatively higher air stability than pentacene TFTs (Zschieschang *et al* 2010).

5.7 DNTT TFT-based digital library

Elsobky *et al* (2017) developed a digital library using DNTT TFTs. This library comprises logic gates, flip flops and shift registers. To minimize power consumption, the complementary logic design is commonly used. However, N-channel organic TFTs still lag considerably behind the P-channel TFTs, both in performance and stability. Therefore, the digital designs exclusively use P-channel DNTT TFTs. Inverters and NAND gates operate with 3 V supply. Static and dynamic type master-slave flip flops are fabricated. In these flip flops, the transmission gates are implemented with P-channel TFTs only. A one stage shift register operates with a maximum clock frequency of 3 kHz.

5.7.1 TFT fabrication

The TFTs have a bottom-gate top-contact structure. The substrate is a bilayer stack of polyimide (PI) with benzocyclobutene (BCB), empirical formula: C_8H_8 and molecular weight 104.15. Its thickness is 40–50 µm. Interconnects are formed by thermally evaporating gold film of thickness 30 nm through a stencil mask in a vacuum coating system. Then a gate electrode is formed by thermal evaporation of aluminum film of thickness 30 nm through another stencil mask. The gate insulator has a two-layer structure. The first layer is an AlO_x film grown in oxygen plasma. It is 3.6 nm thick. The second layer is a SAM made from solution-processed tetradecylphosphonic acid, linear formula $CH_3(CH_2)_{13}P(O)(OH)_2$, molecular weight 278.37. The SAM layer thickness is 1.7 nm. The high capacitance of the gate insulator of ~700 nF cm^{-2} enables operation of the TFT at a low supply voltage. After the gate insulator, the DNTT film is deposited through a mask by sublimation in vacuum. The thickness of the DNTT film is 25 nm. The final step is deposition of source/drain electrodes using a mask. These electrodes are made of gold of thickness 25 nm.

5.7.2 TFT parameters

The TFTs have a threshold voltage of −1 V, the current on–off ratio is 10^6 and the mobility is 1.3 cm^2 V^{-1} s^{-1} (Elsobky *et al* 2017).

5.8 Discussion and conclusions

Organic semiconductors have several positive features: low cost, low temperature processes, inexpensive plastic substrates, fabrication over large areas, and so forth. Although more difficult to process than polymers, the full potential of small organic molecules needs to be harnessed for flexible electronics. Except for the solution-processed case of perylene diimide, the progression of carrier mobility has shown good improvement: 0.3 cm^2 V^{-1} s^{-1}, pentacene (Kato *et al* 2004, Sekitani *et al* 2005); 0.1 cm^2 V^{-1} s^{-1}, pentacene, 0.002 cm^2 V^{-1} s^{-1}, F$_{16}$CuPc (Klauk *et al* 2005); 1.2 × 10^{-5} cm^2 V^{-1} s^{-1} to 3.6 × 10^{-5} cm^2 V^{-1} s^{-1}, perylene diimide (Castro-Carranza *et al* 2012); 0.6 cm^2 V^{-1} s^{-1}, DNTT (Zschieschang *et al* 2010); 1.3 cm^2 V^{-1} s^{-1}, DNTT (Elsobky *et al* 2017).

Review exercises

5.1 In the pentacene TFT on PEN substrate, what materials are used for the following: (i) substrate, (ii) gate electrode, (iii) gate insulator and (iv) source/drain contacts?

5.2 In the pentacene TFT, how is the polyimide gate dielectric formed? How is the pentacene channel layer deposited? How are the electrodes for gate and source/drain contacts formed?

5.3 What is the effect of compressive and tensile strains on the saturated drain–source current? Are these changes reversible up to a radius of 4.6 mm? What happens when TFT is bent to smaller radius?

5.4 For complementary circuit fabrication using organic TFTs, what materials are used for the P-channel and N-channel TFTs? Which TFT shows superior performance? How is this difference in performance accommodated during complementary circuit design with these TFTs?

5.5 What does F$_{16}$CuPc stand for? What type of semiconductor is it, N-type or P-type?

5.6 In the organic complementary circuit, name the materials used for: (i) the substrate, (ii) P-type TFT, (iii) N-type TFT, and (iv) the gate dielectric.

5.7 Arrange the following organic semiconductors in ascending order of carrier mobility: TIPS-pentacene TFT, PTAA TFT and perylene diimide TFT.

5.8 To which type of organic semiconductors, N-type or P-type, do the following materials belong: (i) PTAA, (ii) TIPS-pentacene and perylene diimide?

5.9 Why is a TFT made from an organic semiconductor not able to maintain a stable performance when stored in the atmosphere? In what way does the TFT performance deteriorate? How can the drift in TFT parameters be minimized?

5.10 How is DNTT synthesized? How is the SAM/AlO$_x$ gate insulator of a DNTT TFT made?

5.11 Describe the aging experiments carried out to investigate the stability of a DNTT TFT with respect to time? Make a relative comparison of the stability of parameters of DNTT TFT versus pentacene TFT.

5.12 How do the transfer curves of an inverter made with DNTT TFT shift when it is aged over a period of eight months? What happens to the maximum small signal gain?

5.13 During the design of a digital library with DNTT TFTs, only P-channel devices are used. Why?

5.14 What two-layer gate insulator structure is used in DNTT TFTs? How are these two layers formed?

5.15 Write the formula of tetradecylphosphonic acid. What material does $C_{10}H_7CHO$ represent?

References

Castro-Carranza A, Nolasco J C, Estrada M, Gwoziecki R, Benwadih M, Xu Y, Cerdeira A, Marsal L F, Ghibaudo G, Iñiguez B and Pallarès J 2012 Effect of density of states on mobility in small-molecule n-type organic thin-film transistors based on a perylene diimide *IEEE Electron Device Lett.* **33** 1201–3

Elsobky M, Elattar M, Alavi G, Letzkus F, Richter H, Zschieschang U, Strecker M, Klauk H and Burghartz J N 2017 A digital library for a flexible low-voltage organic thin-film transistor technology *Org. Electron.* **50** 491–8

Kato Y, Iba S, Teramoto R, Sekitani T, Someya T, Kawaguchi H and Sakurai T 2004 High mobility of pentacene field-effect transistors with polyimide gate dielectric layers *Appl. Phys. Lett.* **84** 3789–91

Klauk H, Halik M, Zschieschang U, Eder F, Rohde D, Schmid G and Dehm C 2005 Flexible organic complementary circuits *IEEE Trans. Electron Devices* **52** 618–22

Sekitani T, Kato Y, Iba S, Shinaoka H, Someya T, Sakurai T and Takagi S 2005 Bending experiment on pentacene field-effect transistors on plastic films *Appl. Phys. Lett.* **86** 073511-1–3

Yamamoto T and Takimiya K 2007 Facile synthesis of highly π-extended heteroarenes, dinaphtho [2,3-b:2',3'-f]chalcogenopheno[3,2-b]chalcogenophenes, and their application to field-effect transistors *J. Am. Chem. Soc.* **129** 2224–5

Zschieschang U, Ante F, Yamamoto T, Takimiya K, Kuwabara H, Ikeda M, Sekitani T, Someya T, Kern K and Klauk H 2010 Flexible low-voltage organic transistors and circuits based on a high-mobility organic semiconductor with good air stability *Adv. Mater.* **22** 982–5

IOP Publishing

Flexible Electronics, Volume 2
Thin-film transistors
Vinod Kumar Khanna

Chapter 6

Polymer TFT

Polymeric materials offer facile solution-based TFT processing but suffer from purification difficulties. A P3HT TFT is made on a polycarbonate substrate with a SiO_2 adhesion cum barrier layer by contact printing the poly (3-hexylthiophene) semiconductor with a PDMS stamp on the polyimide–SiO_2 dual layer gate dielectric (Park *et al* 2002). Offset printing is used to deposit the PEDOT:PSS source/drain interdigitated electrodes of PTAA TFT on PET foil with spin coating of PTAA semiconductor and $BaTiO_3$ filled polymer gate dielectric. Electrical conductivity of the PEDOT:PSS ink is enhanced by mixing with glycol (Zielke *et al* 2005). An all-solution-processed PDQT array is realized on a PET substrate. Inkjet printing is used for PEDOT:PSS gate and source/drain electrodes. Before printing, the non–polar surface of the PET substrate is PVA-modified to increase its hydro-philicity. By this treatment, the contact angle is decreased from 107° to 38° ensuring continuity of the line formation with consecutively inkjet printed droplets. The PDQT semiconductor and PMMA gate insulator are spin-coated (Xu *et al* 2016). An eco-friendly TFT is fabricated by green electronics technology on a biodegrad-able cellulose acetate substrate using a decomposable PDPP–PD polymer as the semiconductor. For device release after processing, the substrate is peeled off from the carrier by dissolution of dextran film underlying the substrate layer. Functional logic circuits are demonstrated with this TFT (Lei *et al* 2017). A high mobility TFT is made using a three-layer moisture and gas barrier, FBT–TH_4(1, 4) as the semiconductor and AlO_x–Nd gate insulator with surface modified with ODPA SAM. The highest field-effect mobility is $2.88 \, cm^2 \, V^{-1} \, s^{-1}$. The mobility is crucially dependent upon the ODPA SAM in the absence of which it is appreciably degraded (Sun *et al* 2015).

6.1 Introduction

The greatest convenience with polymeric materials is solution processing. The biggest drawback of these materials is difficulty of purification. Once defects gain entry into the polymer chain, they cannot be removed without decomposition of the whole chain. The distribution of a statistically defined chain length of polymers makes the entry of defects almost inevitable. These defects have the unwholesome effect of lowering the mobility of carriers, which affects the device properties unfavorably.

6.2 P3HT TFT on polycarbonate substrate

Park *et al* (2002) printed polymer TFTs on a plastic substrate (figure 6.1). Salient features of their work are described below.

6.2.1 Substrate

This is made of polycarbonate (polycarbonate of bisphenol A), which is a clear plastic.

6.2.2 Gate dielectric

This is a dual layer structure. The two layers used include polyimide deposited by spin coating and SiO_2 deposited by a low-temperature process.

6.2.3 Semiconductor film

This is poly (3-hexylthiophene) (P3HT).

6.2.4 Process steps

(i) Substrate preparation: The polycarbonate substrate is given a pre-annealing treatment. This treatment is meant to minimize shrinkage of the polymer during the process. After pre-annealing, a SiO_2 film of thickness 50 nm is sputtered onto the polycarbonate. It has a two-fold function, firstly to promote adhesion of the subsequent metal layers and secondly to act as a barrier to gases.

(ii) Gate electrode formation: Al metal is deposited on the substrate at 100 °C. The choice of aluminum as a gate metal is made keeping in view its high coefficient of thermal expansion (25 ppm K^{-1}) and ductile property.

(iii) Polyimide–SiO_2 dual layer gate dielectric deposition: The polyimide film is deposited by spin coating. It is subjected to heat treatment at 150 °C under vacuum. When the pressure inside the vacuum chamber has dropped down to 10^{-6} Torr, electron gun evaporation is used to vaporize SiO_2 powder on the polyimide buffer layer at 10^{-5} Torr; the acceleration voltage is 7 kV and the emission current is 80 mA. The composite gate insulator is 250 nm thick (40 nm polyimide + 210 nm SiO_2). Annealing is done in nitrogen ambience.

(iv) Source/drain metallization: The metal combination selected is Ti/Au with titanium as the adhesion layer and gold is chosen for ohmic contact with the polymer owing to the high work function of Au. The lift-off process is used for patterning source/drain layout.

Sputtering SiO$_2$ on polycarbonate substrate

SiO$_2$ (Gas barrier cum adhesion layer)

Polycarbonate film

Making aluminum gate electrode

Gate electrode (Al)

SiO$_2$ (Gas barrier cum adhesion layer)

Polycarbonate film

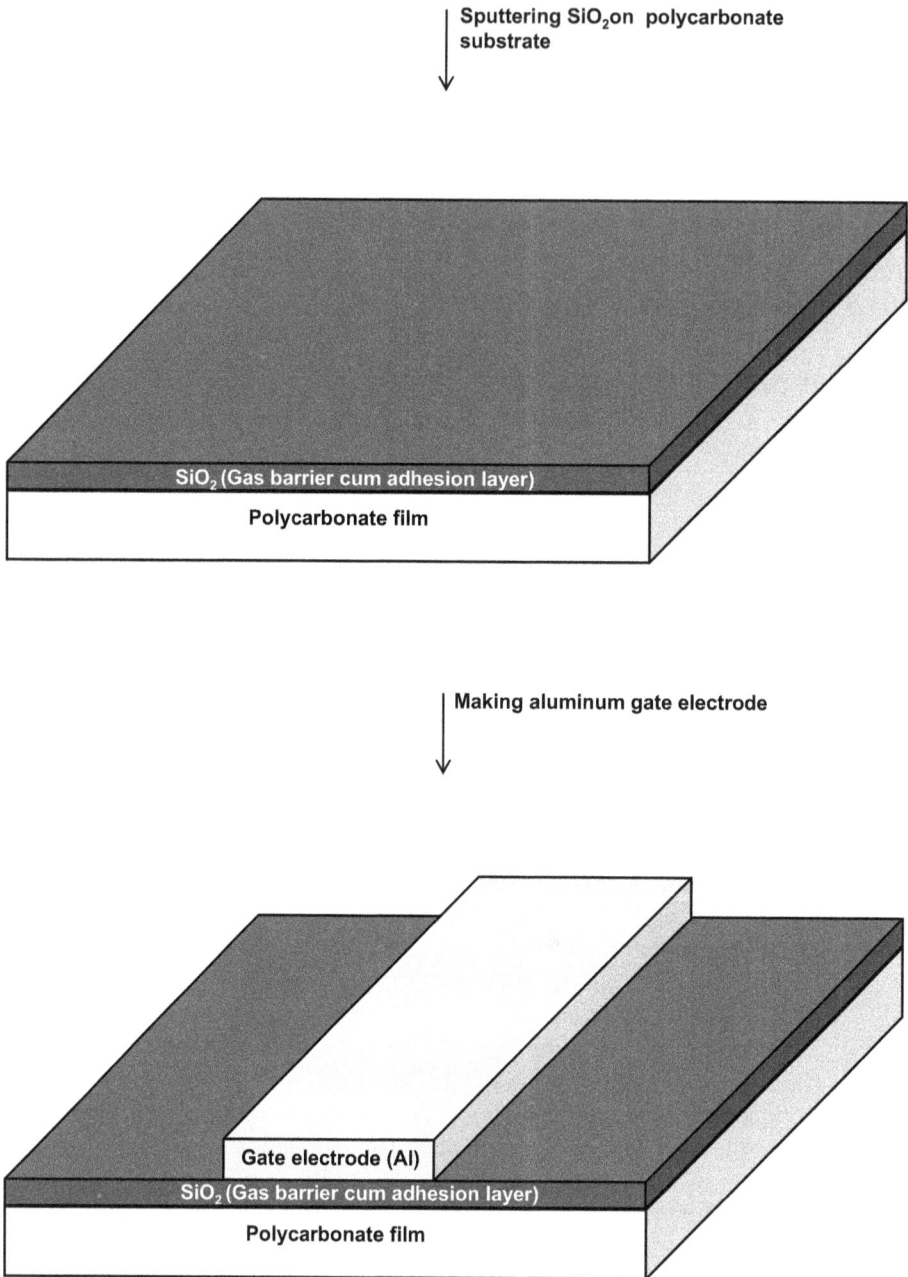

Figure 6.1. TFT made with P3HT organic semiconductor on polycarbonate substrate.

(v) Semiconducting polymer layer deposition: The gate dielectric layer is heat treated with hexamethyldisilazane (HMDS), $(CH_3)_3SiNHSi(CH_3)_3$. Then 1% P3HT solution in chloroform, $CHCl_3$ is deposited by the contact printing technique in atmospheric environment.

Spin coating polyimide

Polyimide film

Gate electrode (Al)

SiO$_2$ (Gas barrier cum adhesion layer)

Polycarbonate film

Evaporating SiO$_2$

SiO$_2$

Polyimide film

Gate dielectric

Gate electrode (Al)

SiO$_2$ (Gas barrier cum adhesion layer)

Polycarbonate film

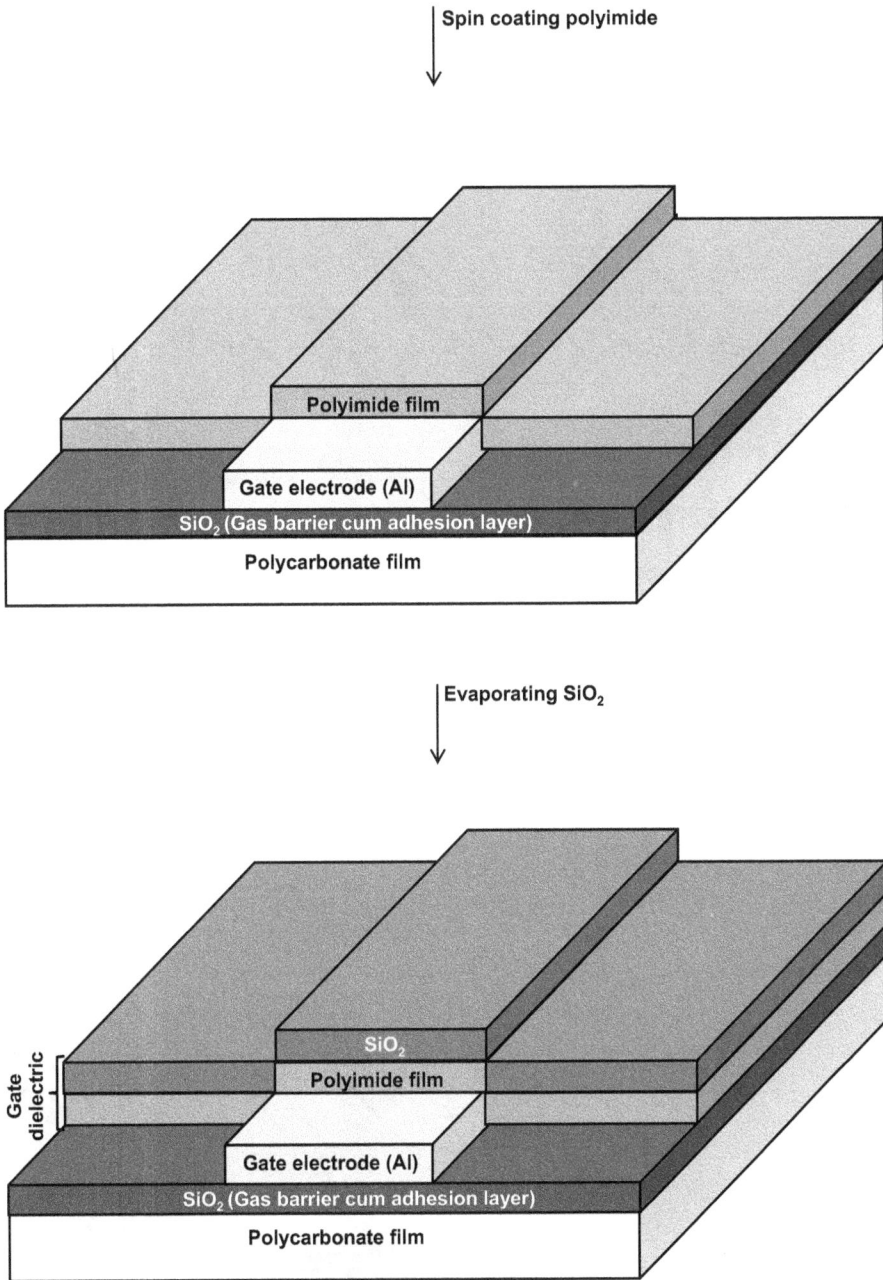

Figure 6.1. (Continued.)

Contact printing is a two-stage process. The first stage of contact printing is the definition of an accurate pattern. It involves the preparation of a stamp using silicone elastomer material. By the micromachining process, a master structure is

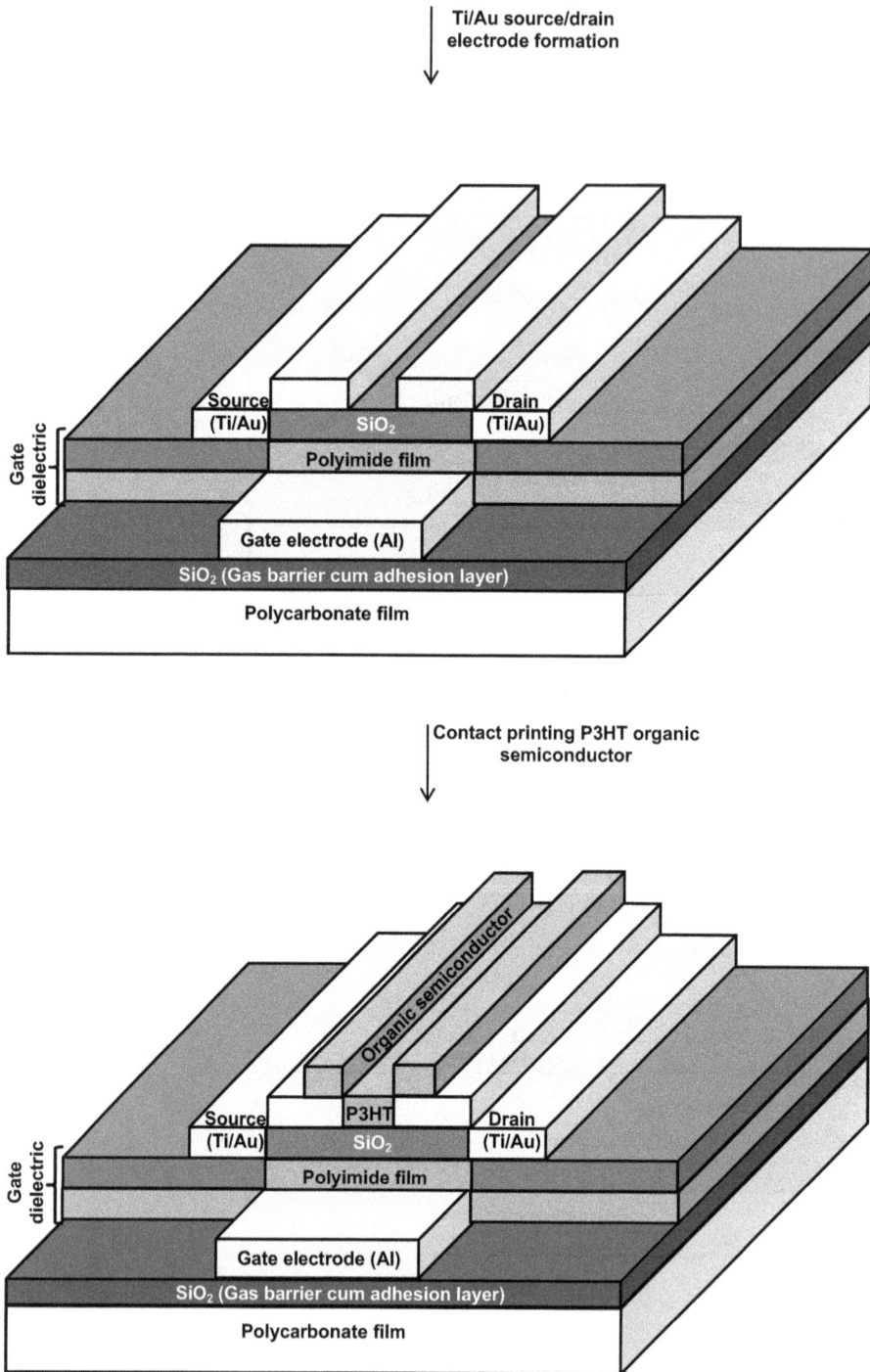

Figure 6.1. (Continued.)

produced using 400 μm thick photoresist. To make the stamp, liquid poly (dimethylsiloxane) (PDMS), $(C_2H_6OSi)_n$ is poured into the master structure. The PDMS is cured at temperatures from 20 °C to 80 °C in ambient conditions for a time duration of 24 h. The stamp obtained has a surface roughness of 1 nm. Its Young's modulus is 3 MPa.

The second stage is printing with the help of this stamp. For printing, a drop of P3HT polymer solution is placed on the stamp for 2 s. Then nitrogen gas is blown over it. A reproducible quantity of polymer solution is left on the stamp. The stamp is pressed against the HMDS-treated gate dielectric layer under ambient conditions. Before polymer deposition, the structure is treated with oxygen plasma at 50 W RF power and 1 torr pressure. This treatment is found to increase the mobility and the drain current values for the fabricated TFT.

Four parameters decide the thickness of the polymer coating and its window size. These parameters are: the quantity of polymer solution dropped on the stamp; the concentration of polymer solution; the time period for which printing is carried out; the pressure applied during printing. After stamping, annealing is done at 120 °C in nitrogen for 3 h. Then 12 h annealing is done under vacuum. The semiconducting polymer film has a thickness of 250–500 nm.

6.2.5 TFT parameters

The W/L ratio of the TFT is 500 μm/25 μm = 20. TFTs exposed to plasma after fabrication of the electrode showed good characteristics. They had a threshold voltage ~2–3 V. The field-effect mobility in saturation mode is 0.02–0.025 $cm^2 V^{-1} s^{-1}$. The current on–off ratio is 10^3–10^4 (Park *et al* 2002).

6.3 PTAA TFT on PET foil

Zielke *et al* (2005) showed the use of offset printing for making source/drain structures of TFT (figure 6.2). Printing technology is attractive because of its capability of providing mass production at affordable prices.

6.3.1 Printing ink formulation

The basis for the conducting printing ink consists of a mixture of poly(3,4-ethylenedioxythiophene) (PEDOT), $(C_6H_4O_2)_n$ and poly(4-styrene sulfonate) (PSS) $[(C_8H_7NaO_3S)_n$, sodium salt] dissolved in water at a concentration of 1%. Evaporation of water thickens the ink and increases its viscosity. The conductivity of the ink formulation is increased by mixing it with glycol, $C_2H_6O_2$ in the ratio 1:1. Then the viscosity of the ink becomes 100 Pa s (Pascal second). Its surface tension is 50 mN m^{-1}. The printed ink after thermal curing at 150 °C has a conductivity of 30 S cm^{-1}.

6.3.2 Formation of interdigitated source/drain structures by offset printing

An interdigitated source/drain geometry is used. The substrate is a PET foil of thickness 50 μm and a width of 140 mm, and is coiled on rolls. The foil has a surface roughness of 75 nm. The foil is given an in-line corona treatment prior to printing.

Making interdigitated source/drain electrodes

Printed drain electrode (PEDOT:PSS conducting ink)

Printed source electrode (PEDOT:PSS conducting ink)

PET foil

Spin coating PTAA

Printed drain electrode (PEDOT:PSS conducting ink)

Organic semiconductor (PTAA)

Printed source electrode (PEDOT:PSS conducting ink)

PET foil

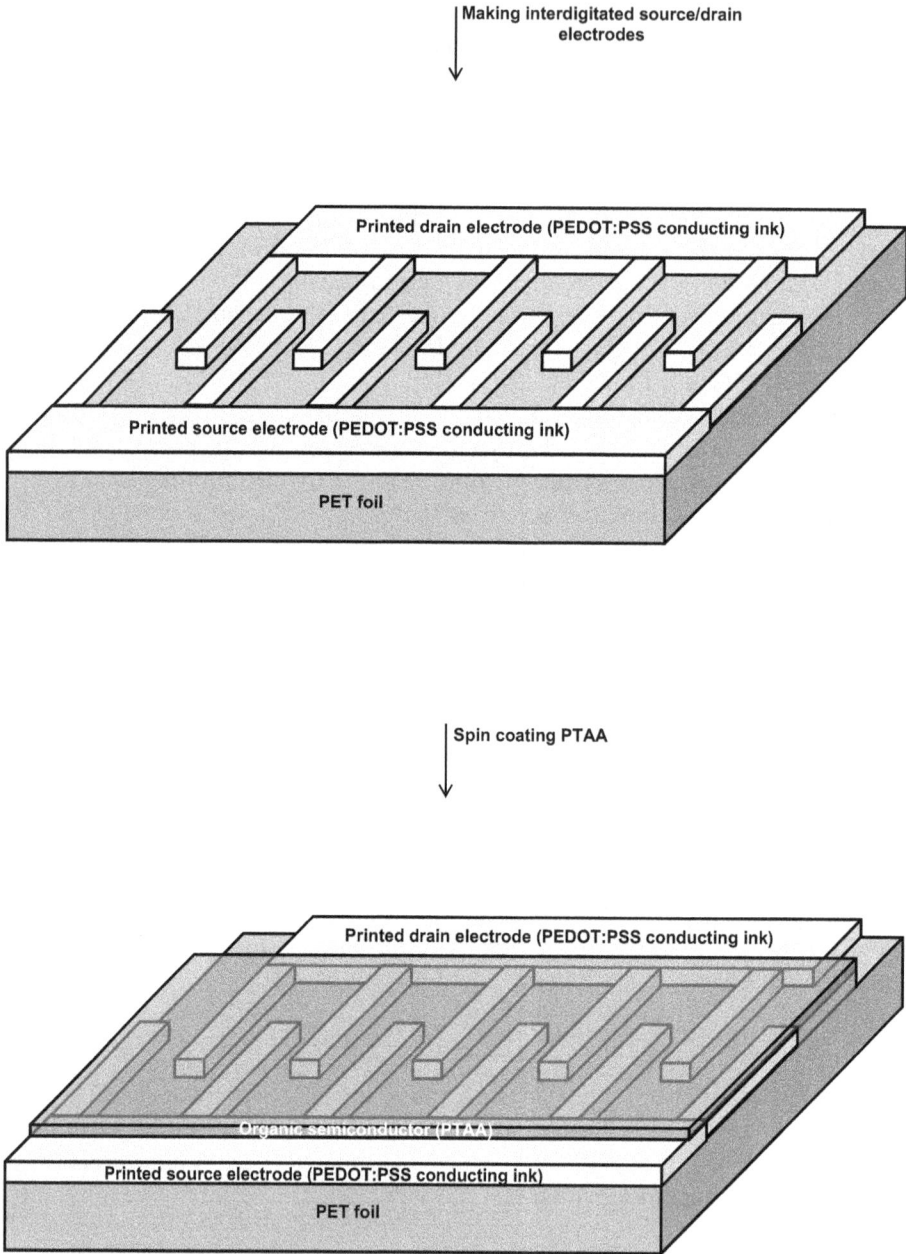

Figure 6.2. TFT with PTAA organic semiconductor on PET substrate.

Offset technology is used to print the source/drain geometry using the PEDOT:PSS-based conducting ink. The printed foil passes through an in-line oven at 150 °C for 8 s. The printing speed is $0.5\,\mathrm{m\,s^{-1}}$. The dried film has a thickness of 600 nm. Its sheet

Spin coating butylene copolymer as
semiconductor/insulator interface optimization
layer

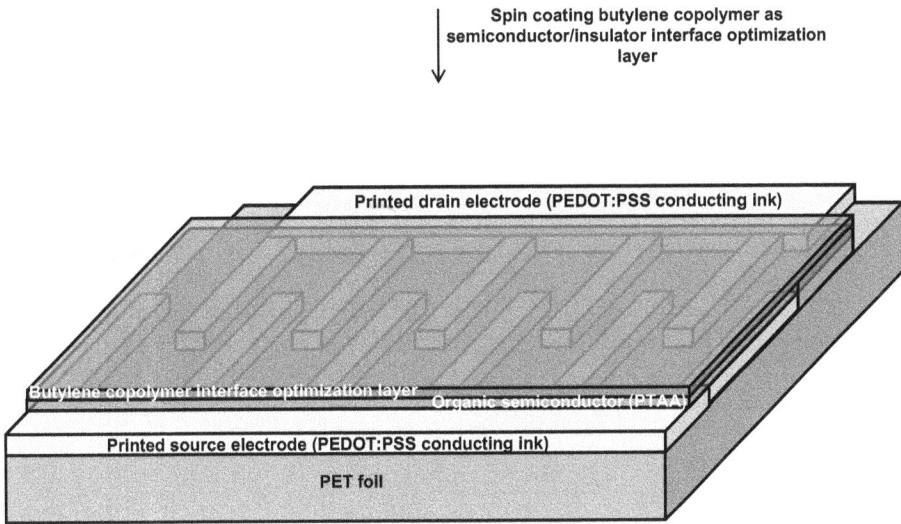

Spin coating BaTiO$_3$-filled polymer as gate
dielectric layer

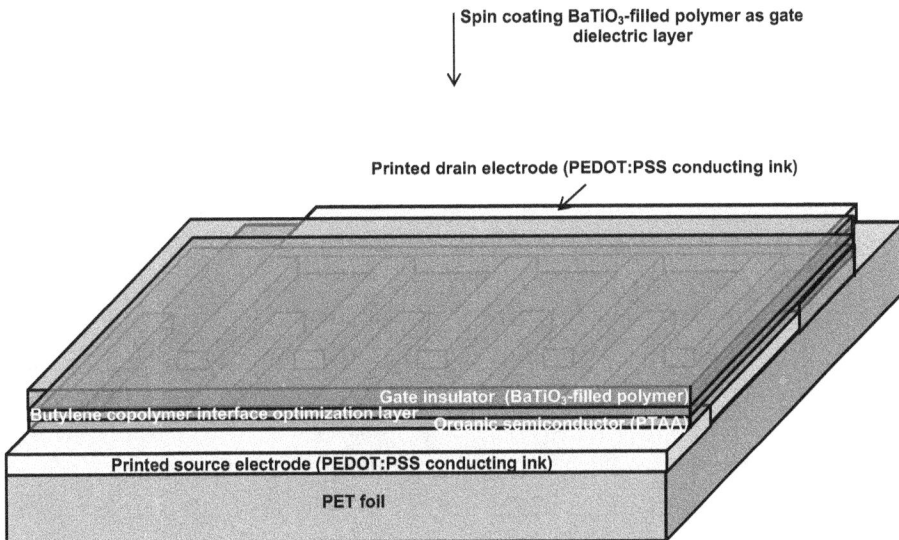

Figure 6.2. (Continued.)

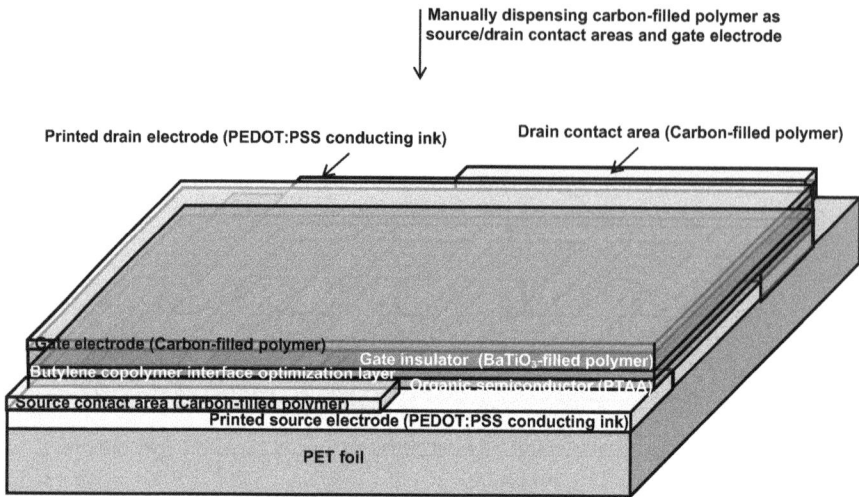

Figure 6.2. (Continued.)

resistance is 13 kΩ/sq. The geometrical parameters of the source/drain structures are: linewidths = 100–400 μm, gap widths = 50–160 μm, and channel width = 40 000 μm.

6.3.3 PTAA semiconductor deposition

A 1% solution of polytriarylamine (PTAA) polymer, linear formula $[C_6H_4N(C_6H_2(CH_3)_3)C_6H_4]_n$ in toluene, $C_6H_5CH_3$ is prepared. This solution is spin coated over the source/drain structures. The semiconducting polymer film is 50 nm thick.

6.3.4 Semiconductor/insulator interface optimization

A butylene (C_4H_8) copolymer with a low relative permittivity of ~2.2 is deposited by spin coating its solution in hexane (C_6H_{14}). The thickness of the copolymer is 100–600 nm.

6.3.5 Deposition of BaTiO$_3$ filled polymer as gate insulator

The BaTiO$_3$ filled polymer is deposited by spin coating. It is 10–15 μm thick. Barium titanate is a high dielectric constant material ($\varepsilon = 100$–1250).

6.3.6 Deposition of carbon-filled polymer on gate and source/drain contact areas

This polymer is manually dispensed. It acts as the gate electrode and source/drain contact areas.

6.3.7 TFT characteristics

The threshold voltage of TFT is 0 V ± 8 V. The mobility is 3×10^{-3} cm^2 V^{-1} s^{-1} and the current on–off ratio is 10^3. Despite the non-ideal threshold voltage, a seven stage ring oscillator driven by −40 V supply voltage is demonstrated with functional inverters. The gain from the seven series-connected inverters is >1 in the voltage range between −5 V and −25 V. The output swing is 7.5 V and the frequency of oscillation is 0.4 Hz (Zielke *et al* 2005).

6.4 PDQT TFT array on PET substrate

The manifold advantages of inkjet printing, such as small consumption of material and fully additive approach, are utilized by Xu *et al* (2016) to fabricate an ultra-short channel TFT; the channel length is 2 μm (figure 6.3). The gate and source/drain electrodes are made of inkjet printed conducting polymer. An all-solution process is employed to fabricate an array of 200 TFTs.

A primary feature of the technology of a PDQT TFT array is that all the constituent process steps are solution-based. Another important feature is that all the steps are implemented at room temperature. The process sequence does not contain any high-temperature step. In addition, the extremely short channel is formed using a simple process. The solution-processed TFT is transparent.

6.4.1 Substrate surface pretreatment and modification

The substrate is a PET foil of thickness 200 μm. It is treated with oxygen plasma for 20 min. The treatment is applied to increase the surface energy.

Then the substrate surface is modified by spin coating polyvinyl alcohol (PVA), $[-CH_2CHOH-]_n$ solution in DI water having concentration of $20 \, mg \, mL^{-1}$, prepared by heating at 90 °C for 5 h. Thus a 50 nm thick buffer layer of PVA is formed on the surface of PET substrate. The reason why this step is necessary is as follows: Upon inkjet printing, the PEDOT:PSS droplet first spreads out owing to its kinetic energy and afterwards contracts due to surface tension effect to acquire a stable shape. When the PEDOT:PSS in aqueous solution (polar solvent) is printed over the non-polar surface of the PET substrate, the large contact angle ~107° does not allow formation of a continuous line with successively inkjet printed droplets. Rather, the individual droplets form isolated islands. For continuity of the line, the hydrophilicity of the PET surface must be increased. The PVA buffer layer decreases the contact angle to 38° so that a continuous inkjet printed line is obtained by merging of the consecutively printed droplets. The wettability provided by the common methods of oxygen plasma or UV/ozone treatments being unstable in air, PVA is a better choice guaranteeing wettability.

6.4.2 Inkjet printing of PEDOT:PSS source/drain electrodes

On the PVA-modified PET surface, the pattern of source/drain electrodes is laid out by inkjet printing of conducting polymer PEDOT:PSS.

6.4.3 Organic semiconductor PDQT film deposition

A solution of poly [3,6-bis(40-dodecyl[2,20]bithiophenyl-5-yl)-2,5-bis(2-hexyldecyl)-2,5-dihy-dropyrrolo[3,4-c]pyrrole-1,4-dione] (PDQT) is made in chloroform,

Modification of surface of PET foil
with polyvinyl alcohol (PVA)

← PVA
← PET

Surface modification layer
Substrate

Making PEDOT:PSS source/drain
electrodes by inkjet printing

← PVA
← PET
← PDOT:PSS

Drain

Source

Surface modification layer
Substrate

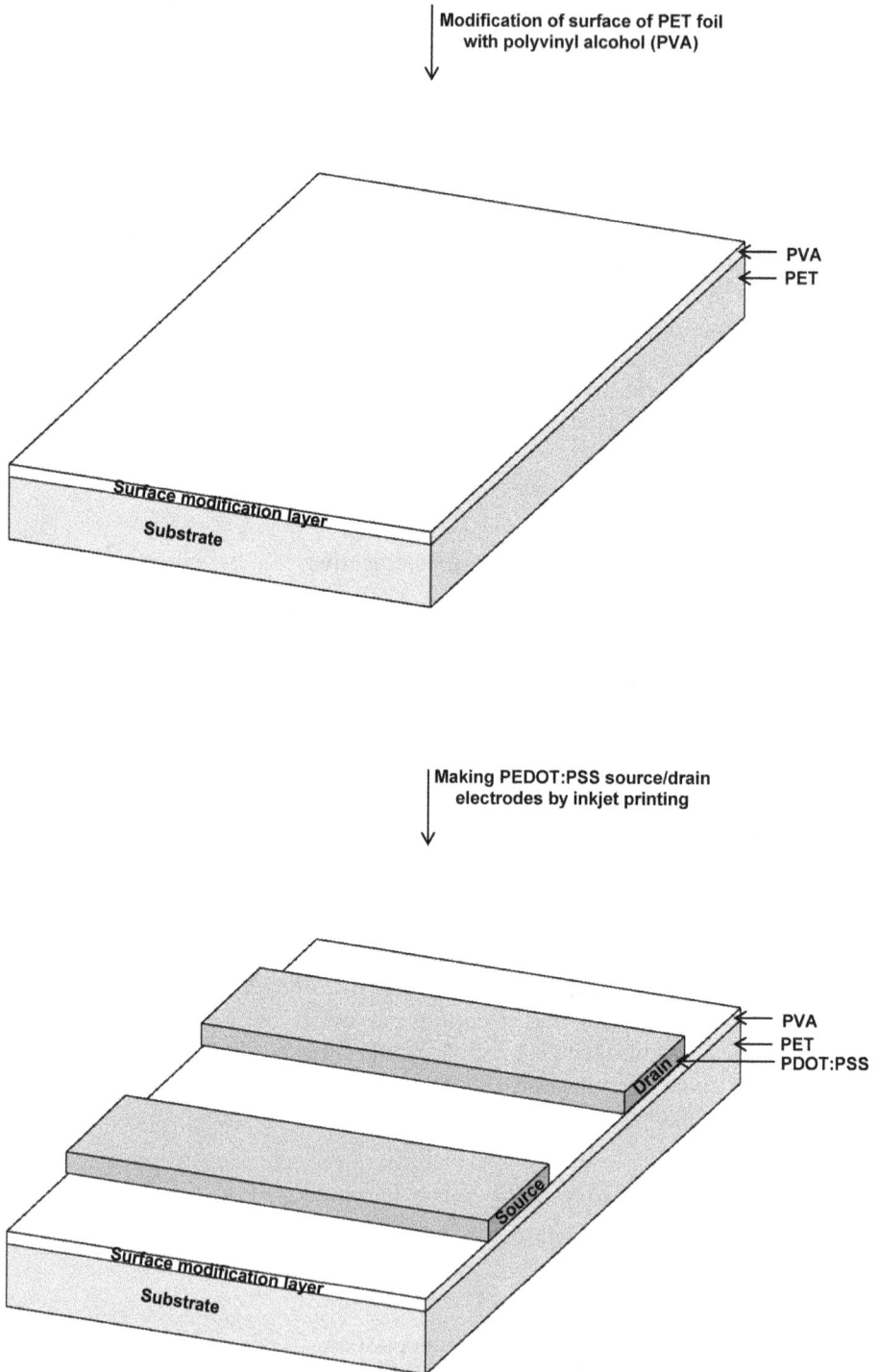

Figure 6.3. TFT with PDQT organic semiconductor on PET substrate.

Spin coating PDQT film

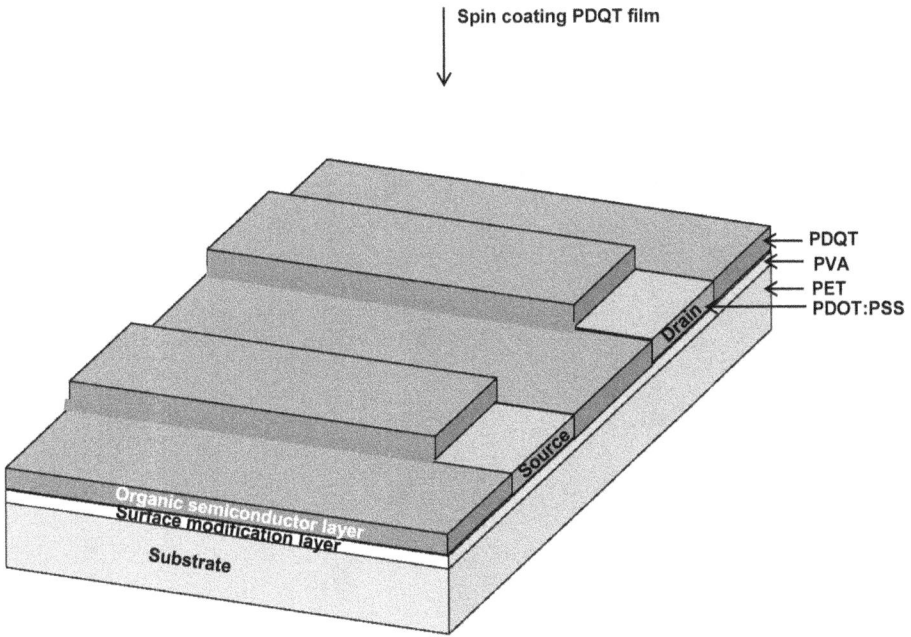

PDQT
PVA
PET
PDOT:PSS

Drain
Source

Organic semiconductor layer
Surface modification layer
Substrate

Spin coating PMMA

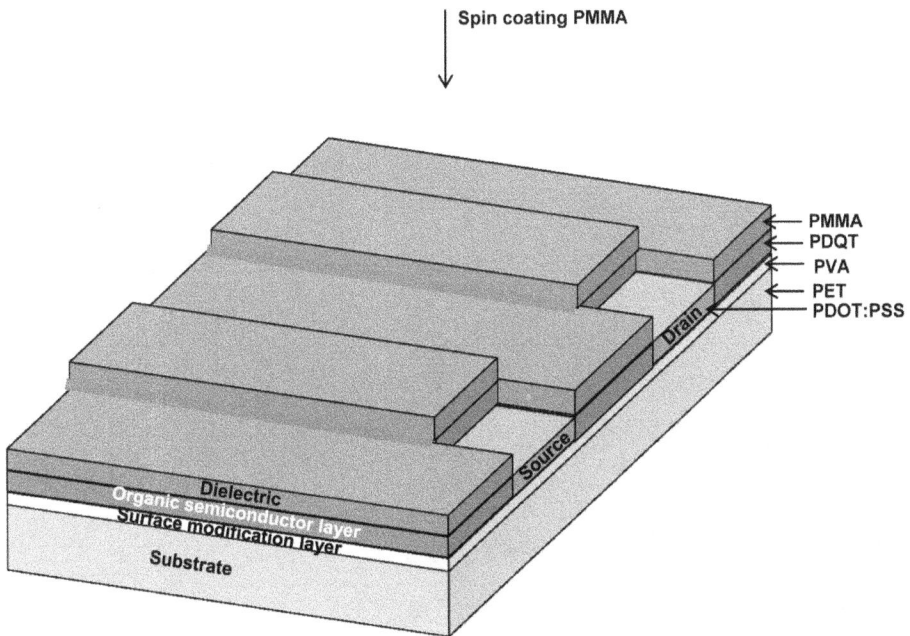

PMMA
PDQT
PVA
PET
PDOT:PSS

Drain
Source

Dielectric
Organic semiconductor layer
Surface modification layer
Substrate

Figure 6.3. (Continued.)

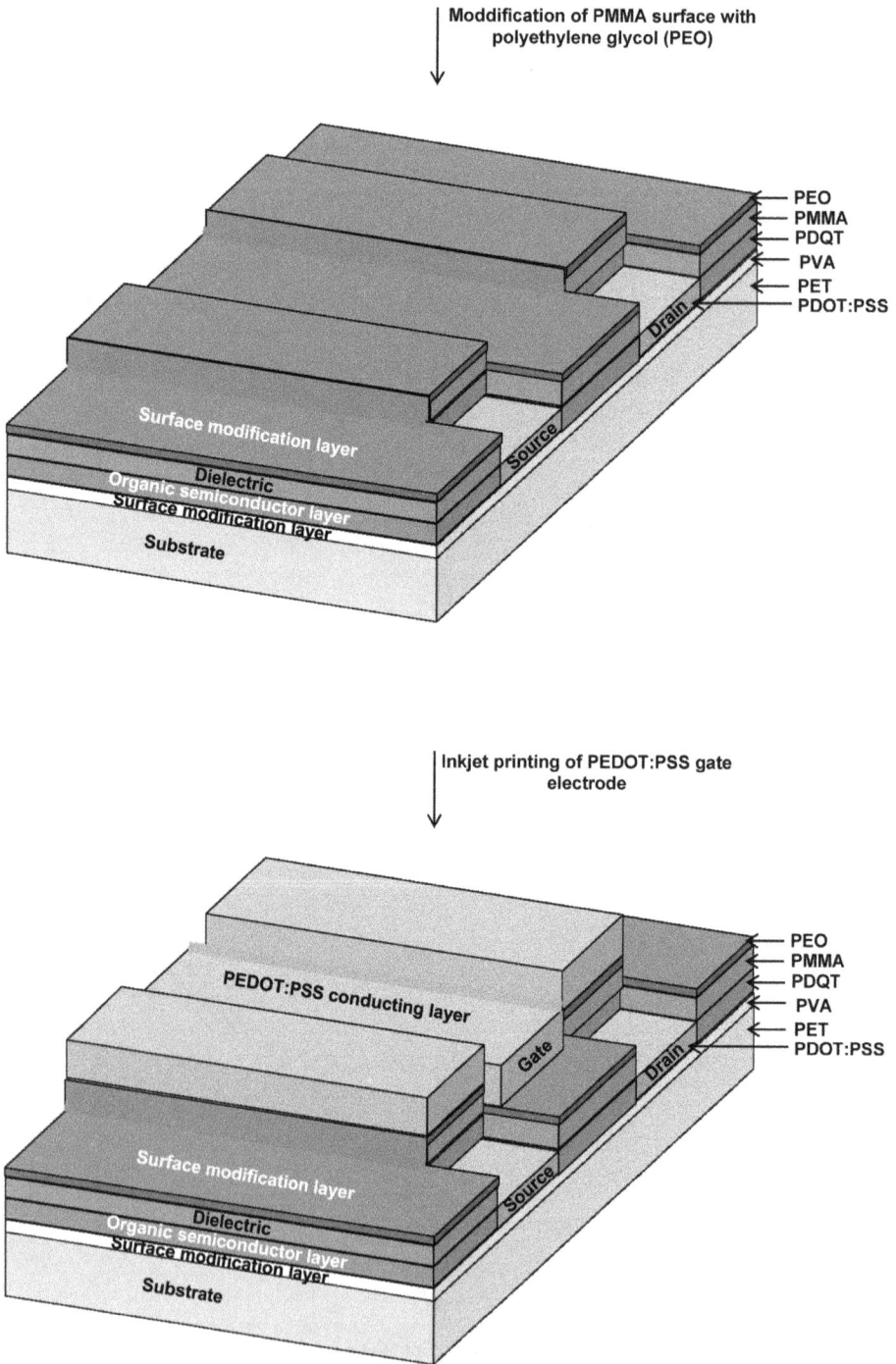

Figure 6.3. (Continued.)

$CHCl_3$. The concentration of the solution is $6 \, mg \, mL^{-1}$. It is deposited by spin coating on the source/drain electrode pattern. The PDQT film thickness is 60 nm.

6.4.4 PMMA gate insulator layer deposition

The gate insulator is poly(methyl methacrylate), linear formula $[CH_2C(CH_3) (CO_2CH_3)]_n$ or $(C_5O_2H_8)_n$, average molecular weight 3 500 000 by gas permeation chromatography (GPC). The $50 \, mg \, mL^{-1}$ solution of PMMA in n-butyl acetate, $CH_3COO(CH_2)_3CH_3$ or $C_6H_{12}O_2$, is spin-coated to form the gate dielectric. The gate dielectric film is 500 nm thick.

6.4.5 Enhancement of wettability of PMMA surface

To increase the wettability, a $5 \, mg \, mL^{-1}$ solution of polyethylene glycol (PEO), H $(OCH_2CH_2)_nOH$, in methanol, CH_3OH, is spin coated over the PMMA surface. The wettability enhancement step is a surface preparation for the next step of PEDOT:PSS printing.

6.4.6 PEDOT:PSS gate printing

The gate electrode is a PEDOT:PSS film formed by inkjet printing over the PMMA surface prepared in the preceding step.

6.4.7 TFT characterization

The TFT has a turn-on voltage of 1.94 V. The current on–off ratio is $\sim 10^4$. The maximum mobility of holes is $0.64 \times 10^{-3} \, cm^2 \, V^{-1} \, s^{-1}$ (Xu *et al* 2016).

6.5 Ultrathin, disintegrable PDPP–PD polymer TFT and logic circuits on cellulose substrate

Lei *et al* (2017) break away from the contemporary practice of manufacturing consumer electronics products from non-decomposable and non-biocompatible substances to follow an environment-friendly path of using biocompatible and disintegrable polymers (figure 6.4). Their approach is pursued to make circuits operating at a low voltage ~ 4 V. Moreover, the circuit is less than a micron thick and also exceedingly light in weight $\sim 2 \, g \, m^{-2}$. Thus, ultrathin, ultra-lightweight green electronics is conceptualized.

6.5.1 Biodegradable substrate

A cellulose substrate of thickness 800 nm is used. The substrate is chemically stable in water and organic solvents. Its glass transition temperature $T_g > 180 \, °C$ shows its thermal stability.

The ultrathin substrate is prepared from trimethylsilyl cellulose (TMSC), which is synthesized by dissolving 2 g microcrystalline cellulose powders in 9% LiCl/*N*, *N*-dimethylacetamide (lithium chloride/DMAc) (Lei *et al* 2017) by heating at 150 °C. DMAc is $CH_3CON(CH_3)_2$ or C_4H_9NO with molecular weight 87.12. After

Spin coating 5% dextran on
glass/silicon carrier substrate

Dextran

Glass or Silicon carrier substrate

Spin coating TMSC in chlorobenze

TMSC (1200 nm)
Dextran

Glass or Silicon carrier substrate

Hydrolyzing TMSC film in 95% acetic
acid vapor

Cellulose film (400 nm)
Dextran

Glass or Silicon carrier substrate

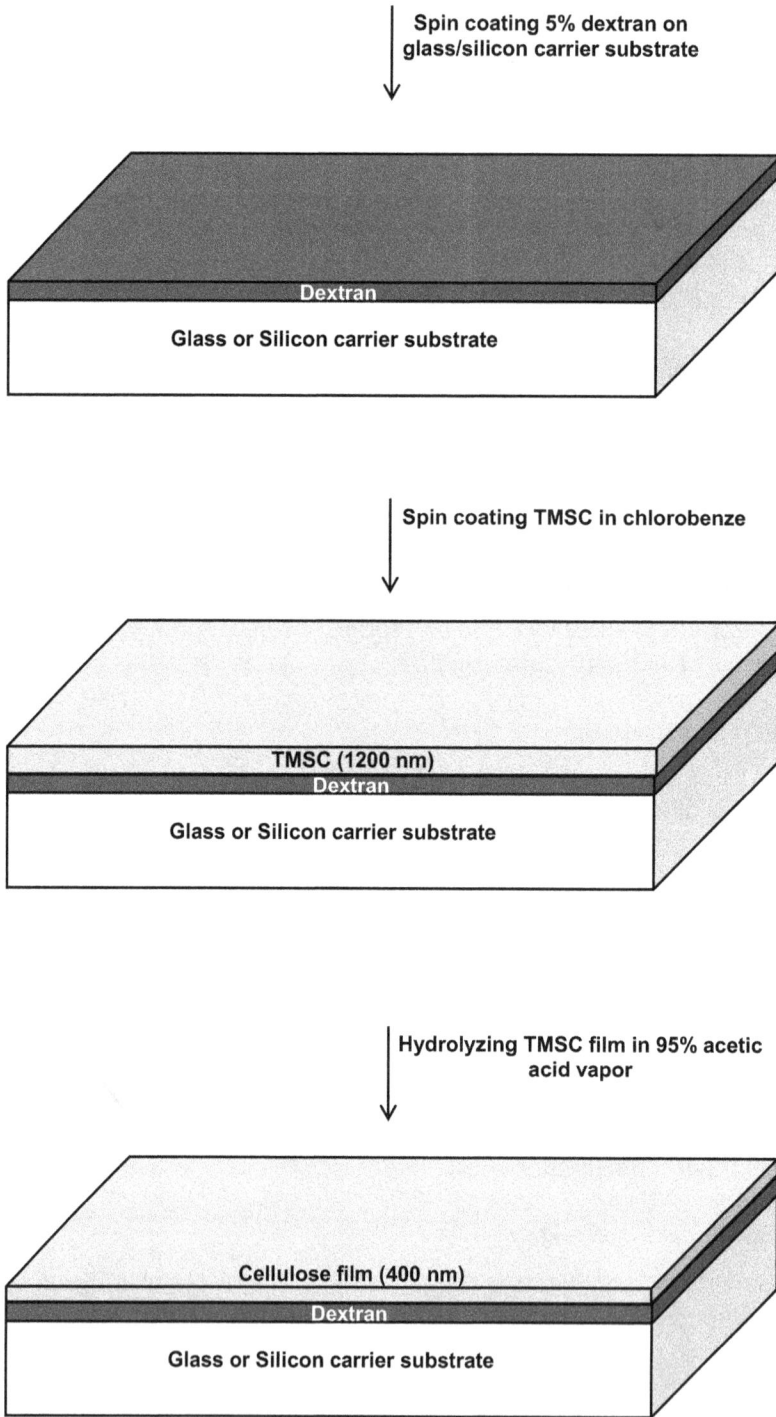

Figure 6.4. Decomposable PDPP–PD polymer TFT on a biodegradable substrate.

RepeatingTMSC spin coating and
hydrolyzing steps till cellulose film is
800 nm thick

Cellulose film (800 nm)

Dextran

Glass or Silicon carrier substrate

Ti/Au/Ti gate electrode formation

Au — Gate

Ti

Cellulose film (800 nm)

Dextran

Glass or Silicon carrier substrate

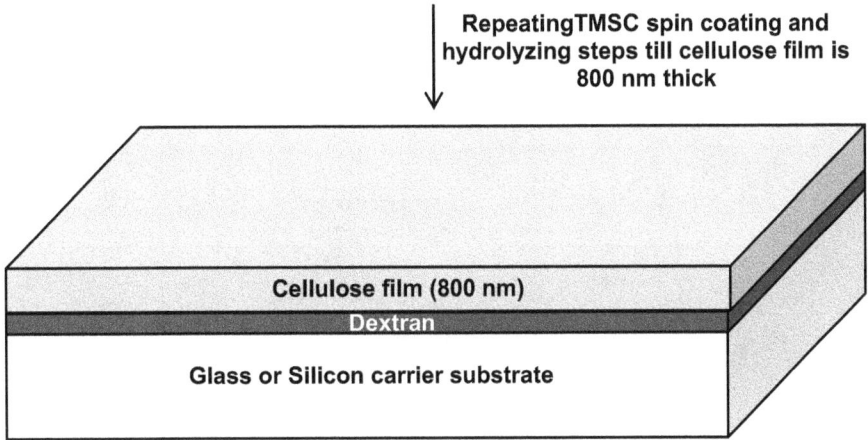

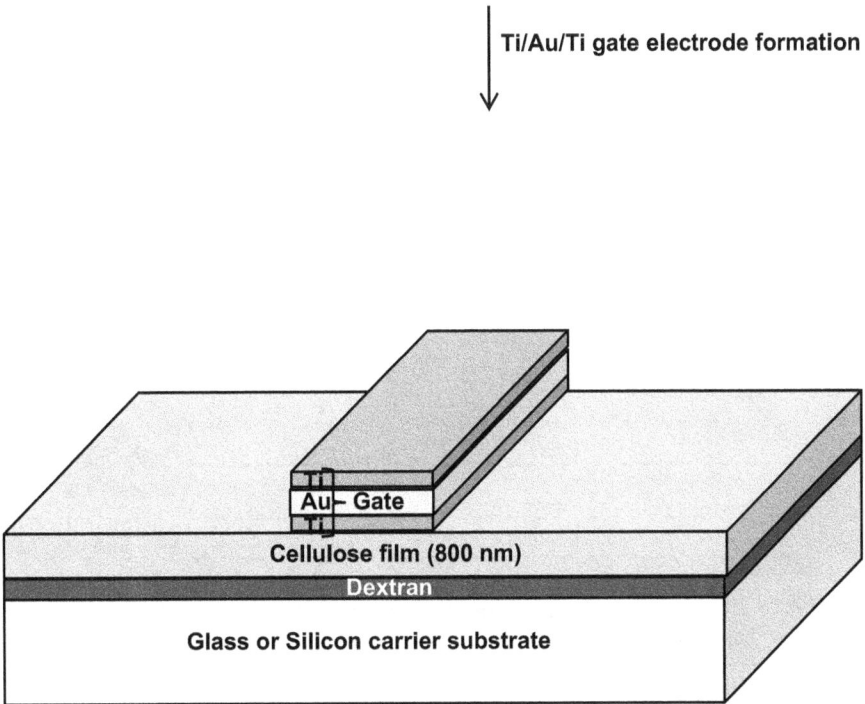

Figure 6.4. (Continued.)

Al₂O₃ gate dielectric by ALD

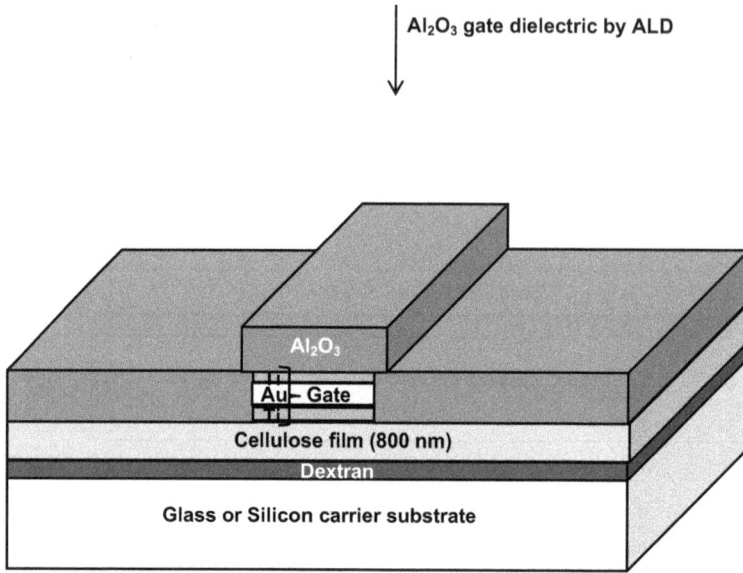

Modifying the surface of Al₂O₃ with
SAM of butylphosphonic acid

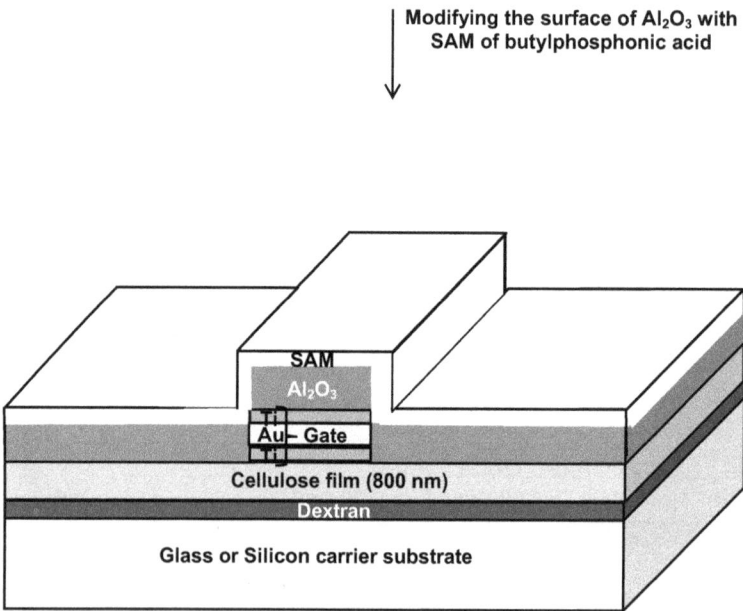

Figure 6.4. (Continued.)

Spin coating decomposable polymer

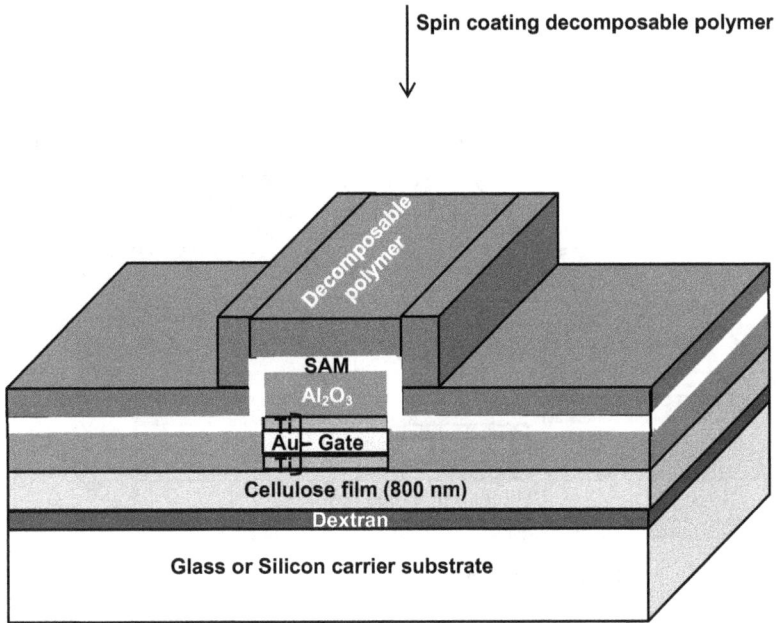

SAM
Al₂O₃
Au ─ Gate
Cellulose film (800 nm)
Dextran
Glass or Silicon carrier substrate

Depositing PTFE over
decomposable polymer

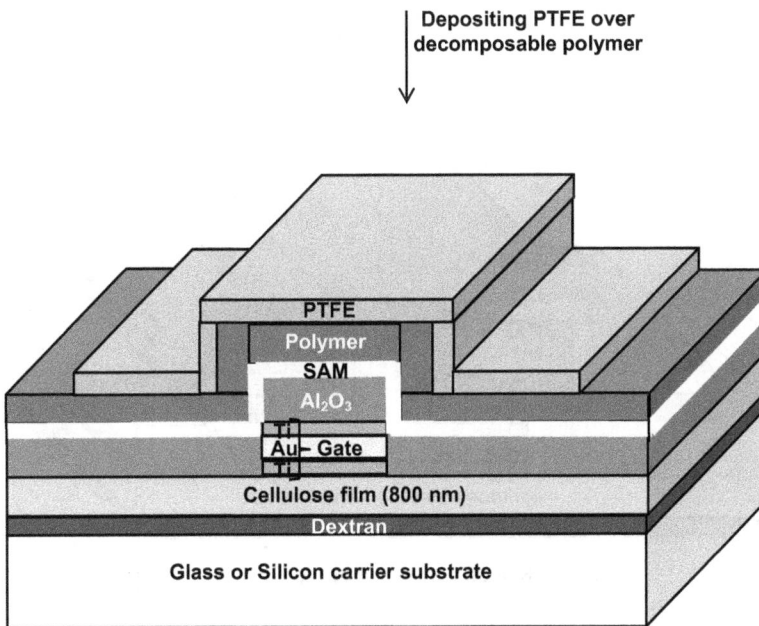

PTFE
Polymer
SAM
Al₂O₃
Au ─ Gate
Cellulose film (800 nm)
Dextran
Glass or Silicon carrier substrate

Figure 6.4. (Continued.)

Depositing copper over PTFE

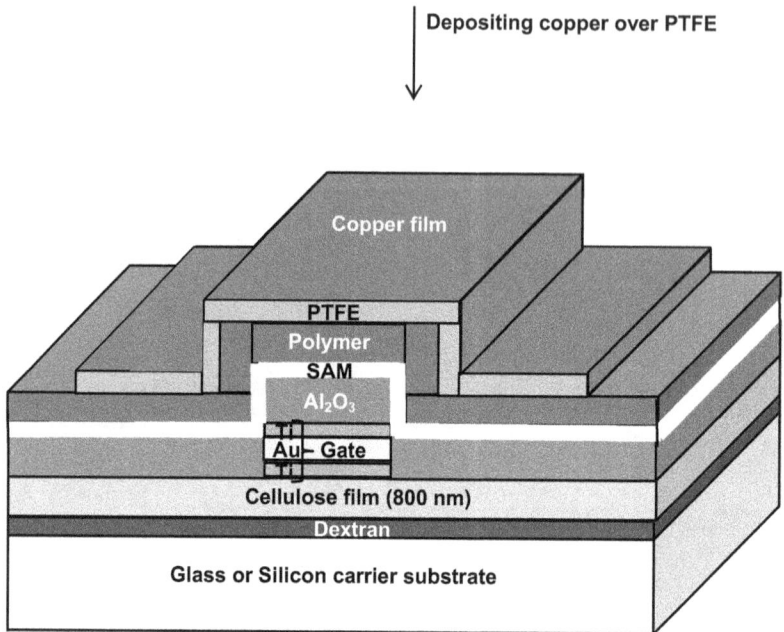

Copper film

PTFE
Polymer
SAM
Al$_2$O$_3$
Au — Gate
Cellulose film (800 nm)
Dextran
Glass or Silicon carrier substrate

Dry etching the polymer in oxygen
plasma with (copper+PTFE) as mask

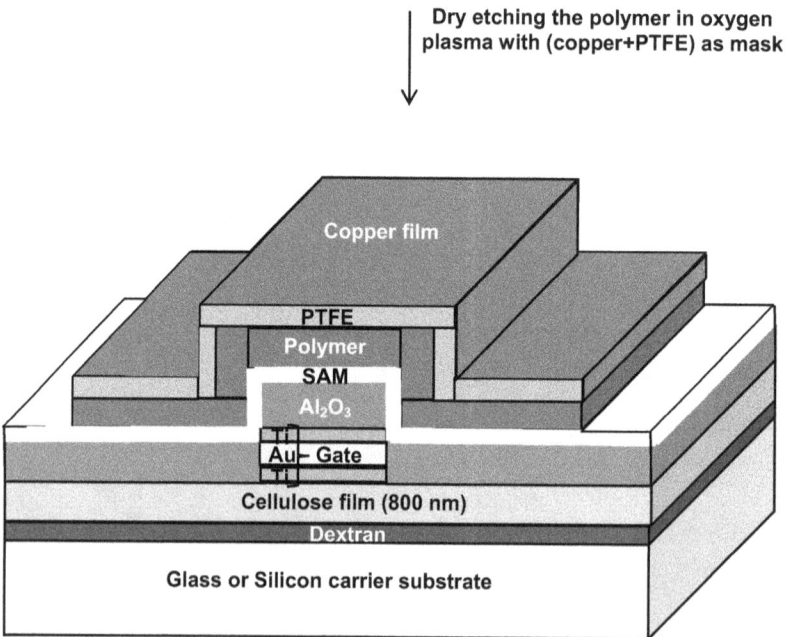

Copper film

PTFE
Polymer
SAM
Al$_2$O$_3$
Au — Gate
Cellulose film (800 nm)
Dextran
Glass or Silicon carrier substrate

Figure 6.4. (Continued.)

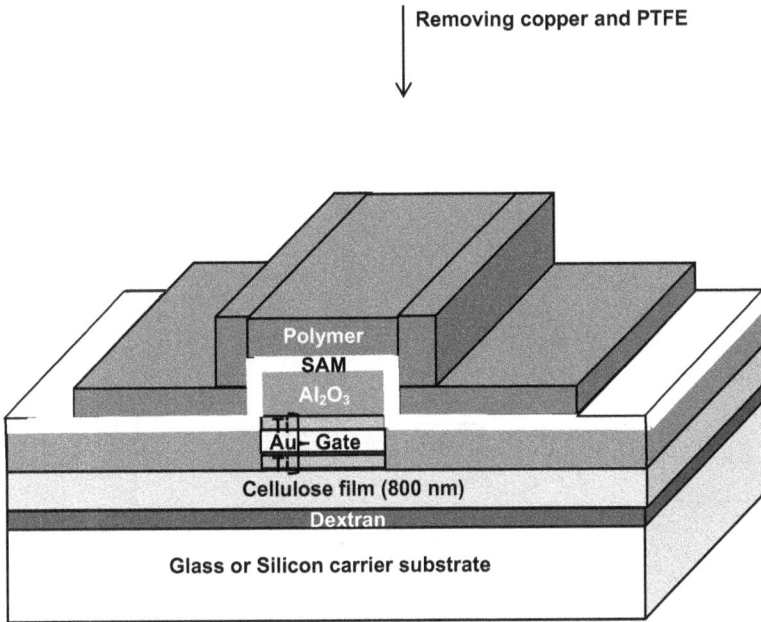

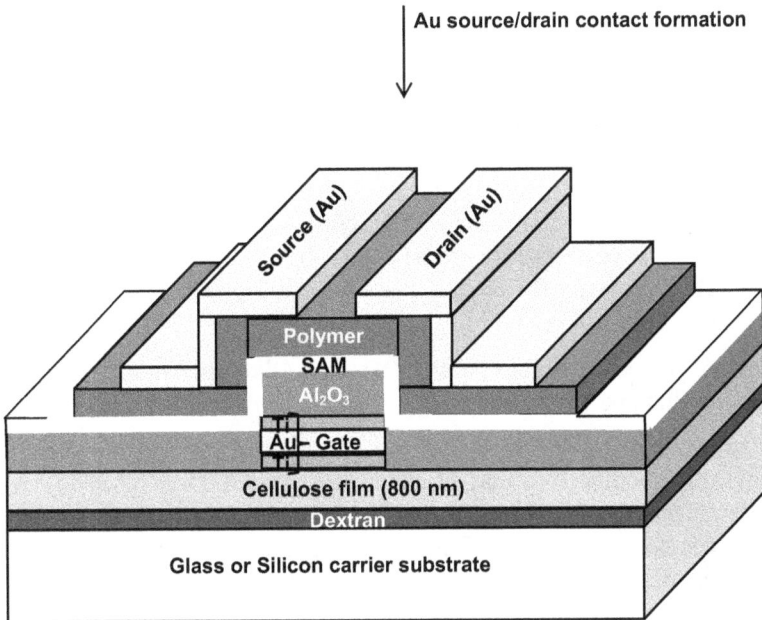

Figure 6.4. (Continued.)

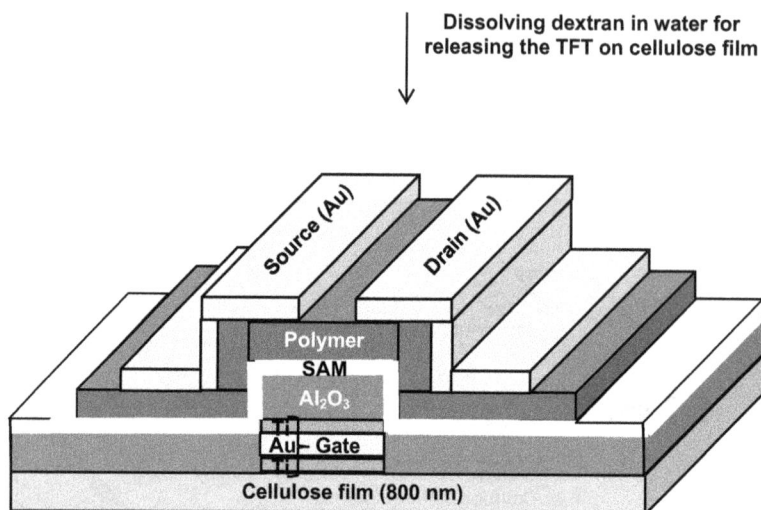

Figure 6.4. (Continued.)

dissolution of TMSC, the solution is heated to 80 °C and hexamethyldisilazane (HMDS), linear formula $(CH_3)_3SiNHSi(CH_3)_3$, molecular weight 161.39 is added dropwise in nitrogen atmosphere. After stirring at 80 °C for 12 h, the mixture is cooled with methanol (CH_3OH) addition to precipitate out TMSC. The crystallized TMSC is filtered, washed in methanol and dried in vacuum.

For making the ultrathin substrate film, a 5% solution of dextran, $C_{18}H_{32}O_{16}$ in DI water is deposited by spin coating on a carrier chip at 2000 RPM for 1 min and baked at 150 °C for 10 min. A 70 mg mL^{-1} solution of TMSC in chlorobenzene, C_6H_5Cl, is spin coated over the dextran film at 2000 RPM for 1 min and baked at 100 °C for 10 min. Hydrolyzation of the TMSC film is done in 95% acetic acid, CH_3COOH vapor for 2 h. The hydrolyzation is followed by annealing at 150 °C for 10 min. TMSC coating and hydrolyzation steps are repeated until a 800 nm film thickness is achieved. The film is mechanically robust to carry out device processing steps after which it is peeled off from the carrier by dissolving dextran in water. The ultrathin cellulose carrying the device floats on water and is picked up.

6.5.2 Synthesis of decomposable PDPP–PD {poly(diketopyrrolopyrrole–phenylenediamine)} polymer

Diketopyrrolopyrrole (DPP), $C_6H_2N_2O_2$ is synthesized as follows (Lei *et al* 2017): Succinic acid, molecular formula: $HOOCCH_2CH_2COOH$ or $C_4H_6O_4$, molecular weight: 118.088 g mol^{-1} is reacted with 2-thiophenecarbonitrile, empirical formula C_5H_3NS, molecular weight 109.15 in tert-amyl alcohol, $CH_3CH_2C(CH_3)_2OH$ or $C_5H_{12}O$. The solubility is increased by attaching branched alkyl chains. The chain is synthesized from octyldodecanol, linear formula $CH_3(CH_2)_9CH[(CH_2)_7CH_3]CH_2OH$, molecular weight 298.55. Thus DPP is obtained. From DPP, DPP–CHO is formed by attaching two aldehyde groups to the DPP monomer. Using

DPP–CHO, the decomposable polymer is synthesized by a condensation reaction between DPP–CHO and p-phenylenediamine, linear formula $C_6H_4(NH_2)_2$ or $C_6H_8N_2$, molecular weight 108.14 under catalysis with p-toluenesulfonic acid (PTSA), $CH_3C_6H_4SO_3H$, or $C_7H_8O_3S$, molecular weight $172.2 \, g \, mol^{-1}$. The addition of a drying agent $CaCl_2$ gives a decomposable polymer of higher molecular weight; its molar mass is 39.6 kDa. The polymer decomposes above 400 °C and is stable under ambient exposure. Its HOMO level is at −5.11 eV and LUMO level at −3.54 eV.

6.5.3 Fabrication of logic circuits

Lei *et al* (2017) proceed further to fabricate PDPP–PD polymer TFT and logic circuits using PDPP–PD polymer TFT as given below:

    (i) Formation of patterned gate electrodes: A triple-layer Ti (2.5 nm)/Au (35 nm)/Ti (2.5 nm) metallization scheme is followed in which the bottom 2.5 nm thick titanium layer is an adhesion promoter. The middle 35 nm gold layer is the contact metal and the uppermost 2.5 nm thick titanium layer has the role of providing nucleation sites for the forthcoming ALD step. Shadow masks are used for defining the electrodes during thermal evaporation of layers.

    (ii) Al_2O_3 gate dielectric layer deposition: The aluminum oxide gate dielectric is formed by atomic layer deposition using the top titanium film of the three-layer metallization scheme as nucleation sites. The ALD process is carried out at 150 °C. The Al_2O_3 film is 25 nm thick.

    (iii) Creation of vertical interconnect holes and interconnects: The vertical interconnect access holes are defined photolithographically. Etching is done in aluminum etchant. The interconnects are formed by thermal evaporation of titanium (2 nm) and gold (40 nm) through a shadow mask. They do not include source/drain electrodes.

    (iv) Surface modification of Al_2O_3 gate dielectric: A solution of butylphosphonic acid, $C_4H_{11}O_3P$ in trichloroethylene, $ClCH=CCl_2$ or C_2HCl_3 is prepared. The solution has a concentration of 2 mM. A self-assembled monolayer (SAM) of butylphosphonic acid is formed over the Al_2O_3 film by spin coating the prepared solution at 3000 RPM, 30 s. The SAM layer is annealed at 100 °C for 10 min.

    (v) Decomposable polymer deposition: A solution of decomposable polymer is made in trichloroethylene, $ClCH=CCl_2$. The concentration of the solution is $5 \, mg \, mL^{-1}$. This solution is coated on the surface-modified substrate obtained from the preceding step by spinning at 1000 RPM for 1 min. The polymer film is thermally annealed at 150 °C in nitrogen ambience.

    (vi) Patterning of the polymer film: On the decomposable polymer, a film of poly[4,5-difluoro-2,2-bis(trifluoromethyl)-1,3-dioxole-*co*-tetrafluoroethylene] (PTFE AF 2400) is deposited by spin coating. The PTFE film has a thickness of 400 nm. Additionally, 80 nm thick copper film is deposited by thermal evaporation through a shadow mask. Dry etching of PDPP–PD layer is done in oxygen plasma at 150 W for 2 min from regions uncovered by copper. After the dry etching, the remaining copper is removed in sodium persulfate, $Na_2S_2O_8$ etchant and the leftover PTFE AF 2400 by immersing in methoxyperfluorobutane, CH_3OCF_2R, R = $-CF_2CF_2CF_3$ or $-CF(CF_3)_2$ for 5 min.

(vii) Source/drain contact formation: Gold is deposited by thermal evaporation through a shadow mask at the locations of source and drain contacts. The gold film is 40 nm thick.

(viii) Dextran dissolution and device release: The dextran film is mechanically abraded from the edge of the chip. After immersing in water, the dextran starts dissolving from the edges towards the center of the chip. The device chip thus released from the carrier starts floating on the surface of water. It is transferred to a chosen target substrate, e.g. PI or PDMS.

(ix) TFT and logic circuit characteristics: The channel width/channel length (W/L) of the TFT is 1000 µm/50 µm = 20. The average value of threshold voltage of the TFT fabricated on cellulose substrate is 4.67 V. The carrier mobility is 0.21 $cm^2\,V^{-1}\,s^{-1}$.

For flexibility testing, the TFTs are transferred to a 25 µm thick PI substrate. The device is bent to a radius of 2 mm and its transfer characteristic after bending is compared with that before bending. The change in characteristic is nominal ~5%.

The pseudo-D inverter circuit shows output voltages close to the supply voltages, viz, 4 V and 0 V. The inverter switches sharply at 1.9 V, which is nearly one-half of 4 V. The noise margin is 1.2 V (60% of 2 V), which is at par with the performance of CNT and organic flexible logic circuits. Rail-to-rail voltage swings between 0 to 4 V are displayed by NOR and NAND gates (Lei *et al* 2017).

6.6 FBT–TH$_4$(1, 4) TFT on PEN substrate

Sun *et al* (2015) fabricated TFT using a low bandgap organic semiconductor FBT–TH$_4$(1, 4) as the active layer (figure 6.5). The gate insulator is AlO_x–Nd. Besides being economical, AlO_x–Nd has several favorable properties. A few of those advantages are: its high dielectric constant in the range 8–12, environmental friendliness and room-temperature deposition process. These advantages lend strong support to its selection. Among its drawbacks are its surface roughness and coverage of its surface with hydroxyl groups. Therefore, the as-deposited AlO_x:Nd film leads to TFT with reduced performance. By modifying its surface with SAM of octadecyl-phosphonic acid (ODPA), empirical formula : $C_{18}H_{39}O_3P$, molecular weight 334.47, this inadequacy is removed.

6.6.1 Substrate preparation

A 50 µm thick PEN film is fixed on a glass carrier so that the process steps on a rigid support can be carried out easily.

Fixing PEN film on glass carrier

PEN Film

Glass carrier

Three layer SiN$_x$/photoresist/SiN$_x$
barrier by PECVD/spin coating/PECVD

SiN$_x$

Photoresist

SiN$_x$

PEN Film

Glass carrier

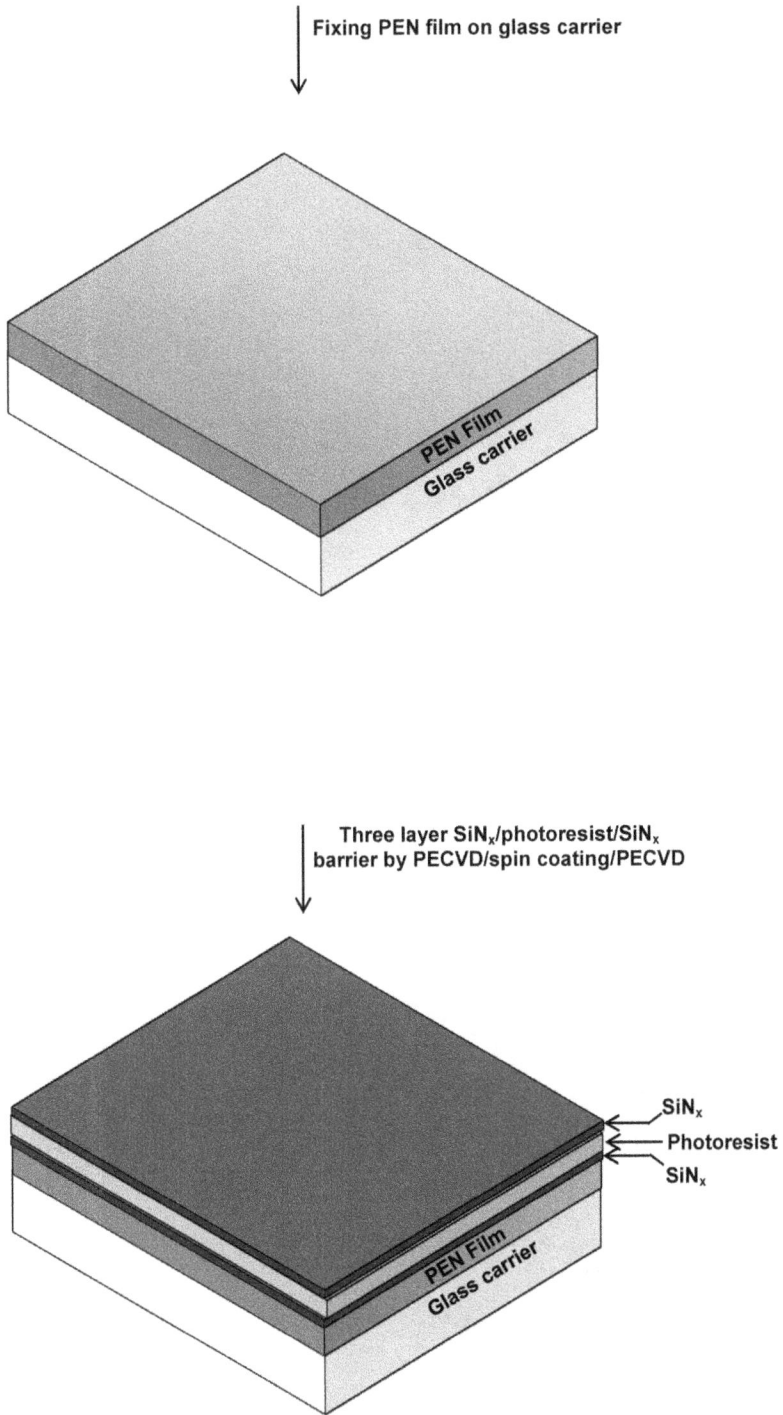

Figure 6.5. TFT with organic semiconductor FBT–TH$_4$(1, 4).

Al:Nd film sputtering

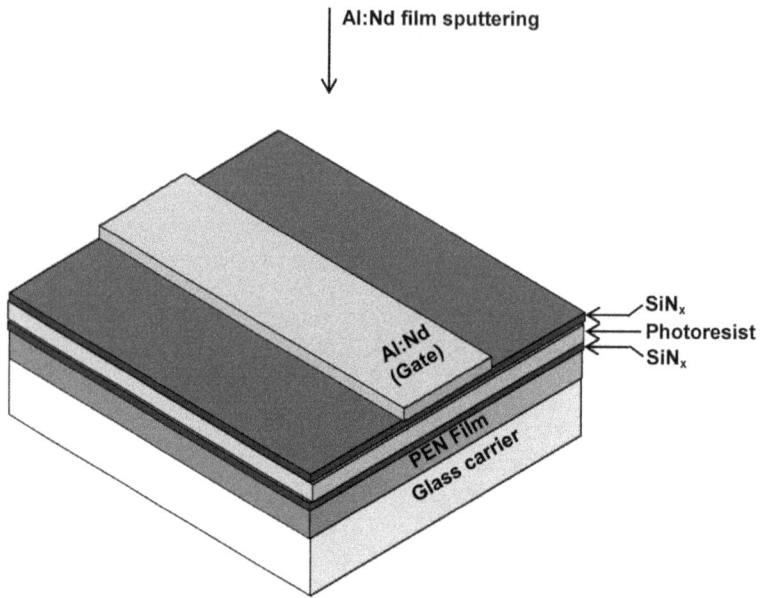

Anodization of Al:Nd to form AlO_x:Nd
on the surface of Al:Nd

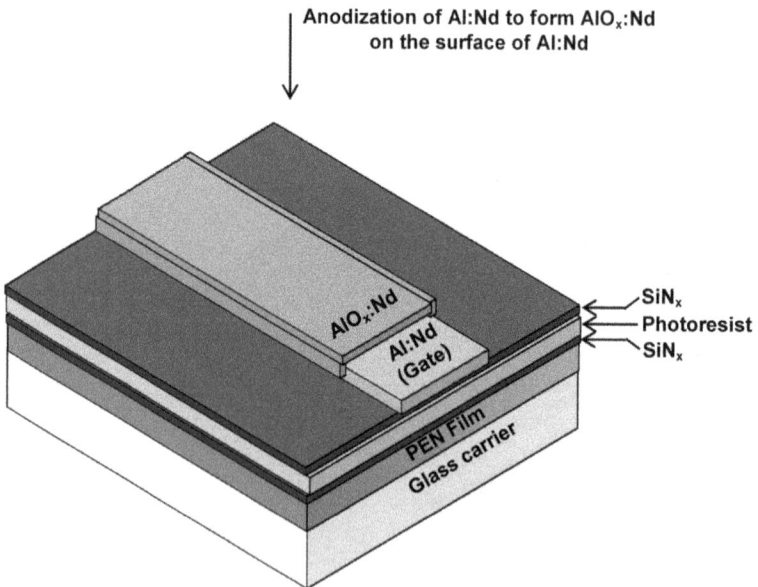

Figure 6.5. (Continued.)

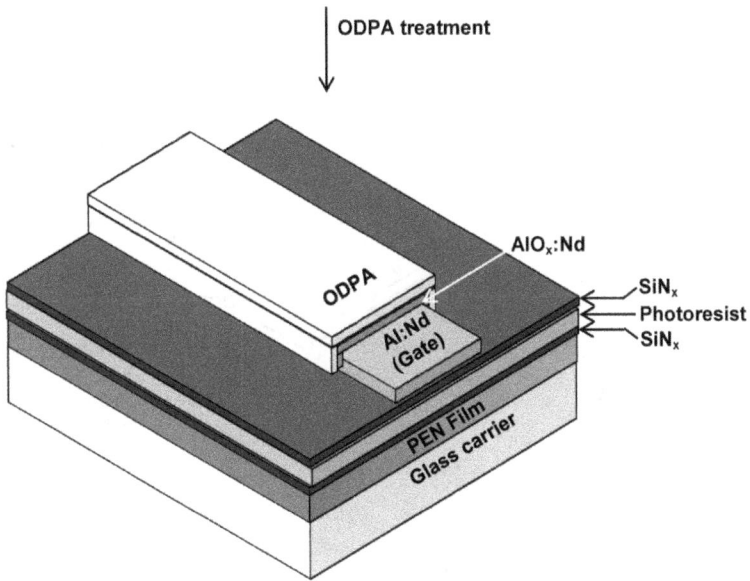

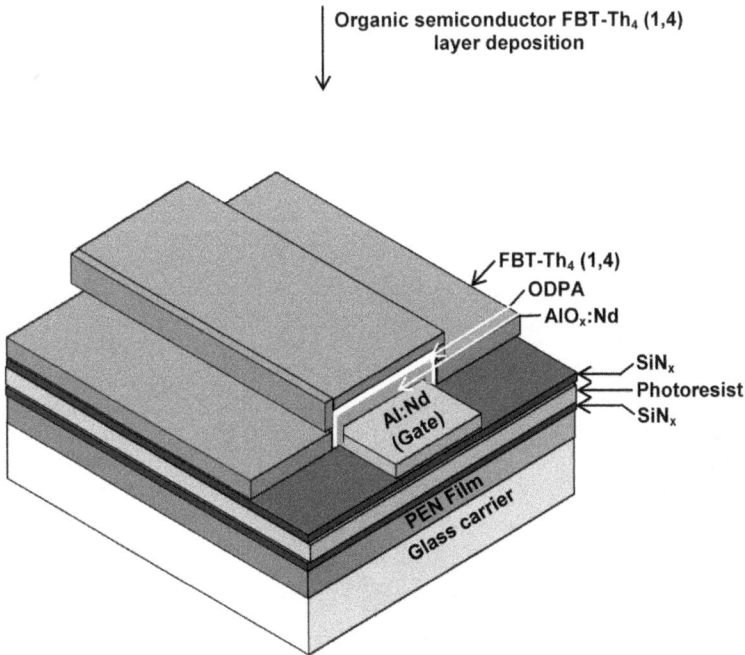

Figure 6.5. (Continued.)

Au source/drain electrode deposition
through shadow mask

Gold
(Drain)

ODPA
AlO$_x$:Nd
FBT-Th$_4$ (1,4)
SiN$_x$
Photoresist
SiN$_x$

Gold
(Source)

Al:Nd
(Gate)

PEN Film
Glass carrier

TFT removal from glass carrier

Gold
(Drain)

ODPA
AlO$_x$:Nd
FBT-Th$_4$ (1,4)
SiN$_x$
Photoresist
SiN$_x$

Gold
(Source)

Al:Nd
(Gate)

PEN Film

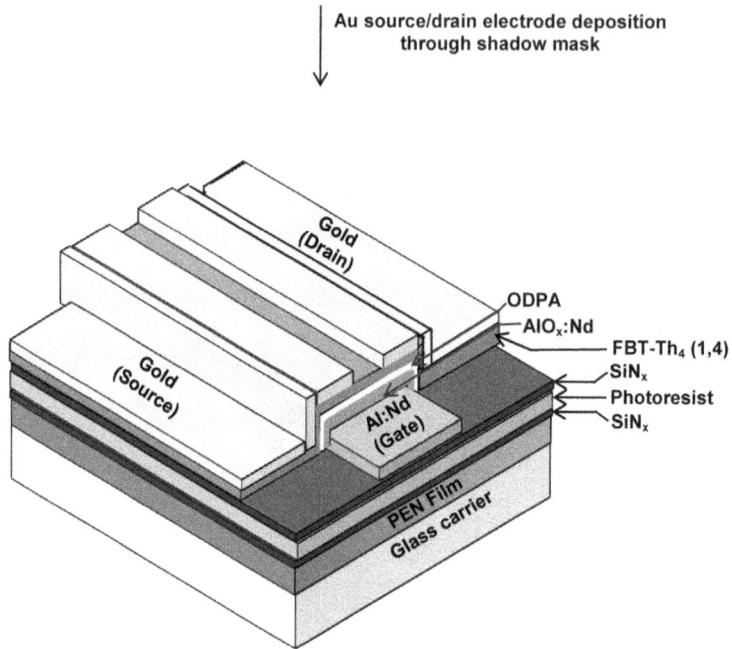

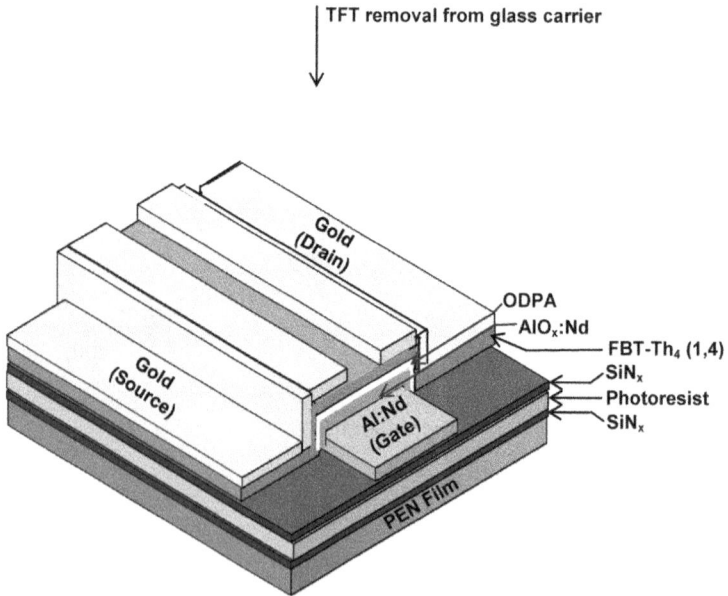

Figure 6.5. (Continued.)

6.6.2 Gas and moisture barrier layer deposition

This consists of three layers: SiN_x/photoresist/SiN_x. The SiN_x film is formed by PECVD at 150 °C; its thickness is 250 nm. The photoresist film is spin coated. It is 2000 nm thick.

6.6.3 Al:Nd film deposition

As the first step to gate dielectric formation, DC sputtering is used to deposit an Al: Nd (3 wt%) film. The thickness of this film is 150 nm. The pattern is defined by photolithography. The incorporation of Nd has a two-fold effect: first, it suppresses the formation of hillocks and reduces surface roughness; second, Nd segregates to form clusters in the film. These clusters release compressive or tensile stresses during bending.

6.6.4 Anodic oxidation of the Al:Nd film

The Al:Nd film is anodized for 1.5 h. By anodization, a 200 nm thick AlO_x:Nd film is formed on the surface of the Al:Nd film. The AlO_x:Nd film serves as the gate insulator while the Al:Nd layer below it acts as the gate electrode.

6.6.5 SAM deposition

ODPA solution of strength 1 mM is made in 2-propanol, $(CH_3)_2CHOH$. The substrate with all the layers formed in the preceding steps is subjected to plasma treatment and then immersed in an ODPA solution for 8 h at room temperature. After 8 h, it is taken out, rinsed thoroughly with 2-propanol and dried by blowing nitrogen.

6.6.6 Organic semiconductor layer deposition

The semiconductor is a low bandgap donor–acceptor (D–A) conjugated polymer. Its A-unit is 5,6-difluoro-2,1,3-benzothiadiazole (FBT), $C_6H_2F_2N_2S$ with molecular weight 172.16. The D-unit is quaterthiophene (Th_4), empirical formula: $C_{16}H_{10}S_4$, molecular weight: 330.51 with alkyl chains fixed on the two terminal thiophene rings to make it soluble.

By dissolution of FBT–Th_4(1,4) in dichlorobenzene, $C_6H_4Cl_2$, a solution of concentration $0.8\ mg\ mL^{-1}$ is made. This solution is deposited by spin coating at 2000 RPM on AlO_x:Nd surface modified with ODPA. The FBT–Th_4(1,4) film is 25 nm thick.

6.6.7 Source/drain electrodes

Using a shadow mask, a gold film is deposited on the FBT–Th_4(1,4) polymer by thermal evaporation. The channel width/length ratio is 500 μm/70 μm = 7.14. The gold film is 100 nm thick.

6.6.8 TFT characteristics and bending effects

The P-channel TFT has a threshold voltage of −5.4 V. The current on–off ratio is 10^5. The mean field-effect mobility is $1.10\,\mathrm{cm^2\,V^{-1}\,s^{-1}}$ while the highest field-effect mobility is $2.88\,\mathrm{cm^2\,V^{-1}\,s^{-1}}$. Without ODPA treatment, the mobility is much lower = $0.24\,\mathrm{cm^2\,V^{-1}\,s^{-1}}$, reducing by a factor of 12, which reveals the significance of this treatment in controlling mobility. It is, therefore, a crucial step to realize high-performance TFTs. The improvement in mobility is caused by the decrease in surface roughness of the gate insulator. In an ODPA-untreated TFT, this roughness introduces disorder in the morphology of the semiconductor film hindering the carrier transport in the interfacial region between the semiconductor film and the gate insulator.

The transfer curves deviate by small amounts at bending curvatures >30 mm. But the device behavior degrades appreciably on bending in the curvature range of 5–20 mm through damage incurred at the semiconductor/insulator interface (Sun *et al* 2015).

6.7 Discussion and conclusions

The performance of polymer TFTs has continuously improved towards achievement of better mobilities. The polymer TFTs described here have mobilities of 0.02–$0.025\,\mathrm{cm^2\,V^{-1}\,s^{-1}}$ (Park *et al* 2002), $3 \times 10^{-3}\,\mathrm{cm^2\,V^{-1}\,s^{-1}}$ (Zielke *et al* 2005), $0.64 \times 10^{-3}\,\mathrm{cm^2\,V^{-1}\,s^{-1}}$ (Xu *et al* 2016), $0.21\,\mathrm{cm^2\,V^{-1}\,s^{-1}}$ (Lei *et al* 2017), $1.10\,\mathrm{cm^2\,V^{-1}\,s^{-1}}$ (mean) and $2.88\,\mathrm{cm^2\,V^{-1}\,s^{-1}}$ (highest of the compared mobilities) (Sun *et al* 2015).

Review exercises

6.1 What is the greatest convenience of using polymeric materials in TNT fabrication? What is the biggest drawback of these materials?

6.2 How do defects in polymers affect TFT parameters?

6.3 With reference to P3HT TFT on polycarbonate, name the: (i) gate dielectric material, (ii) gate electrode metal, and (iii) source/drain metallization.

6.4 Why is the polycarbonate substrate subjected to an annealing treatment before commencing TFT fabrication? What is the intent of sputtering a silicon dioxide layer on the PC substrate after the annealing treatment?

6.5 How is the stamp made for contact printing P3HT polymer? How is the P3HT polymer applied on the gate dielectric using this stamp?

6.6 Name the four parameters determining the thickness of the P3HT polymer coating applied by contact printing.

6.7 How is the electrical conductivity of the PEDOT:PSS ink for conductor printing increased? What is the electrical conductivity of the printed ink after thermal curing?

6.8 What geometry is used for the source/drain electrodes of the PTAA TFT? By what method are these electrodes formed? What is the gate insulator made of?

6.9 What are the salient features of the fabrication technology of PDQT TFT array?

6.10 Why does the PET surface need to be covered with a polyvinyl alcohol buffer layer before inkjet printing of source/drain electrodes with PEDOT: PSS ink?

6.11 What is the contact angle of the PET surface before PVA coating? What is its contact angle after PVA coating?

6.12 How is the PDQT semiconductor film deposited? What is the gate insulator material used? How is it deposited?

6.13 Explain the notion of ultrathin, ultra-lightweight green electronics.

6.14 How is the biodegradable cellulose acetate substrate made? Explain how the substrate film is removed from the carrier after processing.

6.15 How is the decomposable polymer semiconductor synthesized? How is it patterned?

6.16 How does the TFT device with disintegrable polymer semiconductor on biodegradable substrate respond to bending? How does the pseudo-inverter fabricated with this TFT work?

6.17 What are the advantages of AlO_x–Nd as a gate insulator? What is its shortcoming? How is this shortcoming overcome?

6.18 How is the AlO_x–Nd insulating film formed in the FBT–TH$_4$(1, 4) TFT? How is the SAM formed over this film?

6.19 Explain the significance of ODPA SAM in the gate dielectric of FBT–TH$_4$(1, 4) TFT. What happens when this SAM is not deposited?

6.20 What is the effect of bending on the FBT–TH$_4$(1, 4) TFT?

References

Lei T, Guan M, Liu J, Lin H-C, Pfattner R, Shaw L, McGuire A F, Huang T-C, Shao L, Cheng K-T, Tok J B-H and Bao Z 2017 Biocompatible and totally disintegrable semiconducting polymer for ultrathin and ultralightweight transient electronics *Proc. Natl. Acad. Sci. U.S.A.* **114** 5107–12

Park S K, Kim Y H, Han J I, moon D G and Kim W K 2002 High-performance polymer TFTs printed on a plastic substrate *IEEE Trans. Electron Devices* **49** 2008–15

Sun S, Lan L, Xiao P, Chen Z, Lin Z, Li Y, Xu H, Xu M, Chen J, Peng J and Cao Y 2015 High mobility flexible polymer thin-film transistors with an octadecyl-phosphonic acid treated electrochemically oxidized alumina gate insulator *J. Mater. Chem. C* **3** 7062–6

Xu W, Hu Z, Liu H, Lan L, Peng J, Wang J and Cao Y 2016 Flexible all-organic, all-solution processed thin film transistor array with ultrashort channel *Sci. Rep.* **6** 29055

Zielke D, Hübler A C, Hahn U, Brandt N, Bartzsch M, Fügmann U, Fischer T, Veres J and Ogier S 2005 Polymer-based organic field-effect transistor using offset printed source/drain structures *Appl. Phys. Lett.* **87** 123508-1–3

IOP Publishing

Flexible Electronics, Volume 2
Thin-film transistors
Vinod Kumar Khanna

Chapter 7

Organic single-crystal TFT

Rubrene TFT fabrication is done by growing 150 nm thick crystals of size 1 cm ×
1 cm by the physical vapor transport method. The TFT is fabricated by adhering the
crystal to the gate/source/drain electrode pattern by PVP, which also serves as the
gate dielectric. The TFTs are not affected by bending up to a radius of 9.4 mm with
strain of 0.74% (Briseno *et al* 2006a). Anthracene crystal TFT is fabricated by
choosing the higher mobility β-phase of the crystal. The crystal is grown by placing
the seed crystals over the source/drain gold electrodes. The gold electrodes are
treated with thiophenol to lower the contact resistance between the electrodes and
the semiconductor crystal. Bending experiments carried out up to 1 mm radius
reveal that the TFT can sustain the bending with initial mobility reduction up to 20
bending times and constant mobility up to 300 bending times (Cai *et al* 2014). An
important finding that OTS domains act as nucleation centers for growth of organic
semiconductor crystals enables the growth of crystals at precise predefined locations.
These locations are marked by applying OTS with a PDMS stamp over the pattern
of gold source/drain electrodes. The method obviates the difficulty of manually
attaching prefabricated crystals over the source/drain electrodes. The method offers
versatility and is applied to fabrication of pentacene and rubrene TFTs (Briseno *et al*
2006b). A complementary inverter circuit is made with P-type CuPc TFT and
N-type F_{16}CuPc TFT. These TFTs are first made on an OTS-modified silicon
substrate. The CuPc and F_{16}CuPc single crystals are synthesized in a quartz tube
furnace and the polystyrene gate dielectric is spin-coated. The fabricated TFTs are
transferred to a gold wire to assemble the inverter for wearable electronics
applications (Zheng *et al* 2016).

7.1 Introduction

In analogy with silicon single TFTs, organic single-crystal TFTs too are high-
performance promising devices. Single-crystal materials have fewer grain boundaries.
Their long-range ordered structure helps to gain access to the intrinsic properties of

the material. Physical insights into their carrier transfer mechanisms are provided. A better understanding of the limits of performance capabilities of these materials is acquired. On the other hand, impure materials with structural defects have besmirched properties. Particularly, the presence of any structural imperfections and impurity content in a material greatly limits its charge carrier properties.

7.2 Rubrene single-crystal TFT

Briseno *et al* (2006a) fabricated organic single-crystal TFT using the rubrene $C_{42}H_{28}$ crystal. The channel length is 45 μm and the channel width is 978 μm giving a *W/L* ratio = 978 μm/45 μm = 21.73.

7.2.1 Growth of rubrene single crystal

The growth is done by physical vapor transport (Kloc *et al* 1997, Laudise *et al* 1998). It involves three stages: source vaporization, transport of source vapors and crystalline growth of source material; the source is made of the material whose crystals are to be grown. For experimental convenience, a horizontal tube is used for vapor transport and crystal growth (figure 7.1). The reactor tube contains two tubes: a source tube and a crystal growth tube. This arrangement makes insertion of the starting source material easier. Moreover, after crystal growth, the crystals formed as well as the left-over source material are easy to take out. The source tube is nearer to the inlet of the reactor tube than the crystal growth tube, which is placed farther away. As the crystals grow, they are deposited along the crystal growth tube. A resistance wire-wound furnace provides the heating.

Commercial rubrene material is used. The crystal is grown in an inert environment of argon gas. The gas from the inlet flows over the rubrene source to the crystal growth tube and exits through the outlet. Flat crystal flakes, both small in thickness (150 nm) and large in size (1 cm × 1 cm) are realized by rapid growth in 20 min to 1 h. The rapid growth process is chosen keeping in consideration that the slow

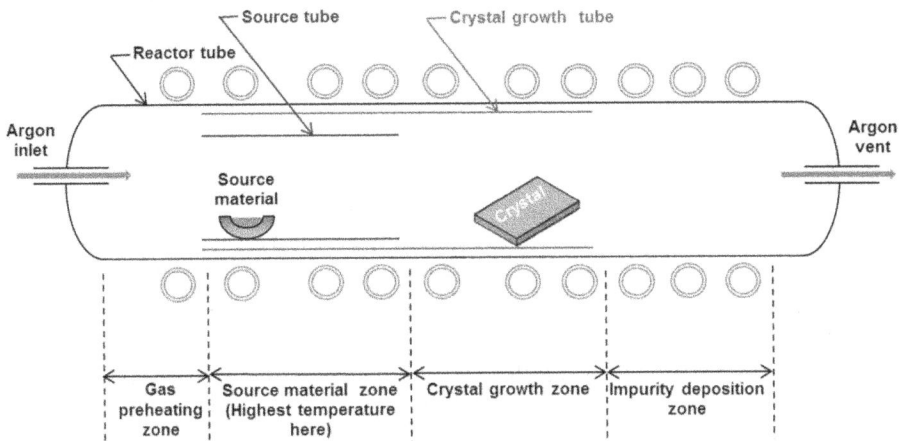

Figure 7.1. Apparatus for crystal growth by physical vapor transport.

growth method leads to the production of large thick crystals. Rapid crystal growth is initiated by applying a temperature in the range 280 °C–300 °C for sublimation of rubrene. The vaporized rubrene is carried down the tube by the inert gas flowing through the tube. Here, resolidification of rubrene occurs in crystalline form along a temperature gradient. The method of setting up the thermal gradient varies with the equipment. A second resistive heating element is sometimes used. As soon as rubrene nucleation is noticed on the inner walls of the tubes, the furnace temperature is raised to 330 °C. The gas flow rate is kept at 100 mL min^{-1}.

7.2.2 Source/drain/gate electrodes

The device is fabricated in a bottom contact structure (figure 7.2). The electrodes are formed by evaporating chromium and gold through a shadow mask on a Kapton substrate of thickness 140 μm. The chromium film has a thickness of 5 nm. The gold film is 50 nm thick. Another substrate used is a transparent PET sheet coated with ITO. The electrodes are formed in the ITO film.

7.2.3 Adhering the rubrene crystal to the electrode pattern

The thin rubrene crystal is adhered to the electrode pattern with a poly-4-vinyl-phenol (PVP), $[CH_2CH(C_6H_4OH)]_n$ thin film. The PVP thin film serves the double role of: (i) glue for attachment of the rubrene crystal, and (ii) gate dielectric. The adherent dielectric solution is made using PVP (22 wt%) and poly(melamine-co-formaldehyde) methylated, $(C_4H_8N_6O)_n$ (8 wt%). PVP has a weight-average molecular weight of 20 000 while the same for poly(melamine-co-formaldehyde) methylated is 511. Cross-linking is done by curing the substrates at 100 °C for 10 min. This is followed by another curing at 200 °C for 10 min.

7.2.4 Electrical characteristics of the TFT

The switch-on voltage of the TFT is −2.1 V. The normalized subthreshold swing is 0.9 nF per decade per cm^2. The drain current is 280 μA at a gate voltage of −60 V. The current on–off ratio is 10^6. The carrier mobility is 4.6 cm^2 V^{-1} s^{-1}.

7.2.5 Tolerance to bending

During bending experiments, the TFTs are fastened to curved cylinders. The TFT substrate is kept in a bent position for 1 h for each bending radius. It is found that the performance of the device is not affected up to a bending radius of 9.4 mm, and strain = 0.74%. At a radius of 7.4 mm, the mobility decreases to 0.078 cm^2 V^{-1} s^{-1}. Nevertheless, the current on–off ratio is still 10^5. At 5.9 mm radius, strain = 1.18%, the mobility is only 0.0065 cm^2 V^{-1} s^{-1}; the on–off ratio is 10^3. However, when the strain of 1.18% is removed, the mobility recovers to 91.3% of its value in original flat condition. The on–off ratio too is restored to the value of 10^6, the value at the beginning. Recovery from such strenuous conditions is made possible by the ultra-thin rubrene crystal used (Briseno et al 2006a).

Taking Kapton or PET substrate

Kapton or PET substrate

Cr/Au gate electrode for PI or ITO
electrode for PET

Cr/Au or ITO gate electrode

Kapton or PET substrate

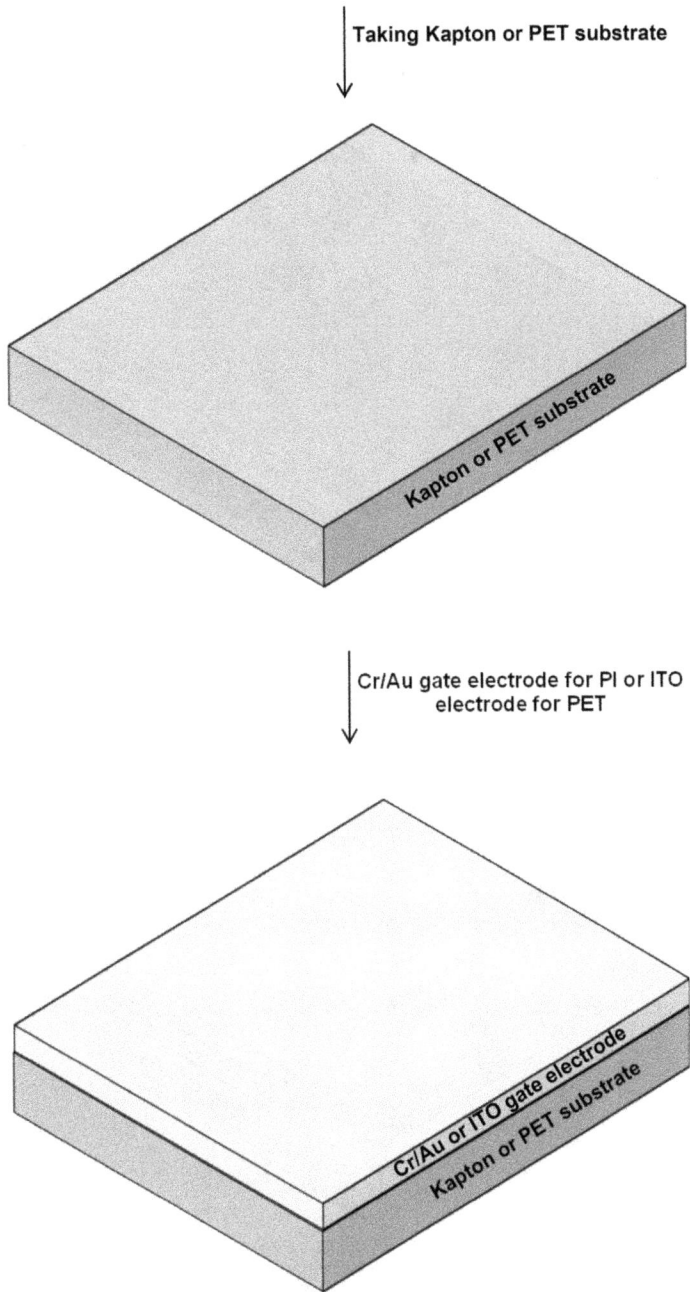

Figure 7.2. TFT using rubrene single crystal for semiconductor channel.

PVP gate dielectric deposition

PVP gate dielectric
Cr/Au or ITO gate electrode
Kapton or PET substrate

Cr/Au source/drain electrode
deposition

Cr/Au drain
electrode

Cr/Au source
electrode

PVP gate dielectric
Cr/Au or ITO gate electrode
Kapton or PET substrate

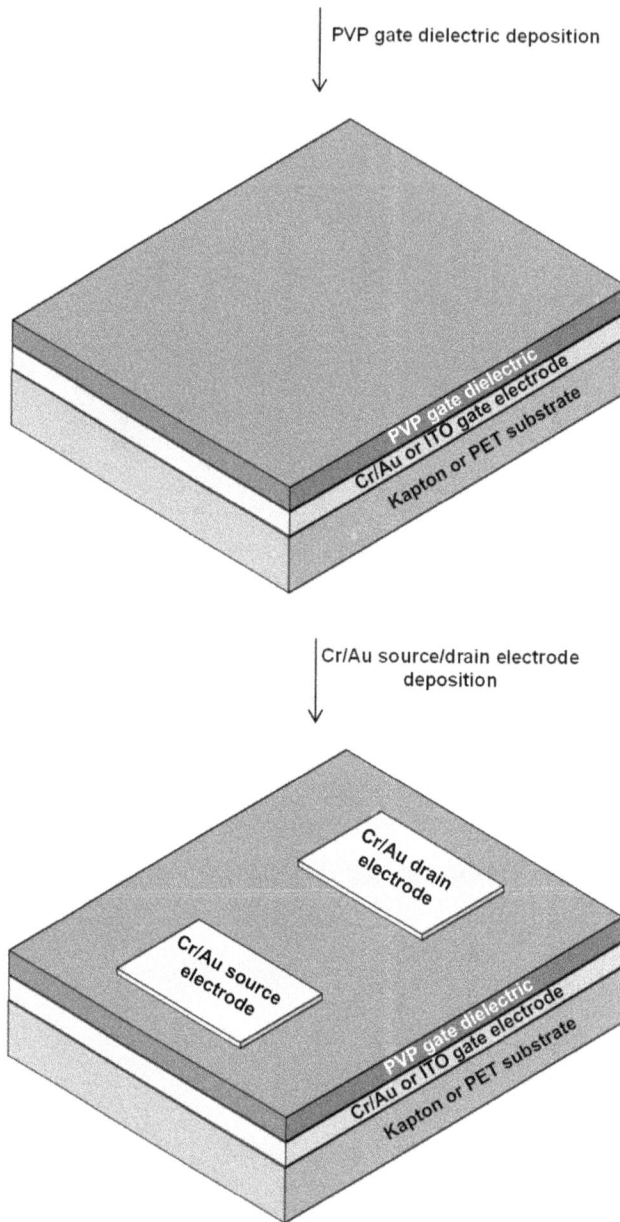

Figure 7.2. (Continued.)

Pasting the rubrene crystal to PVP

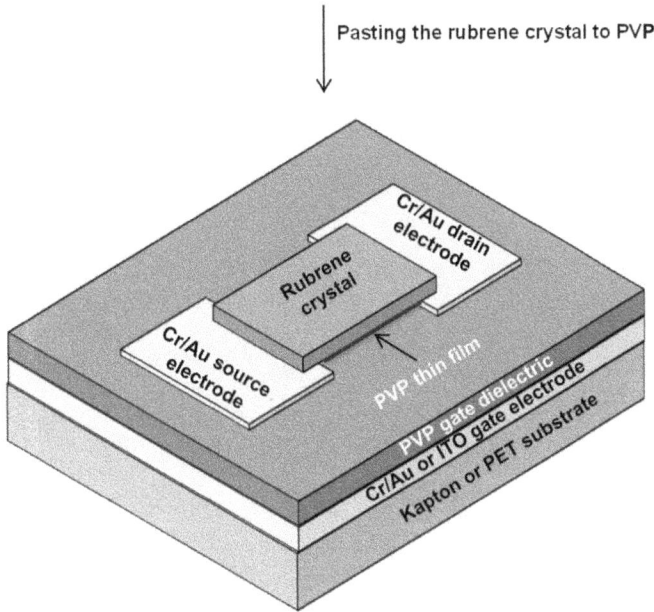

Figure 7.2. (Continued.)

7.3 BPEA single-crystal TFT

Cai *et al* (2014) fabricated TFT using a single-crystal ribbon of an organic semiconductor named 9,10-bis (phenyl ethynyl) anthracene; synonyms: BPEA, anthracene; empirical formula $C_{30}H_{18}$, molecular weight 378.46; an aromatic hydrocarbon with high mobility of holes and fluorescence properties. This bottom gate, bottom contact TFT uses PI as the gate dielectric. BPEA crystallizes in two phases: α-phase crystal and β-phase crystal. As the β-phase crystal shows poorer hole mobility than the α-phase crystal, the α-phase BPEA crystal is selected for TFT fabrication. A solution-based process is used for crystal growth. It is called seed-assisted crystal growth from solution. The crystal growth takes place under the guidance and regulation of a seed crystal in a securely closed ambient environment. A splendid long ribbon of single-crystal materials extending up to hundreds of microns can be formed by this method.

The process is developed on an ITO conductive glass substrate by spin coating PI precursor of polyamic acid, $(C_{12}H_{12}N_2O.C_{10}H_2O_6)_x$ and cross-linking at high temperature (300 °C for 2 h). The thickness of PI film is 800 nm. The process carried out on ITO glass is then repeated using a PI sheet as the substrate. The steps for PI substrate are given below (figure 7.3):

7.3.1 Formation of gate electrode

The substrate is a polyimide sheet which also serves as the gate dielectric. Ni or Al is deposited by thermal evaporation for gate contact.

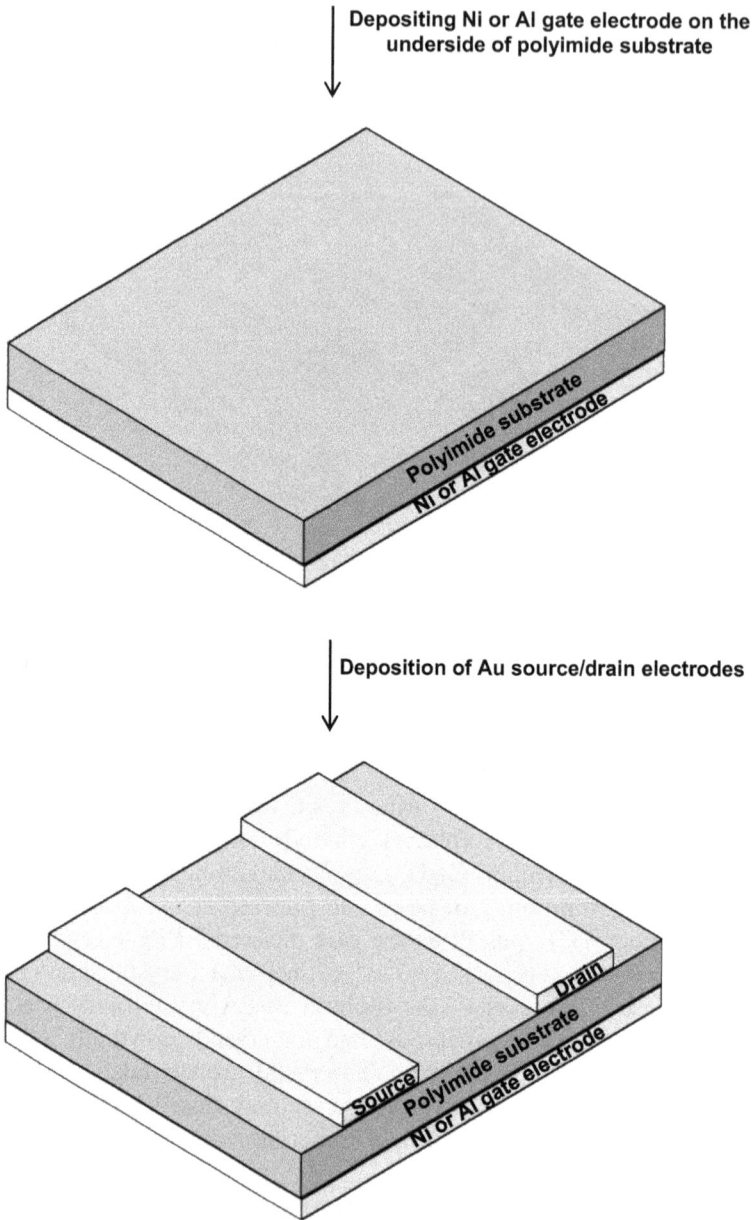

Depositing Ni or Al gate electrode on the
underside of polyimide substrate

Polyimide substrate

Ni or Al gate electrode

Deposition of Au source/drain electrodes

Drain

Source

Polyimide substrate

Ni or Al gate electrode

Figure 7.3. TFT with BPEA single-crystal ribbon as organic semiconductor.

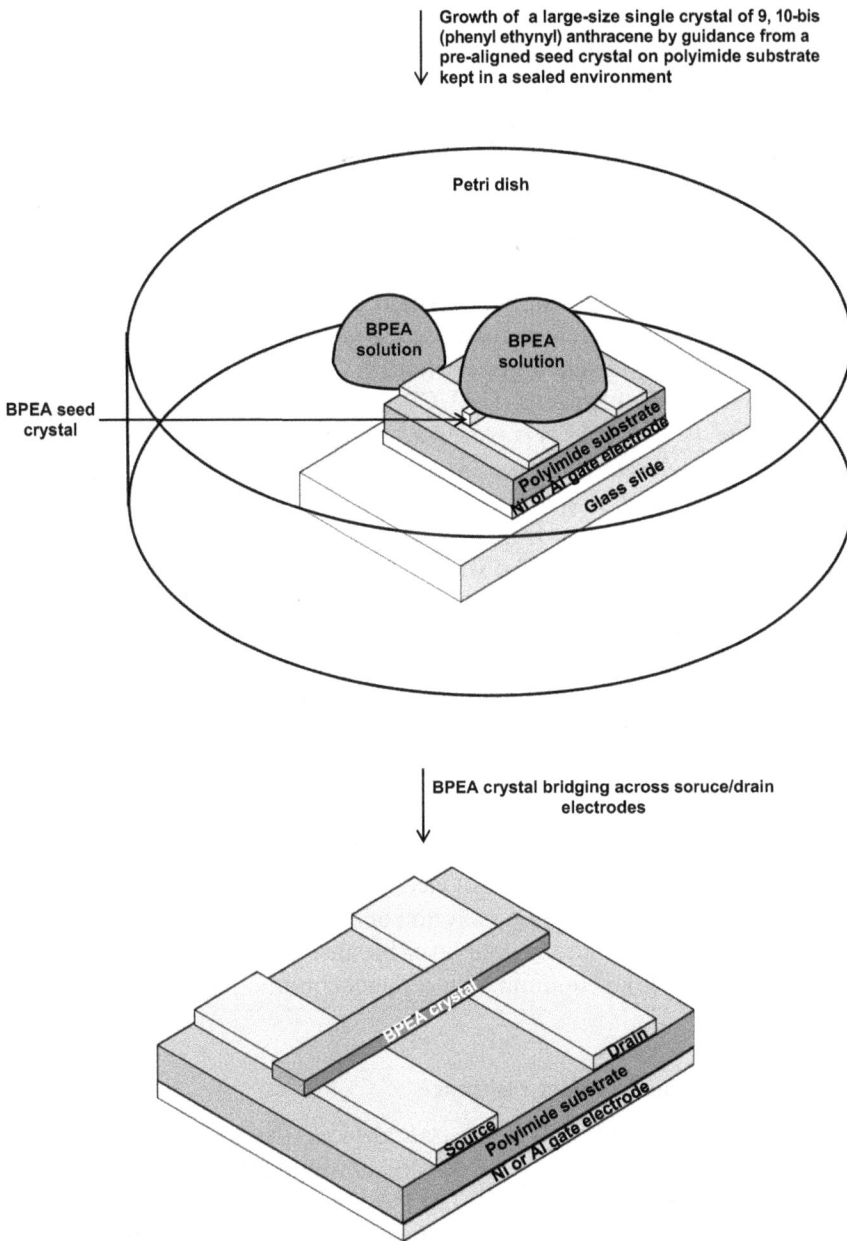

Growth of a large-size single crystal of 9, 10-bis (phenyl ethynyl) anthracene by guidance from a pre-aligned seed crystal on polyimide substrate kept in a sealed environment

Petri dish

BPEA solution

BPEA solution

BPEA seed crystal

Polyimide substrate

Ni or Al gate electrode

Glass slide

BPEA crystal bridging across soruce/drain electrodes

BPEA crystal

Drain

Source

Polyimide substrate

Ni or Al gate electrode

Figure 7.3. (Continued.)

7.3.2 Formation of source/drain electrodes

For lift-off photolithography, the pattern of source/drain electrodes is defined in photoresist on the opposite side of PI substrate on which there is no Ni or Al film. Over this photoresist pattern, gold is deposited by thermal evaporation. The thickness of gold film is 25 nm. When the photoresist is dissolved in acetone, the gold film over the photoresist portions is washed off while the remaining gold film on non-photoresist coated portions is retained to create the metal pattern.

7.3.3 Growth of BPEA single-crystal ribbon

A suspension of BPEA crystals in ethanol, C_2H_5OH is prepared by sonicating previously-prepared BPEA crystals in ethanol (Cai *et al* 2014). This suspension is dropped at the correct location on the source/drain gold electrode pattern over the PI gate dielectric film. The ethanol is allowed to evaporate.

After the ethanol completely evaporates, the aligned BPEA seed crystals on the PI gate dielectric are placed inside a glass petri dish. The crystal growth is carried out in this petri dish and chlorobenzene (C_6H_5Cl) solution of BPEA crystals (with BPEA concentration of 1–2 mg mL^{-1}) is placed adjoining the aligned BPEA seed crystals. The petri dish is enclosed inside a cool and dry atmosphere. After 2 days, when all the solvent has vanished, it is found that BPEA crystal has grown along a straight line in the shape of a ribbon from the seed crystal. The crystal is 100 μm long. Its height is typically 600 nm. The width is 800 nm. The ribbon shaped crystal is adequate to provide a large area for contact. Further, the BPEA crystalline ribbon is firmly fixed to the PI surface due to the clean BPEA/PI interface available during *in situ* crystal growth.

A single ribbon of BPEA crystal is used for each device. The devices are 20–30 μm long and 500–800 nm wide. The device width depends on that of the individual ribbon used.

7.3.4 Treatment of gold electrodes with thiophenol

The contact resistance between the Au electrode and the semiconductor is lowered by immersing in thiophenol (C_6H_6S) solution of concentration 1 mmol mL^{-1} for 24 h. The thiophenol sticks only to the gold layer because of the presence of the thiol end group (Myny *et al* 2006). It optimizes the contact condition bringing about performance improvement.

7.3.5 Hole mobility and contact resistance

From measurements on ITO glass substrate devices, the TFTs exhibit an averaged mobility of 2.47 cm^2 V^{-1} s^{-1} with 57% of devices showing a mobility >2.5 cm^2 V^{-1} s^{-1} and maximum mobility of 3.2 cm^2 V^{-1} s^{-1}. The maximum current on–off ratio is 10^5. The thiophenol treatment decreases the averaged contact resistance from 90–300 kΩ cm to 38 kΩ cm, which is still quite large. A five stage ring oscillator fabricated on a single ribbon of BPEA crystal shows a low oscillation frequency of 512 Hz, which is ascribed to the high contact resistance.

7.3.6 Effect of bending on TFT

The PI-substrate TFT is bent down to a curvature radius of 1 mm. During the first 20 bending times, the mobility decreases from 3.2 cm^2 V^{-1} s^{-1} to 2.4 cm^2 V^{-1} s^{-1} but thereafter it becomes stable and remains constant at the value of 2.1 cm^2 V^{-1} s^{-1} for 300 bending times. The experiment shows that this TFT has good endurance to bending (Cai *et al* 2014).

7.4 Speedier process of building large arrays of organic single crystals

Briseno *et al* (2006b) noted that the process of selection of individual organic single crystals and their precise placement at pre-stipulated locations is a lengthy, cumbersome, labor-intensive procedure which can indeed be carried out in a research laboratory but is certainly inapplicable to the production of large numbers of devices in bulk manufacturing. They appreciated this difficulty and came up with a process of patterning arrays of organic single crystals, which is very useful for large-scale production of devices with a high throughput. This method allows fabrication of arrays of single crystals directly on the pre-created pattern of source/drain electrodes of the TFTs. By this method, arrays of crystals of a wide range of organic semiconductors can be grown over patterned source/drain electrode designs.

Based on this method, Briseno *et al* (2006b) fabricated TFTs by growing single crystals of different organic semiconductors on flexible substrates. For the rubrene TFT specimen on PI substrate, the threshold voltage, current on–off ratio and carrier mobility are ~1.5 V, 10^4 and 0.9 cm^2 V^{-1} s^{-1} respectively. In the case of pentacene TFT on PI, the highest mobility value is 0.11 cm^2 V^{-1} s^{-1}. The current on–off ratios are ~10^3. These rubrene and pentacene TFTs did not incur any perceivable performance degradation up to a bending radius of 6 mm. This experiment shows that the TFTs are tolerant towards bending. Apart from these two semiconductors, the above authors also realized TFTs of tetracene and anthracene on PET substrates by this method.

7.4.1 The principle

The key to this method is that domains of octadecyltriethoxysilane (OTS), empirical formula C$_{24}$H$_{52}$O$_3$Si, molecular weight 416.75, formed on the target film by micro-contact printing with a PDMS stamp act as the controlling center of the nucleation of organic single-crystal film from the vapor phase. The PDMS stamp contains a relief structure in the pattern according to which printing is to be done.

7.4.2 The substrate

The substrate is Kapton (polyimide) (figure 7.4). Its thickness is 140 μm. One side of the substrate is coated with a gold film of thickness 100 nm. This gold film is used as the gate electrode.

7.4.3 Coating the PVP dielectric layer on the gold film over the substrate

This coating is done over the gold film on the PI substrate. The dielectric solution is prepared by dissolving 22 wt% PVP, [CH$_2$CH(C$_6$H$_4$OH)]$_n$ and 8 wt% poly(mela-mine-*co*-formaldehyde) methylated, (C$_4$H$_8$N$_6$O)$_n$ in propylene glycol monomethyl

Au gate electrode deposition on
kapton substrate

Gate electrode (Gold)
Kapton (Polyimide) substrate

PVP gate dielectric deposition on
the gold film

Gate dielectric (PVP)
Gate electrode (Gold)
Kapton (Polyimide) substrate

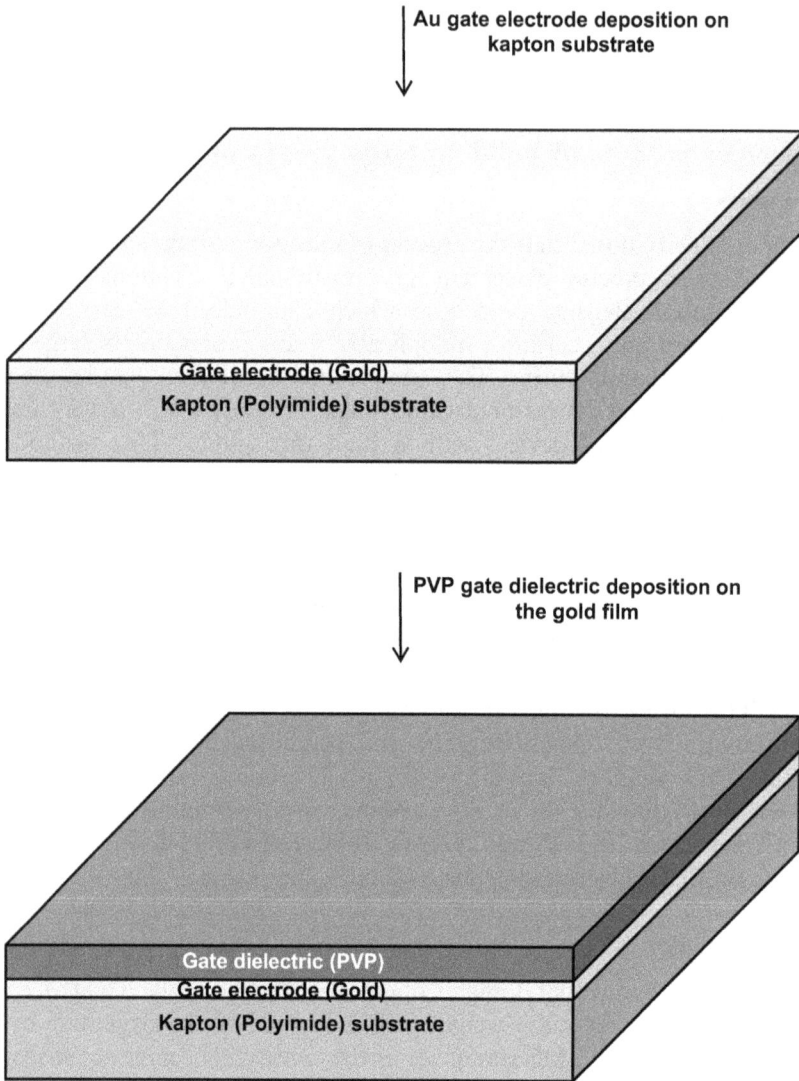

Figure 7.4. Building large arrays of organic single-crystal TFTs.

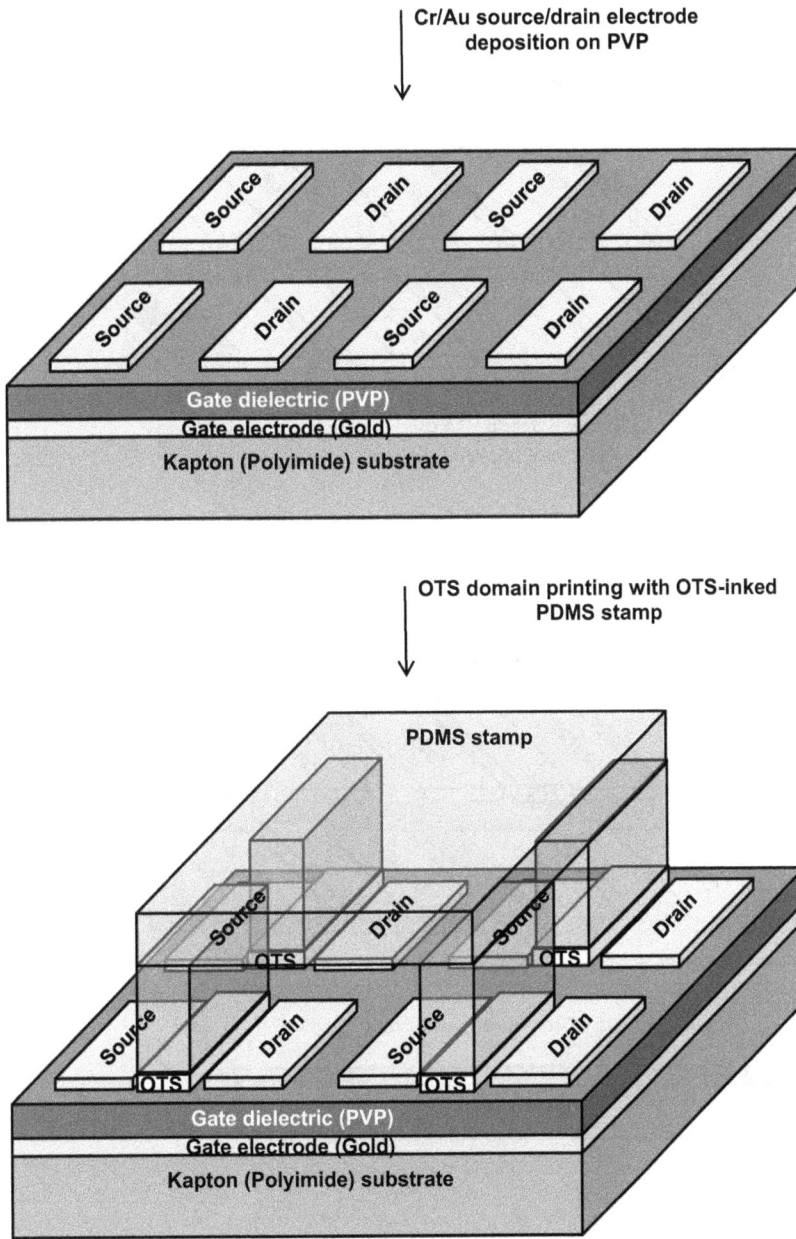

Figure 7.4. (Continued.)

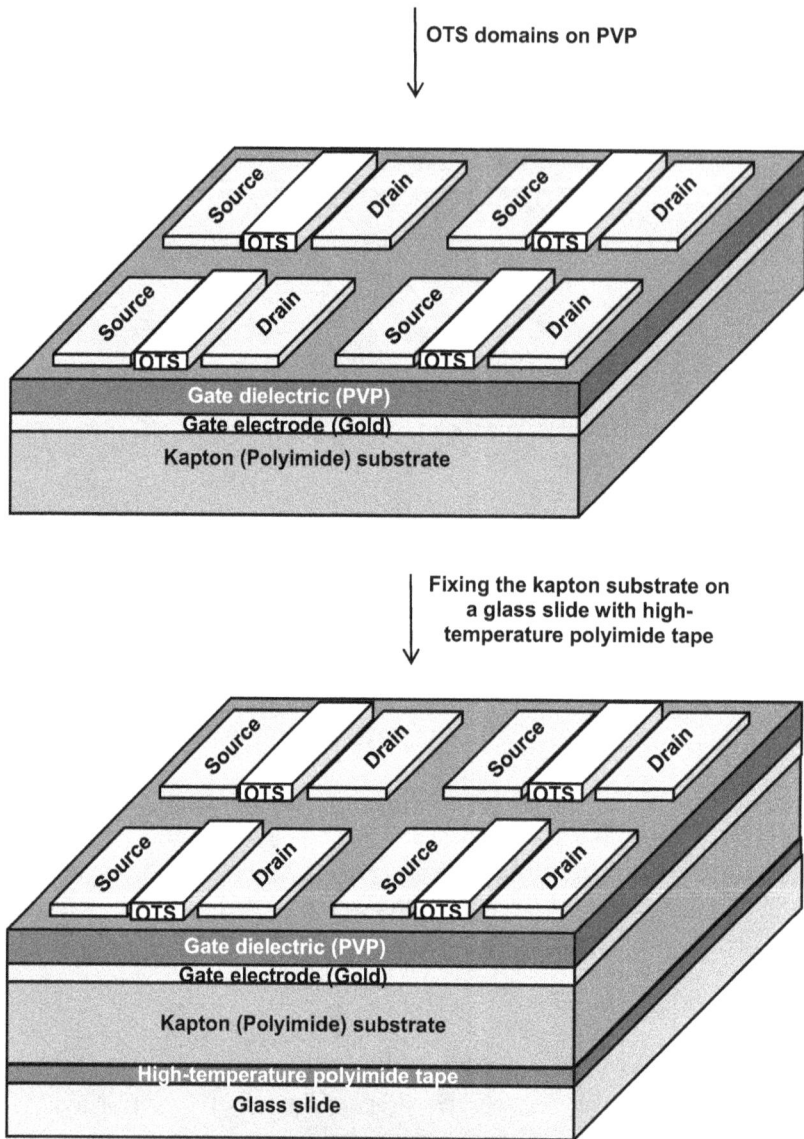

Figure 7.4. (Continued.)

Sealing the OTS-inked substrate along with organic semiconductor inside a glass tube in vacuum and keeping in reactor

Temperature-gradient sublimation reactor

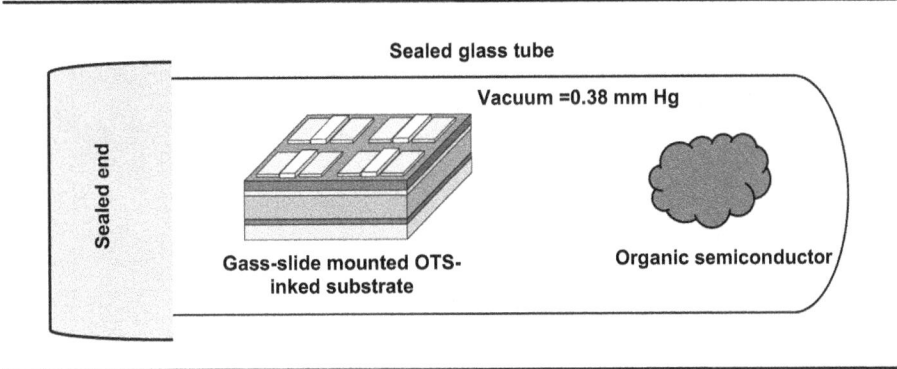

Sealed glass tube

Vacuum =0.38 mm Hg

Sealed end

Gass-slide mounted OTS-inked substrate

Organic semiconductor

Rubrene crystals in a definite pattern at OTS defined locations

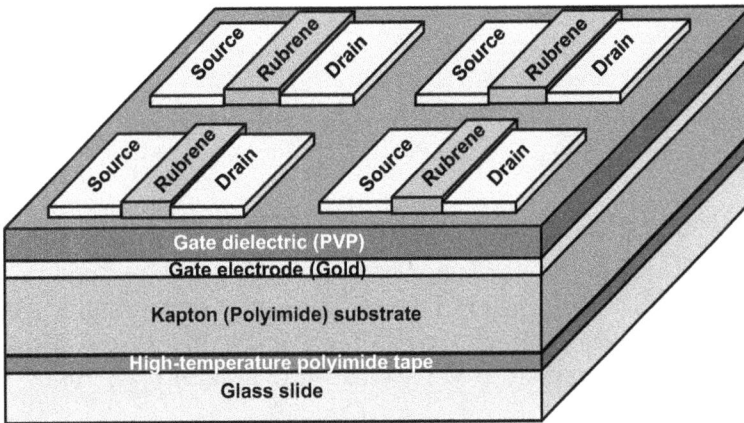

Source Rubrene Drain   Source Rubrene Drain

Source Rubrene Drain   Source Rubrene Drain

Gate dielectric (PVP)
Gate electrode (Gold)
Kapton (Polyimide) substrate
High-temperature polyimide tape
Glass slide

Figure 7.4. (Continued.)

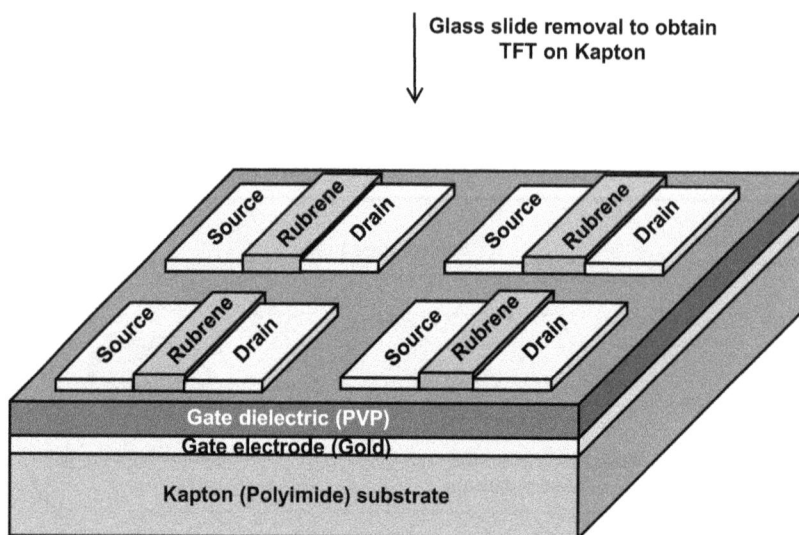

Glass slide removal to obtain
TFT on Kapton

Gate dielectric (PVP)
Gate electrode (Gold)
Kapton (Polyimide) substrate

Figure 7.4. (Continued.)

ether acetate (PGMEA), $CH_3CO_2CH(CH_3)CH_2OCH_3$ or $C_6H_{12}O_3$. This dielectric solution is deposited over the gold film by spin coating at 2000 RPM. Baking is done in two steps: first at 100 °C for 10 min and then at 200 °C for 10 min. By baking, the polymer is crosslinked. The thickness of the dielectric film is 1.5 μm.

7.4.4 Fabrication of source/drain electrodes

Thermal evaporation of two metals is done through a shadow mask: 1.5 nm thick chromium and 50 nm thick gold.

7.4.5 Inking the PDMS stamp with OTS

The OTS solution is prepared in toluene. The solution has a high OTS concentration typically = 100 mM–1 M. The OTS solution is applied to the PDMS stamp with a cotton applicator. The inked stamp is dried in a stream of air from an air gun. The surplus solvent is driven away. The thickness of OTS layer is ~13 nm.

7.4.6 Formation of OTS domains on the source/drain electrode pattern

Before printing, the substrate is given a 30 s exposure to UV-ozone (Briseno *et al* 2006b). The OTS-inked PDMS stamp is contacted with the PVP dielectric film on the gold-coated PI substrate. The stamp and the PVP dielectric surfaces remain in contact for a duration of 5 min, after which the stamp is taken away. As the organic crystal must be formed in such a way that it accurately bridges the source and drain electrodes of the transistors in the array, it is necessary to correctly contact the inked stamp with the PVP dielectric surface. To ensure that this alignment is properly done, a square-shaped registration bar is made on the stamp and an identical registration bar on the substrate. Both the PDMS stamp and the TFT array are predefined using photolithographic procedures. With the help of registration bars, the OTS printing is done in areas overlying the source/drain electrodes of the separate transistors. Hence, OTS areas are formed connecting the source and drain electrodes of each transistor. The accuracy of alignment of the stamp with the electrode pads determines the yield of functional transistors from a batch. The OTS domains formed by the stamp show a rough surface.

7.4.7 Growth of organic single crystal

The OTS-inked substrate is taped on a pre-cut microscopic slide of glass (Briseno *et al* 2006b). High-temperature polyimide tape is used for fixing the substrate. The glass slide-mounted substrate is put on a clip-shaped coil of nichrome wire and pushed into a 16 mm diameter glass tube open at one end. It is kept at a distance of several centimeters away from 10 mg of relevant organic semiconductor material lying in the glass tube at the closed end. Following its evacuation to 0.38 mm mercury, the glass tube containing the substrate and the source material is sealed with an oxygen-aided flame.

The vacuum-sealed tube is introduced into a furnace tube, referred to as a temperature gradient sublimation reactor. In this reactor, there is a temperature gradient in accordance with the particular organic semiconductor whose crystals are to be formed. The source temperature for pentacene is ~260 °C; for rubrene, 250 °C–270 °C; for C_{60}, 425 °C–500 °C. The sublimation temperature for anthracene is very low ~120 °C owing to its high vapor pressure. Substrate temperatures, i.e. temperatures of crystal-forming zones, are lower than source temperatures and the difference between source and substrate temperatures depends on the distance between them. For pentacene, the nearest edge of the substrate is 2 cm away from the source if the substrate temperature is 250 °C. The nucleation zone stretches up to a distance of 5 cm from the source. The temperature at 5 cm distance is 220 °C. The rate of crystal growth is increased when the source temperature is raised. At the same time, the distance between the source and the crystal-forming region becomes larger and, therefore, the substrate must be positioned farther away from the source.

The crystal growth is done by the vapor transport method. This method is very versatile from the viewpoint that crystals of several organic semiconductors, P-type as well as N-type, can be grown by applying it. Notable examples of P-type organic semiconductors include rubrene, $C_{42}H_{28}$; tetracene, $C_{18}H_{12}$; and pentacene, $C_{22}H_{14}$. Some examples of N-type semiconductors are; $F_{16}CuPc$, $C_{32}CuF_{16}N_8$; Buckminsterfullerene, C_{60}; and tetracyanoquinodimethane, $C_{12}H_4N_4$. Times of sublimation of the organic semiconductor vary widely among the materials. It is ~5 min for pentacene and ~2 h for C_{60}.

The crystal growth is confined to the rough surfaces of OTS domains, whereas in the smoother regions between two OTS domains, no crystals are found to grow. Selectivity of crystal growth is enhanced by choosing a higher substrate temperature. At a typical difference of temperatures by 20 °C, with the substrate cooler than the source by this amount, the desorption rate in the rough OTS domains is less severe than in the smoother regions whereby there is more likelihood of nucleation to take place on the OTS domains than in the smoother areas. Hence, crystal growth is selectively restricted to OTS domains.

A pertinent question relates to the fact that if any gold surfaces of source/drain electrodes are covered with OTS by its spreading, how does the crystal establish contact with gold? The reason that there is no contacting issue is that the OTS layer consists of sparse pillars that do not fully cover the gold surface. Moreover, during crystal growth, some of the OTS evaporates because of the high temperature. Further, the growing crystal pushes away a portion of OTS outside the stamping field, thereby increasing the area of contact between the organic semiconductor crystal and the gold surface of the electrode (Briseno *et al* 2006b).

7.5 CuPc and F_{16}CuPc TFTs on 15 µm diameter Au wire

Hexadecahydrogen copper phthalocyanine (CuPc) is a P-type semiconductor and hexadecafluoro copper phthalocyanine (F_{16}CuPc) is an N-type semiconductor. These materials display good flexibility and superb field-effect properties. They are highly stable in air, both thermally and chemically.

Zheng *et al* (2016) devised a jigsaw puzzle method for device fabrication. In their method (figure 7.5), there is a pre-fabrication step in which the device is first made on a stiff, flat substrate. Subsequently, it is transferred to an ultra-fine gold wire. The extremely small size TFT mounted on a thin Au wire is very useful for applications in wearable and implantable electronics.

The gold wire onto which the device is transferred serves as the gate electrode. The gate dielectric is made of polystyrene (PS), $(C_8H_8)_n$. Gold is used to make the source/drain electrodes.

7.5.1 PS dielectric deposition on OTS (octadecyltrichlorosilane)-modified silicon substrate

The solution of polystyrene in toluene has a strength of 2%. This solution is spun on an OTS-coated Si substrate at 4000 RPM for 30 s; OTS is $CH_3(CH_2)_{17}SiCl_3$. Thermal baking is done at 90 °C for 20 min. The thickness of PS film is 130 nm.

Modification of silicon substrate with OTS is necessary because OTS treatment renders the Si surface hydrophobic, which makes peeling off the PS film from the silicon substrate easier. Therefore, the PS dielectric layer along with the device built upon it can be easily peeled off from the silicon substrate at a later stage for mounting on gold wire.

7.5.2 Synthesis of single-crystal CuPc and F_{16}CuPc nanowires

The set up for crystal growth consists of a quartz tube inserted into a furnace. This is a two-zone furnace consisting of a high-temperature zone (typically 425 °C) and a low-temperature zone (150 °C) (Tang *et al* 2006). In the high-temperature zone the source material, either CuPc or F_{16}CuPc, is placed. When this material vaporizes, its vapors are carried away by an inert gas flowing through the tube towards the

Silicon substrate surface modification
with OTS

OTS
Silicon substrate

Spin coating polystyrene insulating
layer on OTS

PS gate dielectric
OTS
Silicon substrate

(a)

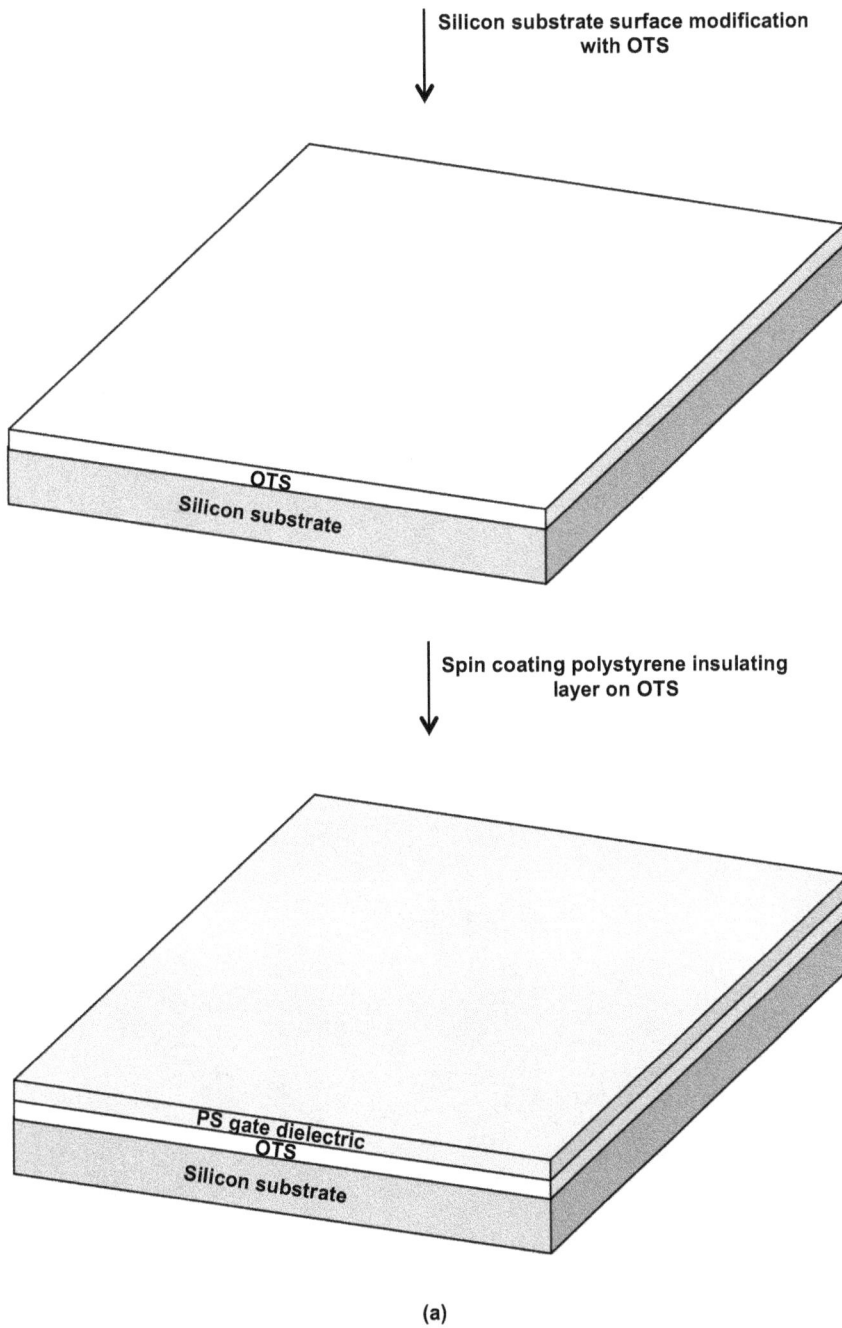

Figure 7.5. CuPc or F_{16}CuPc TFT on gold wire: (a) substrate preparation for nanowire placement and (b) remaining steps.

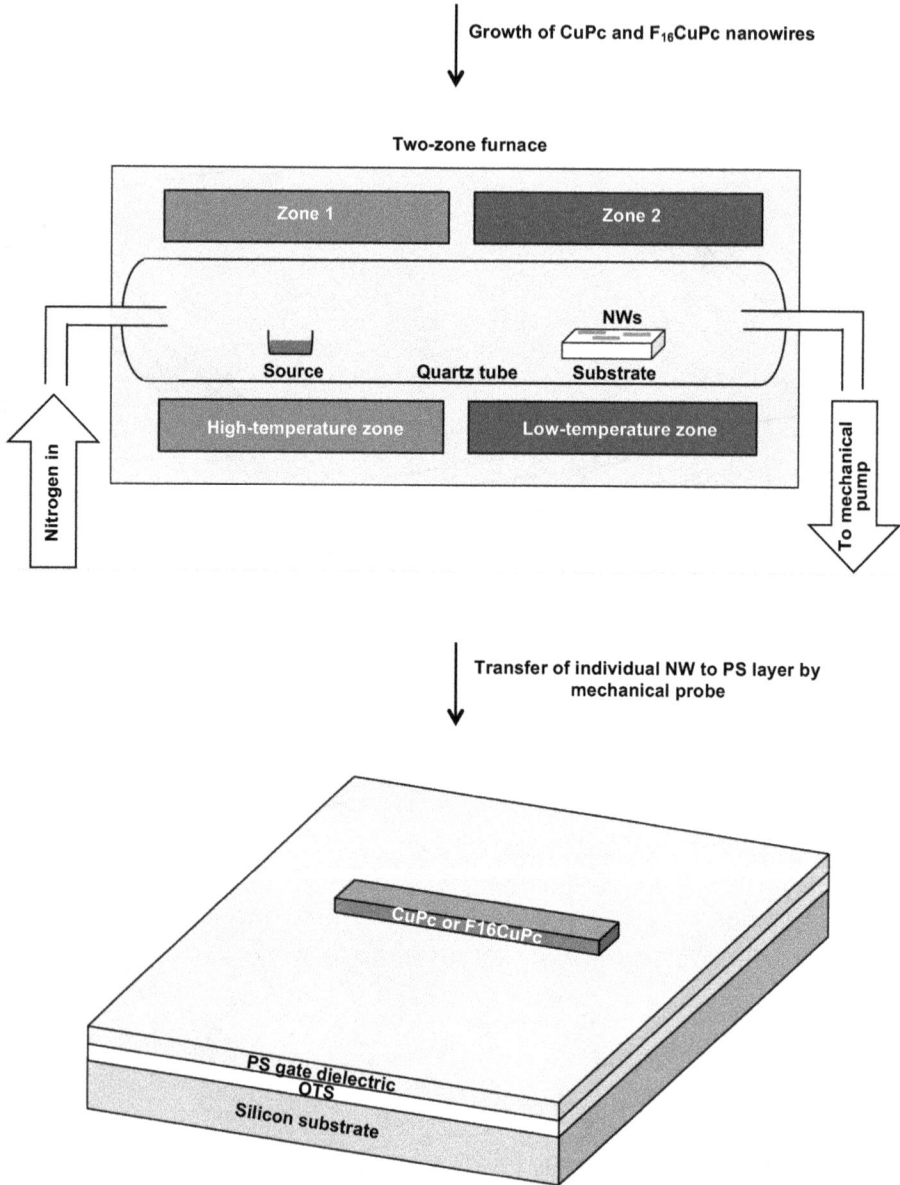

Figure 7.5. (Continued.)

Source/drain electrode formation

Peeling away the OTS layer from silicon substrate

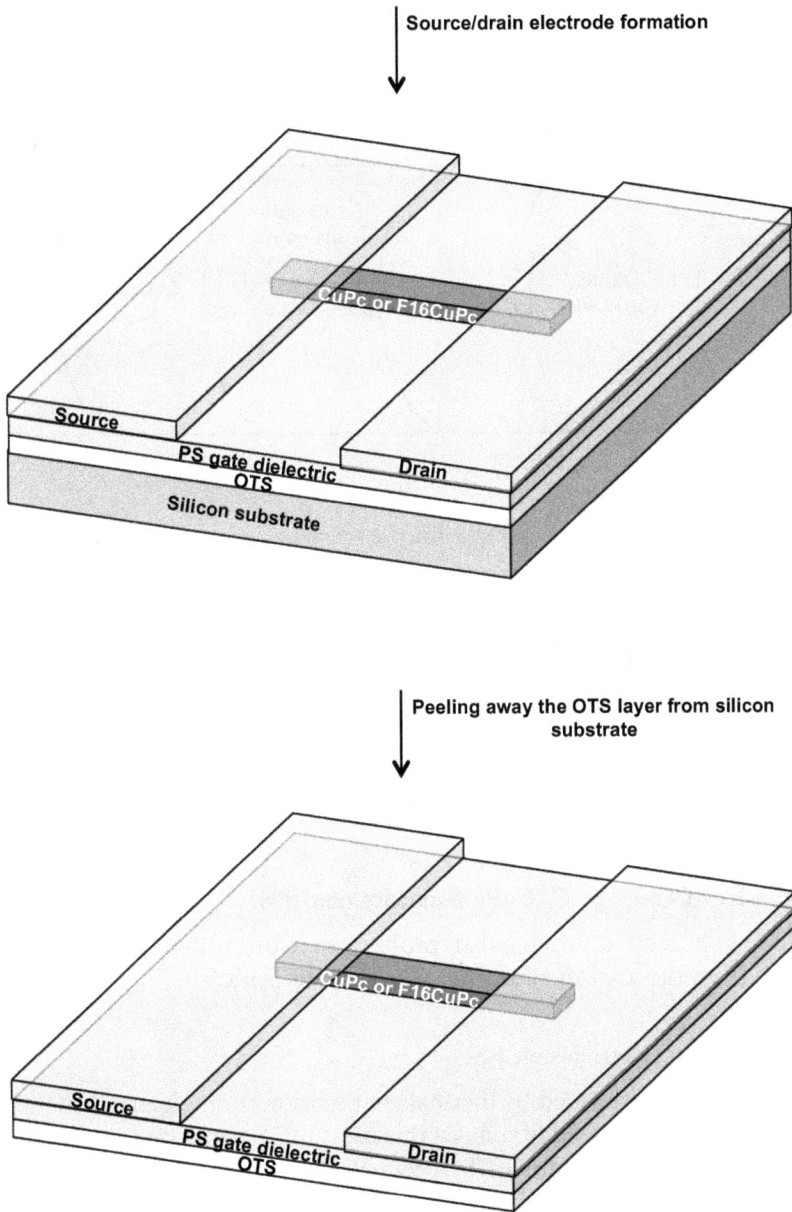

Figure 7.5. (Continued.)

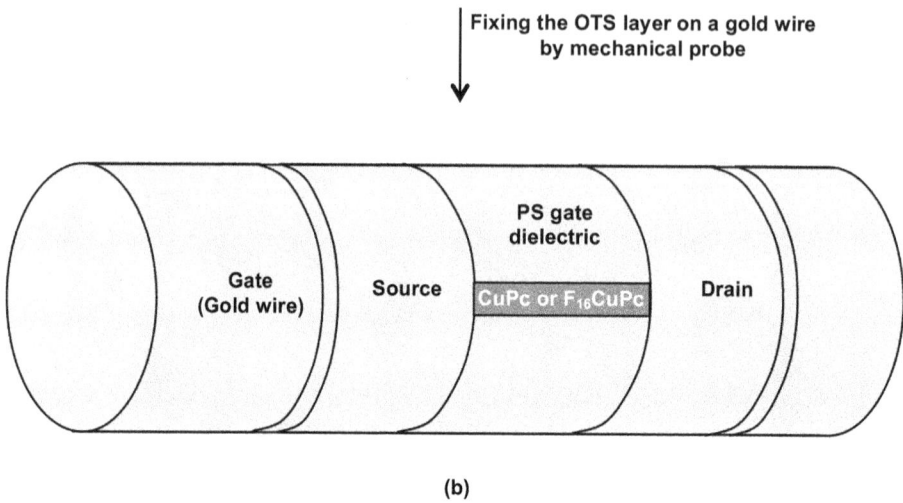

Fixing the OTS layer on a gold wire by mechanical probe

Gate (Gold wire)   Source   CuPc or F$_{16}$CuPc   PS gate dielectric   Drain

(b)

Figure 7.5. (Continued.)

low-temperature zone, where the material condenses and solidifies into the crystalline state on a suitable substrate.

Powdered CuPc or F$_{16}$CuPc is used as the source material. For CuPc the temperature of the high-temperature zone is kept at 300 °C. For F$_{16}$CuPc, it is 290 °C. In both cases, the vapor carrier gas is ultra-pure nitrogen. The gas flow rate is 25 sccm. The chamber pressure is 0.225 Torr. The crystal growth time is 2 h (Zheng *et al* 2016).

7.5.3 Transfer of CuPc or F$_{16}$CuPc nanowire onto PS

Manipulating it with a mechanical probe, the CuPc or F$_{16}$CuPc nanowire is transferred from the growth substrate over PS gate dielectric.

7.5.4 Source/drain electrodes on PS

These electrodes are deposited by thermal evaporation through a gold shadow mask of thickness 80 nm. The gold film has a thickness of 35 nm. After gold evaporation, the mask is mechanically removed by peeling off from the substrate.

7.5.5 Cleaning of Au wire

The gold wire is cleaned in acetone. It is polished with hydrochloric acid, HCl to provide a smooth surface. It is then treated with oxygen plasma for 5 min.

7.5.6 Transfer of the TFT to Au wire

With the help of a mechanical probe, the planar TFT device fabricated on the OTS-modified silicon substrate is peeled off from the silicon substrate. It is fixed on the gold wire using a mechanical probe. No additional treatment is required for fixation of the PS layer to the Au wire. The adhesion between the two is excellent.

7.5.7 CuPc and F_{16}CuPc TFT characteristics

CuPc fiber-shaped TFT has a threshold voltage of +2.97 V. The field-effect mobility is 0.07 cm^2 V^{-1} s^{-1}. The F_{16}CuPc TFT shows a threshold voltage of −6.16 V; the best mobility value is 0.04 cm^2 V^{-1} s^{-1}.

7.5.8 Complementary inverter

Taking P-type CuPc TFT and N-type F_{16}CuPc TFT, a complementary inverter circuit is assembled on an Au wire. The inverter has a voltage gain of 8.2 at V_{DD} = +20 V; the gain at V_{DD} = −20 V is 7.7 (Zheng *et al* 2016).

7.6 Discussion and conclusions

Crystals of organic semiconductors are mainly grown by a physical vapor transport process in horizontal reactor tubes for the sake of convenience. A solution growth method is also described. During the initial stages of development, the TFTs are fabricated by growing the crystals separately in furnaces and placing them manually at the intended locations. With the knowledge that OTS-stamped sites can act as nucleation centers for crystal growth, a faster method of TFT fabrication was proposed. The method is generic and is widely applicable to various organic semiconductors. It is well received because it eliminates the errors and inconvenience of manually placing the crystal at the desired site. For wearable electronics circuits, the P-type and N-type TFTs are made on OTS-modified silicon substrate and subsequently transferred to a gold wire to make a complementary inverter.

The mobilities achieved with different TFTs are: 4.6 cm^2 V^{-1} s^{-1}, rubrene (Briseno *et al* 2006a); average 2.47 cm^2 V^{-1} s^{-1} and maximum mobility of 3.2 cm^2 V^{-1} s^{-1}, anthracene (Cai *et al* 2014); 0.9 cm^2 V^{-1} s^{-1}, rubrene, and 0.11 cm^2 V^{-1} s^{-1}, pentacene (Briseno *et al* 2006b); 0.07 cm^2 V^{-1} s^{-1}, CuPc and 0.04 cm^2 V^{-1} s^{-1}, F_{16}CuPc (Zheng *et al* 2016). These values stand up well against the competitive devices.

Review exercises

7.1 Write the formulae of rubrene and anthracene.

7.2 What are the three stages of the physical vapor transport process? What are the two tubes inside the reactor tube used for?

7.3 How are rubrene crystals grown? How thick are the crystals? What is the typical size of the crystals?

7.4 How is the rubrene crystal adhered to the gate/source/drain electrode pattern?

7.5 Up to what bending radius and strain level does the performance of the rubrene crystal TFT remain unaffected by bending? By what percentage does the mobility recover on withdrawal of the strain of 1.18%?

7.6 What are the two phases in which BPEA crystallizes? Which of these phases is chosen for TFT fabrication?

7.7 What is the method used for growing BPEA crystals? Where and how are the seed crystals placed? From which solution do the crystals grow? What is the shape of the BPEA crystal? What are its dimensions?

7.8 Why are the gold electrodes treated with thiophenol?

7.9 Is the BPEA single-crystal TFT tolerant to bending? Discuss its performance under bending conditions.

7.10 How do you overcome the difficulty of fabricating individual organic single crystals and placing them at the precise locations on flexible substrates?

7.11 On what property of OTS is the method for growing organic single crystals at defined locations based?

7.12 How are the OTS domains formed on the pattern of source/drain electrodes?

7.13 What method is used for growing crystals over OTS domains formed on source/drain electrodes with a PDMS stamp? Is this method generally applicable for different materials? What source temperatures are used for growing pentacene and rubrene crystals?

7.14 For the surfaces of source/drain electrodes being covered with OTS, how does the crystal grown over OTS establish contact with the electrodes?

7.15 How are single-crystal CuPc and F_{16}CuPc nanowires grown on OTS-modified silicon substrate? In what form is the source material introduced? Into how many zones is the furnace divided? What happens in these zones?

7.16 What is the gate dielectric in CuPc and F_{16}CuPc TFTs? How is it deposited?

7.17 How are the TFTs removed from the silicon substrate and mounted on the gold wire?

References

Briseno A L, Mannsfeld S C B, Ling M M, Liu S, Tseng R J, Reese C, Roberts M E, Yang Y, Wudl F and Bao Z 2006b Patterning organic single-crystal transistor arrays *Nature* **444** 913–7

Briseno A L, Tseng R J, Ling M-M, Falcao E H L, Yang Y, Wudl F and Bao Z 2006a High-performance organic single-crystal transistors on flexible substrates *Adv. Mater.* **18** 2320–4

Cai X, Ji D, Jiang L, Zhao G, Tan J, Tian G, Li J and Hu W 2014 Solution-processed high-performance flexible 9, 10-bis(phenylethynyl)anthracene organic single-crystal transistor and ring oscillator *Appl. Phys. Lett.* **104** 063305-1–5

Kloc C, Simpkins P G, Siegrist T and Laudise R A 1997 Physical vapor growth of centimeter-sized crystals of α-hexathiophene *J. Cryst. Growth* **182** 416–27

Laudise R A, Kloc C, Simpkins P G and Siegrist T 1998 Physical vapor growth of organic semiconductors *J. Cryst. Growth* **187** 449–54

Myny K, De Vusser S, Steudel S, Janssen D, Müller R, De Jonge S, Verlaak S, Genoe J and Heremans P 2006 Self-aligned surface treatment for thin-film organic transistors *Appl. Phys. Lett.* **88** 222103-1–3

Tang Q, Li H, He M, Hu W, Liu C, Chen K, Wang C, Liu Y and Zhu D 2006 Low threshold voltage transistors based on individual single-crystalline submicrometer-sized ribbons of copper phthalocyanine *Adv. Mater.* **18** 65–8

Zheng L, Tang Q, Zhao X, Tong Y and Liu Y 2016 Organic single-crystal transistors and circuits on ultra-fine Au wires with diameters as small as 15 μm via jigsaw puzzle method *IEEE Electron Device Lett.* **37** 774–7

IOP Publishing

Flexible Electronics, Volume 2
Thin-film transistors
Vinod Kumar Khanna

Chapter 8

Electrolyte-gated organic FET (EGOFET) and organic electrochemical FET (OECFET)

The EGOFET benefits enormously from the high double layer capacitance formed at the electrolyte–semiconductor interface, which lowers the gate driving voltage to <1 V and increases the drain current appreciably (Wang *et al* 2016). In contrast to EGOFET, OECFET exemplified by the familiar PEDOT:PSS transistor, works by doping and de-doping of the organic semiconductor by ion injection into and withdrawal from it. In OECFET, the combined action of electronic and ionic charges together with utilization of the full volume of the semiconductor instead of a thin surface channel layer leads to a higher transconductance than EGOFET (Rivnay *et al* 2018). Among the water-gated pentacene EGOFET, P3HT EGOFET and PEDOT:PSS OECFET fabricated on a PEN substrate, the PEDOT:PSS OECFET exhibits the best overall characteristics in terms of high signal-to-noise ratio and small switching time (de Oliveira *et al* 2016). Three different architectural designs of EGOFETs with polyelectrolyte gate and pBTTT-C14 semiconductor are fabricated, viz, bottom gate with top contacts, top gate with bottom contacts and planar gate. All these designs are useful for flexible electronics but the planar gate structure is easiest to fabricate because it allows the patterning of gate and source/drain electrodes in a single step (Dumitru *et al* 2013). A multi-stack structure OECFET with a vertical architecture is fabricated on both sides of paper and PET substrates by punching via holes in the substrates and using inkjet printing on PET and screen printing on paper for defining the PEDOT:PSS channel. Better current on–off ratios and switching times are achieved on PET than paper (Kawahara *et al* 2013). A fiber-embedded EGOFET/OECFET is fabricated by creating source/drain microgaps using fibers woven into a mesh, depositing P3HT semiconductor across the microgaps as the channels and sewing gold wires onto the P3HT film as gate electrodes; an ionic liquid-based electrolyte is used. The device works in two modes: as an EGOFET in the first mode and as an OECFET in the second mode, as depicted by the initial fast current rise in <3 ms followed by a gradual current ascent up to 2 s (Hamedi *et al* 2009).

8.1 Introduction

A major challenge faced with organic TFTs is to lower the operating voltage. When efforts to decrease the voltage with traditional insulating materials like silicon dioxide, polymeric insulators and even high-κ dielectrics yielded limited success, electrolytic gating emerged as a useful solution. The electrolyte-gated FET works in the capacitive mode exploiting the high capacitance of the electrical double layer at the electrolyte–semiconductor interface. A similar device with electrolyte at the gate works on a different principle. Here, injection of ions into the semiconductor and their withdrawal from it modulates the drain–source current instead of capacitive coupling.

8.2 Principle of electrolyte–gate organic FET (EGOFET)

8.2.1 Differentiating EGOFET from traditional OFET

The main difference between a conventional OFET and an EGOFET (figure 8.1) lies in the gate dielectric (Wang *et al* 2016). In the OFET, the gate dielectric is a non-electrolytic

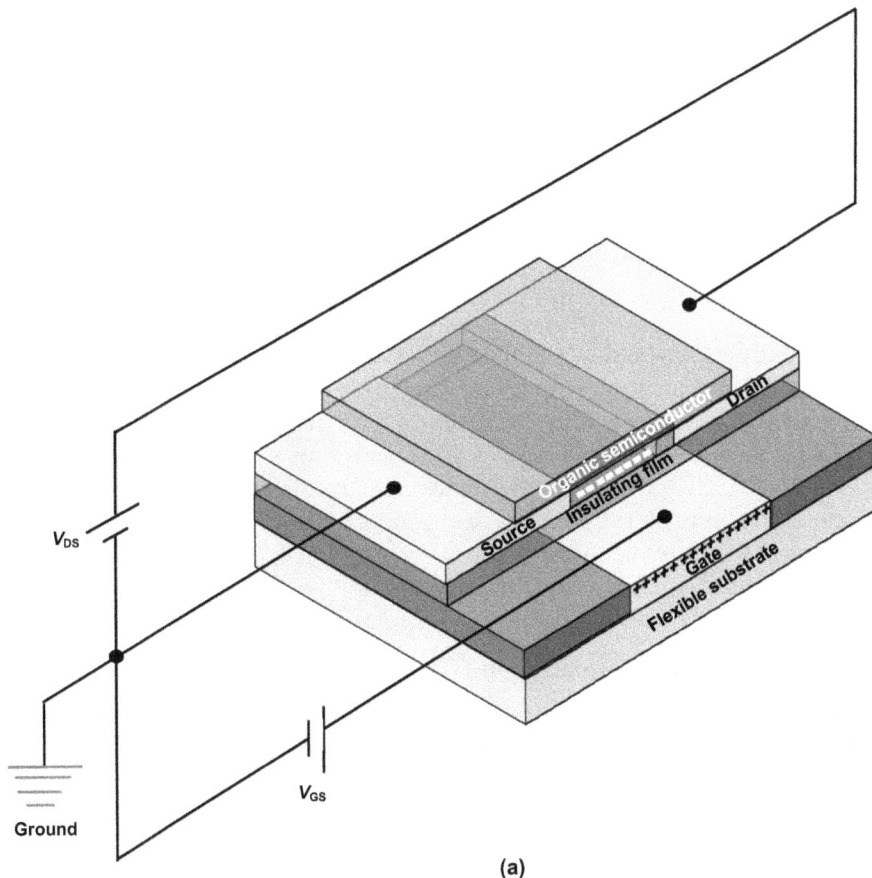

Figure 8.1. Comparison between two types of transistors: (a) an organic field-effect transistor (OFET) and (b) electrolyte-gated organic field-effect transistor (EGOFET). Note the absence of insulating film in (b).

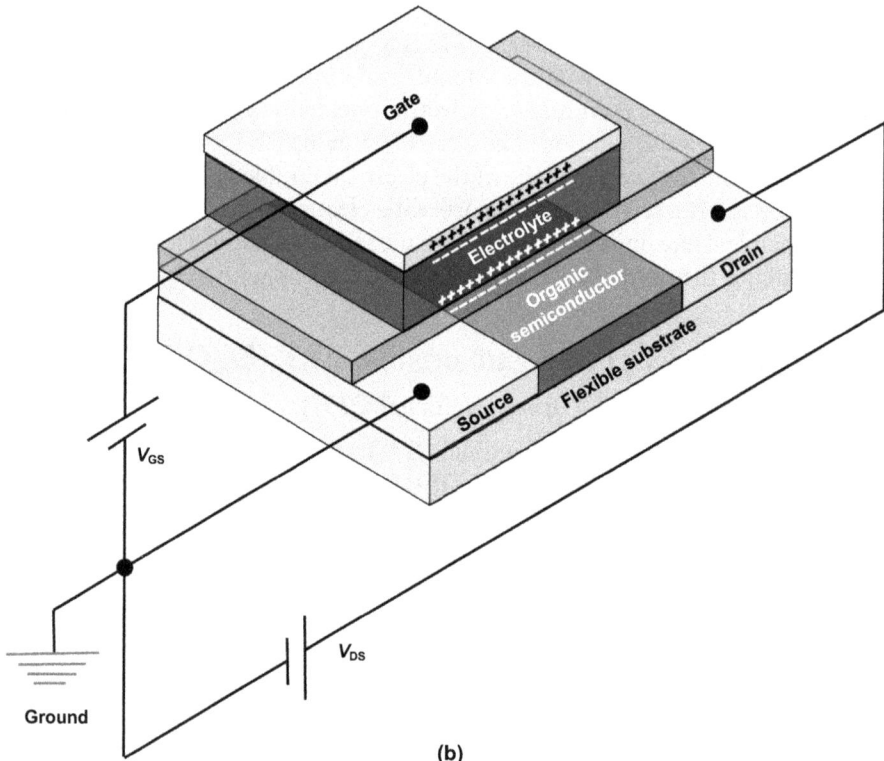

Figure 8.1. (Continued.)

insulating film, e.g. poly(vinyl alcohol) (PVA), poly(4-vinyl phenol) (PVP), parylene C, polymethylmethacrylate, polystyrene, octadecyltrichlorosilane (OTS), hexamethyldisi-lazane (HMDS), etc. In the EGOTFT, the gate dielectric is an electrolytic solution, either liquid, solid or semi-solid, e.g. a salt suspended in a polymeric matrix, an ionic liquid or ionic gel. The gate electrode is immersed in this electrolytic solution. The source and drain electrodes contact the organic semiconductor, and the channel is formed in the semiconductor between source and drain contacts, as usual. The source/drain electrodes are carefully isolated from the electrolytic solution.

8.2.2 Capacitive principle of operation

Replacement of the conventional gate dielectric by an electrolytic solution does not alter the operating principle of the EGOFET. Like the OFET, it works on the capacitive principle. Charge accumulation in the electrolyte is equivalent to the generation of a gate bias. This gate bias modulates the current flow between source and drain by electrostatic field effect.

On applying a positive voltage to the gate electrode (figure 8.1(b)), the negative ions in the electrolyte are attracted towards the positively charged gate electrode. As a result, an electrical double layer is formed at the electrolyte–gate interface. This double layer consists of positive electronic charges in the gate and negative ionic charges in the electrolyte.

Furthermore, the positive ions in the electrolyte are pushed away by the positively charged gate electrode from the neighborhood of the gate towards the surface of the semiconductor. Accumulation of these positive ions in the electrolyte near the surface of the semiconductor leads to attraction of the negatively charged electrons in the semiconductor towards the semiconductor surface. As a result, an electrical double layer is formed at the electrolyte–semiconductor interface between the positive ionic charges in the electrolyte and the negative electronic charges in the semiconductor. The electron population buildup near the semiconductor surface forms a conducting channel between the source and drain along which the current starts flowing.

Likewise, the effect of applying a negative bias to the gate electrode can be understood.

8.2.3 Effect of the high capacitance of the electrical double layer

The electrical double layer (EDL) consists of two parts: a Helmholtz layer (HL) and a diffuse layer (DL). The HL comprises a monolayer of ions, whereas the diffuse layer is a layer in which the ionic concentration, whether positive ions or negative ions, is greater than in the bulk electrolyte solution.

Consequent upon the high capacitance (20–500 μF cm^{-2}) of the very thin electrical double layer (thickness in sub-nm range) formed at the electrolyte–semiconductor interface, the EGOFET exhibits a very high gate capacitance, which is ~1000 times higher than the gate capacitance available with traditional non-electrolytic gate dielectrics. The advantage gained by this high capacitance is that EGOFETs need much lower gate voltages <1 V for modulating the conductivity of the underlying channel by electrical field effect in comparison to the much higher gate bias values demanded by competing OFETs, which are >10 V or even of larger magnitudes. Therefore, the EGOFET provides a high drive current at a lower operating voltage. Thus, the EDL confers upon the transistor the double blessings of a low running voltage and a high output current.

8.3 Organic electrochemical TFT (OECFET)

8.3.1 Distinguishing OECFET from EGOFET

The OECFET differs from the EGOFET in its operating principle (Rivnay et al 2018). The EGOFET is based on an interfacial effect causing charge accumulation or depletion in the surface of the semiconductor adjoining the electrolyte. Whereas OECFET works by a capacitive effect, the OECFET is based on injection and withdrawal of ions from the electrolytic gate into and from the semiconductor to increase and decrease the conductance. Hence, its mechanism relies on semiconductor doping (injection of ions) and de-doping (ion withdrawal) effects. In contrast to OECFET, there is no movement of ions into and from the channel in the EGOFET. The advantage of ionic migration into/from the semiconductor is observed in the form of a higher transconductance of OECFET than EGOFET. The reason is that charge movement in a bulk semiconductor by way of the coupling action of electronic and ionic charges in the full volume of the channel provides a higher on-state current than charge motion restricted to a thin channel adjacent to the gate–electrolyte interface. In EGOFET, the charge carriers are electrons/holes while an OECFET uses both electrons/holes and ions as charge carriers.

8.3.2 OECFET example

A familiar example of OECFET is the transistor made using an electrolyte with poly (2,3-dihydrothieno-1,4-dioxin)–poly(styrenesulfonate) (PEDOT:PSS) (figure 8.2). Here, PEDOT is the organic semiconductor and PSS is the dopant. In the normal state, negative sulfonate ions of the PSS extract electrons from PEDOT leaving holes behind and rendering PEDOT P-type; the sulfonate contains the functional group $R-SO_3^-$ where R is an organic group. So, the PEDOT:PSS material shows a high P-type conductivity due to excess holes. Thus in the absence of a gate voltage, the PEDOT:PSS material is conducting through holes and the TFT is on-state. The pristine PEDOT:PSS film has a conductivity ~ 0.2 S cm^{-1} but can be increased to >2100 S cm^{-1} by treatment with organic solutions (Yu *et al* 2016).

Let us see what happens on application of a positive bias to the gate electrode. It causes the positive ions of the electrolyte to be pushed away from the gate electrode. These ions gain entry into the PEDOT:PSS layer. Here, the positive ions of the electrolyte neutralize the negatively charged sulfonate ions of PSS which had created the holes in PEDOT and thereby enhanced its conductivity. As more and more

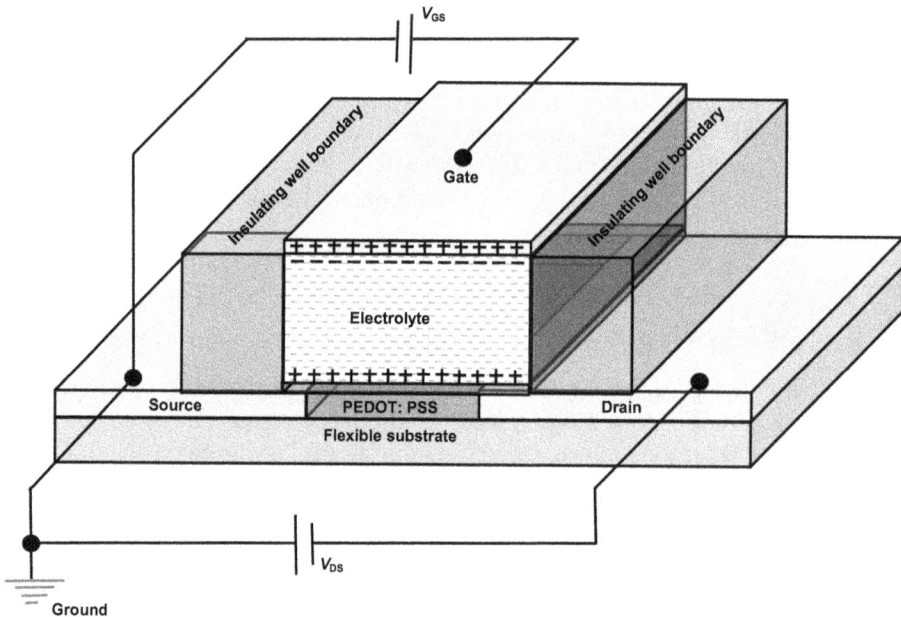

Figure 8.2. Common example of OECFET using PEDOT:PSS.

sulfonate ions are compensated by the positive ions of the electrolyte, the holes produced by them are lost. As a result, the hole population decreases and, therefore, the current flow also declines. Ultimately, the current falls to a very low level and the TFT is switched off. Conversely, by applying a negative voltage to the gate, the reverse sequence of events starts, eventually turning the TFT on.

8.4 EGOFET and OECFET with water as gate dielectric

de Oliveira *et al* (2016) fabricated two types of water-gated EGOFETs and one type of water-gated OECFET. In these devices, the substrate is poly(ethylene 2,6-naphthalate) (PEN) plastic film of thickness 150 μm and RMS roughness 6–10 nm. In the two types of EGOFETs, the semiconductor materials are pentacene and poly(3-hexylthiophene-2,5 diyl) (P3HT). In the OECFET, the conducting polymer poly (3,4-ethylenedioxythiophene):poly(styrenesulfonate) (PEDOT:PSS) is used.

8.4.1 Formation of test patterns of source/drain electrodes on the PEN substrate

Cr/Au metals are deposited on PEN substrates (figure 8.3). By photolithography, test patterns of FETs having different channel lengths and (channel width/channel length) ratios are etched out to form source/drain electrodes. Over the electrodes thus defined, the organic semiconductor films will be deposited in the following steps to fabricate the FET devices.

8.4.2 Fabrication of pentacene-based EGOFET

The test patterns of source/drain electrodes defined on the PEN substrates in the preceding step are cleaned in ethanol, CH_3CH_2OH and isopropanol, $CH_3CHOHCH_3$ or C_3H_8O. Sonication is done in each of these solvents for 15 min. Pentacene, $C_{22}H_{14}$ is deposited over the source/drain electrode pattern in a vacuum of 7.5×10^{-8} Torr with the substrate kept at room temperature. The deposition rate is 0.35 nm min^{-1}. The thickness of the pentacene film is 15 nm, which corresponds to the 10 monolayers of the material.

8.4.3 Fabrication of P3HT-based EGOFET

A solution of P3HT in chloroform, $CHCl_3$, is prepared. The concentration of the solution is 1 mg mL^{-1}. Over the cleaned source/drain electrode pattern, the P3HT solution is spin coated at 2000 RPM for 1 min. The sample is baked at 90 °C for 30 min. The thickness of P3HT film is 150 nm.

8.4.4 Fabrication of PEDOT:PSS-based OECFET

The source/drain pattern is cleaned in air plasma for 3 min. This plasma treatment is necessary for proper adhesion of the PEDOT:PSS film over the pattern. The solution of PEDOT:PSS is prepared in dimethyl sulfoxide (DMSO), $(CH_3)_2SO$ or C_2H_6OS and 3-glyci-doxypropyltrimethoxysilane, $C_9H_{20}O_5Si$ (Silquest). The solution contains 0.2% v/v Silquest and 5% v/v DMSO. It is deposited over the source/drain electrode pattern by spin coating. The coating process consists of three steps:

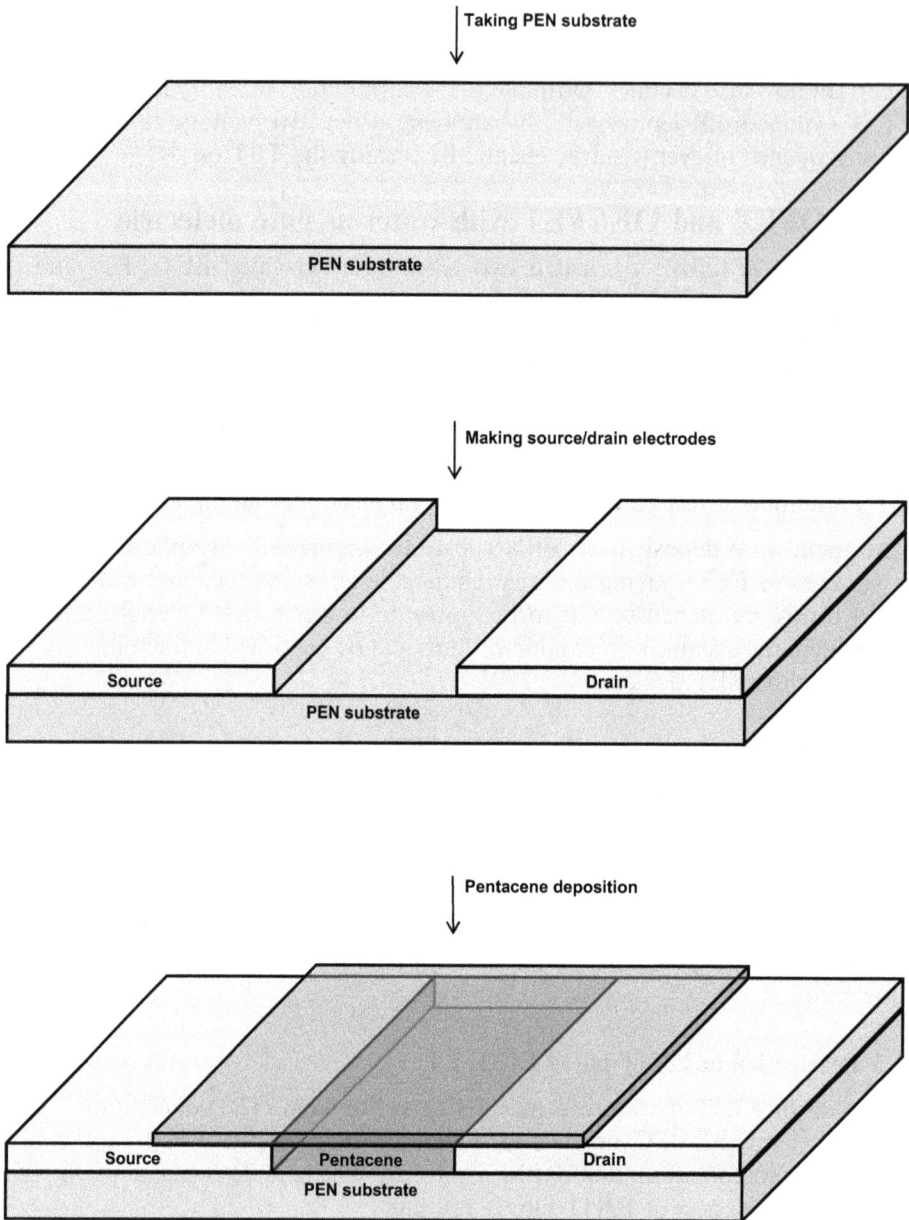

Figure 8.3. Pentacene-based EGOFET with water as gate dielectric.

Mounting the PDMS reservoir
containing the electrolyte

Electrolyte (water)

PDMS reservoir

Source
Pentacene
Drain
PEN substrate

Making the electrical connections

V_{DS}

V_{GS}

Gate (Platinum wire)

Electrolyte (water)

PDMS reservoir

Source
Pentacene
Drain
PEN substrate

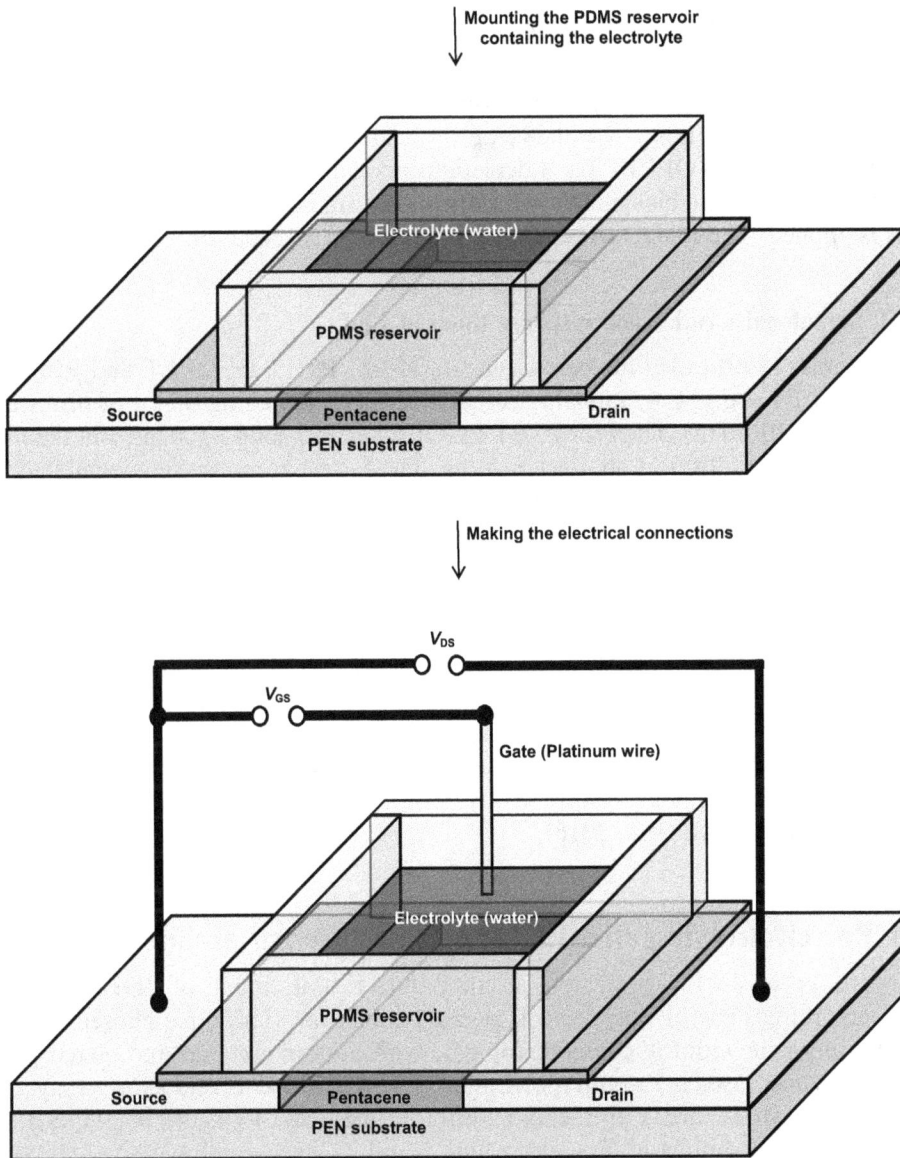

Figure 8.3. (Continued.)

500 RPM for 5 s, 1000 RPM for 10 s and finally 2000 RPM for 35 s. Baking is done at 50 °C for 30 min. The thickness of PEDOT:PSS layer is 100–200 nm.

8.4.5 Formation of PDMS reservoir on all devices

A PDMS reservoir is mounted on the top surface of all FETs. It encloses the electrolyte (here water) and borders it so as to cover the active gate area. A platinum wire is dipped inside the electrolyte to apply the gate voltage. It has a diameter of 1 mm.

8.4.6 Threshold voltage, transconductance and mobility values for different FETs

For the pentacene EGOFET, the threshold voltage V_{Th} is -0.21 V, the trans-conductance g_m is 0.012 µS and the mobility μ is 5.9×10^{-5} cm^2 V^{-1} s^{-1}.

For P3HT EGOFET, $V_{Th} = 0.28$ V, $g_m = 0.003$ µS and $\mu = 2 \times 10^{-5}$ cm^2 V^{-1} s^{-1}.

The PEDOT:PSS OECFET is a depletion-mode device which acts as a conductor in the absence of gate bias but shows semiconducting behavior when a positive gate bias is applied. The PEDOT:PSS OECFET has a $g_m = 2000$ µS.

8.4.7 Signal–noise ratios and response times of FETs

Signal–noise (S/N) ratios for pentacene EGOFET, P3HT EGOFET and PEDOT:PSS OECFET are 194, 4.5 and 111 respectively. The switching time τ of pentacene EGOFET is 10–20 ms. The τ for P3HT EGOFET is very long >100 ms. But the τ for PEDOT:PSS OECFET is shortest $= 4$ ms. The fastest response time exhibited by OECFET in comparison to EGOFETs is contrary to expectations because a disadvantage of OECFET is its sluggish response time relative to EGOTFT owing to slower movements of heavy ions as compared to faster motion of lightweight electrons. However, it is believed that ionic transport in hydrated PEDOT:PSS takes place much like the transport in water (Stavrinidou *et al* 2013). Furthermore, the higher charge density in OECFET leads to larger electronic conduction. As a result, the resistance R of the resistance–capacitance (RC) circuit decreases. Since the RC circuit determines the time response, the τ for PEDOT:PSS OECFET is lowest. Thus, judged on the above figures of merit, the PEDOT:PSS OECFET exhibits the best overall performance followed by pentacene EGOFET and then P3HT EGOFET (de Oliveira *et al* 2016).

8.5 Polyelectrolyte-gated EGOTFTs of different architectures

Dumitru *et al* (2013) presented a comparative assessment of three different architectural designs of polyelectrolyte-gated transistor. The three chosen designs are: bottom gate with top contacts, top gate with bottom contacts and planar gate.

The substrate used is Kapton (polyimide). The organic semiconductor is poly(2,5-bis(3-tetradecylthiophen-2-yl)thieno[3,2-b]thiophene) (pBTTT-C14), $(C_{42}H_{62}S_4)_n$. It is a P-type semiconductor. The electrolyte is poly(4-styrenesulfonic acid) (PSSH), $(C_8H_8O_3S)_n$. It is a polyanionic proton conductor.

8.5.1 Fabrication of the structure: bottom gate with top contacts

This structure is: PI substrate/gate electrode (Au)/electrolyte (PSSH)/organic semiconductor (pBTTT-C14)/source/drain electrodes (Au) (figure 8.4). The gate electrode is a gold strip produced by thermal evaporation of gold on PI. The electrolyte is deposited by spin coating at 2000 RPM for 1 min on the gold strip. The organic semiconductor too is spun over the electrolyte layer. The source/drain electrodes are deposited by thermal evaporation through a shadow mask.

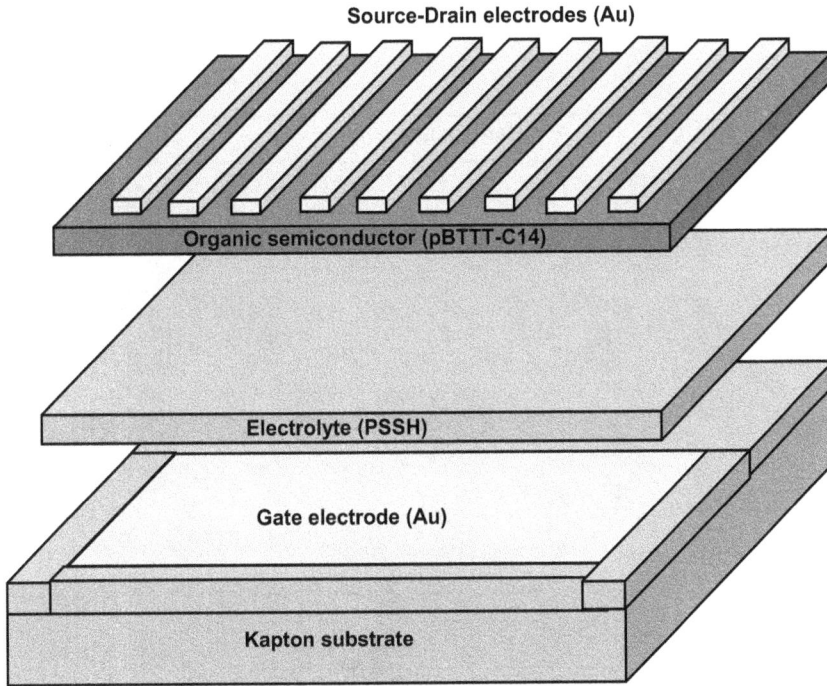

Figure 8.4. TFT with bottom gate and top contacts.

8.5.2 Fabrication of the structure: top gate with bottom contacts

This has the structure: PI substrate/source/drain electrodes (Au)/organic semiconductor (pBTTT-C14)/electrolyte (PSSH)/gate electrode (Au) (figure 8.5). Different layers are formed in exactly the same way as in the previous case of the structure having bottom gate with top contacts. Only the sequence of deposition of layers is altered.

8.5.3 Fabrication of planar gate structure

Its structure is: PI substrate/source/drain electrodes (Au) and gate electrode (Au)/organic semiconductor (pBTTT-C14)/electrolyte (PSSH) (figure 8.6). All the electrodes: source, drain and gate are deposited in one step, followed by spinning of organic semiconductor, baking at 110 °C on a hot plate for 1 min, and finally drop-casting the electrolyte over the gate and source/drain electrodes with doctor blading.

8.5.4 Threshold voltage, on–off current ratio and field-effect mobility for the three structures

For the bottom gate, top contacts structure, the threshold voltage V_{Th} is -0.27 V, the on–off current ratio I_{on}/I_{off} is 100 and the mobility is 1.07 cm^2 V^{-1} s^{-1}.

For the top gate, bottom contacts structure, $V_{Th} = 0.161$ V, $I_{on}/I_{off} = 10$ and the mobility is 4.95×10^{-3} cm^2 V^{-1} s^{-1}.

For the planar structure, $V_{Th} = -0.21$ V, $I_{on}/I_{off} = 10$ and the mobility is 0.37 cm^2 V^{-1} s^{-1}.

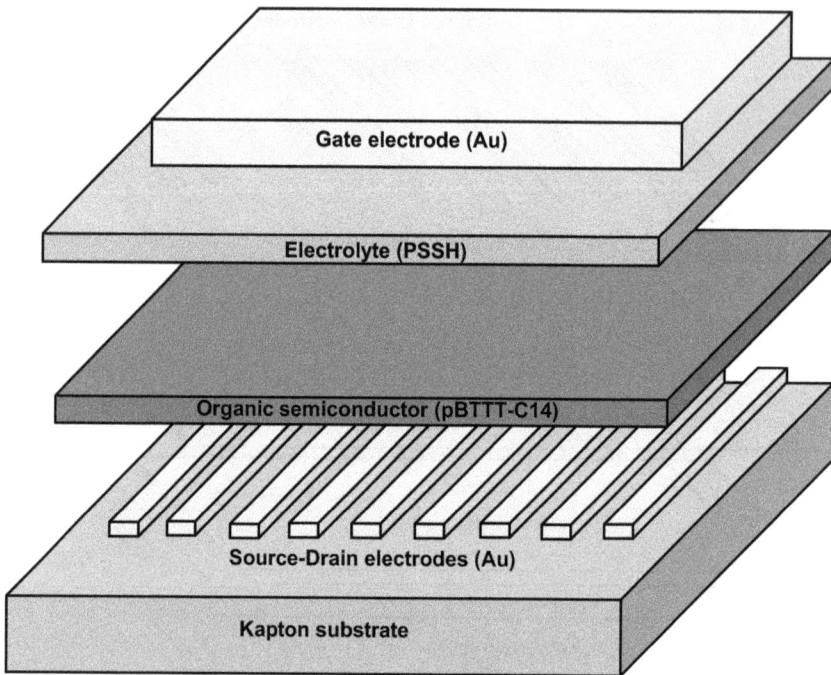

Figure 8.5. TFT with top gate and bottom contacts.

It is found that the three structures are promising for flexible electronics. Particularly, the planar gate structure considerably simplifies the fabrication process because the source/drain and gate electrodes can be patterned in a single step (Dumitru *et al* 2013).

8.6 Vertical architecture OECFET

Kawahara *et al* (2013) reported a novel vertical architecture OECFET made by printing techniques. They adopt the configuration of a multi-stack structure. This OECFET is fabricated on both sides of plastic/paper substrates (figure 8.7). On the top surface of the substrate lie the source, PEDOT:PSS layer, the electrolyte and the gate. The drain contact is located on the bottom surface of the substrate. Between the gate electrode and the PEDOT:PSS layer lies the electrolyte. Electrical vias are made with carbon paste.

The OECFET is made by printing techniques and is a depletion-mode device. The plastic substrate is a PET film. Its thickness is 50 μm. Polyethylene $(C_2H_4)_n$-coated photo paper is used as the paper substrate. PEDOT:PSS is the organic semiconductor. The electrolyte is poly[sodium(quaternary ammonium)]. It is a water-based polyelectrolyte.

8.6.1 OECFET fabrication on PET substrate

A 30–70 μm diameter hole is drilled in the PET substrate by a carbon dioxide laser drilling machine. The hole is filled with carbon paste with a pipette. The drain

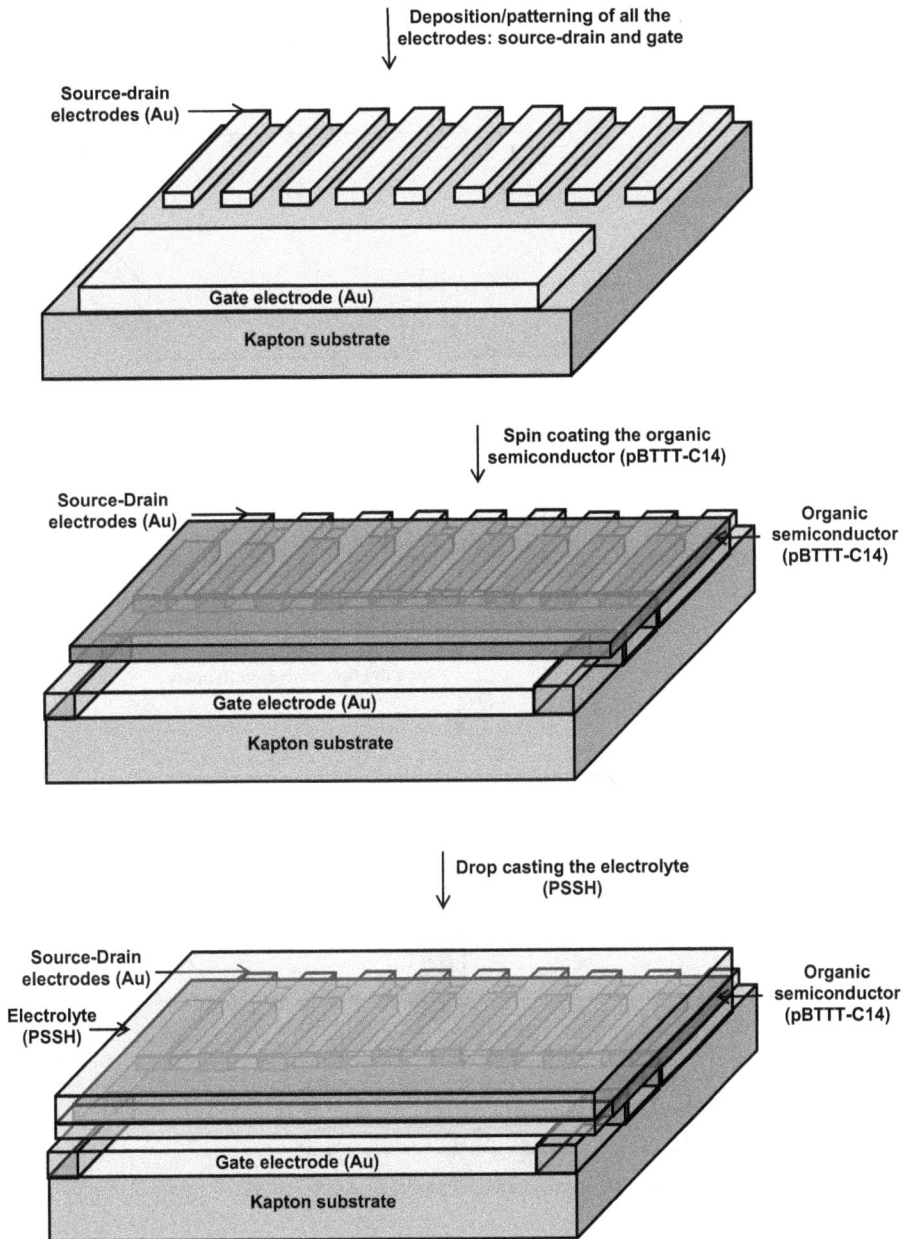

Figure 8.6. Planar TFT.

electrode is defined on the bottom surface of the substrate. A 100 μm × 500 μm rectangle of PEDOT:PSS is defined by inkjet printer on the top surface of the substrate using PEDOT:PSS inkjet printing ink. The rectangle overlaps the carbon paste-filled via hole making contact with the drain electrode below. The PEDOT: PSS ink is dried at 120 °C for 5 min. Carbon paste is used to make the source

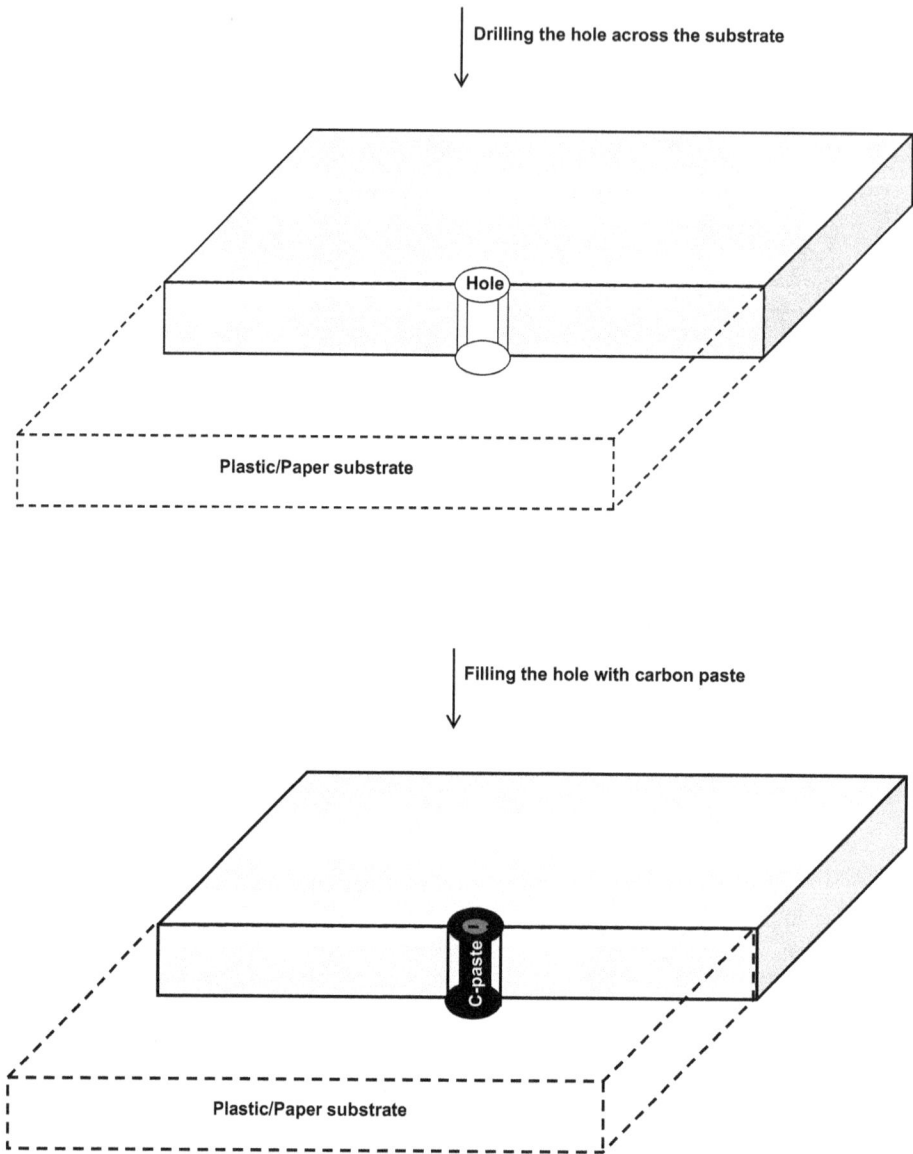

Figure 8.7. Vertical architecture electrochemical TFT.

Defining the drain electrode on the bottom surface of the substrate with C-paste

C-paste

Drain electrode (C)

Plastic/Paper substrate

Printing a PEDOT:PSS rectangle on the top surface of the substrate

C-paste

Drain electrode (C)

PEDOT:PSS (Organic semiconductor)

Plastic/Paper substrate

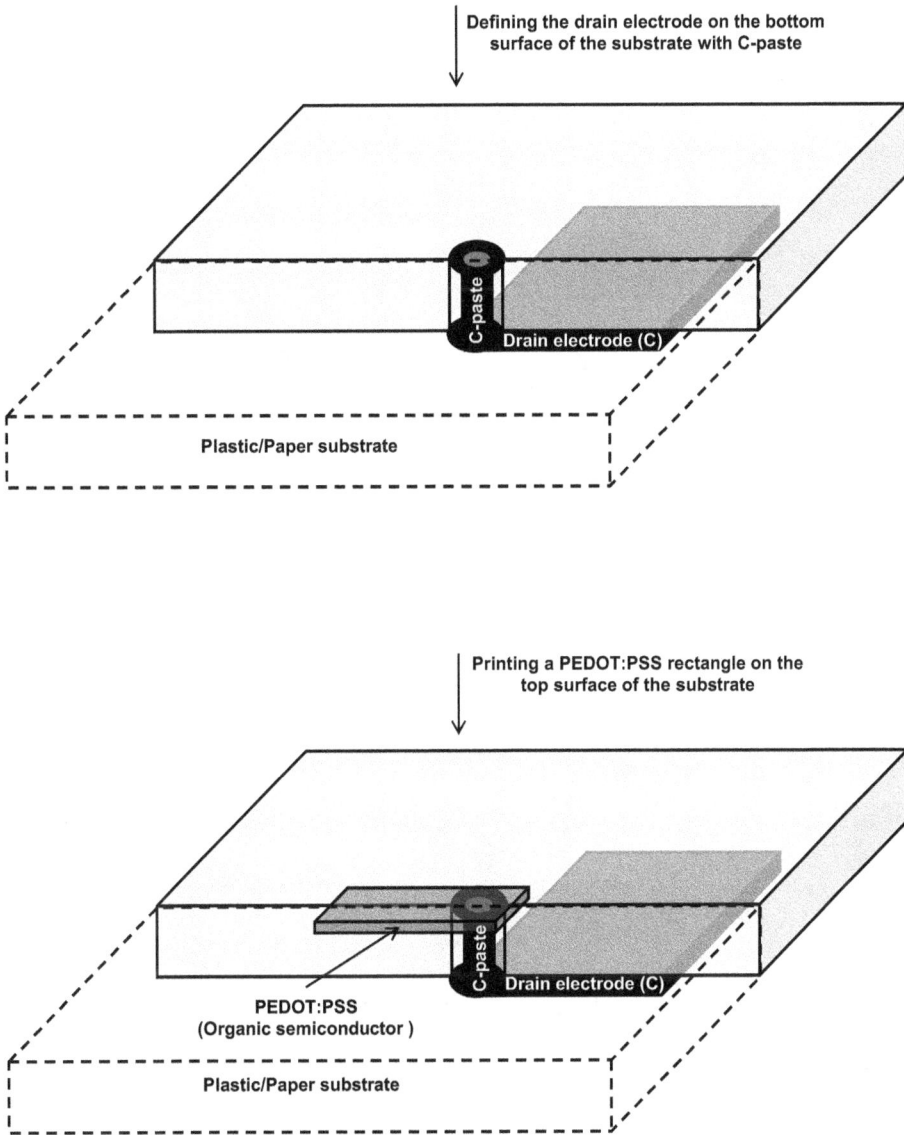

Figure 8.7. (Continued.)

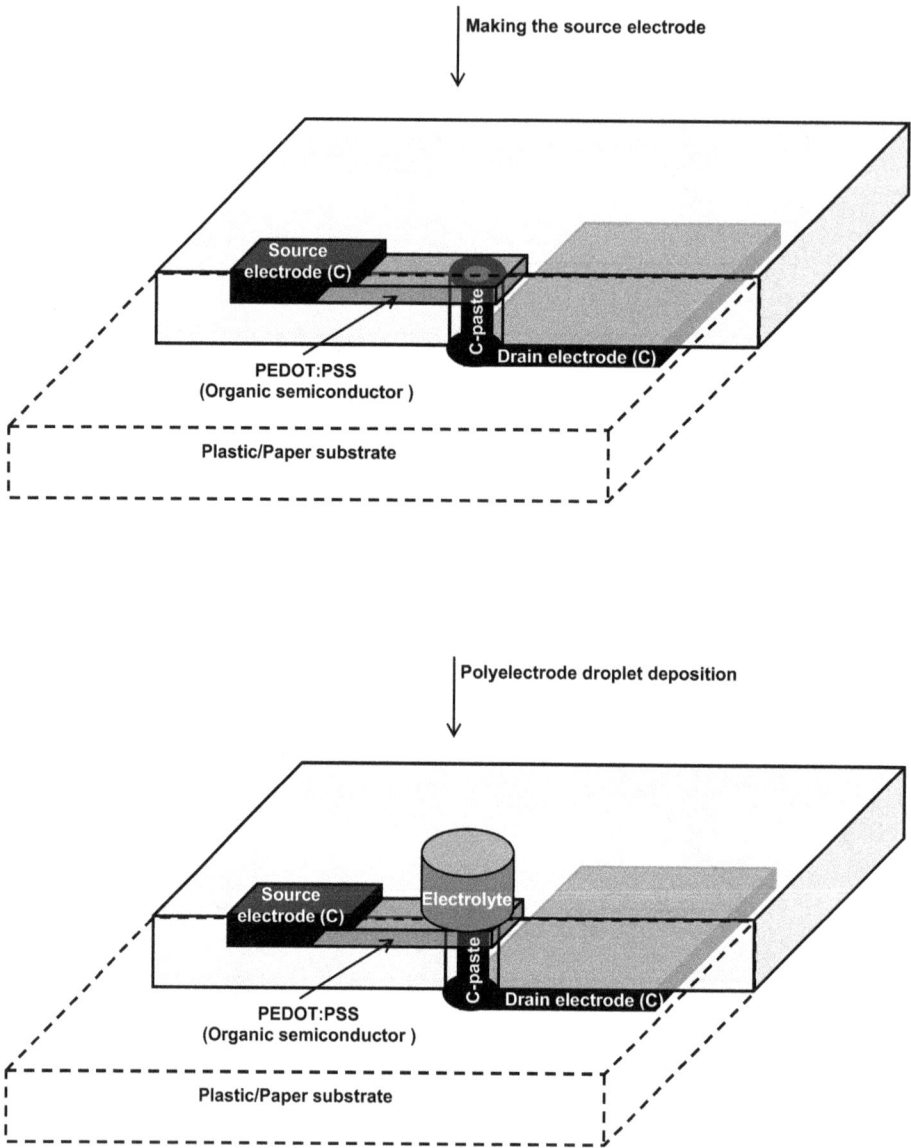

Figure 8.7. (Continued.)

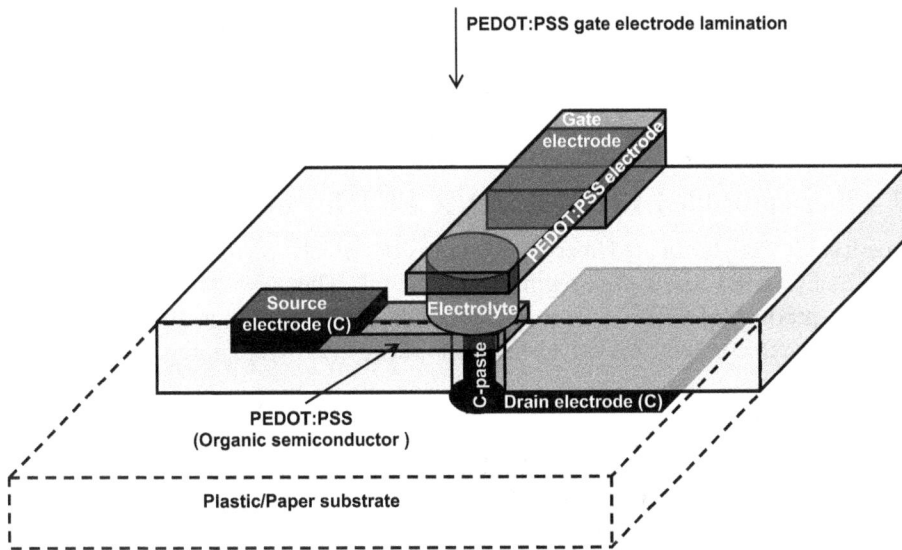

Figure 8.7. (Continued.)

electrode on the top surface near the edge of the PEDOT:PSS rectangle away from the drain electrode on the bottom surface. Over the PEDOT:PSS layer, a droplet of polyelectrolyte measuring 2–3 µL is deposited. It is dried at 60 °C for 1 min yielding a dried electrolyte droplet of size 200–300 µm. Thereby complete coverage of the via hole by the electrolyte is guaranteed. Further, it is assured that the electrolyte and the carbon source contact pad are 100–200 µm apart. A commercial PEDOT:PSS conductive polymer electrode is laminated on the electrolyte droplet. It serves as the gate electrode.

8.6.2 OECFET fabrication on paper substrate

The via hole, made by punching with a metallic pin, has a diameter of 200–300 µm. The same procedure is followed as for the PET substrate except that the PEDOT:PSS layer is formed by screen printing and the polyelectrolyte droplet size is increased to 10–15 µL because of the larger size of via hole.

8.6.3 Switching times and current on–off ratios

For the OECFET made on a PET substrate, the time taken to turn off the device from the on-state is 5 ms while time necessary to turn it on again is ~20 ms. In the case of OECFET made on a paper substrate, the on-to-off switching time is 10–20 ms while off-to-on time is 50–60 ms. The longer switching times for the device on a paper substrate as compared to those on a PET substrate arise from the larger diameter via hole so that the effective volume of PEDOT:PSS channel is increased. Moreover, on the PET substrate OECFET geometry is inkjet printed and, therefore, has a comparatively thinner channel than the screen printed OECFET geometry on a paper substrate.

Like the switching times, the current on–off ratio for the OECFET on a paper substrate (280) is inferior to that on a PET substrate (4000). Nonetheless, the on current level is the same for both the devices being an inherent property of PEDOT: PSS material (Kawahara *et al* 2013).

8.7 Fiber-embedded EGOFET/OECFET for e-textiles

Hamedi *et al* (2009) fabricated fiber-based transistors which operate in two regimes: a fast regime as EGOFET and a slow regime as OECFET. They give a new idea of unifying electronic production with textile manufacturing. The substrate is a fiber (figure 8.8). The organic semiconductor is poly(3-hexylthiophene-2,5-diyl) regioregular, $(C_{10}H_{14}S)_n$.

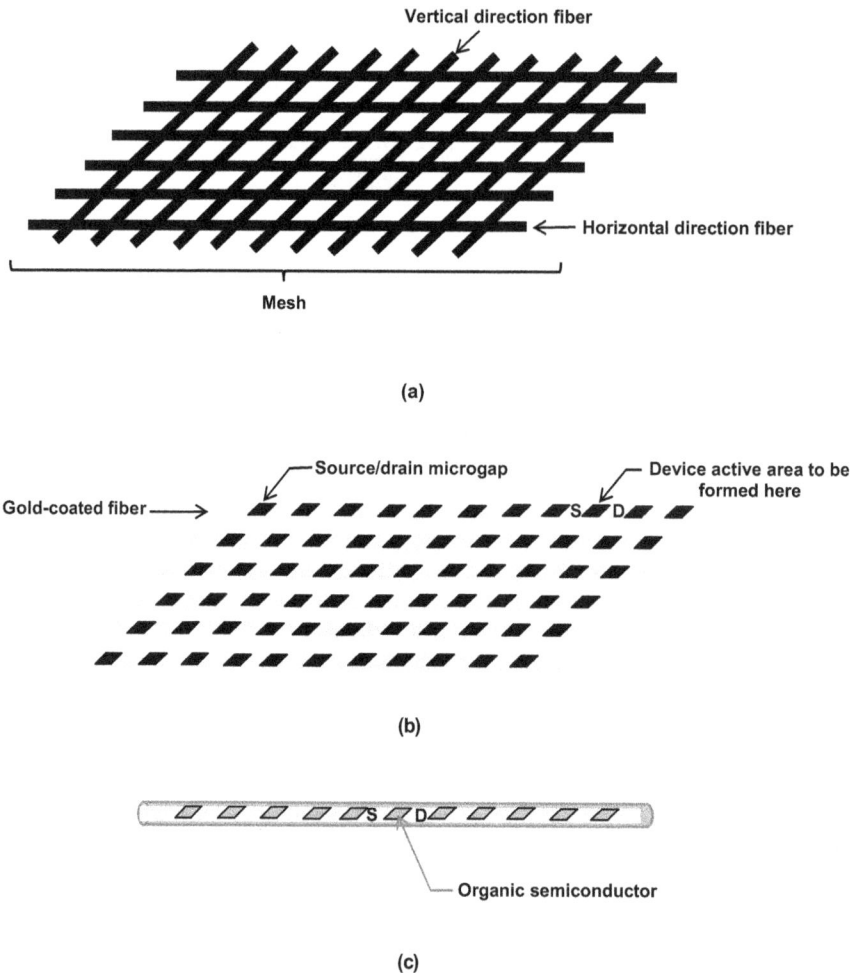

(a)

(b)

(c)

Figure 8.8. Fiber-embedded EGOFET: (a) weaving the mesh, (b) gold evaporation using vertical direction fibers as shadow masks, (c) P3HT coating over source/drain microgaps, (d) sewing gold wire as gate electrode, and (e) electrolyte droplet placement.

(d)

(e)

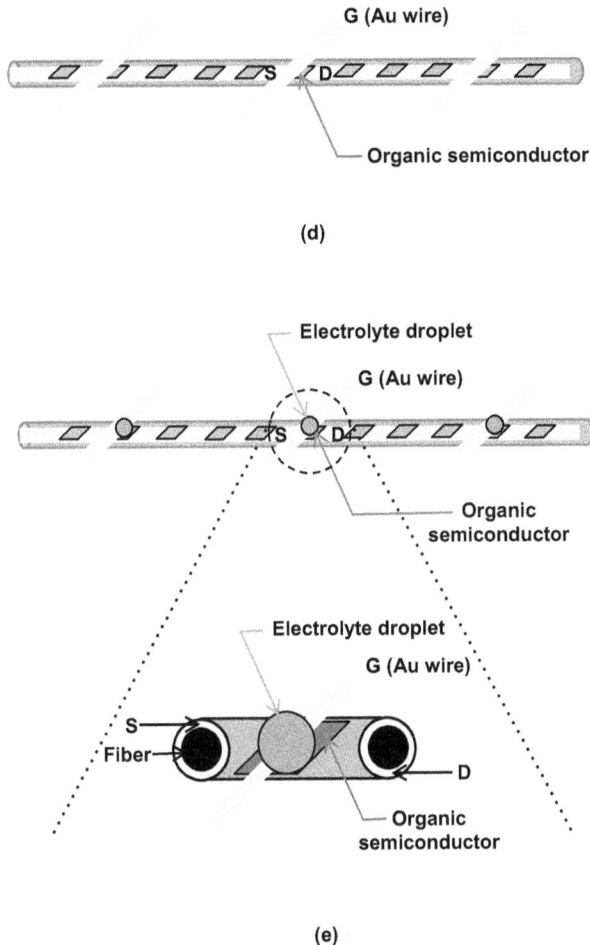

Figure 8.8. (Continued.)

Imidazolium ($C_3H_5N_2^+$) ionic liquid-based electrolyte is used which is a solid ionic liquid made up of a mixture of 1-butyl-3-methylimidazolium bis(trifluoromethylsulfonyl)imide $C_{10}H_{15}F_6N_3O_4S_2$ [(bmim)(Tf2N)] with poly(1-vinyl-3-methylimidazolium) bis(tri-fluo-romethanesulfonimide) [poly(ViEtIm)(Tf2N)] (Marcilla *et al* 2006). The source, drain and gate electrodes are made of gold.

8.7.1 Creation of source/drain microgaps

The fibers are woven into a mesh with one series of fibers overlying the other. Thus a mesh is formed by the intersection of fibers. Suppose the fibers are placed along horizontal and vertical directions with the vertical direction fibers lying over the horizontal direction fibers, then the vertical direction fibers will constitute a shadow mask when gold is evaporated over the mesh. After evaporation, the horizontal direction fibers lying below the vertical ones will be covered with gold but there will be microgaps at the points of location of vertical direction fibers. The gap regions will be active areas of the devices where the organic semiconductor film will be deposited and the gold covered fiber regions on opposite sides of the gaps will serve as the source/drain electrodes. The gold thickness is 100 nm.

8.7.2 Dip coating of P3HT over the source/drain gaps

A solution of P3HT, regioregular in chloroform, $CHCl_3$ is prepared. Its strength is 4 mg mL^{-1}. By dip coating, the P3HT layer is formed over the predefined source/drain gaps. The thickness of the P3HT film is 100 nm.

8.7.3 Sewing gold wires on the P3HT film

Gold wires are sewn at right angles to the Au/P3HT coated fibers with the gold wire positioned exactly above the P3HT film applied on the gold film on the microgap. These gold wires act as the gate electrodes.

8.7.4 Placement of electrolyte

A drop of the electrolyte is poured over the junction of the gold wire with the P3HT film. Thus the EGOFET structure is complete: fiber/Au film (100 nm)/P3HT layer (100 nm)/Au wire/solid polymeric ionic liquid electrolyte.

8.7.5 Two stages of channel formation

The EGOFET works as a P-channel enhancement mode device. Its operating voltage is <−1 V. It gives a high saturation current (−30 μA). Current on–off ratio is ~1000.

The temporal response of the device is determined by applying a step voltage to the gate electrode from 0 V to −1 V. The channel is established in <3 ms. However, after reaching the first plateau at 0.2 μA, the current continues to increase slowly, and finally attains the value of 30 μA in 2 s. This observation reveals that the device works in two modes. The initial fast current rise is ascribed to the EGOFET mode due to the electrical double layer capacitance ~5 μF cm^{-2}. The subsequent slow current upswing is attributed to the OECFET mode. It takes about 2 s to set up equilibrium as anions penetrate into the P3HT layer.

The use of ionic liquid electrolyte makes the transistor characteristics stable with respect to time. A functional circuit with two transistors is demonstrated. The circuit is fully woven (Hamedi *et al* 2009).

8.8 Discussion and conclusions

Using an electrolyte as the gate dielectric solves the high operating voltage problem of an organic TFT. Many different types of devices have been fabricated in the EGOFET configurations. Recalling the threshold voltages: for the pentacene EGOFET, V_{Th} is −0.21 V, for P3HT EGOFET, $V_{Th} = 0.28$ V (de Oliveira *et al* 2016); for the bottom gate, the top contact structure, $V_{Th} = -0.27$ V, for the top gate, the bottom contact structure, $V_{Th} = 0.161$ V, for the planar structure, $V_{Th} = -0.21$ V (Dumitru *et al* 2013).

The closely resembling OECFET devices have also been demonstrated. In particular, water-gated TFTs are very effective for the conversion of biochemical responses into electrical signals. For unification of the device with clothing, a fiber-embedded TFT is developed. This effort provides a seamless and comfortable integration of device with wearable clothing in contrast with the cruder approach of attaching available discrete devices into clothes for health monitoring.

Review exercises

8.1 What is the gate dielectric in: (i) EGOFET and (ii) conventional OFET? Where is the gate electrode placed in EGOFET?

8.2 Does replacement of the conventional gate dielectric film change the principle of operation of an EGOFET? Describe in detail how an EGOFET works?

8.3 What are the two parts of the electrical double layer formed at an electrolyte/semiconductor interface? What is the typical thickness and capacitance range of the electrical double layer?

8.4 What is the advantage offered by the electrical double layer capacitance over the capacitance of insulating film used in a traditional MOS device?

8.5 Does an OECFET operate on capacitive action? How does the working principle of an OECFET differ from that of an EGOFET?

8.6 Is current flow in an OECFET restricted to a thin channel layer adjacent to the electrolyte/semiconductor interface or does the bulk semiconductor also participate in current conduction?

8.7 Which has a higher transconductance: OECFET or OEGFET? Why?

8.8 In PEDOT:PSS material, which is the organic semiconductor? Which is the dopant?

8.9 In the normal state, does PEDOT:PSS show P-type or N-type conductivity? Explain giving the reason.

8.10 What happens when a positive bias is applied to the gate electrode of a PEDOT:PSS transistor? What happens with a negative gate bias?

8.11 How is PEDOT:PSS OECFET switched off and switched on?

8.12 Are pentacene and P3HT deposited on the source/drain electrode pattern by the same method? How is PEDOT:PSS deposited?

8.13 Is PEDOT:PSS OECFET a depletion or enhancement mode device?

8.14 Which has the highest signal-to-noise ratio: pentacene EGOFET, P3HT EGOFET or PEDOT:PSS OECFET?

8.15 Which has the shortest switching time: pentacene EGOFET, P3HT EGOFET or PEDOT:PSS OECFET?

8.16 Why is the PEDOT:PSS OECFET fastest in response even though ions move more slowly than electrons?

8.17 Name the three different architectural designs fabricated with pBTTT-C14 semiconductor? Which of these designs simplifies the fabrication? Why?

8.18 For the pBTTT-C14 semiconductor TFT, name: (i) the substrate and (ii) the electrolyte used.

8.19 Highlight the novelty and special features of the vertical architecture OECFET. Is it a normally-on or normally-off device?

8.20 For the vertical architecture OECFET, name the materials used for: (i) the substrate, (ii) the electrolyte, (iii) the semiconductor and (iv) the source/drain electrodes, and (v) the gate electrode.

8.21 Describe the fabrication process of a vertical architecture OECFET on paper and PET substrates. Which of the two OECFETs, the one on paper or the one on PET substrate exhibits inferior characteristics? Why?

8.22 In the fiber-embedded EGOFET/OECFET, name the: (i) substrate, (ii) semiconductor, (iii) ionic liquid-based electrolyte, (iv) metal used for gate and source/drain electrodes.

8.23 How are the source/drain microgaps created for the fabrication of fiber-embedded EGOFET/OECFET?

8.24 How are gold wires sewn on the P3HT film to act as gate electrodes of the fiber-embedded EGOFET/OECFET?

8.25 Describe the two stages of channel formation observed in the fiber-embedded EGOFET/OECFET. What do these stages signify? Explain.

8.26 How do you infer that fiber-embedded EGOFET/OECFET works in two modes? What are these two modes?

References

de Oliveira R F, Casalini S, Cramer T, Leonardi F, Ferreira M, Vinciguerra V, Casuscell V, Alves N, Murgia M, Occhipinti L and Biscarini F 2016 Water-gated organic transistors on polyethylene naphthalate films *Flex. Print. Electron.* **1** 025005

Dumitru L, Manoli K, Magliulo M and Torsi L 2013 Comparison between different architectures of an electrolyte-gated organic thin-film transistor fabricated on flexible Kapton substrates *5th IEEE International Workshop on Advances in Sensors and Interfaces (IWASI) (Bari, 13–14 June 2013)* pp 91–4

Hamedi M, Herlogsson L, Crispin X, Marcilla R, Berggren M and Inganäs O 2009 Fiber-embedded electrolyte-gated field-effect transistors for e-textiles *Adv. Mater.* **21** 573–7

Kawahara J, Ersman P A, Katoh K and Berggren M 2013 Fast-switching printed organic electrochemical transistors including electronic vias through plastic and paper substrates *IEEE Trans. Electron Devices* **60** 2052–6

Marcilla R, Alcaide F, Sardon H, Pomposo J A, Pozo-Gonzalo C and Mecerreyes D 2006 Tailor-made polymer electrolytes based upon ionic liquids and their application in all-plastic electrochromic device *Electrochem. Commun.* **8** 482–8

Rivnay J, Inal S, Salleo A, Owens R M, Berggren M and Malliaras G G 2018 Organic electrochemical transistors *Nat. Rev. Mater.* **3** 17086

Stavrinidou E, Leleux P, Rajaona H, Khodagholy D, Rivnay J, Lindau M, Sanaur S and Malliaras G G 2013 Direct measurement of ion mobility in a conducting polymer *Adv. Mater.* **25** 4488–93

Wang D, Noël V and Piro B 2016 Electrolytic gated organic field-effect transistors for application in biosensors—A review *Electronics* **5** 9

Yu Z, Xia Y, Du D and Ouyang J 2016 PEDOT:PSS films with metallic conductivity through a treatment with common organic solutions of organic salts and their application as a transparent electrode of polymer solar cells *ACS Appl. Mater. Interfaces* **8** 11629–38

Chapter 9

2D-material TFT

A graphene TFT is fabricated by synthesizing graphene on copper foil and transferring it to a polyimide film by the PMMA-assisted wet transfer process. The polyimide film along with the gate electrode and exfoliated h-BN dielectric layer is already pulled away from a Si/SiO$_2$ wafer by a capture and release process. Mobilities >2300 cm^2 V^{-1} s^{-1} are achieved in graphene TFT (Lee *et al* 2013). The 260 nm channel length graphene TFT on a PEN substrate is fabricated by applying the LPCVD graphene on copper foil over the HfO$_2$ gate dielectric by a dry-transfer method. The low-field mobility is 1000 cm^2 V^{-1} s^{-1}. The extrinsic unity-power-gain frequency is 7.6 GHz, which degrades by <20% of unstrained value during 0%–2% straining (Petrone *et al* 2015). Special qualities of flexible glass, notably higher glass transition temperature and thermal conductivity with lower surface roughness than polyimide or PET are exploited in the graphene TFT on a flexible glass substrate which uses nanoscale polyimide as the gate dielectric, yielding electron (hole) mobility of 4540 (1100) cm^2 V^{-1} s^{-1} and intrinsic cut-off frequency of 95 GHz (Park *et al* 2016). MoS$_2$ TFT devices with low-field mobility of 30 cm^2 V^{-1} s^{-1} are fabricated on Kapton by extrafoliating MoS$_2$ from commercial crystals onto the Al$_2$O$_3$ or HfO$_2$ gate dielectric. Al$_2$O$_3$-gated devices fail at a bending radius of 2 mm while HfO$_2$-gated devices work up to 1 mm (Chang *et al* 2013). WS$_2$ TFT is fabricated on a polyimide film by PMMA-aided transferring of WS$_2$ film synthesized by sulfiding WO$_3$ powder over the Al$_2$O$_3$ gate dielectric. The maximum on-current of the TFT diminishes by 20% after 30 000 bending cycles and by 30% after 50 000 cycles at 2 mm radius (Gong *et al* 2016). An all-2D WSe$_2$ TFT is fabricated on a PET substrate with h-BN as gate insulator and electrodes for gate, source and drain made of graphene. The drain/source current of this TFT remains practically undisturbed up to 2% strain (Das *et al* 2014).

9.1 Introduction

Graphene, a 2D semiconducting carbon monolayer sheet, is an ideal material for flexible electronics because of its exceptional electronic and mechanical properties: high

doi:10.1088/2053-2563/ab0d18ch9

mobility >10 000 cm^2 V^{-1} s^{-1}, ambipolar charge transport, carrier saturation velocity 1–5 × 10^7 cm s^{-1}, and mechanical strain limit up to 25%.

The greatest shortcoming of graphene is the lack of a bandgap. Consequently, low-power switching transistors are not realizable with graphene. This weakness of graphene has compelled researchers to look for layered atomic sheets with sufficient bandgaps. Semiconducting transition metal dichalcogenides (TMDs), e.g. molybdenum disulfide (MoS$_2$), tungsten disulfide (WS$_2$), molybdenum diselenide (MoSe$_2$), tungsten diselenide (WSe$_2$), etc provide an answer.

9.2 Graphene TFT on polyimide

Lee *et al* (2013) implemented graphene TFTs by choosing hexagonal boron nitride (h-BN), an isomorph of graphene with B and N atoms in a hexagonal lattice, as the material for the gate dielectric film (figure 9.1). The justifying reasons for choosing h-BN are:

  (i) its small lattice mismatch ~1.8% relative to graphene;
  (ii) reduction of charge fluctuations by virtue of its surface flatness whereby nearly ideal graphene properties are achieved without complex suspended structures;
  (iii) cleavage and exfoliation in the same manner as graphene resulting from strong intra-plane bonding and frail inter-plane bonding through Van der Waals forces.

9.2.1 Capture–release process for PI film with gate electrode and h-BN dielectric

The h-BN film exfoliated on thermally grown SiO$_2$ (285 nm) on Si wafer, is annealed at 300 °C in N$_2$ for 1 h to smoothen its surface and remove any leftover adhesive traces. On the h-BN/SiO$_2$/Si structure thus obtained, gate electrodes are defined. For this purpose, e-beam lithography, evaporation and lift-off techniques are used. On the gate electrode patterned surface, liquid polyimide is spin-coated. On curing the polyimide in nitrogen, the polyimide layer captures the device structures within itself. The PI film with the embedded gate electrode and h-BN gate dielectric is released by etching away SiO$_2$ in buffered oxide etchant. The above process is called the capture–release process. To ensure a clean surface for receiving the graphene the PI film with gate electrode and h-BN dielectric is annealed again

9.2.2 Synthesis of graphene film

First, a copper foil of thickness 25 μm is taken. The graphene growth is done in a hot-wall furnace containing a fused silica tube (Li *et al* 2009). To grow the graphene, the Cu foil is placed inside the silica tube, and the tube is loaded into the furnace. After evacuation, the tube is filled with hydrogen and is heated to 1000 °C. The pressure of hydrogen gas is maintained at 40 mTorr with a flow rate of 2 sccm. The Cu foil is stabilized at the desired temperature up to 1000 °C. Then, methane gas is introduced for the required period of time. The total pressure is 500 Torr and the methane flow rate is 35 sccm. Finally, the furnace is cooled down until room temperature is attained. Graphene film grows on the copper foil.

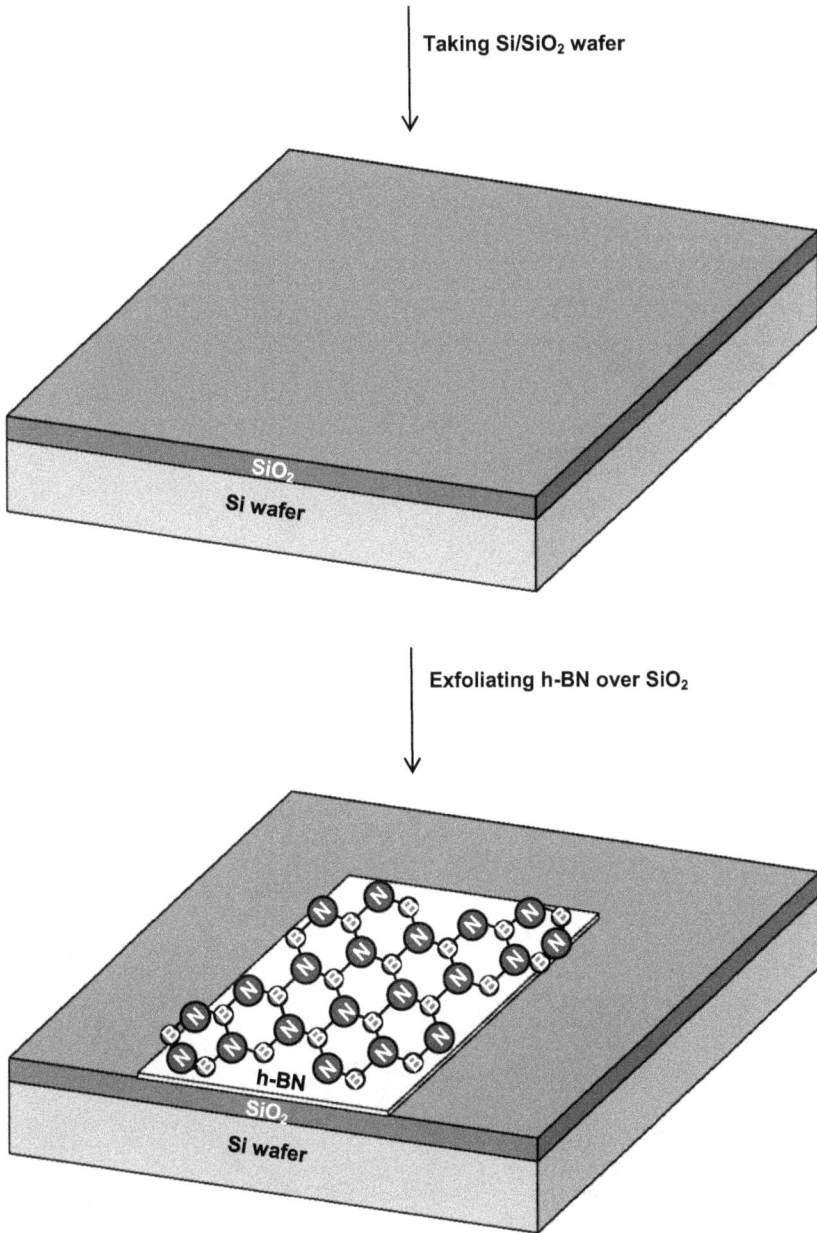

Figure 9.1. TFT on polyimide with a graphene semiconducting layer.

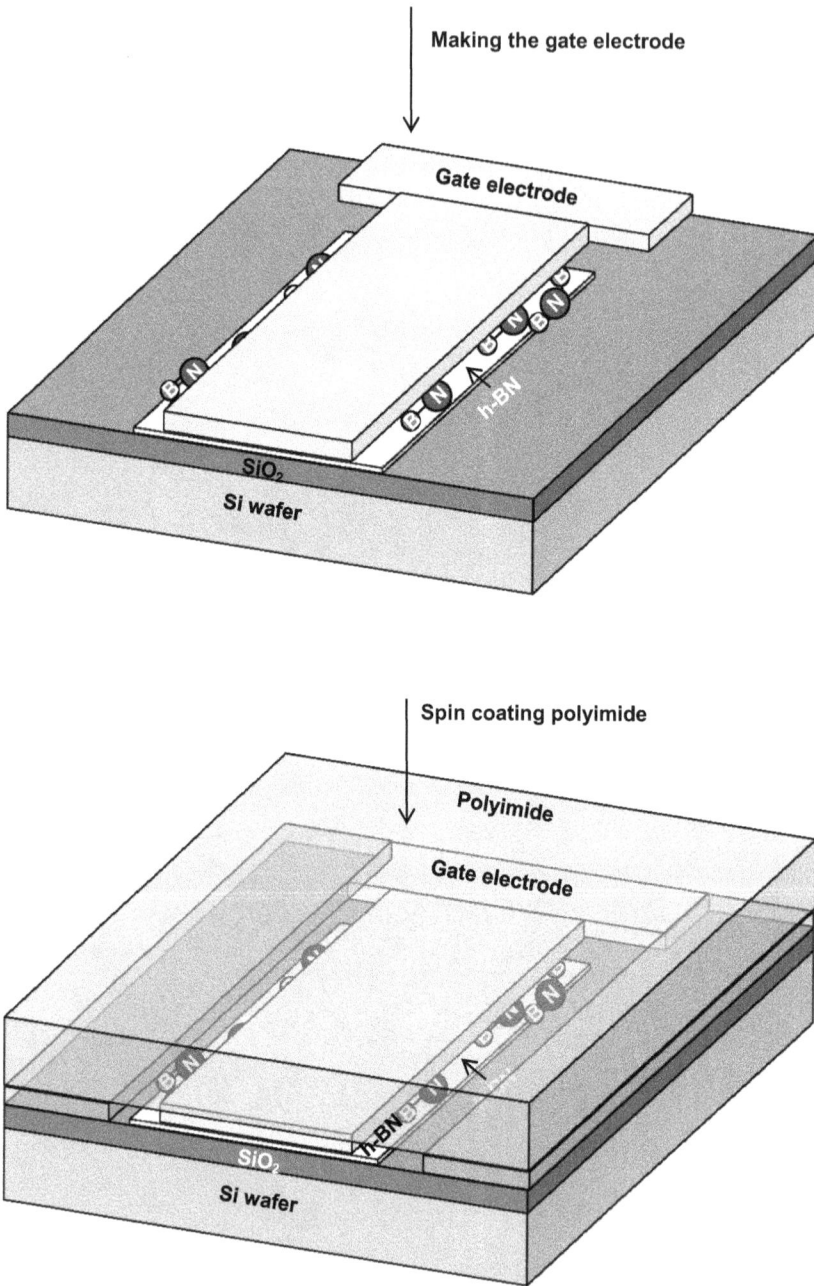

Figure 9.1. (Continued.)

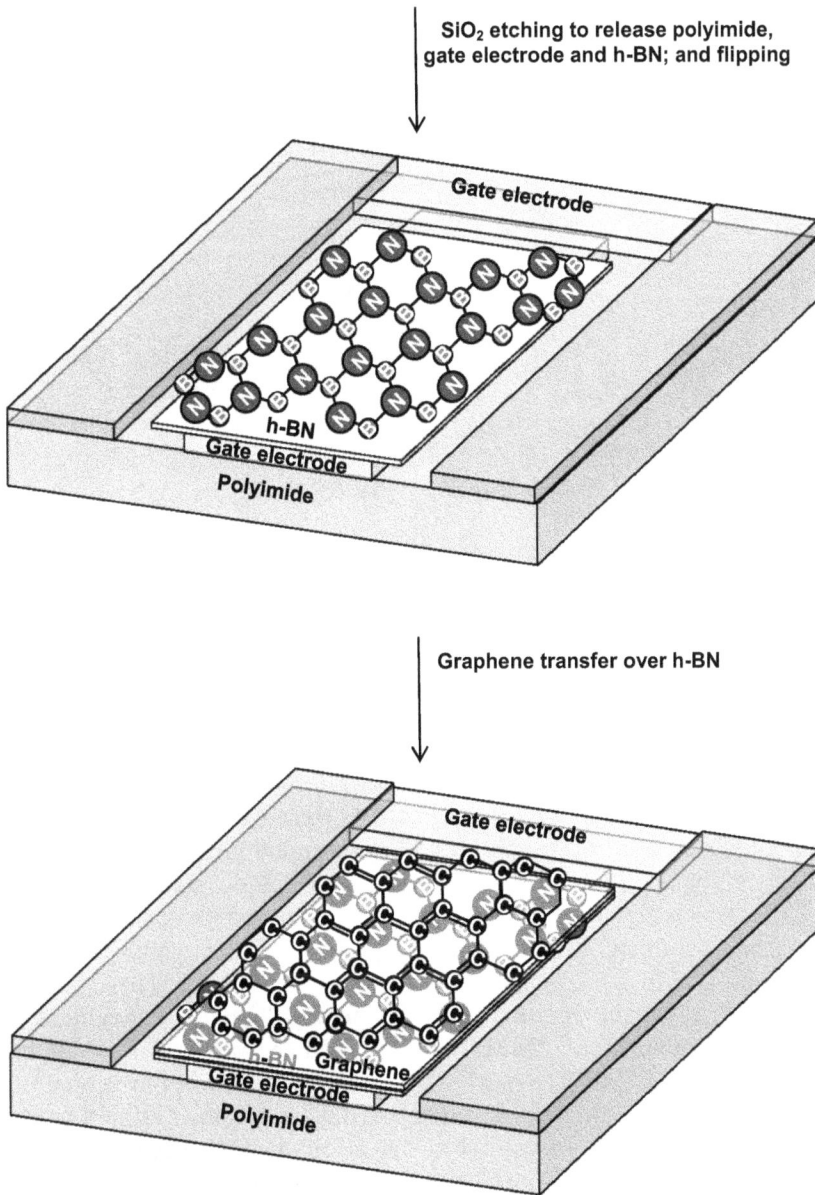

Figure 9.1. (Continued.)

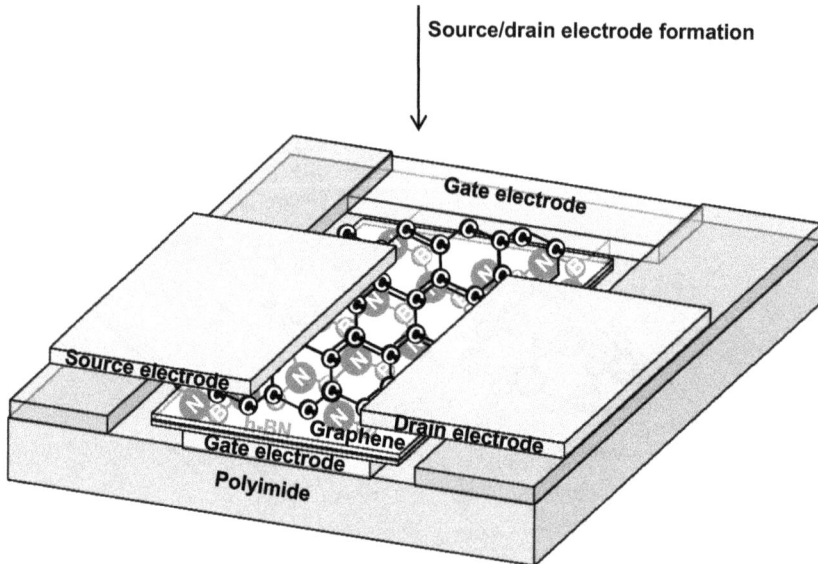

Figure 9.1. (Continued.)

9.2.3 Transferring graphene film from copper foil to polyimide by PMMA-assisted wet transfer process

Graphene is transferred with the help of PMMA. Hence, the process is referred to as the PMMA-assisted wet transfer process. Graphene film grows on both sides of the Cu foil. The graphene film on one side of the Cu foil is drop-coated with PMMA, and the PMMA is cured at 180 °C for 1 min. The graphene film on the opposite side of Cu foil is removed by polishing, thereby exposing the underlying Cu foil, which is removed by etching in ammonia persulfate etchant for copper. Then the graphene film is transferred to the chosen substrate, here polyimide. Subsequent to this transfer, a small quantity of PMMA is coated on the previous PMMA layer of PMMA/graphene structure. The new PMMA film is cured at room temperature for 30 min. Finally, full PMMA is removed by dissolution in acetone. The second PMMA layer dissolves the previous PMMA layer, thereby mechanically relaxing the graphene below it and hence improving the contact of graphene with the substrate. If the second PMMA layer is not applied, full contact between graphene and substrate is not established leaving some gaps so that the unattached regions break easily leading to the formation of cracks.

9.2.4 Source/drain electrode deposition and electrical testing

Superfluous area of graphene is removed by plasma etching in oxygen and source–drain electrodes are deposited.

9.2.5 TFT parameters

The extracted electron mobility is 2307 cm^2 V^{-1} s^{-1}; for holes, it is 2324 cm^2 V^{-1} s^{-1}. The drive current is 300 µS µm^{-1}. The conversion gain (CG) of the frequency doubler circuit is −29.5 dB with an output power = −22.2 dBm. The high CG is enabled by the low impurity level of the interface between graphene and h-BN. In addition, the low inter-plane thermal conductivity of h-BN protects polyimide from deformations by thermal effects as well as damage arising from high fields (Lee *et al* 2013).

9.3 Graphene TFT on transparent PEN substrate

9.3.1 Bottom gate patterning and dielectric deposition

In the process developed by Petrone *et al* (2015), the PEN substrate is mounted on silicon handle wafers using PDMS adhesive (figure 9.2). The PDMS film is 6 µm thick. A dual-finger design of bottom gate is defined on the substrate with e-beam lithography. After evaporation of 1 nm Cr/20 nm Au–Pd alloy (60–40 wt%), the unwanted metal is removed by the lift-off process. The gate dielectric is a hafnium oxide (HfO$_2$) layer of thickness 6 nm with dielectric constant 13. It is done by atomic layer deposition (ALD).

9.3.2 Graphene growth

The semiconductor channel layer is graphene obtained by LPCVD (Petrone *et al* 2012):

(i) Large-grain patches of graphene are grown on the interior surface of a Cu foil (25 µm) folded into a pocket shape with the edges crimped tightly. The pocket is annealed at 1030 °C, 1 mTorr pressure in a hydrogen atmosphere for 15 min. By flowing methane at 1 sccm and hydrogen at 2 sccm over the Cu foil at 10 mTorr, 1000 °C for 30 min to 2 h, graphene patches of size 20–250 µm are grown. The graphene layer is cooled down to room temperature under flowing (methane + hydrogen).

(ii) Sheets of small-grain polycrystalline graphene are grown on Cu foil (25 µm), annealed at 800 °C, 50 mTorr pressure in hydrogen at for 12 h. By flowing methane at 35 sccm and hydrogen at 2 sccm over the Cu foil at 300 mTorr, 1000 °C for 30 min, continuous graphene sheets are grown. The process ends with cooling to ambient temperature.

9.3.3 Dry-transfer method for applying graphene over the gate

The graphene film is spin-coated with PMMA (Petrone *et al* 2012), figure 9.3. This PMMA layer provides mechanical support to graphene. Thus a copper/graphene/PMMA stack is formed. The stack is cut into square-shaped sections. The stack sections are glued to polyimide tape with the copper side of the stack sections placed over windows cut in the tape, thereby exposing copper. The copper film in the stack section is removed by etching in ammonium persulphate. We are now left with the graphene/PMMA membrane. The membrane is suspended across the window in the tape. After rinsing in water and then dipping in isopropanol, the membrane is dried in nitrogen. The graphene is applied to the gate dielectric film and the supporting tape is

Applying PDMS adhesive on silicon handle wafer

PDMS adhesive

Silicon handle wafer

Pasting the PEN substrate with silicon handle wafer

PEN substrate

PDMS adhesive

Silicon handle wafer

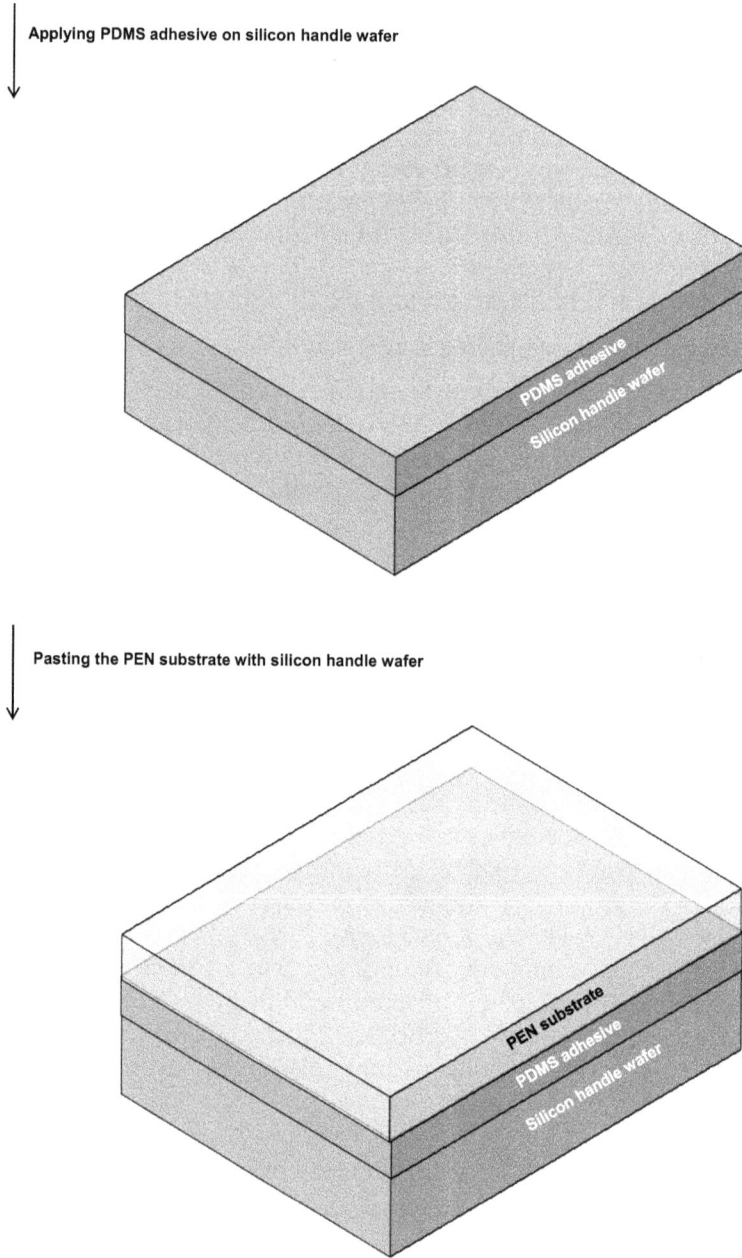

Figure 9.2. TFT on a PEN substrate using graphene as the active semiconductor channel.

Dual finger gate electrode deposition/patterning

Hafnium oxide gate dielectric by ALD

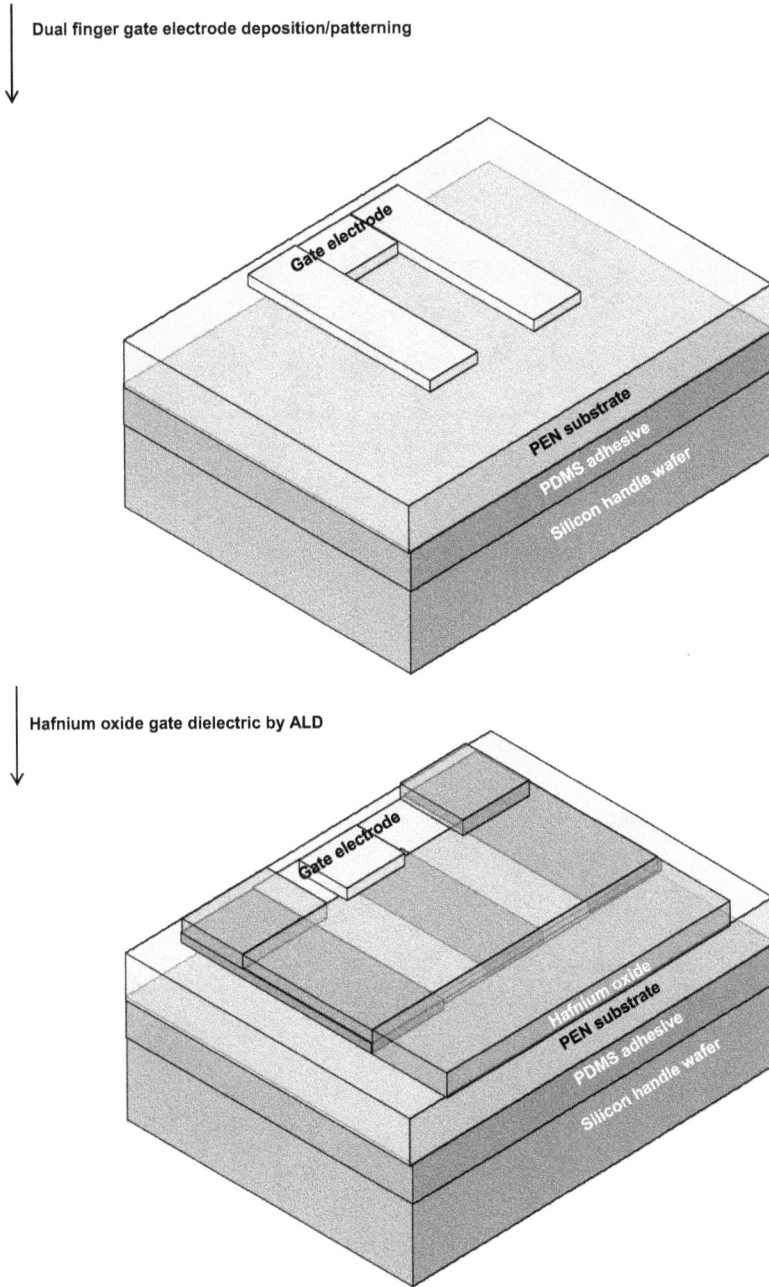

Figure 9.2. (Continued.)

Graphene transfer on HfO₂

Source/drain electrode
deposition/patterning

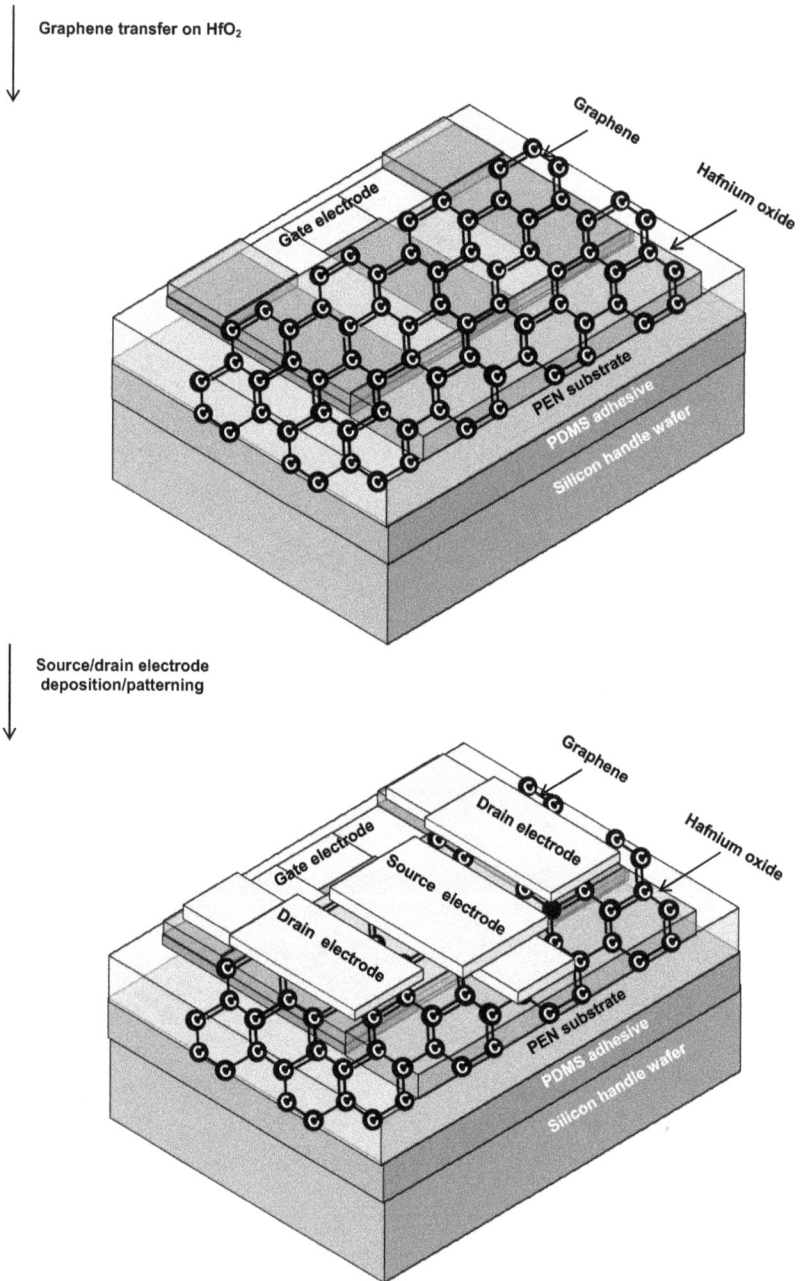

Figure 9.2. (Continued.)

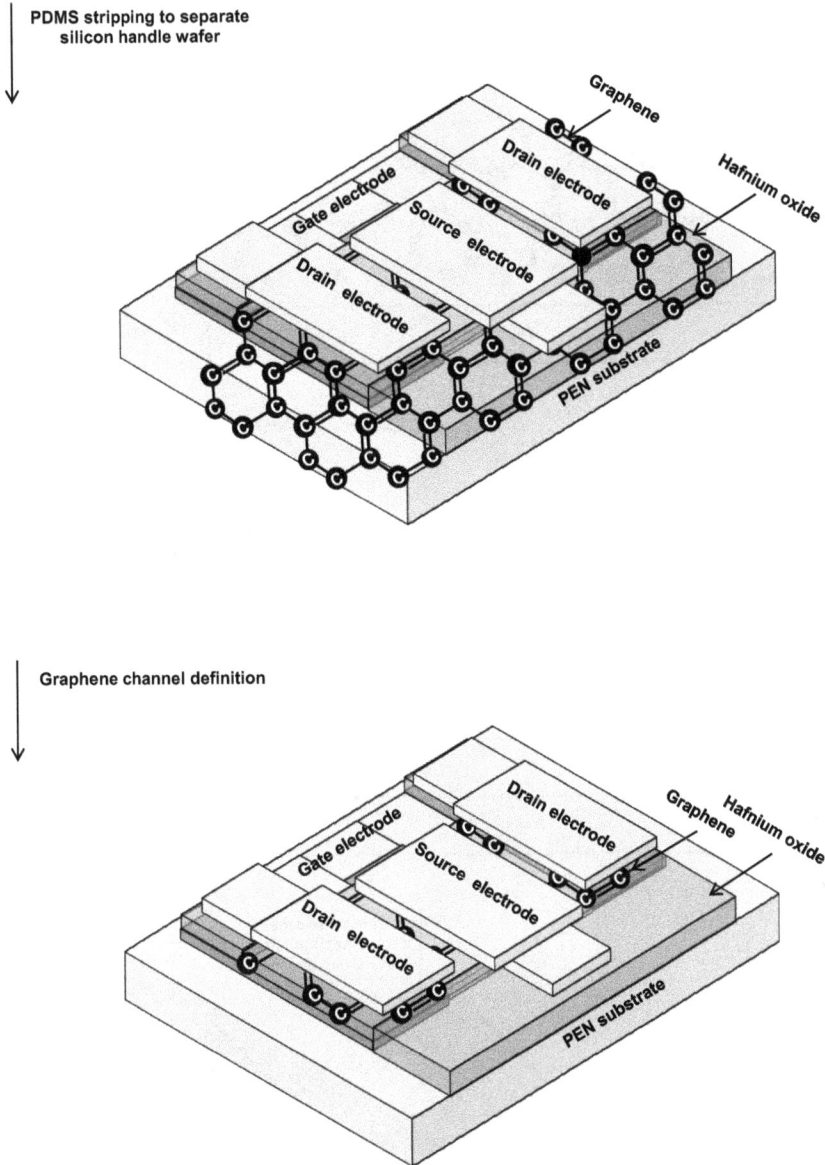

PDMS stripping to separate silicon handle wafer

Graphene channel definition

Figure 9.2. (Continued.)

removed. The PMMA is stripped off in acetone. Electron-beam lithography is used to define the graphene channel. The graphene etching is done in oxygen plasma.

9.3.4 Source/drain electrodes and measurements

Trilayer Cr/Pd/Au metallization with layer thickness of 1, 20, 110 nm is done. The channel length of the device is 260 nm and the effective channel width is 20 µm. The channel aspect ratio is 20 000 nm/260 nm = 76.92. At low fields, the field-effect

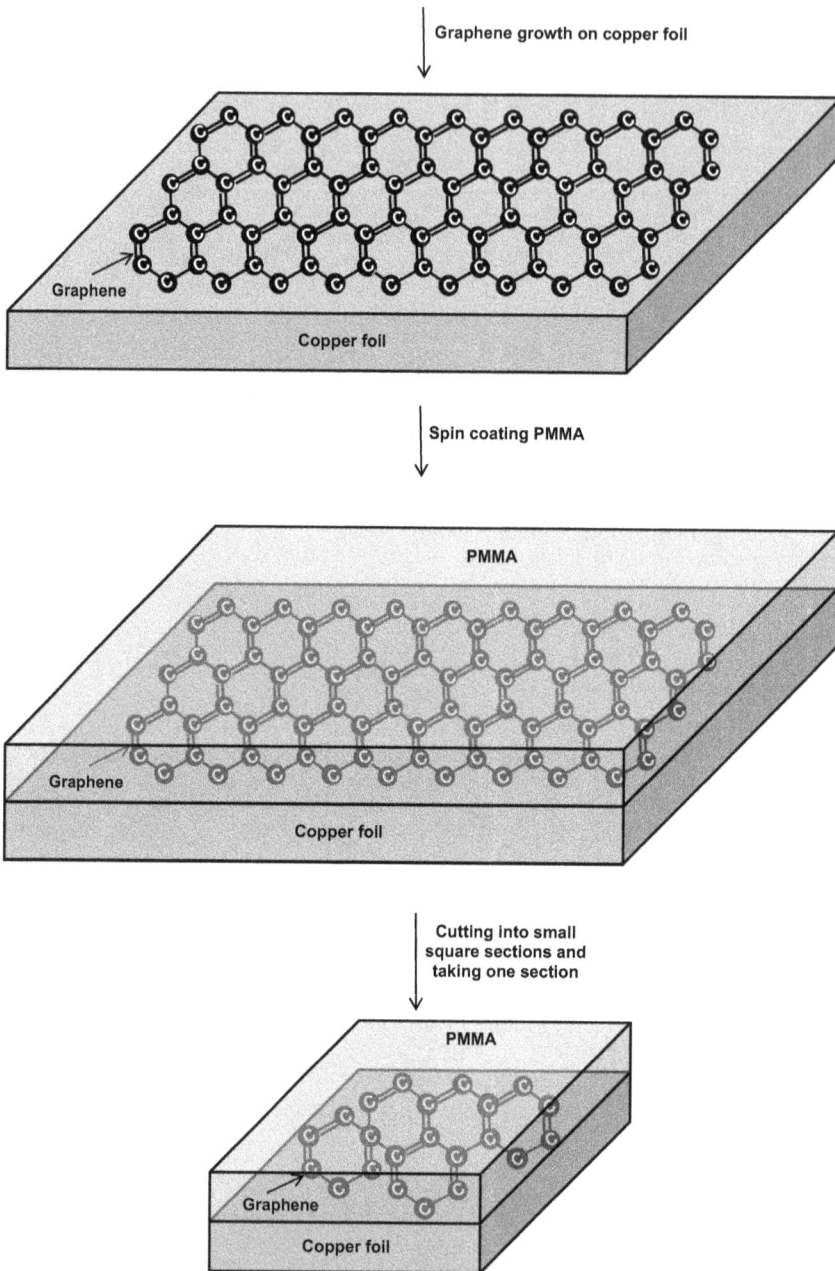

Figure 9.3. Fixing graphene on the gate dielectric of TFT by the dry-transfer method.

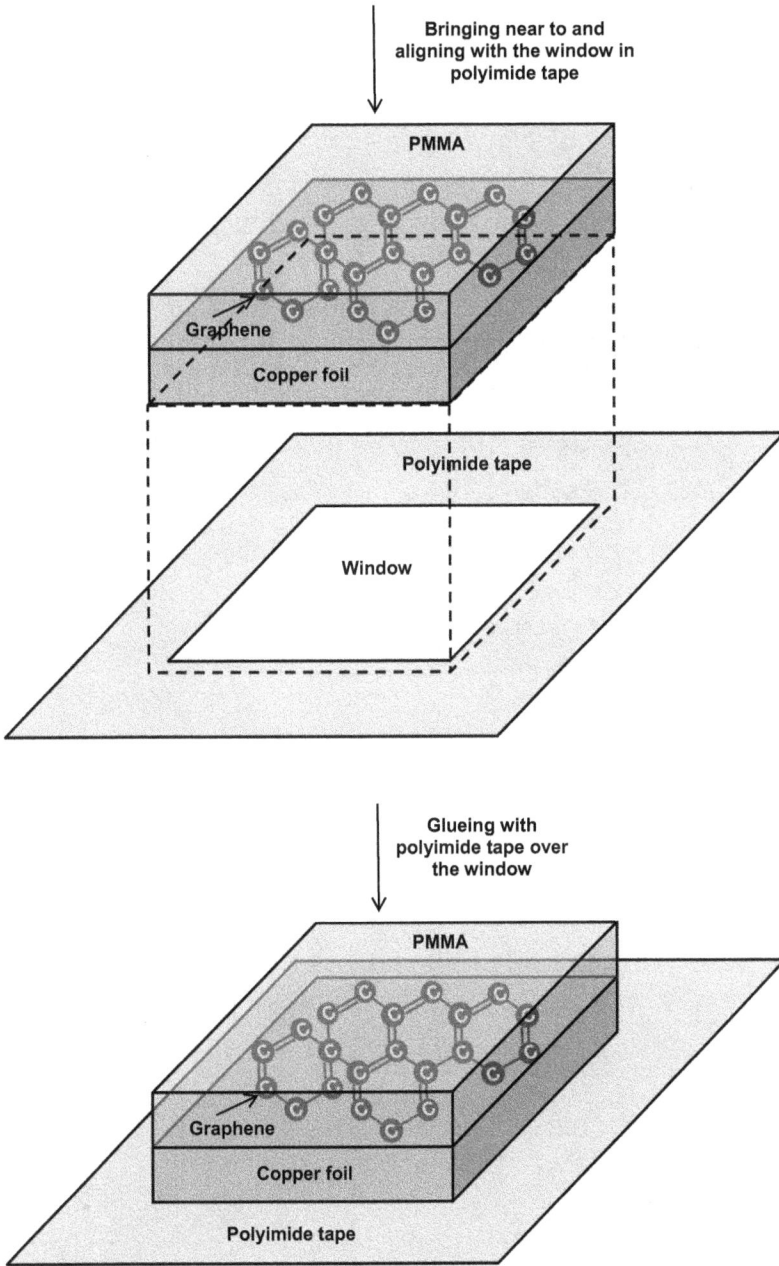

Figure 9.3. (Continued.)

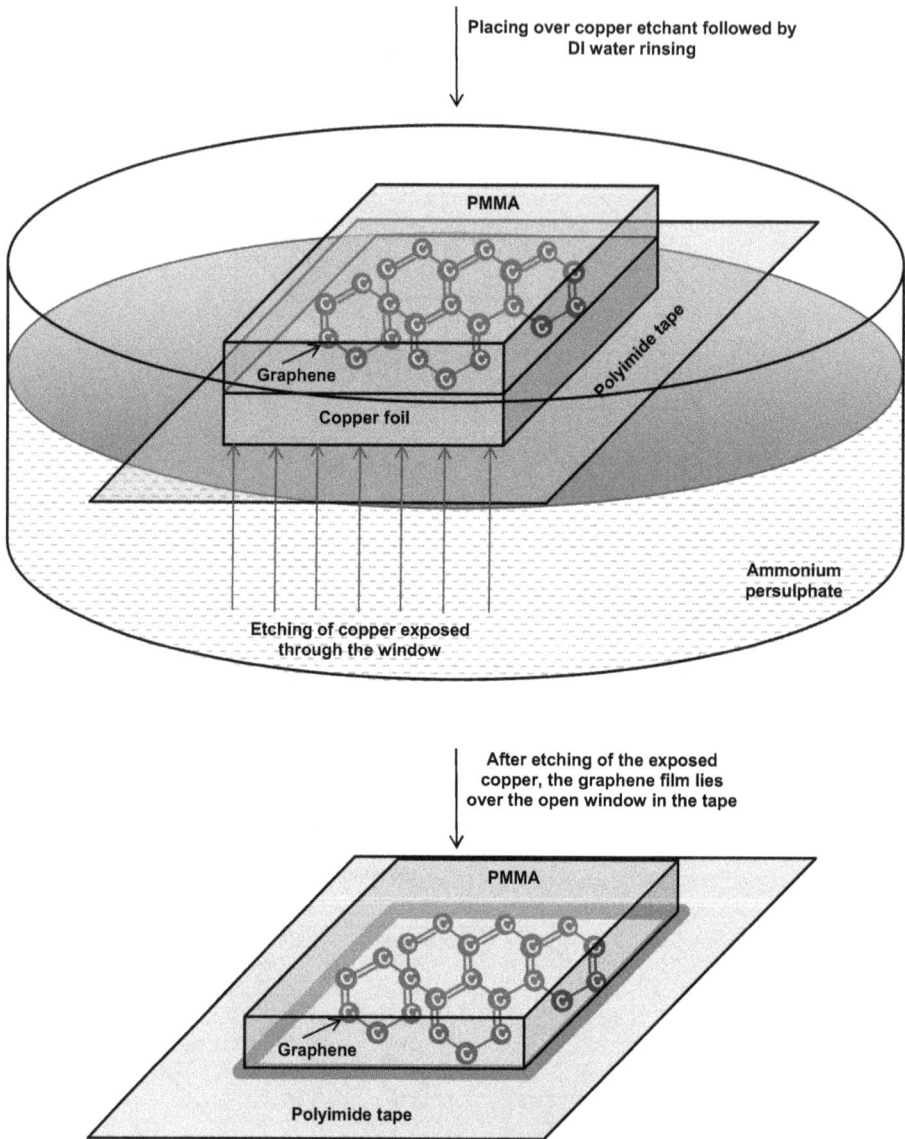

Figure 9.3. (Continued.)

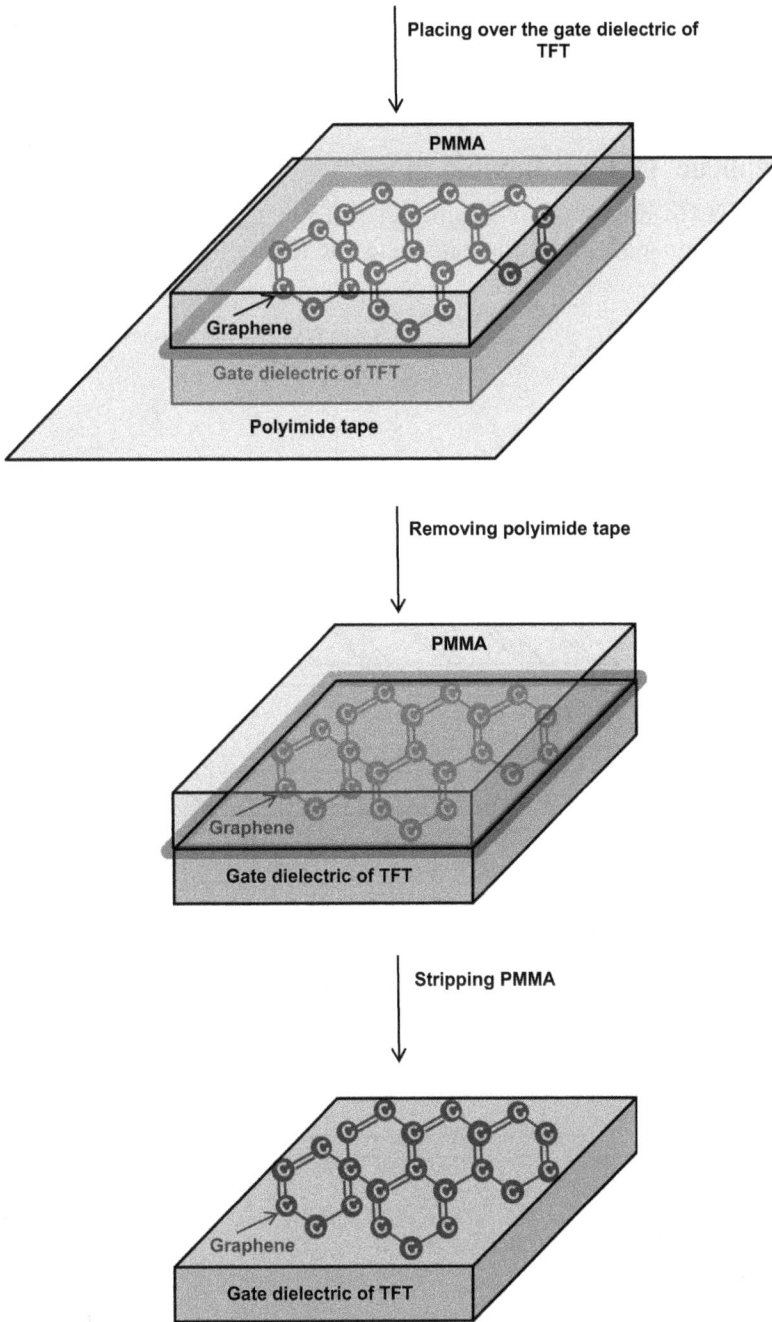

Figure 9.3. (Continued.)

mobility is $1000 \text{ cm}^2 \text{ V}^{-1} \text{ s}^{-1}$. For the as-fabricated TFT, intrinsic $f_T = 23.6$ GHz and $f_{\max} = 6.5$ GHz. Extrinsic $f_T = 38.7$ GHz and $f_{\max} = 7.6$ GHz. For strain levels from 0%–2%, f_T degradation is <35% while $f_{\max}$ degradation is <20% (Petrone et al 2015).

9.4 Graphene TFT on flexible glass

Park et al (2016) used Willow flexible glass from Corning as the substrate and nanoscale polyimide (NPI) as the gate dielectric (figure 9.4).

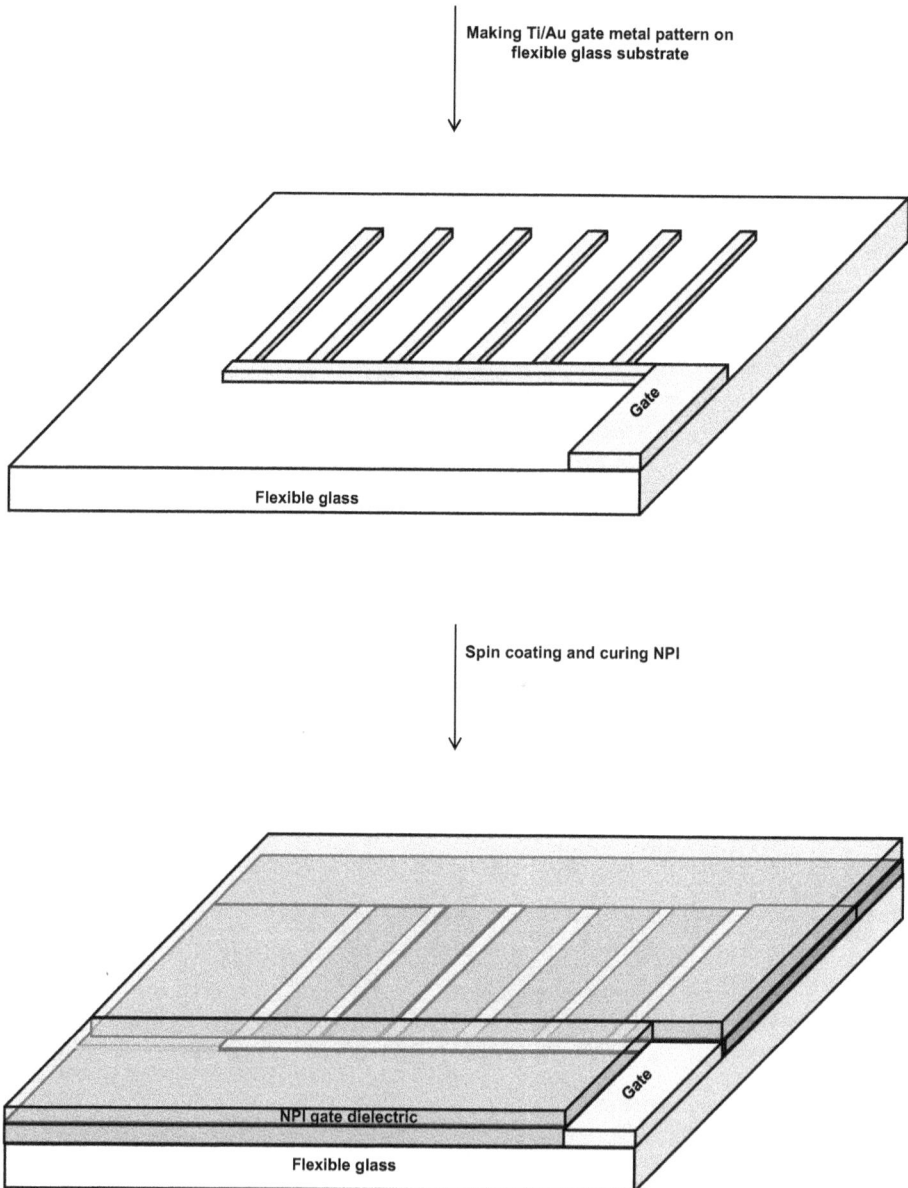

Figure 9.4. TFT with graphene semiconductor on flexible glass substrate.

Attaching graphene over NPI

Graphene

C C
C C C C C C C C C C
C C C C C C C C C C C
C C C C C C C C C C
C C C C C C C C C C C
C C C C C C C C C C
C C C C C C C C C C

Gate

NPI gate dielectric

Flexible glass

Ti/Au source/draincontact
formation

Drain

Graphene

C C
C C C
C C C
C C C
C C C
C C C

Source

Gate

NPI gate dielectric

Flexible glass

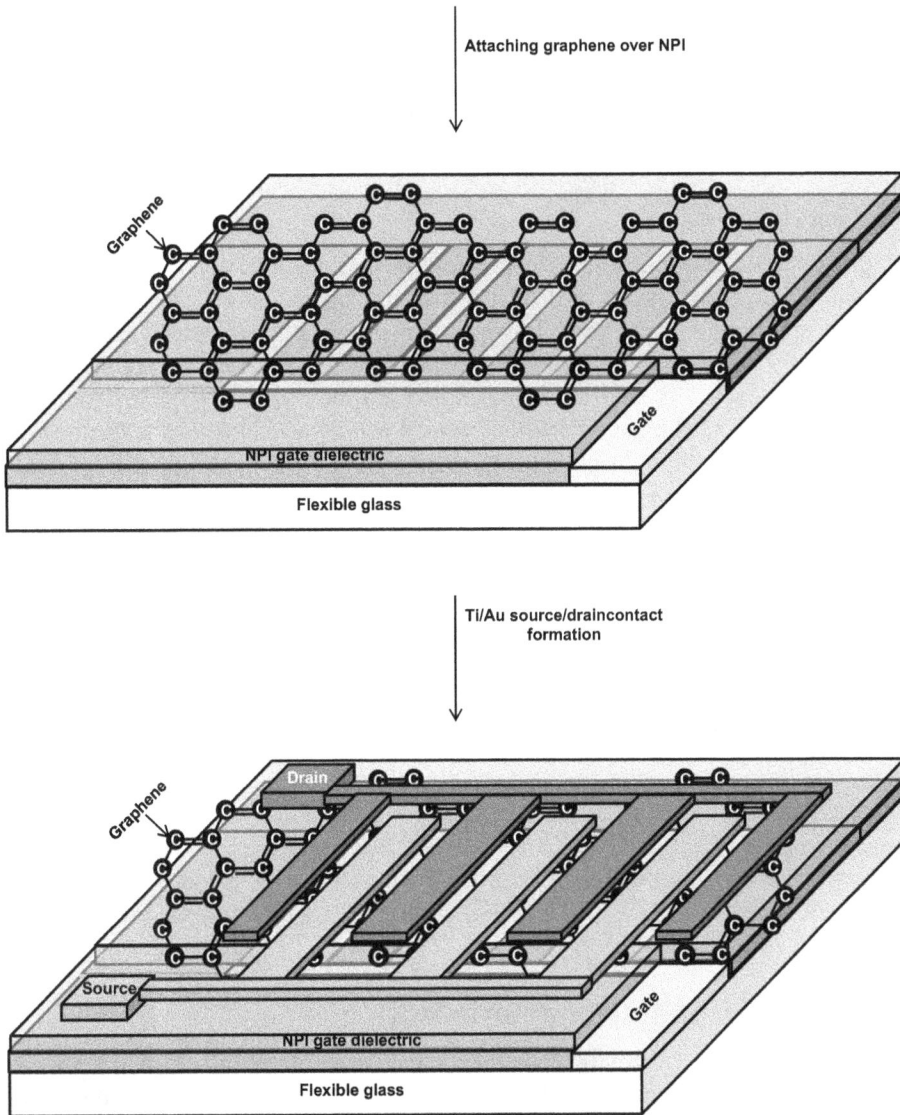

Figure 9.4. (Continued.)

9.4.1 Specialties of flexible glass

Besides transparency and conformability, the flexible glass has many unique properties, e.g. it is ultra-thin and lightweight, and provides a defect-free surface. From a processing viewpoint, its glass transition temperature >750 °C is much higher than that of competing polymers: PI <400 °C and PET <200 °C. Hence, a much wider range of processing temperature opportunities is available with flexible glass. Comparing thermal conductivities values, they are 1, 0.12 and 0.15 W mK^{-1} for flexible glass, PI and PET respectively. Owing to the higher thermal conductivity value of flexible glass, Joule heating effects for TFTs on flexible glass are less severe than on plastic substrates, greatly enhancing the heat dissipation capability for these TFTs. Additionally, the <0.5 nm surface roughness for flexible glass is much better than <1 nm for PI and <2 nm for PET. As a result, thermomechanical failure chances are less for TFTs on flexible glass.

9.4.2 Fabrication process

The gate metallization pattern is formed on the flexible glass substrate using e-beam evaporated Ti (2 nm)/Au(38 nm). Over this pattern, nanoscale polyimide (NPI) is coated by spinning. Curing the NPI film at 250 °C in nitrogen yields an embedded gate structure with NPI thickness = 60 nm. The low surface roughness of NPI film <0.5 nm leads to less disordering in the graphene layer to be laid over NPI. Since the channel of the device will be formed in this graphene layer, the transport properties of TFT will be significantly improved. For this device, the monolayer graphene film is produced by CVD and attached over NPI film by PMMA-assisted wet transfer, as already explained in sections 9.2.2 and 9.2.3. Using electron-beam lithography and oxygen plasma, the edge-injection contact patterns for source–drain metallization are defined. Edge-injection contacts provide lower contact resistance. The source–drain metal film formed by electron-beam evaporation includes 1 nm Ti and 45 nm Au layers. The W/L ratio is 60 µm/140 nm = 60 000 nm/140 nm = 428.57.

9.4.3 TFT parameters

The extracted mobility from the electron (hole) branch is 4540 (1100) cm^2 V^{-1} s^{-1}; the corresponding contact resistance is 1140 (720) Ω cm. The intrinsic cut-off frequency of the TFT is 95 GHz. This is 196% higher than on polymeric substrates. Microwave performance characterization up to 30 GHz revealed that the intrinsic power frequency of the TFT is >30 GHz. The extracted effective saturation velocity of carriers is found to be 8.4 × 10^6 cm s^{-1} (Park et al 2016).

9.5 MoS$_2$ TFT on Kapton (polyimide)

Monolayer MoS$_2$ has a bandgap of ~1.8 eV; bandgap of bulk MoS$_2$ is ~1.3 eV. Together with a carrier mobility of 200 cm^2 V^{-1} s^{-1}, it is a suitable nanomaterial for high-speed TFTs.

9.5.1 TFT fabrication

Chang et al (2013) take a commercial polyimide (Kapton) substrate (76 µm) (figure 9.5). Spin coating Kapton with 26 µm thick layer of liquid polyimide helps to reduce its surface roughness. Liquid polyimide is cured at 300 °C for 1 h. The substrate is now ready for device fabrication. Over the substrate, bottom gate electrode metallization is done. The metal combination is Ti (2 nm)/Pd (50 nm). Metal deposition is done by electron-beam evaporation. Upon the bottom gate electrode thus formed, HfO$_2$ or Al$_2$O$_3$ film is deposited as a gate insulator. Atomic layer deposition (ALD) is used for gate dielectric formation. After the gate dielectric comes the MoS$_2$ semiconductor layer, which is

Spin coating liquid polyimide on kapton

Making Ti/Pd gate electrode

HfO$_2$ or Al$_2$O$_3$ gate insulator by ALD

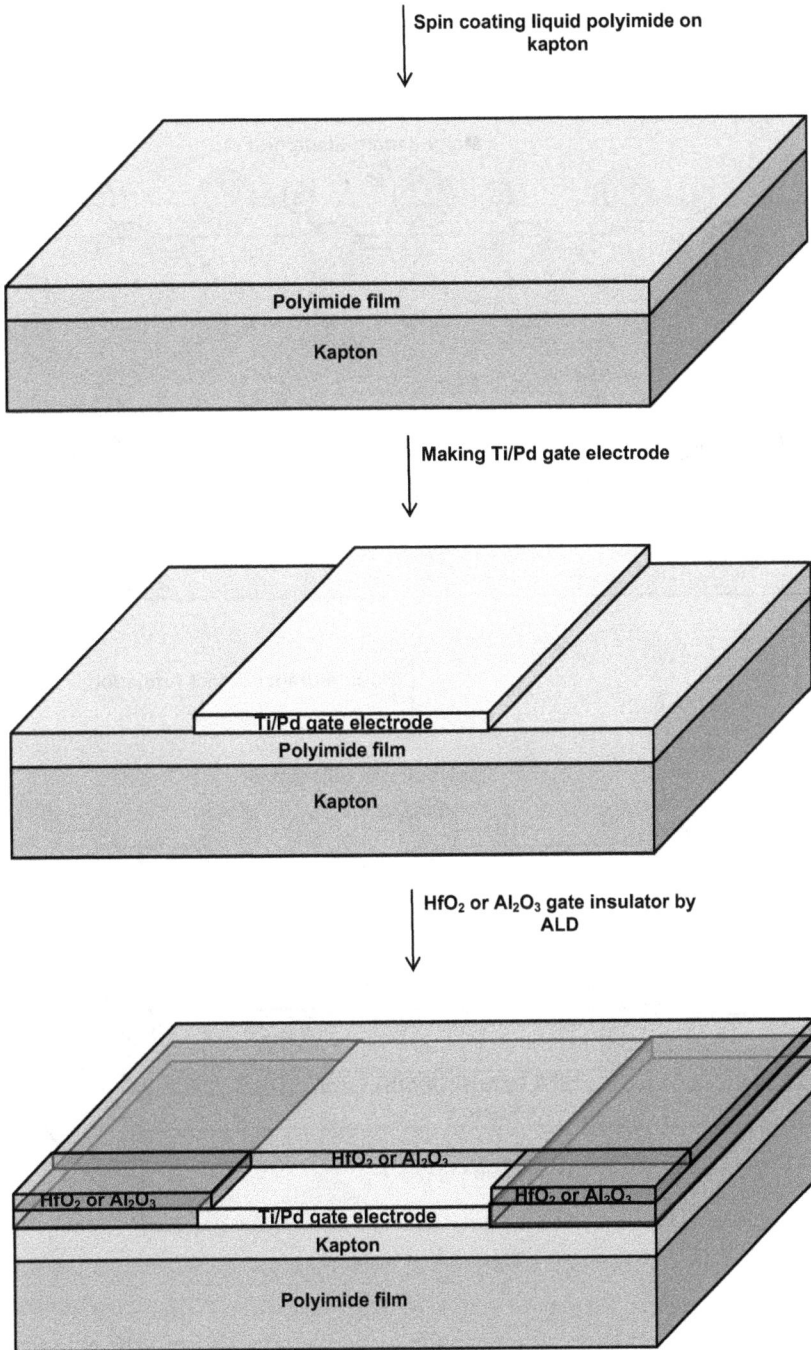

Figure 9.5. TFT with MoS$_2$ semiconductor using Kapton substrate.

Laying down exfoliated MoS$_2$ on
gate insulator

Molybdenum disulphide

HfO$_2$ or Al$_2$O$_3$

HfO$_2$ or Al$_2$O$_3$

HfO$_2$ or Al$_2$O$_3$

Ti/Pd gate electrode

Kapton

Polyimide film

Source/drain contact formation

Molybdenum disulphide

Ti/Au source
electrode

Ti/Au drain
electrode

HfO$_2$ or Al$_2$O$_3$

HfO$_2$ or Al$_2$O$_3$

HfO$_2$ or Al$_2$O$_3$

Ti/Pd gate electrode

Kapton

Polyimide film

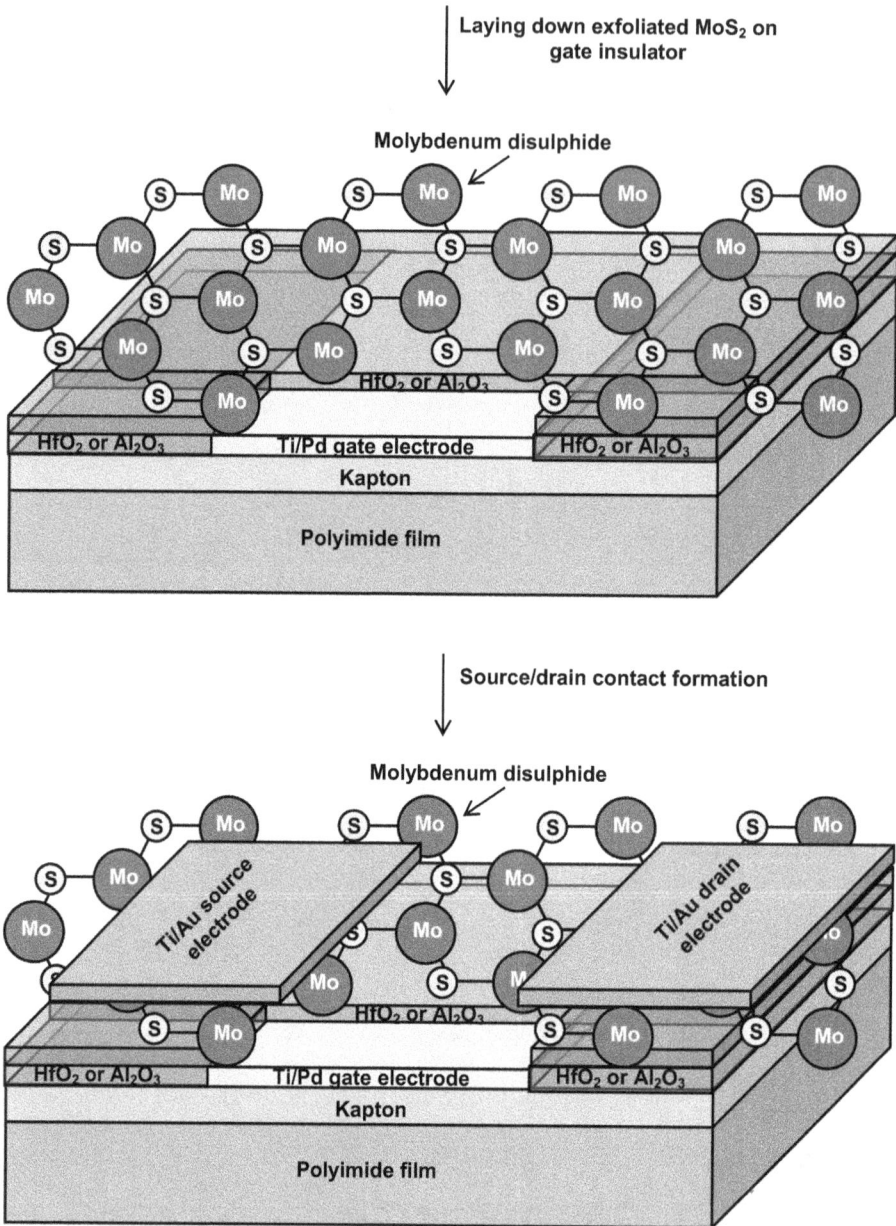

Figure 9.5. (Continued.)

mechanically exfoliated from commercially available crystals onto the gate dielectric. For making source–drain contacts, the contact pattern is defined by photolithography. Over this design, the contact metals Ti (2 nm)/Au (50 nm) are deposited by electron-beam evaporation. Then the source–drain metallization pattern is delineated by the lift-off technique.

9.5.2 TFT characteristics and bending effects

The current on–off ratio of TFT is 10^7, the subthreshold slope is ~82 mV/decade, intrinsic gain is >100, and low-field mobility of carriers is 30 cm^2 V^{-1} s^{-1}. Uniaxial straining studies revealed that devices are functional up to the bending radius of 1 mm for HfO$_2$ and 2 mm for Al$_2$O$_3$ devices. The current on–off ratio remains >10^4 for HfO$_2$-gated TFTs up to 1 mm radius. Failure of HfO$_2$ and Al$_2$O$_3$-gated devices under deformation showed that the cracking of the dielectric film is the primary cause of malfunction with Al$_2$O$_3$ film device failing earlier due to the higher Young's modulus of Al$_2$O$_3$ which increases the velocity of crack propagation. Since HfO$_2$ has higher relative permittivity than Al$_2$O$_3$, it is superior to Al$_2$O$_3$ both electrically and mechanically (Chang *et al* 2013).

9.6 WS$_2$ TFT on solution-cast PI substrate

Gong *et al* (2016) reported TFT with monolayer WS$_2$ film in a bottom gate, top source–drain electrodes configuration (figure 9.6).

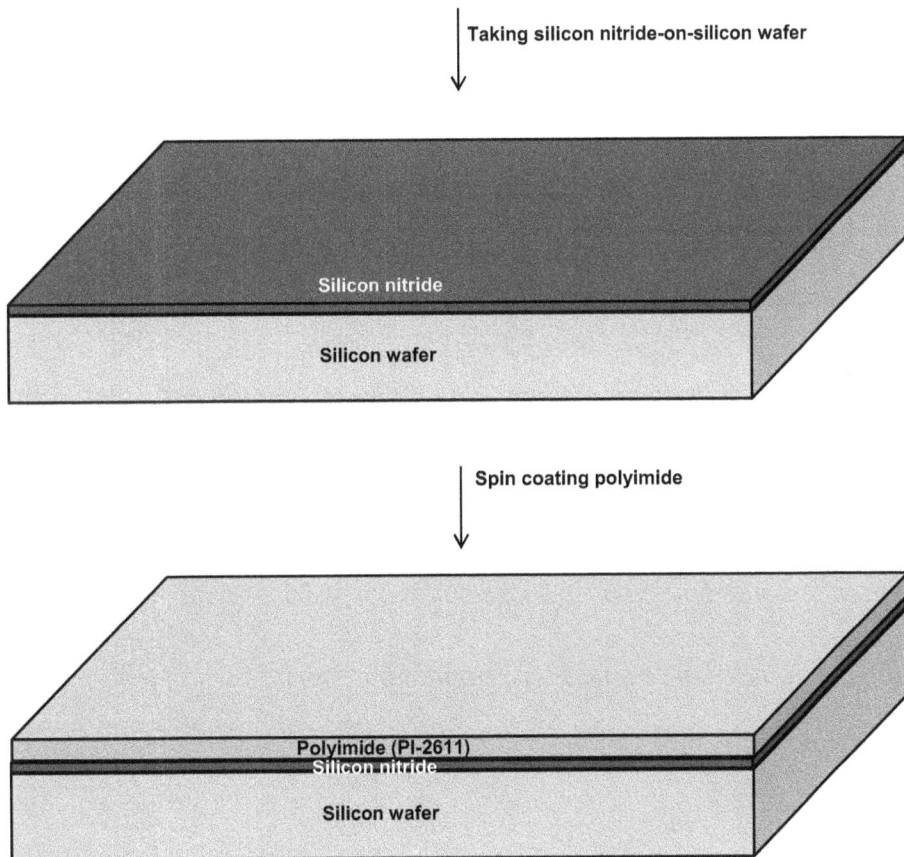

Figure 9.6. TFT with WS$_2$ semiconductor channel.

Titanium gate electrode patterning

Gate (Ti)
Polyimide (PI-2611)
Silicon nitride
Silicon wafer

Al_2O_3 gate dielectric
deposition/patterning

Al_2O_3 gate insulator
Al_2O_3 gate insulator
Al_2O_3 gate insulator
Gate (Ti)
Polyimide (PI-2611)
Silicon nitride
Silicon wafer

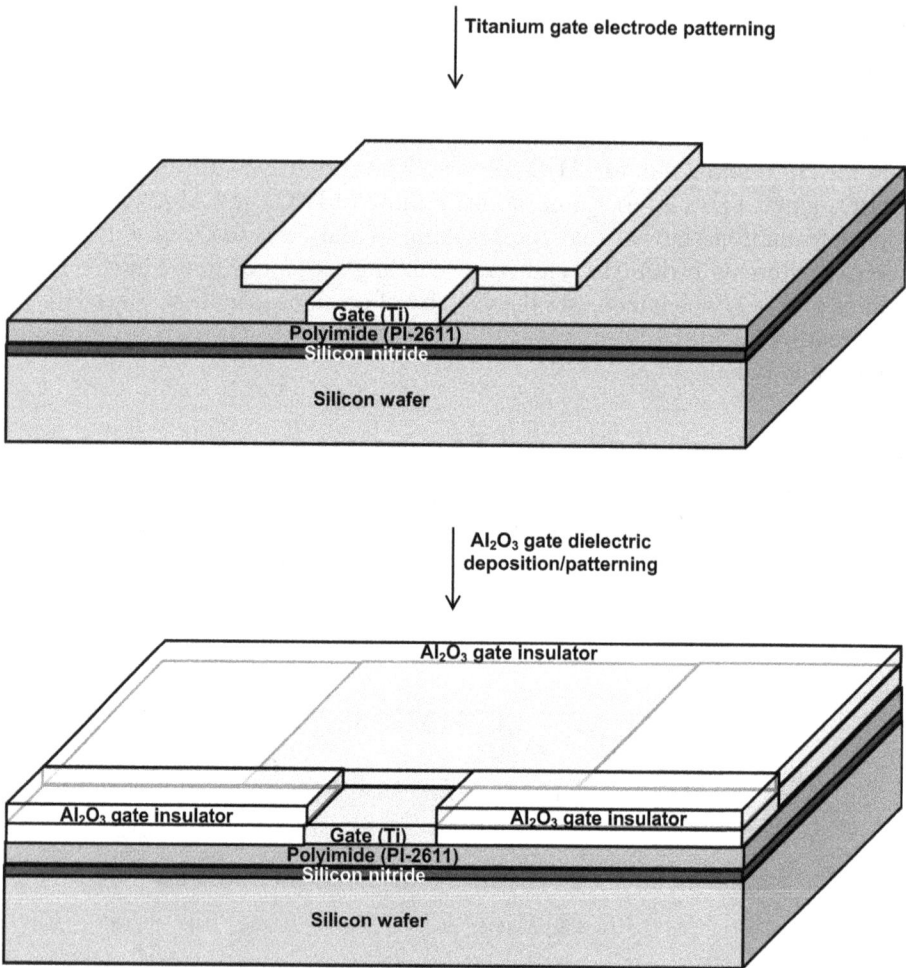

Figure 9.6. (Continued.)

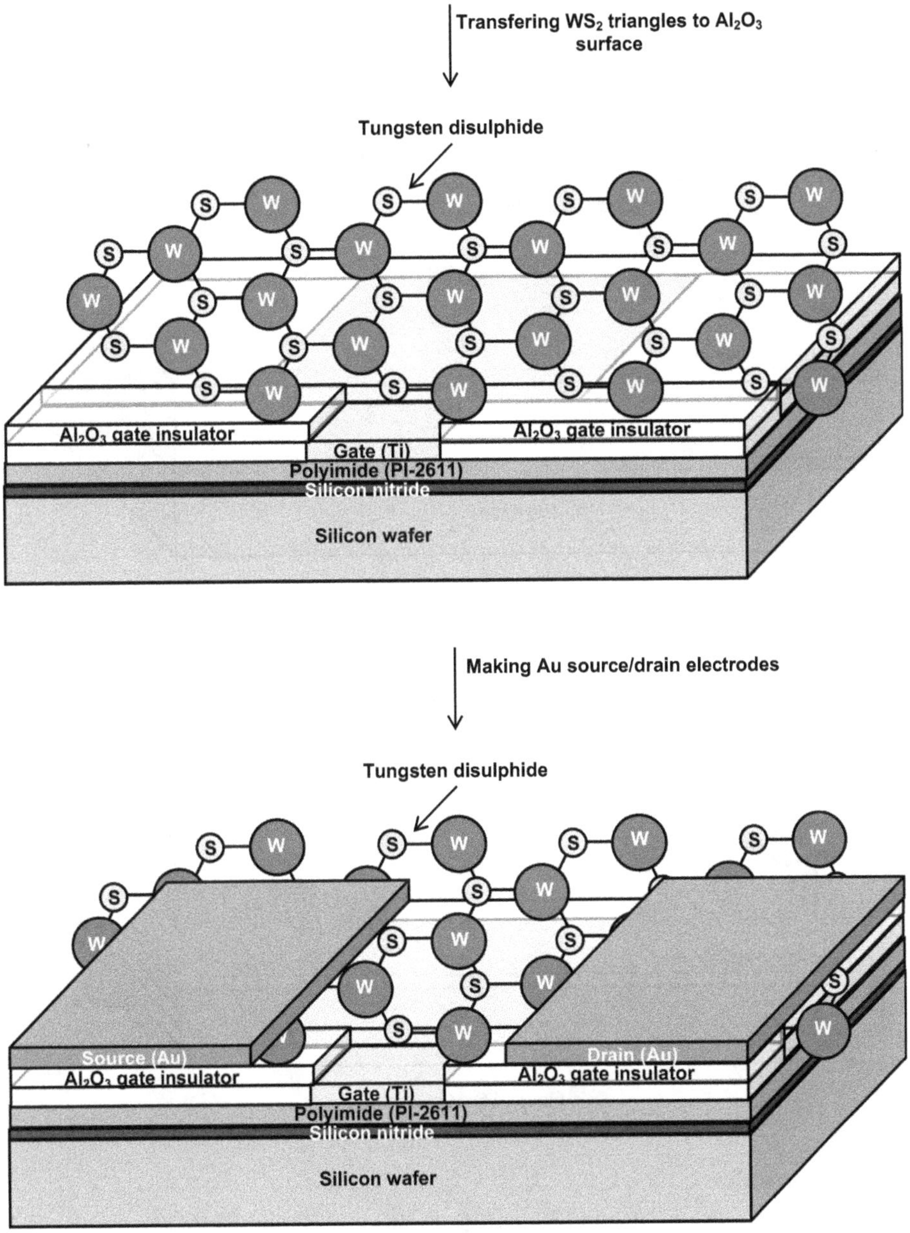

Transfering WS$_2$ triangles to Al$_2$O$_3$ surface

Tungsten disulphide

Al$_2$O$_3$ gate insulator
Al$_2$O$_3$ gate insulator
Gate (Ti)
Polyimide (PI-2611)
Silicon nitride
Silicon wafer

Making Au source/drain electrodes

Tungsten disulphide

Source (Au)
Drain (Au)
Al$_2$O$_3$ gate insulator
Al$_2$O$_3$ gate insulator
Gate (Ti)
Polyimide (PI-2611)
Silicon nitride
Silicon wafer

Figure 9.6. (Continued.)

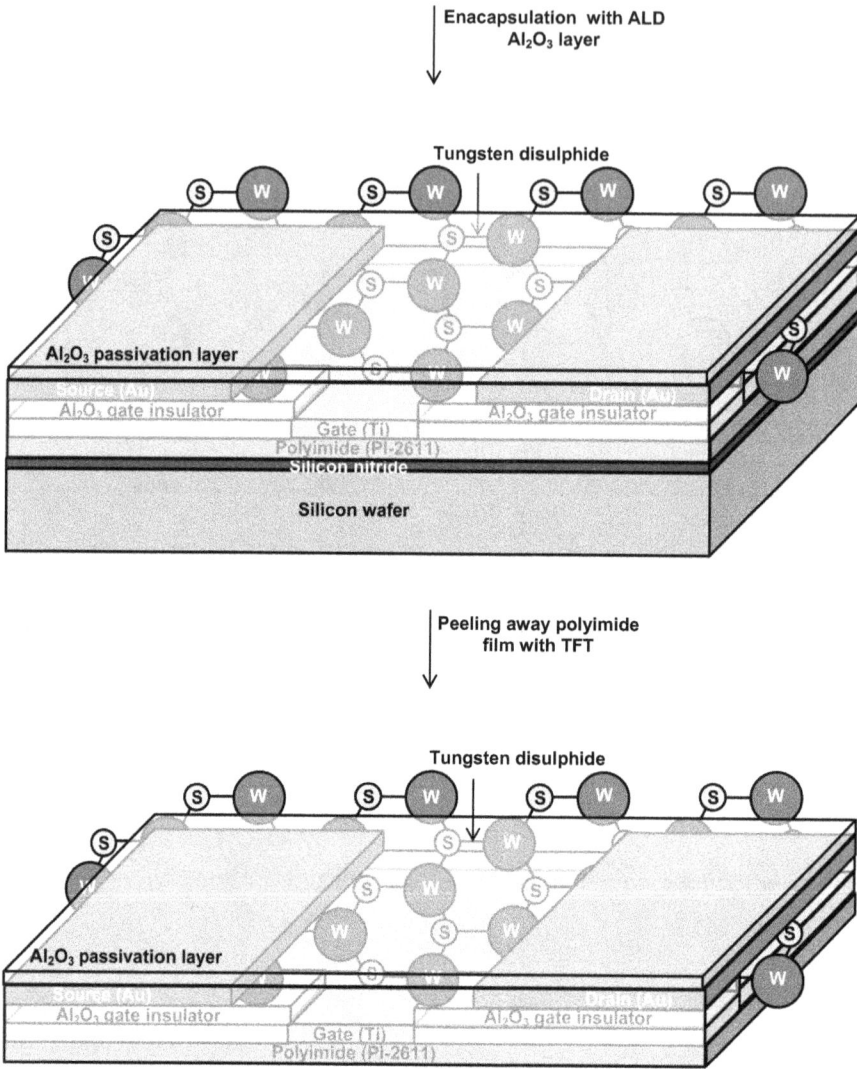

Figure 9.6. (Continued.)

9.6.1 Carrier wafer

This is a silicon wafer with a 50 nm thick coating of silicon nitride.

9.6.2 Substrate

This is a thin polyimide film of thickness 4.8 μm. The substrate film is formed by spin coating the polyimide precursor on the silicon nitride layer. The polyimide film is hardened by curing at 300 °C for 5 h. Its RMS surface roughness is <2 nm. This film is much thinner than the commonly used free-standing PI substrates which have thicknesses in the range 25–100 μm.

9.6.3 Deposition of gate electrode array

The gate electrode is made of titanium. Lift-off photolithography is used to create the gate electrode pattern. So, the photoresist pattern is first defined on the PI layer by photolithography. Then titanium is sputtered over the photoresist. On stripping the photoresist, the titanium adheres to the PI wherever it is directly in contact with PI. These are the gate electrode areas. It is swept away with the photoresist in all regions where it was deposited over the photoresist. The thickness of the titanium gate electrode film is 100 nm.

9.6.4 Gate dielectric deposition

The gate dielectric is aluminum oxide, Al_2O_3. The deposition technique used for the gate dielectric is plasma-enhanced atomic layer deposition (ALD). The ALD process is carried out at 200 °C which is safe for PI. The thickness of the Al_2O_3 film is 30 nm. At this stage, the structure produced is: $Si/SiO_2/PI/Ti/Al_2O_3$.

9.6.5 Synthesis of WS_2 for the active semiconductor layer

The WS_2 film is synthesized by chemical vapor deposition based on the sulfidation of WO_3 powder at 700 °C at atmospheric pressure (figure 9.7). A Si/SiO_2 wafer is taken with SiO_2 film thickness 300 nm. The wafer is cleaned by ultrasonic agitation in acetone and isopropanol (50/50). After drying by blowing nitrogen over it, a small quantity of WO_3 powder is placed over the SiO_2 layer. The wafer is heated at 700 °C in an inert environment generated by flowing argon gas. Afterwards, sulfur vapors are produced from sulfur powder placed upstream in a lower temperature zone at 300 °C, where the temperature is controlled independently. Introduction of the sulfur vapors causes a reaction between sulfur and WO_3 powder resulting in deposition of a sparse distribution of WS_2 flakes on the SiO_2 layer. The WS_2 film is a monolayer thick and WS_2 flakes are triangular-shaped.

9.6.6 PMMA-assisted transferring of the WS_2 triangles to the Al_2O_3 layer on the carrier wafer

The WS_2 flake covered SiO_2 layer is coated with PMMA at 3000 RPM for 30 s (Gutiêrrez *et al* 2012). After curing the PMMA overnight at room temperature, the

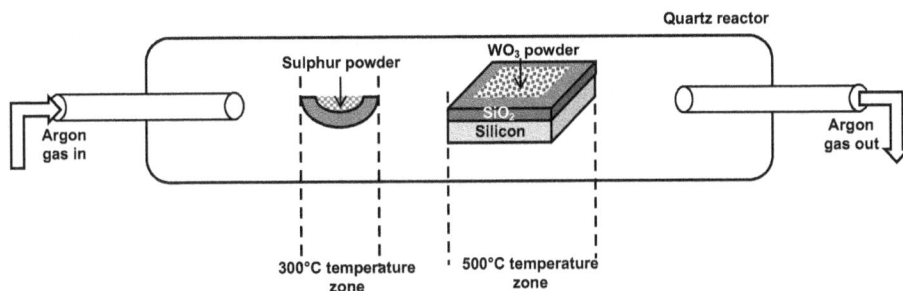

Figure 9.7. WS_2 synthesis using WO_3 and sulfur powders.

edge of the Si/SiO$_2$/WS$_2$/PMMA structure is scratched with a blade to expose the Si/SiO$_2$ interface, and it is immersed in 15 M KOH solution. The KOH solution attacks and etches the SiO$_2$ layer, thereby releasing the WS$_2$/PMMA structure, which is fished out on the Al$_2$O$_3$ insulator on the carrier wafer and washed with DI water. The PMMA is removed in acetic acid, acetone and isopropanol.

After this transfer of WS$_2$ flakes, we have reached the stage: Si/SiO$_2$/PI/Ti/Al$_2$O$_3$/WS$_2$.

9.6.7 Source/drain electrodes

Lift-off photolithography is done. So, the source–drain geometrical design is defined in the photoresist layer on the WS$_2$ film. Pattern definition is followed by thermal evaporation of the gold film on the photoresist. The thickness of the gold film is 50 nm. On dissolving the photoresist in acetone, the source–drain gold electrodes are formed over the WS$_2$ film.

9.6.8 Encapsulation

The TFTs are protected from environmental effects by coating with a thin film of aluminum oxide having a thickness of 30 nm. The capping Al$_2$O$_3$ film is formed by atomic layer deposition. Thus, we get the structure: Si/SiO$_2$/PI/Ti/Al$_2$O$_3$/WS$_2$/Al$_2$O$_3$.

9.6.9 Mechanically peeling off the PI film with the overlying TFT

The TFT has a W/L ratio = 25 μm/8 μm = 3.125. The carrier mobility in the linear region is typically 2–10 cm^2 V^{-1} s^{-1}. The current on–off ratio is 10^6. The peeling step involves bending by a small radius <0.5 mm and influences the turn-on voltage by around a 1 V negative shift. It also increases the subthreshold slope.

9.6.10 Static and multi-cycle bending effects

The effect of tensile strain on TFT is studied by binding the device on a curved surface. The radius of this surface is 5 mm causing a strain of 0.05%. The transfer characteristics of the TFT are found to be only trivially affected when bound in this way. In a more rigorous bending test, the stability of the TFT against repeated bending is investigated. During repeated bending tests, the TFT is bent to a radius of curvature of 2 mm and is then flattened. A linear actuator produces the bending, and the flat–bend–flat cycle duration is 2.5–4 s. After 30 000 bending/flattening cycles, a 20% decrease in maximum on-current is recorded while after 50 000 cycles, the diminution is 30%. Nevertheless, the gate current remains <10^{-10} A throughout (Gong *et al* 2016).

9.7 WSe$_2$ TFT on a PET substrate

Das *et al* (2014) fabricated an all-2D transparent TFT having a total thickness of 10 atomic layers using WSe$_2$ as the semiconducting channel, h-BN as the gate dielectric and graphene source–drain–gate electrodes (figure 9.8).

9.7.1 Mechanical exfoliation of WSe$_2$ flakes

WSe$_2$ flakes are mechanically exfoliated using Scotch tape on PET substrate.

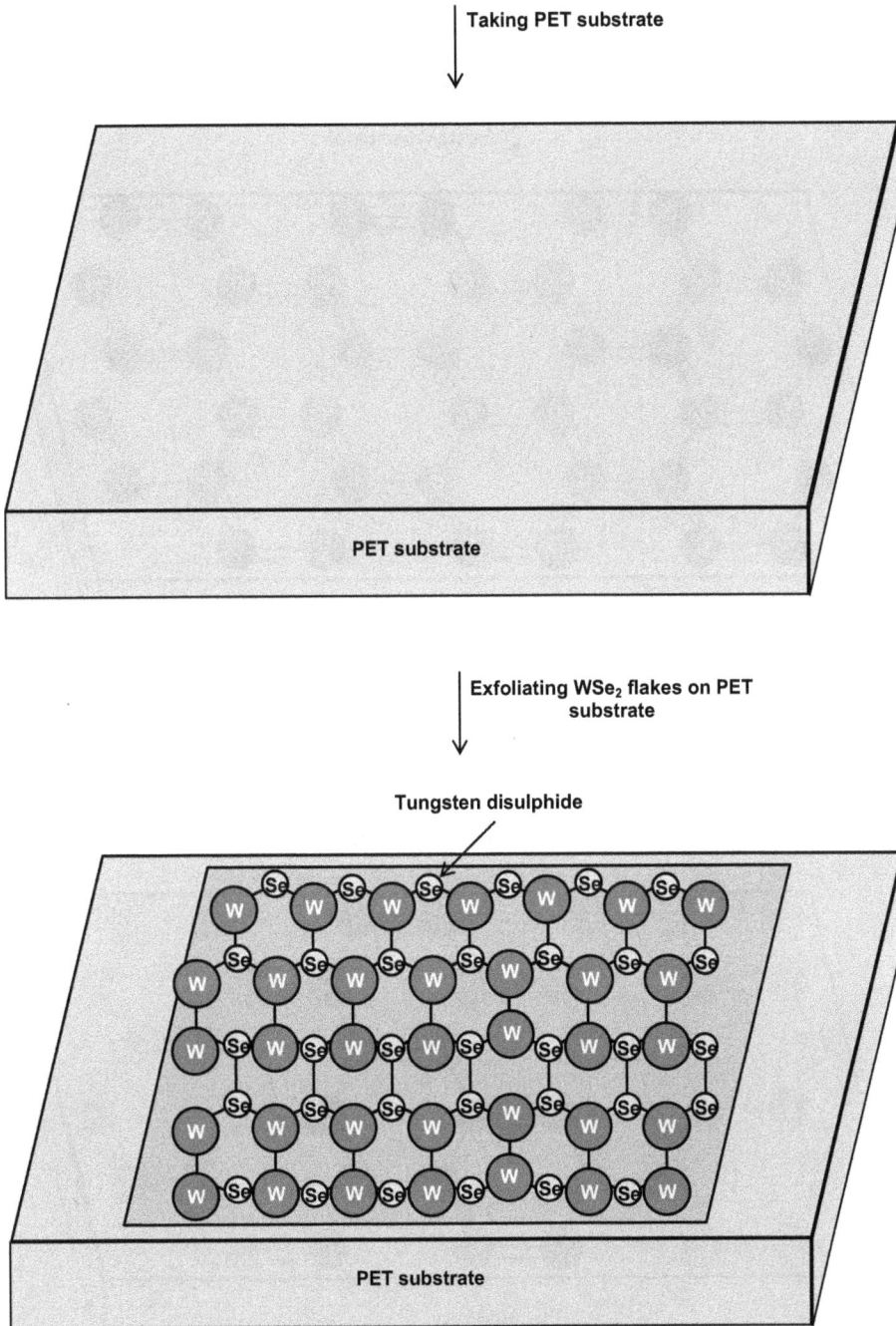

Taking PET substrate

PET substrate

Exfoliating WSe₂ flakes on PET substrate

Tungsten disulphide

PET substrate

Figure 9.8. Thin all-2D WSe$_2$ TFT.

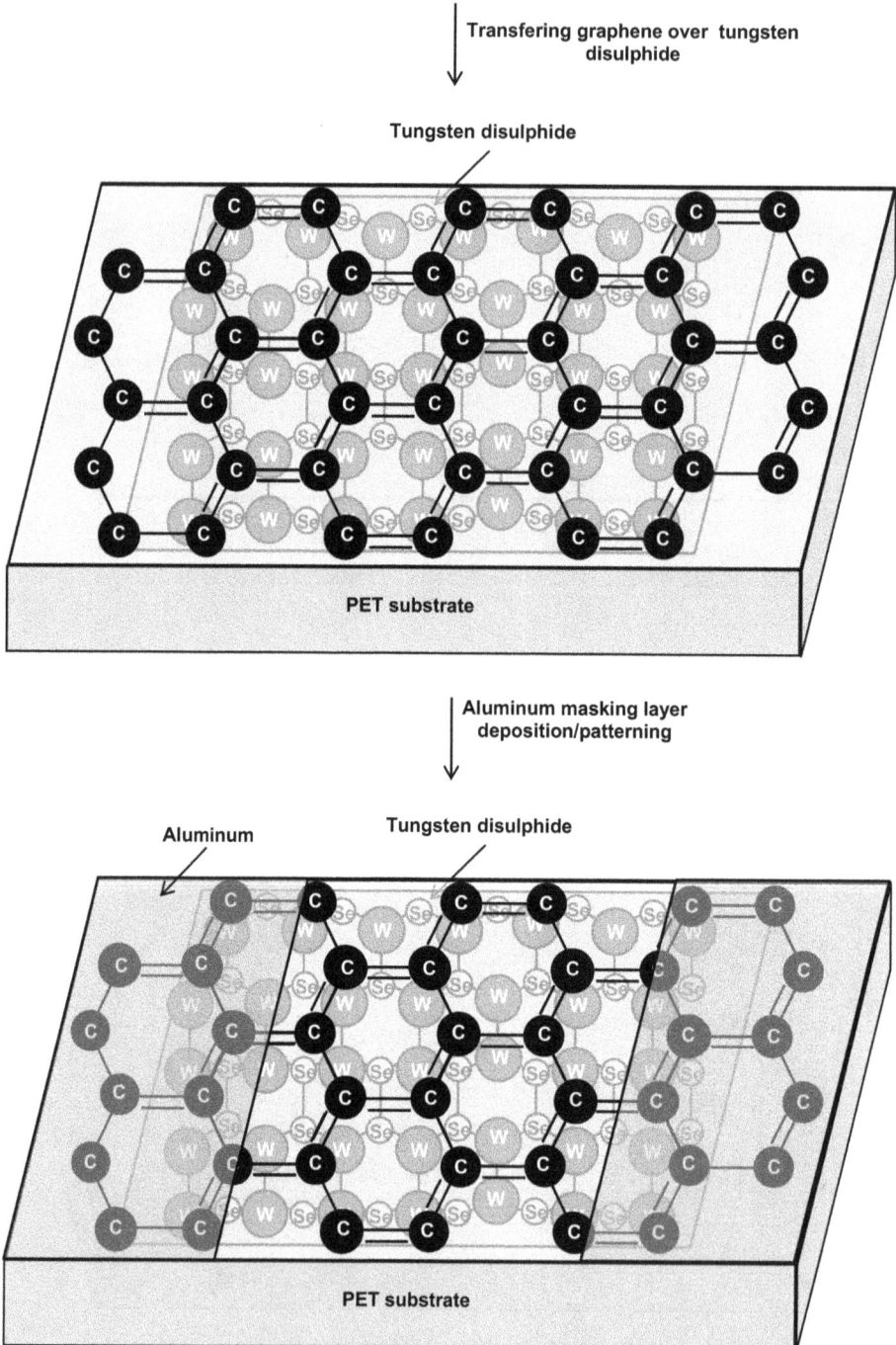

Figure 9.8. (Continued.)

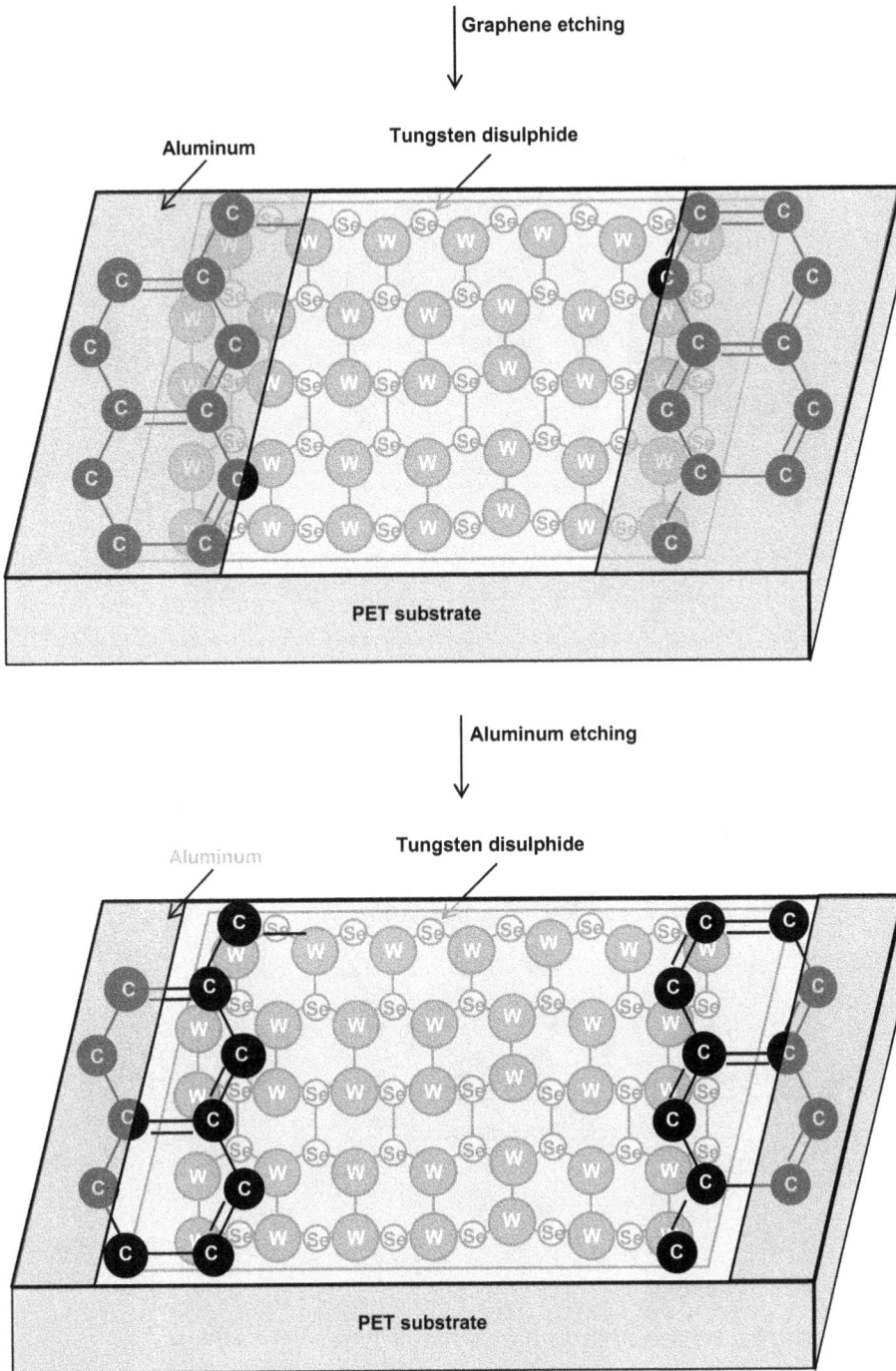

Figure 9.8. (Continued.)

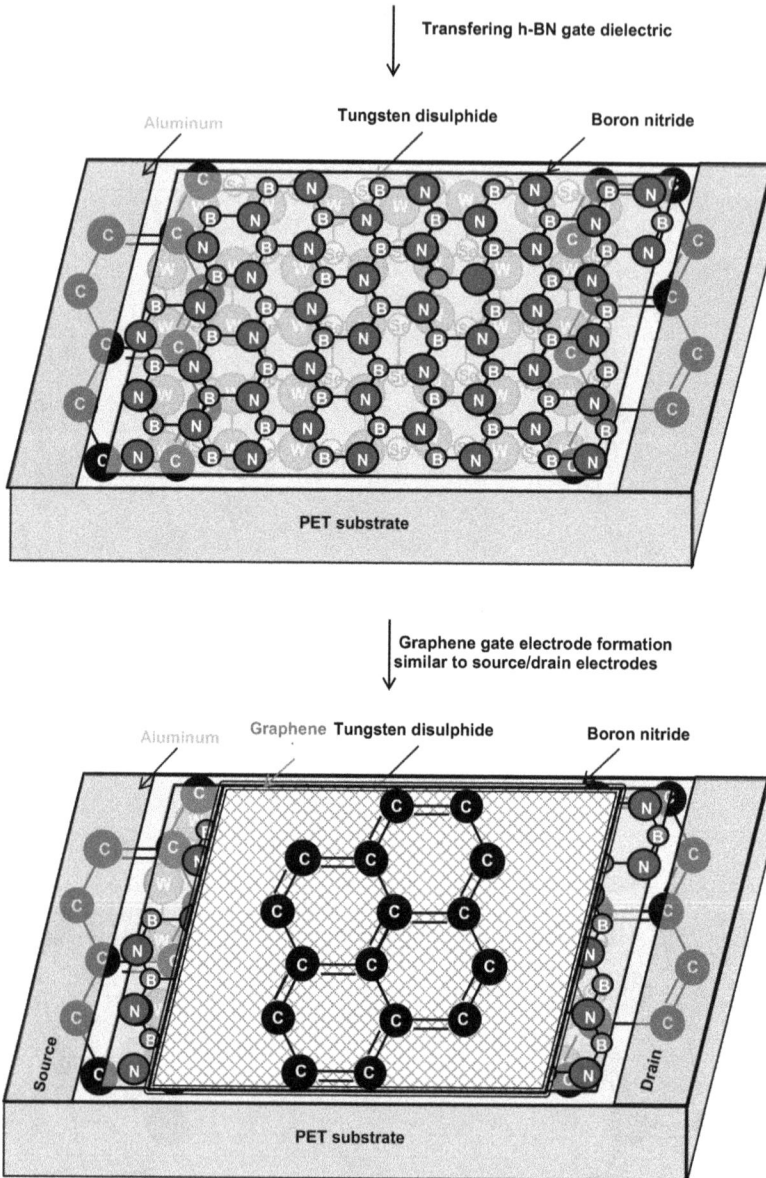

Figure 9.8. (Continued.)

9.7.2 Graphene synthesis

A graphene monolayer is grown on copper foil by CVD technique.

9.7.3 Monolayer graphene transfer over WSe₂ flakes

Monolayer graphene is transferred to the surface of WSe_2 flakes by traditional technique.

9.7.4 Creation of source–drain contact pads and hard mask for graphene etching

The lift-off photolithography technique is applied. The pattern of source–drain contact pads is defined in photoresist. Aluminum is deposited by electron-beam evaporation. The photoresist is stripped off when aluminum is left only in the source–drain contact pad areas. The aluminum film does not cover the channel region and is, therefore, used as a mask for removing graphene from this region.

9.7.5 Dry etching of graphene in oxygen plasma

Oxygen plasma removes graphene from all over the device except at the source–drain contact pads. As oxygen plasma does not affect WSe_2, the semiconductor channel region remains intact.

9.7.6 Photolithography to open windows for aluminum etching from graphene–WSe_2 contact regions

This establishes contact of graphene with WSe_2 only by removing aluminum from regions where trilayer aluminum/graphene/WSe_2 combinations exist. Now the aluminum pads are connected to graphene only.

9.7.7 Transfer of a few layers of h-BN to act as gate dielectric

A few layers of h-BN grown on copper foil by CVD are transferred over the WSe_2 film. The h-BN film constitutes the gate insulator.

9.7.8 Monolayer graphene transfer over h-BN for gate electrode formation

Monolayer/bilayer graphene grown on copper foil is transferred on h-BN gate dielectric.

9.7.9 Photolithography for creating aluminum hard mask for gate electrode patterning

The gate electrode pattern is defined in photoresist. Over this pattern, aluminum is deposited by electron-beam evaporation. By the lift-off technique, a hard mask of aluminum is created on the graphene surface.

9.7.10 Dry etching of graphene

Graphene is etched in oxygen plasma from the unwanted portions.

9.7.11 Removal of aluminum hard mask

The aluminum hard mask is etched away. Thus, the gate electrode of pure graphene is realized.

9.7.12 TFT characteristics

The p-FET has a contact resistance of 1.4 kΩ μm. For the hole branch, the subthreshold slope is 180 mV/decade, whereas for the electron branch, it is 340 mV/decade. Without strain, the drain/source current I_{DS} is ~1.6 × 10^{-4} μA μm^{-1} at $V_{GS} = 0$ V and ~1.2 μA μm^{-1} at $V_{GS} = -1$ V. With 2% in-plane mechanical strain, the corresponding currents are ~1.5 × 10^{-4} μA μm^{-1} at $V_{GS} = 0$ V and ~1.0 μA μm^{-1} at $V_{GS} = -1$ V. These results indicate that the effect of strain on-current is practically trivial. The device is thermally stable in the temperature range -196 °C to $+127$ °C. Moreover, it has a transparency of 88% over the visible spectrum. The WSe$_2$ flakes have an absorbance <5%.

The device exhibits ambipolar characteristics with 34 cm^2 V^{-1} s^{-1} electron mobility and 45 cm^2 V^{-1} s^{-1} hole mobility extracted from an identical device on a Si/SiO$_2$ substrate (Das *et al* 2014).

9.8 Discussion and conclusions

Although graphene possesses outstanding electronic and mechanical properties, the absence of an energy bandgap in graphene has prompted research in alternative materials, notably TMDs.

Graphene TFTs are fabricated on polyimide (Lee *et al* 2013), PEN (Petrone *et al* 2015) and flexible glass (Park *et al* 2016) substrates. The MoS$_2$ TFT is made on a Kapton substrate (Chang *et al* 2013), WS$_2$ TFT is made on a solution-cast polyimide film of thickness 4.8 μm (Gong *et al* 2016), and WSe$_2$ TFT is made on a PET substrate (Das *et al* 2014). Some of these devices have undergone and qualified strain tests to different degrees. A few TFTs have shown high cut-off frequencies. Transparency of TFT is also a useful feature. Looking at the various TFT types, two directions of research trends in 2D TFTs are evident, viz, building transparent bendable systems and integration of wireless communications into flexible electronics.

Review exercises

9.1 What properties of graphene make it an ideal flexible electronics material? What is its drawback?
9.2 Give the names of three transition metal dichalcogenides (TMDs).
9.3 State the reasons for choosing hexagonal boron nitride as the gate dielectric material for graphene TFT.
9.4 For making the graphene TFT on polyimide, how is the polyimide film with gate electrode and h-BN dielectric formed by capture–release process?
9.5 How is graphene sysnthesized on a copper foil? What is the temperature used for graphene synthesis? What gases are used?
9.6 How is graphene transferred to polyimide by the PMMA-based wet transfer process? After graphene has been transferred, why is it necessary to coat a small amount of fresh PMMA on the previous PMMA layer of

PMMA/graphene structure? What happens if this new PMMA is not coated?

9.7 What is the mobility extracted from graphene TFT characteristics? What is the extracted hole mobility?

9.8 In the graphene TFT on PEN substrate, what is the gate metallization? What material is used as the gate dielectric?

9.9 Describe the dry-transfer method used for applying graphene over the gate. How does it differ from the wet transfer process?

9.10 What is the mobility of carriers in the graphene TFT on a PEN substrate? What is the intrinsic cut-off frequency and maximum oscillation frequency of TFT?

9.11 Compare flexible glass with polyimide and PET in terms of surface roughness, glass transition temperature, thermal conductivity and optical transparency.

9.12 Why are Joule heating effects on flexible glass less severe than on polyimide or PET?

9.13 What material is used as the gate dielectric in the graphene TFT on flexible glass?

9.14 What is the extracted mobility for graphene TFT on flexible glass? What is the intrinsic cut-off frequency?

9.15 In the MoS_2 TFT on Kapton, what gate dielectric materials are used?

9.16 In the MoS_2 TFT on Kapton, which of the two gate material devices, Al_2O_3 or HfO_2 are more robust against bending? Compare the bending radii at which the devices using these two materials fail.

9.17 What is the thickness of the solution-cast substrate PI film of WS_2 TFT?

9.18 How is WS_2 synthesized and transferred to the Al_2O_3 gate dielectric? What is the transfer method called?

9.19 How much bending is experienced during the peeling of the WS_2 TFT from the carrier? How does this bending affect the turn-on voltage of TFT?

9.20 In the repetitive bending test of the WS_2 TFT, how much does the maximum on-current change after 30 000 cycles? How much change occurs in 50 000 cycles?

9.21 What is meant by all-2D TFT? What are the gate–source–drain electrodes, the gate dielectric and the semiconductor film in the all-2D TFT made of?

9.22 How are the source–drain contact pads of the all-2D TFT defined using aluminum as a hard mask?

9.23 Is all-2D TFT optically transparent? Comment on its ambipolar conduction behavior.

References

Chang H-Y, Yang S, Lee J, Tao L, Hwang W-S, Jena D, Lu N and Akinwande D 2013 High-performance, highly bendable MoS_2 transistors with high-K dielectrics for flexible low-power systems *ACS Nano* **7** 5446–52

Das S, Gulotty R, Sumant A V and Roelofs A 2014 All two-dimensional, flexible, transparent, and thinnest thin film transistor *Nano Lett.* **14** 2861–6

Gong Y, Carozo V, Li H, Terrones M and Jackson T N 2016 High flex cycle testing of CVD monolayer WS$_2$ TFTs on thin flexible polyimide *2D Mater.* **3** 021008-1–6

Gutiêrrez H R, Perea-López N, Elías A L, Berkdemir A, Wang B, Lv R, López-Urías F, Crespi V H, Terrones H and Terrones M 2012 Extraordinary room-temperature photoluminescence in triangular WS$_2$ monolayers *Nano Lett.* **13** 3447–54

Lee J, Ha T-J, Parrish K N, Sk F C, Tao L, Dodabalapur A and Akinwande D 2013 High-performance current saturating graphene field-effect transistor with hexagonal boron nitride dielectric on flexible polymeric substrates *IEEE Electron Device Lett.* **34** 172–4

Li X, Zhu Y, Cai W, Borysiak M, Han B, Chen D, Piner R D, Colombo L and Ruoff R S 2009 Transfer of large-area graphene films for high-performance transparent conductive electrodes *Nano Lett.* **9** 4359–63 Supporting Online Material (Materials and Methods Figs. S1, S2, S3): www.sciencemag.org

Park S, Shin S H, Yogeesh M N, Lee A L, Rahimi S and Akinwande D 2016 Extremely high-frequency flexible graphene thin-film transistor *IEEE Electron Device Lett.* **37** 512–5

Petrone N, Dean C R, Meric I, van der Zande A M, Huang P Y, Wang L, Muller D, Shepard K L and Hone J 2012 Chemical vapor deposition-derived graphene with electrical performance of exfoliated graphene *Nano Lett.* **12** 2751–6

Petrone N, Meric I, Chari T, Shepard K L and Hone J S 2015 Graphene field-effect transistors for radio-frequency flexible electronics *J. Electron Devices Soc.* **3** 44–8

IOP Publishing

Flexible Electronics, Volume 2
Thin-film transistors
Vinod Kumar Khanna

Chapter 10

CNT FET

In macroelectronics, a single CNT is not to be aligned at the correct location. Instead, a random 2D network of CNTs is deposited to form the semiconductor film of the TFT. In a TFT formed on spin-coated polyimide film, SWCNTs are deposited on a silanized SiO_2 gate dielectric surface by soaking in SWCNT solution. High mobility (150 cm^2 V^{-1} s^{-1}) is achieved but the on–off current ratio is low ~70 due to contamination with metallic CNTs (Snow *et al* 2005). In another TFT on polyimide film, semiconductor-enriched SWCNTs are deposited on a poly-L-lysine-modified SiO_x surface yielding higher on–off ratio ~10^4 with mobility of 20 cm^2 V^{-1} s^{-1}. For stretchability, the PI film is cut into a honeycomb mesh structure. When pulled lengthwise, the device continues to work up to 3 mm displacement corresponding to 11.5% stretchability (Takahashi *et al* 2011). A still higher on–off ratio ~10^5 is achieved in TFT on a Kapton substrate with CNTs deposited on an APTES-modified Al_2O_3 surface from solution with mobility 10–35 cm^2 V^{-1} s^{-1} (Chandra *et al* 2011). In an inkjet printed SWCNT TFT on a PES substrate, the on–off ratio is 3×10^4 and mobility is 9.76 cm^2 V^{-1} s^{-1} (Lee *et al* 2016). The all-inkjet printed CNT TFT on Kapton polyimide film shows an on–off ratio of 138 and operating frequency of 5 GHz (Vaillancourt *et al* 2008). Gravure-printed SWCNT TFT-based D flip–flop has an on–off ratio of 10^4 with mobility of 0.2 cm^2 V^{-1} s^{-1} for the drive TFT and an on–off ratio of 10^2 with mobility of 0.5 cm^2 V^{-1} s^{-1} for the load TFT (Noh *et al* 2011a). A TFT is made on a PET substrate with inverse gravure printing and solution-deposited SWCNTs yielding an on–off ratio of 5.7×10^5 and mobility of 9.13 cm^2 V^{-1} s^{-1}. The transfer characteristics are not altered on bending up to 1 mm (Lau *et al* 2013). The all-CNT TFT made on a PEN substrate using photosensitive dry film shows a mobility of 33 cm^2 V^{-1} s^{-1} and an on–off ratio more than 10^5; the CNTs are solution-deposited (Chen *et al* 2018).

10.1 Introduction

CNTs possess the desirable mechanical flexibility, stretchability and electrical conductivity favoring their utilization in bendable electronics. Individual SWCNTs show mobility >10 000 cm^2 V^{-1} s^{-1} and can work at frequencies beyond 1 GHz. Considerable progress has been made in the purification of CNTs and in the separation of metallic CNTs from semiconducting CNTs. The percentage of semiconducting CNTs has reached 95%–99% using the density gradient ultracentrifugation technique. All these supportive advances have given impetus to TFT fabrication using CNTs.

10.2 High-mobility SWCNT TFT on spin-coated PI substrate

A common issue in nanoelectronic device fabrication is the non-availability of a high throughput manufacturable process for placement of SWCNTs at precise locations. But in large-area macroelectronics, the same issue is not as serious because a 2D random network of SWCNTs represents the averaged properties of a large number of SWCNTs. So, standard processing technology can be applied to fabricate reliable devices. Snow *et al* (2005) fabricated SWCNT TFT on a PI substrate (figure 10.1).

10.2.1 Substrate preparation

The substrate is a polyimide thin film of thickness 10 μm. For easy processing, the PI film is deposited by spin coating on a silicon wafer. The PI film is baked in a nitrogen atmosphere in stages of 100 °C steps up to 300 °C by curing for 1 h at each stage.

10.2.2 Formation of gate metal electrodes and contact pads

This pattern is defined in photoresist. Titanium is deposited by electron-beam evaporation. The Ti film is 25 nm thick. The lift-off technique is used for creating the titanium pattern.

10.2.3 Gate dielectric deposition

The gate dielectric is silicon dioxide. It is formed by electron-beam evaporation. The SiO$_2$ film thickness is 100 nm. Lift-off photolithography is used to define the silica pattern.

10.2.4 Surface modification of SiO$_2$ film for SWCNT deposition

A 3%–5% solution by volume of (3-aminopropyl)trimethoxysilane, linear formula H$_2$N(CH$_2$)$_3$Si(OCH$_3$)$_3$ or C$_6$H$_{17}$NO$_3$Si, molecular weight 179.29 is used. The wafer is immersed in this solution for 1 h. Then it is dried by blowing.

10.2.5 Preparation of SWCNT solution

Sodium dodecyl sulfate (SDS), linear formula CH$_3$(CH$_2$)$_{11}$OSO$_3$Na or NaC$_{12}$H$_{25}$SO$_4$, molecular weight 288.38 (1% by weight) is mixed with water. The mixture is ultra-sonicated until the SDS completely dissolves. Into the SDS solution, 1 mg mL^{-1} of SWCNT powder is added followed by ultrasonification at 10 W for 45 min. Centrifugation of the solution is done at 12 000 g for 1 h. The liquid is decanted leaving the sediment behind. Centrifugation and decantation steps are repeated until the cessation of sediment formation.

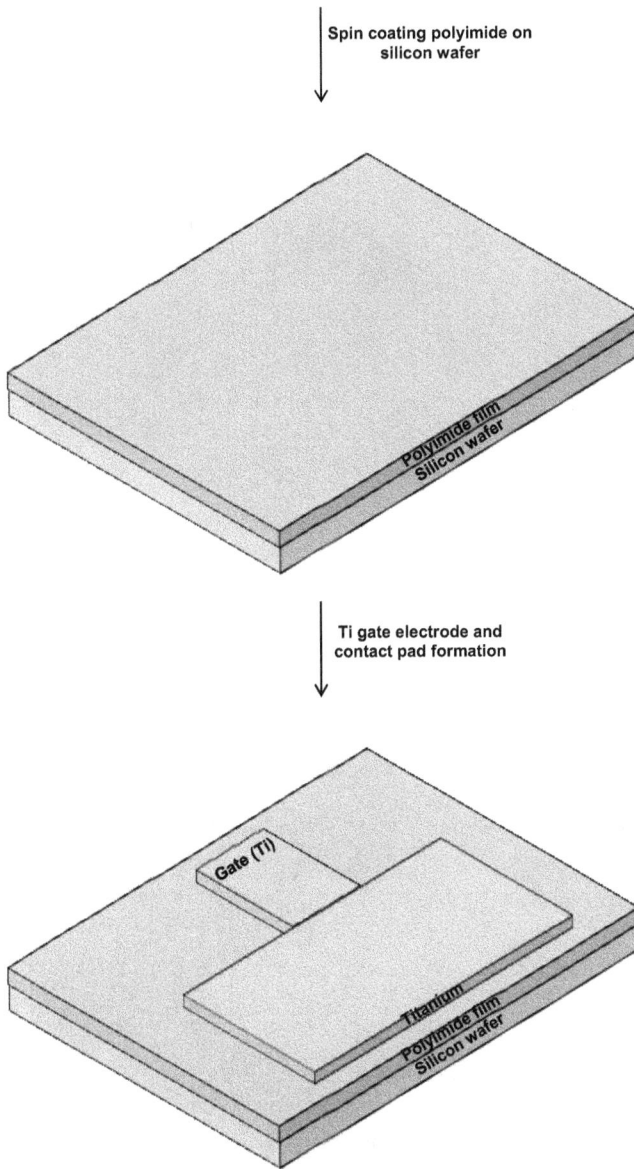

Spin coating polyimide on
silicon wafer

Polyimide film
Silicon wafer

Ti gate electrode and
contact pad formation

Gate (Ti)

Titanium
Polyimide film
Silicon wafer

Figure 10.1. SWCNT TFT on a PI substrate.

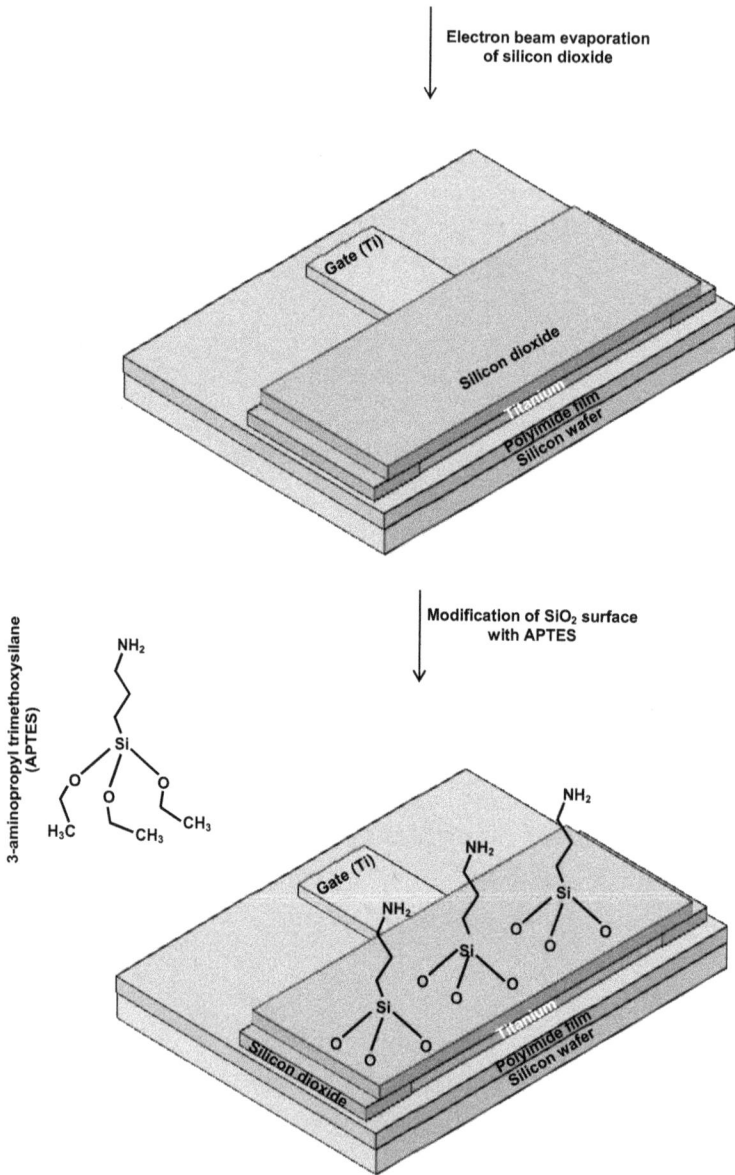

Figure 10.1. (Continued.)

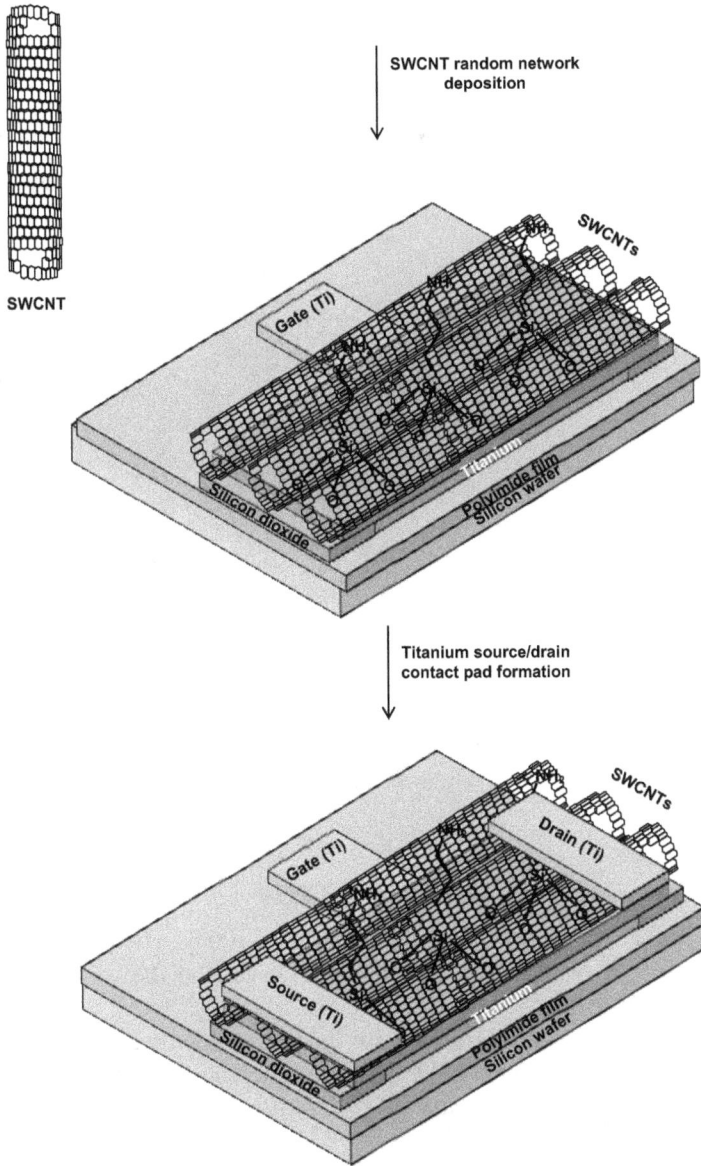

SWCNT random network deposition

SWCNTs

Gate (Ti)

Silicon dioxide

Titanium

Polyimide film

Silicon wafer

SWCNT

Titanium source/drain contact pad formation

SWCNTs

Gate (Ti)

Drain (Ti)

Source (Ti)

Silicon dioxide

Titanium

Polyimide film

Silicon wafer

Figure 10.1. (Continued.)

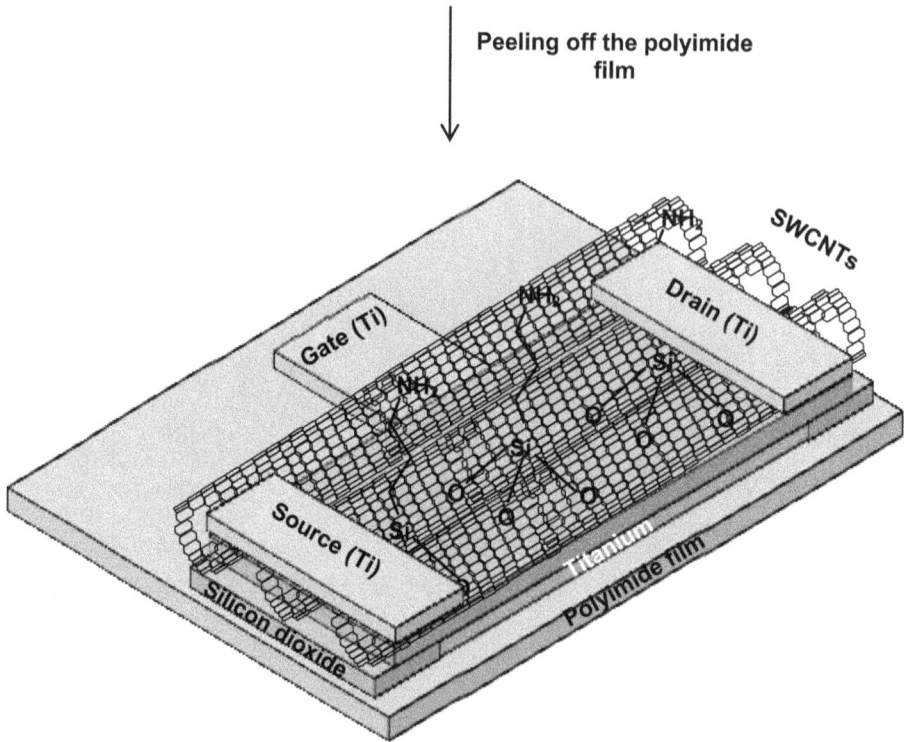

Peeling off the polyimide film

Figure 10.1. (Continued.)

10.2.6 Deposition of random network of SWCNTs

The wafer is soaked in the SWCNT solution for 50–100 h. After taking out from the SWCNT solution, it is blown dry. Remnant SDS left on the wafer is removed by rinsing in DI water for 1 h.

The SWCNT deposition is assessed with the help of identically treated test pieces. These pieces are dipped into the SWCNT solution. They are periodically taken out and their sheet resistances are measured to monitor SWCNT deposition.

10.2.7 Formation of source/drain contact pads

The source/drain contact pattern is formed in photoresist. Titanium is deposited over the patterned photoresist. Ti film thickness is 100 nm. When the photoresist is lifted off, titanium is left only on the regions of source/drain contact pads.

10.2.8 Peeling off the PI film

After processing, the PI film is peeled off from the silicon wafer so that the devices are supported on a free-standing PI film.

10.2.9 TFT characteristics

The W/L ratio of the TFT is 130 μm/7 μm = 18.57. At $V_D = -1.5$ V, the normalized transconductance (g_m/W) of the TFT is 0.5 mS mm^{-1}. The field-effect mobility is 150 cm^2 V^{-1} s^{-1} at $V_D = 0.01$ V. The current on–off ratio is typically 70 with values varying between 50 to 400. The high off-current is a consequence of the presence of metallic and small bandgap SWCNTs in the network. These SWCNTs do not undergo full depletion and, therefore, conduct current. The TFT performance can be significantly ameliorated by purification of SWCNTs to remove the metallic SWCNTs (Snow *et al* 2005).

10.3 Semiconductor-enriched CNT-based TFT on spin-coated PI substrate for active-matrix backplane

Takahashi *et al* (2011) developed flexible/stretchable active-matrix backplanes using TFT arrays formed with SWCNT networks, and utilized them to prepare an artificial electronic skin (e-skin). This skin provides spatial mapping of touch profiles.

10.3.1 Substrate used

The substrate is a polyimide film of 24 μm thickness (figure 10.2). For ease of processing, the PI film is coated by spinning twice on SiO$_2$/Si handle wafer (100 mm diameter), each time at 2000 RPM for 1 min.

10.3.2 Gate structure

The gate electrode is made of nickel. The nickel film is deposited by thermal evaporation over the PI substrate. The gate dielectric comprises three layers: SiO$_x$ (bottom layer: 10 nm)/Al$_2$O$_3$ (20 nm)/SiO$_x$ (top layer: 15 nm). The SiO$_x$ layers are obtained by electron-beam deposition while the Al$_2$O$_3$ layer is formed by atomic layer deposition (ALD). The bottom SiO$_x$ layer serves as the nucleation layer for Al$_2$O$_3$. The top SiO$_x$ layer promotes adhesion of SWCNTs to be laid over the gate oxide. The necessity of this layer arises from the poor adhesion of Al$_2$O$_3$ with SWCNTs. The presence of fixed charges in the as-grown Al$_2$O$_3$ film is the most likely cause for non-adherence. In this way, a sandwich structure with Al$_2$O$_3$ interposed between SiO$_x$ layers is formed.

10.3.3 SWCNT network deposition

The next layer over the gate oxide is the semiconductor channel layer formed by depositing highly dense SWCNT networks with good uniformity. In preparation for SWCNT deposition, the surface of the top SiO$_x$ layer is modified by treatment with oxygen plasma at 30 W for 1 min. Then poly-L-lysine treatment is given for 5 min by solution casting. In solution casting, the poly-L-lysine is dissolved in the solution, the solution is coated on the SiO$_x$ layer and the solvent is dried leaving the poly-L-lysine on the SiO$_x$.

Spin coating polyimide on SiO$_2$/Si wafer

Polyimide
SiO$_2$
Silicon

Thermal evaporation and patterning of nickel gate electrode

Gate (Ni)
Polyimide
SiO$_2$
Silicon

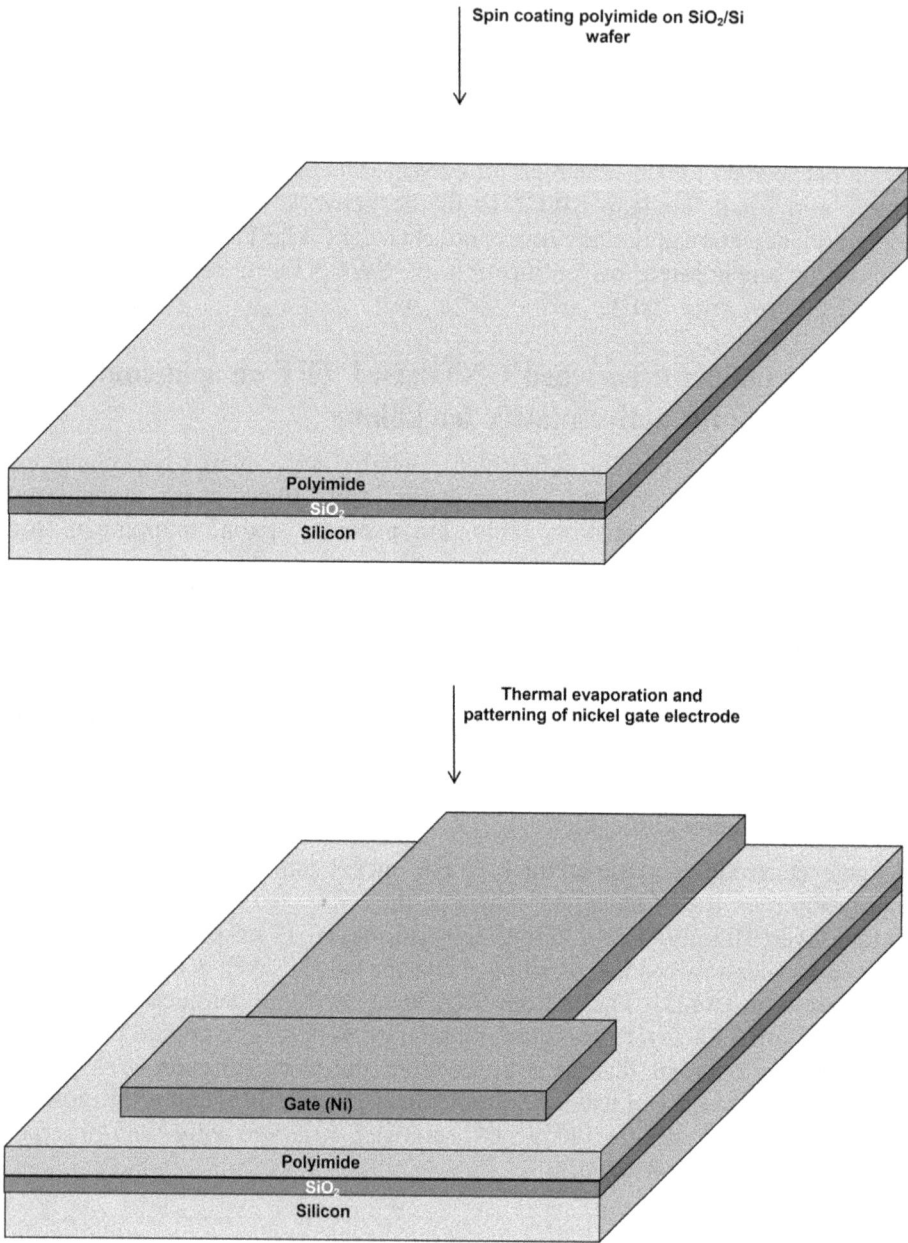

Figure 10.2. Semiconductor-enriched CNT TFT.

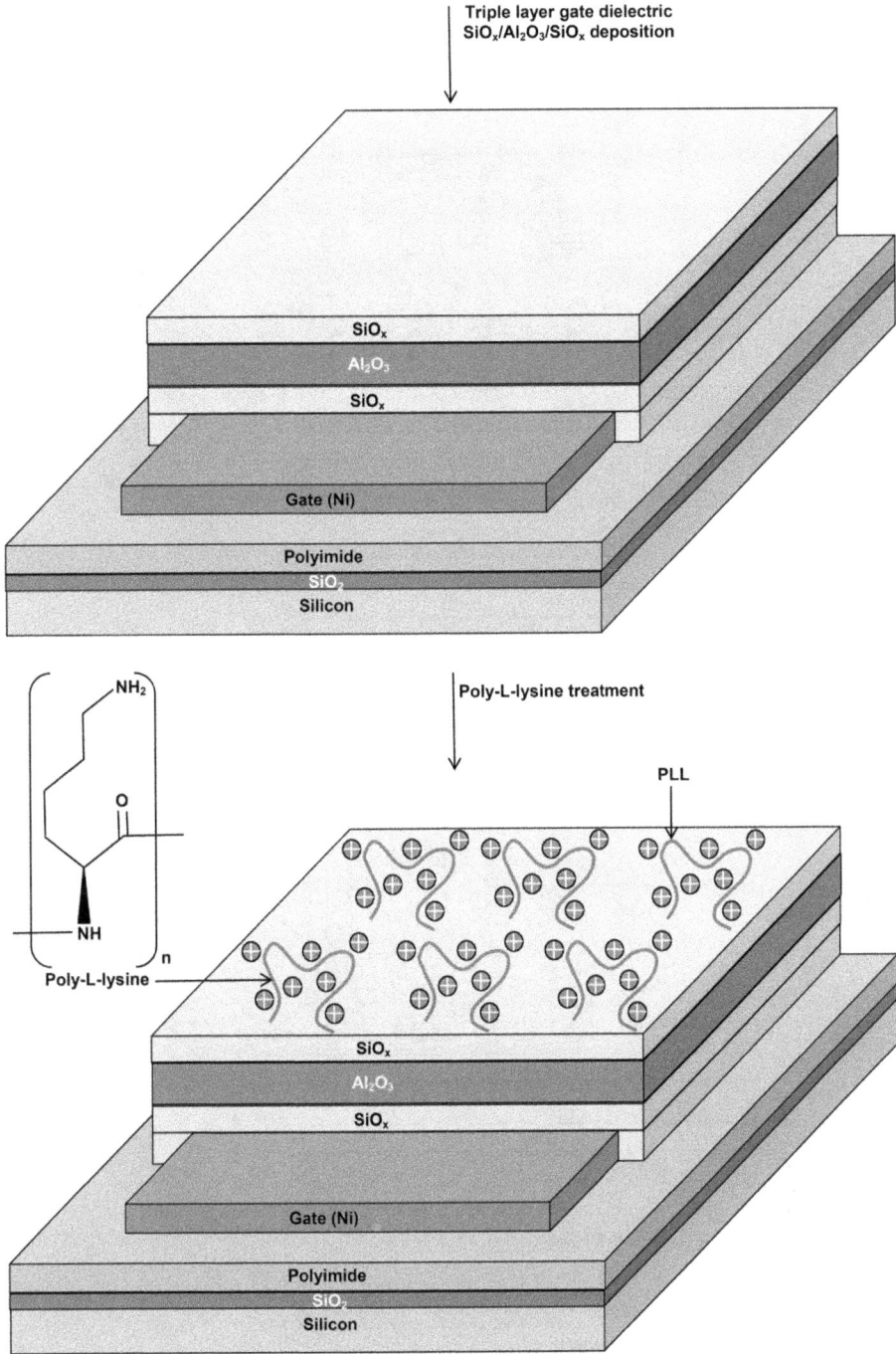

Figure 10.2. (Continued.)

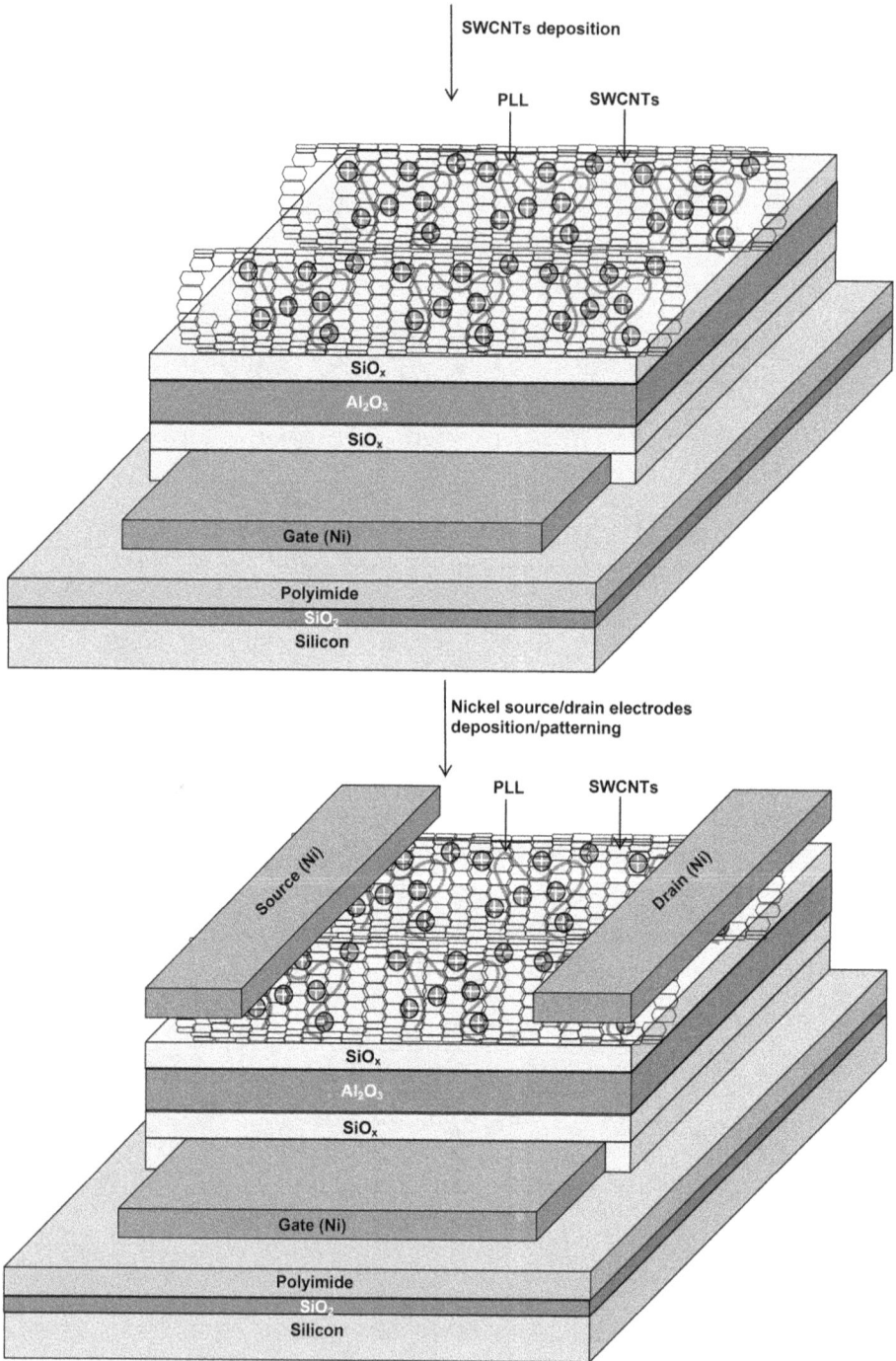

Figure 10.2. (Continued.)

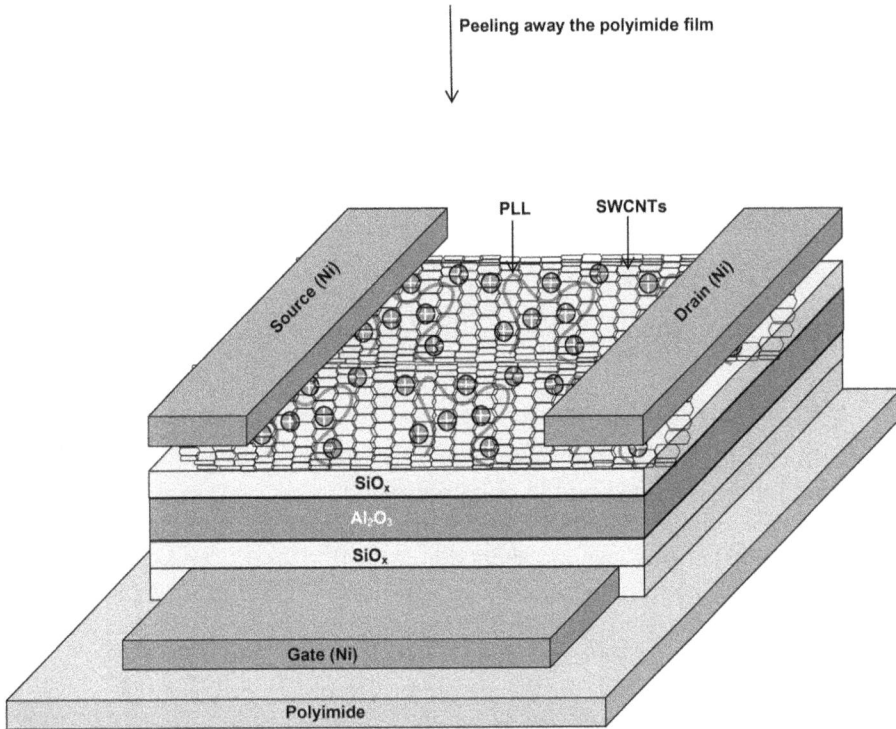

Figure 10.2. (Continued.)

After DI water rinsing, 99% semiconductor-enriched SWCNTs are deposited by solution casting over the gate oxide. Then the SWCNTs are rinsed with DI water. The density of SWCNTs is governed by the surface preparation and the nanotube surfactants. Vacuum annealing is done at 200 °C for 1 h. This annealing ensures removal of the residues of the surfactants. The vacuum annealing also helps to improve the transconductance value and ratio of on-state current to off-state currents for the TFTs.

10.3.4 Source–drain metallization and encapsulation

Ohmic contacts are made with the valence band of SWCNTs. The source/drain electrodes are made of palladium (Pd), with a thickness of 35 nm. The lift-off process is used for defining the metallization pattern.

For packaging, a 500 nm thick parylene-C layer is deposited over the device. This layer not only provides mechanical support to the device but also makes it chemically resistant to environmental degradation agents such as adsorbed water molecules. After photolithography and etching in oxygen plasma, via holes are made in the parylene-C layer to reach the contact pads. The channel is 3 μm long and 240 μm wide. The channel aspect ratio is 240 μm/3 μm = 80.

10.3.5 Removing the PI film from the wafer

After the full process has been completed, the edge of the PI layer is cut with a razor blade and the PI film with the fabricated device(s) upon it is removed from the SiO$_2$/ Si handle wafer by peeling it away. The peel-off process does not pose any difficulty because PI has a poor adhesion with SiO$_2$. In this way a mechanically flexible device is obtained on a PI film.

10.3.6 Electrical properties of TFT

The mobility is ~20 cm^2 V^{-1} s^{-1}. The ratio I_{ON}/I_{OFF} is 10^4.

10.3.7 Stretchability of PI substrate

The as-formed PI film does not allow extension or elongation to the desired extent. The stretchability is defined as the maximum engineering strain that the film can tolerate before breakage. To make the PI film stretchable, it is cut into a honeycomb mesh structure by cutting hexagon-shaped holes with a laser. The holes are made at a pitch of 3.3 mm. The side length of the hexagons is varied from 1 to 1.85 mm. With increasing hole side length from 0 to 1.85 mm, the stretchability is found to increase from 0 to 60%. This increase is enabled by the creation of twisting ability in the PI bridges connecting the holes. Further, structural symmetry makes the honeycomb mesh invariant towards every rotation by 60°.

If the active devices with W/L = 200 µm/5 µm = 40 are positioned at the intersection points on the mesh, the performance degradation remains non-significant when the structure is mechanically bent up to a radius of curvature of 2 mm. This robustness against bending is made possible by properly designing the device structure taking the neutral bending plane of PI substrate into account, the ultra-small dimensions of SWCNTs and their mechanical strength. If the substrate is pulled lengthwise while performing electrical measurements on TFTs, the device remains functional up to a displacement of 3 mm, for which the stretchability is 11.5%. Packaged TFTs are thermally stable from 0 °C to 100 °C in air with minimal change in threshold voltage.

An artificial electrical skin is demonstrated by fabricating an active-matrix backplane. This backplane contains 12 × 8 pixels of size 6 × 4 cm^2 with each pixel controlled by a TFT. After lamination of parylene-C passivation layer with a pressure sensitive rubber (PSR), connection of the drain of each transistor to PSR and grounding through an Al-foil, the device is used for spatial pressure detection and mapping (Takahashi *et al* 2011).

10.4 CNT TFT with high current on–off ratio on a Kapton substrate

Chandra *et al* (2011) fabricated CNT TFTs with on–off ratio of 10^5 using a bottom-gate configuration (figure 10.3).

Applying PMMA on silicon wafer

PMMA adhesive
Silicon carrier

Fixing kapton film on silicon carrier

Kapton film
PMMA adhesive
Silicon carrier

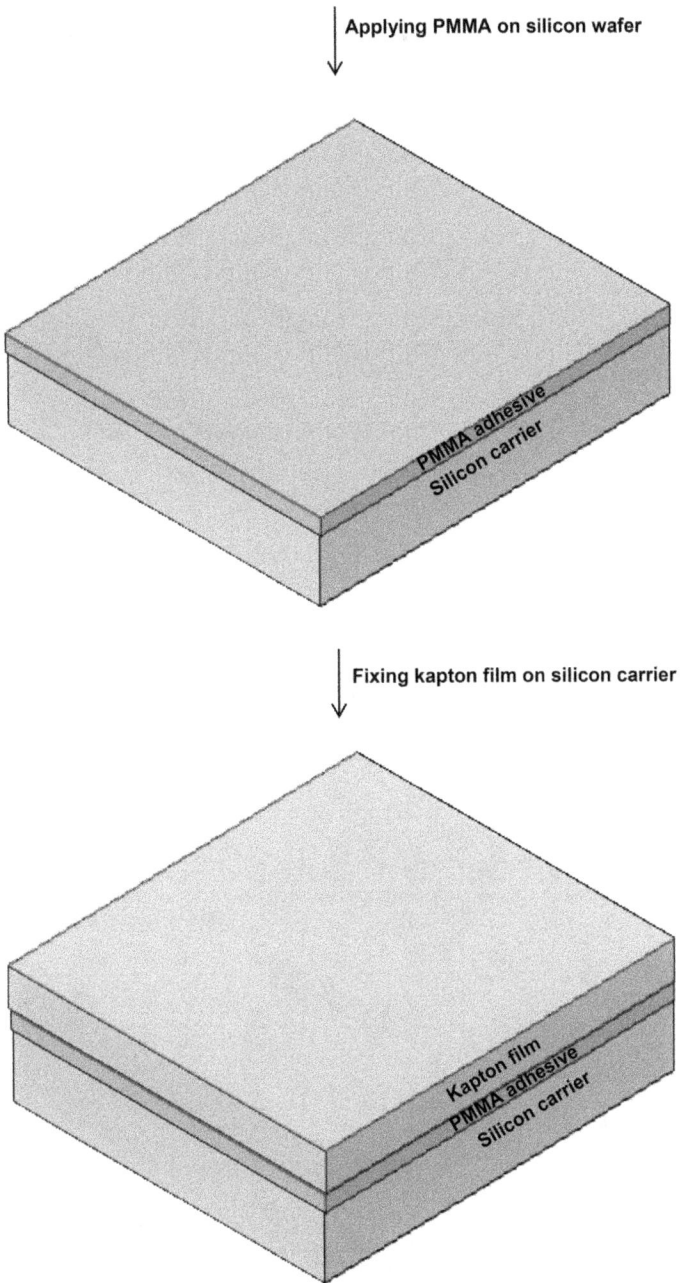

Figure 10.3. CNT FET with high current on–off ratio.

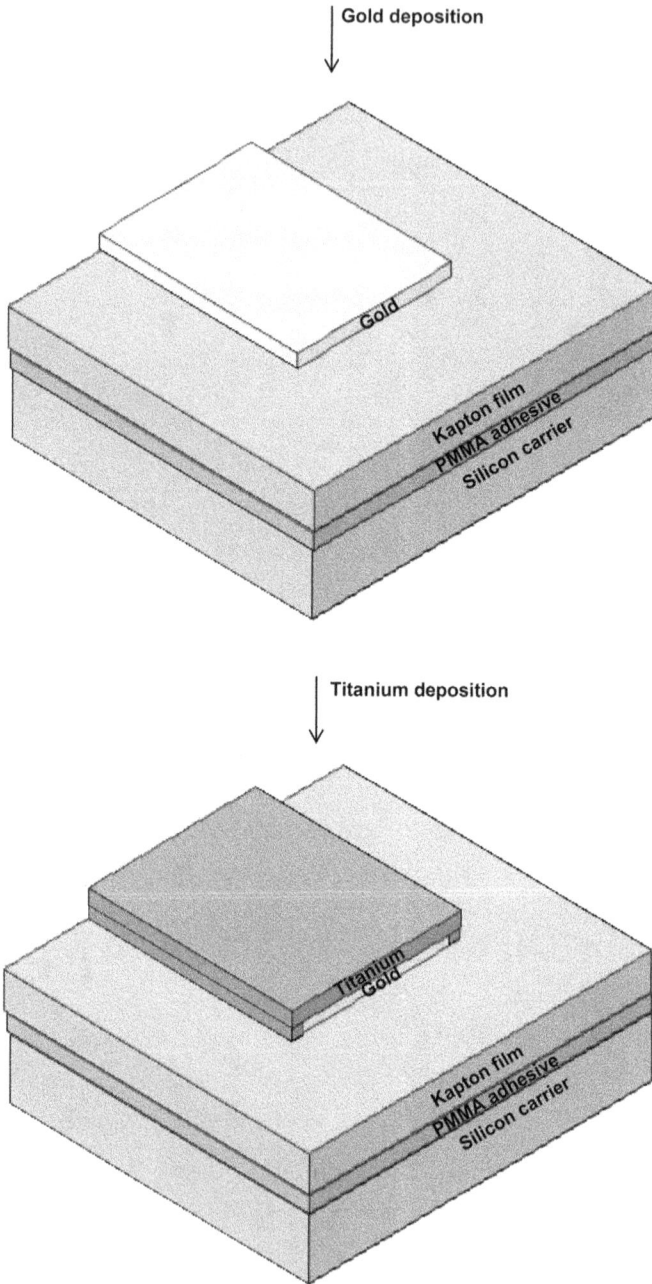

Gold deposition

Gold

Kapton film
PMMA adhesive
Silicon carrier

Titanium deposition

Titanium
Gold

Kapton film
PMMA adhesive
Silicon carrier

Figure 10.3. (Continued.)

Al$_2$O$_3$ deposition

APTES deposition

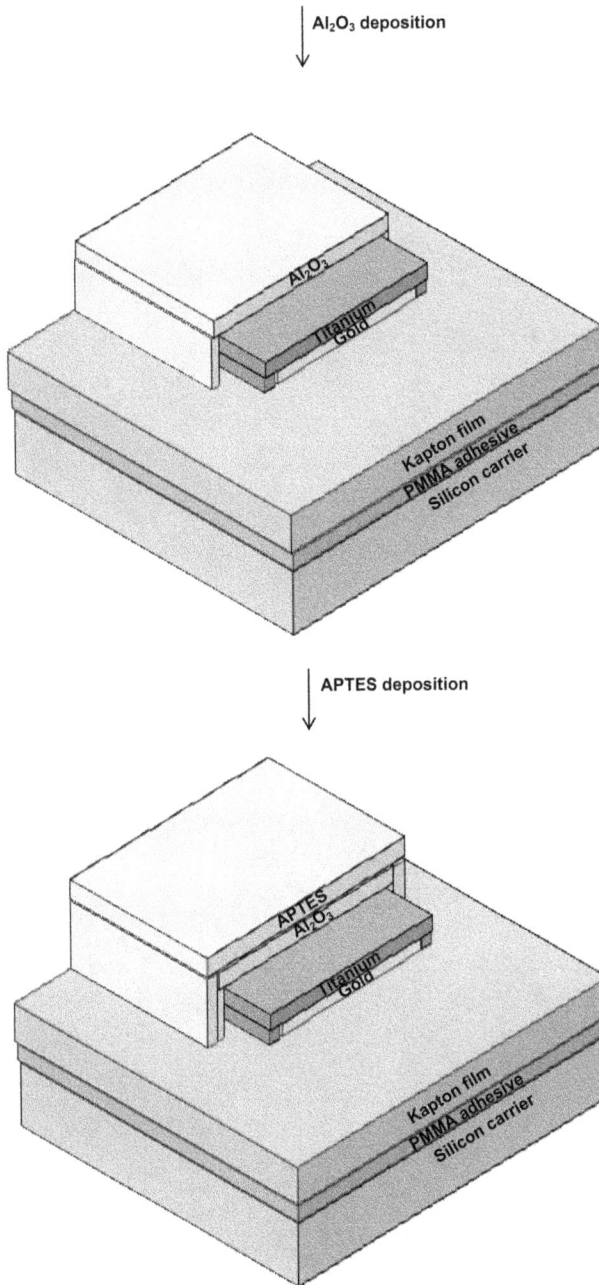

Figure 10.3. (Continued.)

CNTs deposition

Ti/Pd/Au source/drain metallization

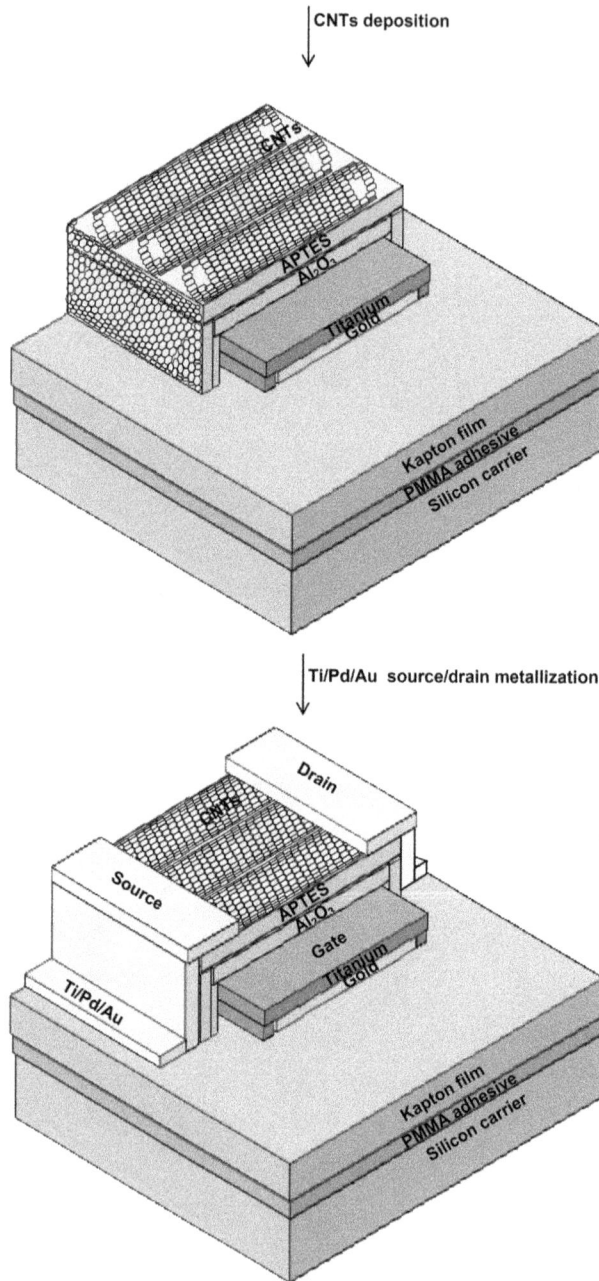

Figure 10.3. (Continued.)

Silicon carrier removal

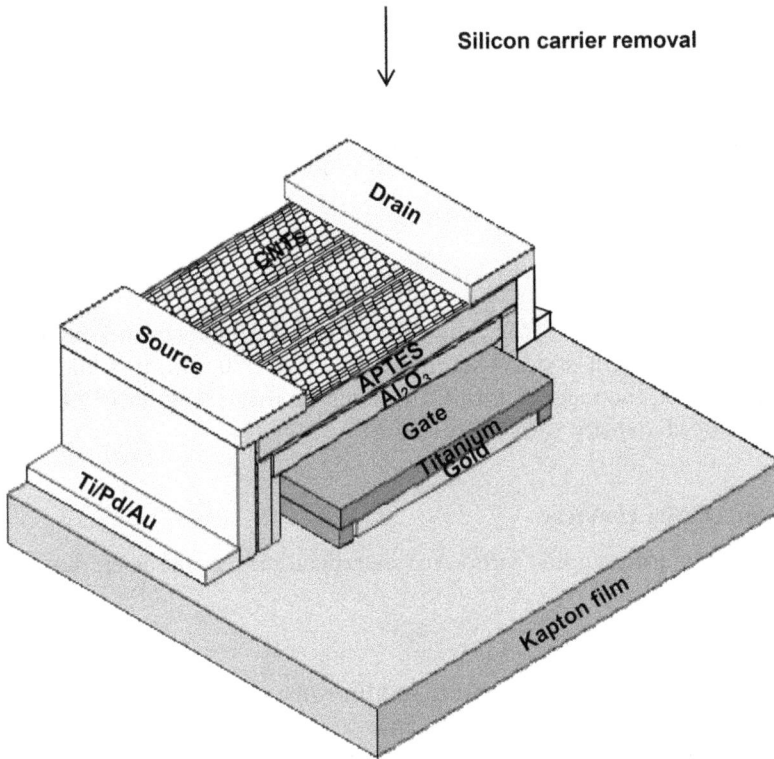

Figure 10.3. (Continued.)

10.4.1 Substrate preparation and mounting

2 cm × 2 cm pieces of Kapton film are used as substrates. Any possible processing-induced distortions of substrate are minimized by subjecting the Kapton pieces to several cycles of heating to 200 °C followed by cooling down.

In order that the substrate does not curl up during processing, it is fixed on a silicon wafer using PMMA as an adhesive, taking care to avoid trapping air bubbles at the substrate–PMMA interface because that will cause loss of planarity.

10.4.2 Gate electrode formation

First gold is deposited and then titanium over the gold.

10.4.3 Gate dielectric deposition

The gate dielectric is high-quality aluminum oxide. Its thickness is 40 nm. It is formed by atomic layer deposition at 150 °C on titanium. Titanium oxidizes in air to form titanium dioxide. So, a thin native oxide layer grows on the titanium surface. Aluminum oxide is selectively deposited on this titanium dioxide film. It is not deposited on Kapton which is hydrophobic.

10.4.4 Al$_2$O$_3$ surface modification

The wafer-mounted Kapton substrate with the gate electrode and gate dielectric film is immersed in a 1% v/v solution of 3-aminopropyltriethoxysilane (APTES), linear formula H$_2$N(CH$_2$)$_3$Si(OC$_2$H$_5$)$_3$ or C$_9$H$_{23}$NO$_3$Si, molecular weight 221.37 in DI water for 30 min. A monolayer of APTES is formed which serves as an adhesive for CNTs.

10.4.5 CNT deposition

The Kapton substrate with APTES-modified Al$_2$O$_3$ surface is immersed in a solution of 95% CNTs in sodium dodecyl sulfate (SDS). After 30 min, it is taken out of the solution and rinsed thoroughly with DI water. The mean length of CNTs is 0.9 μm. The average CNT density is 6–7 CNTs μm^{-2}.

10.4.6 Source/drain contacts

The titanium/palladium/gold (Ti/Pd/Au) metallization scheme is employed.

10.4.7 TFT behavior

CNTs become hole-doped on exposure to atmosphere. The TFT exhibits P-channel characteristics with threshold voltage close to 0 V. The on-current per unit gate width is 0.1 μA μm^{-1}. The TFT on-current varies by ±5% when subjected to repeated cycles of flexing and relaxation. The current on–off ratio is 10^5. The range of the highest carrier mobilities is 10–35 cm^2 V^{-1} s^{-1} (Chandra *et al* 2011).

10.5 Inkjet printed SWCNT TFT on PES substrate

Printed electronics are both scalable and economical. Lee *et al* (2016) used poly (1,3,5-trimethyl-1,3,5-trivinyl cyclotrisiloxane) (pV3D3) as the gate dielectric to fabricate a top gate SWCNT TFT (figure 10.4).

10.5.1 Substrate functionalization

The plasma-assisted chemical vapor deposition (PECVD) technique is used to deposit a SiO$_x$ film of thickness 50 nm on the surface of the polyethersulfone (PES) substrate. The deposition temperature is 150 °C. An amine-terminated adhesion layer is formed on the surface of the PES substrate using 0.1% w/v aqueous poly-L-lysine solution.

10.5.2 SWCNT deposition

The PES substrate with a surface-modified SiO$_x$ layer is immersed in 99.9% semiconducting enriched SWCNT solution for 7 h. Then it is taken out and rinsed with DI water.

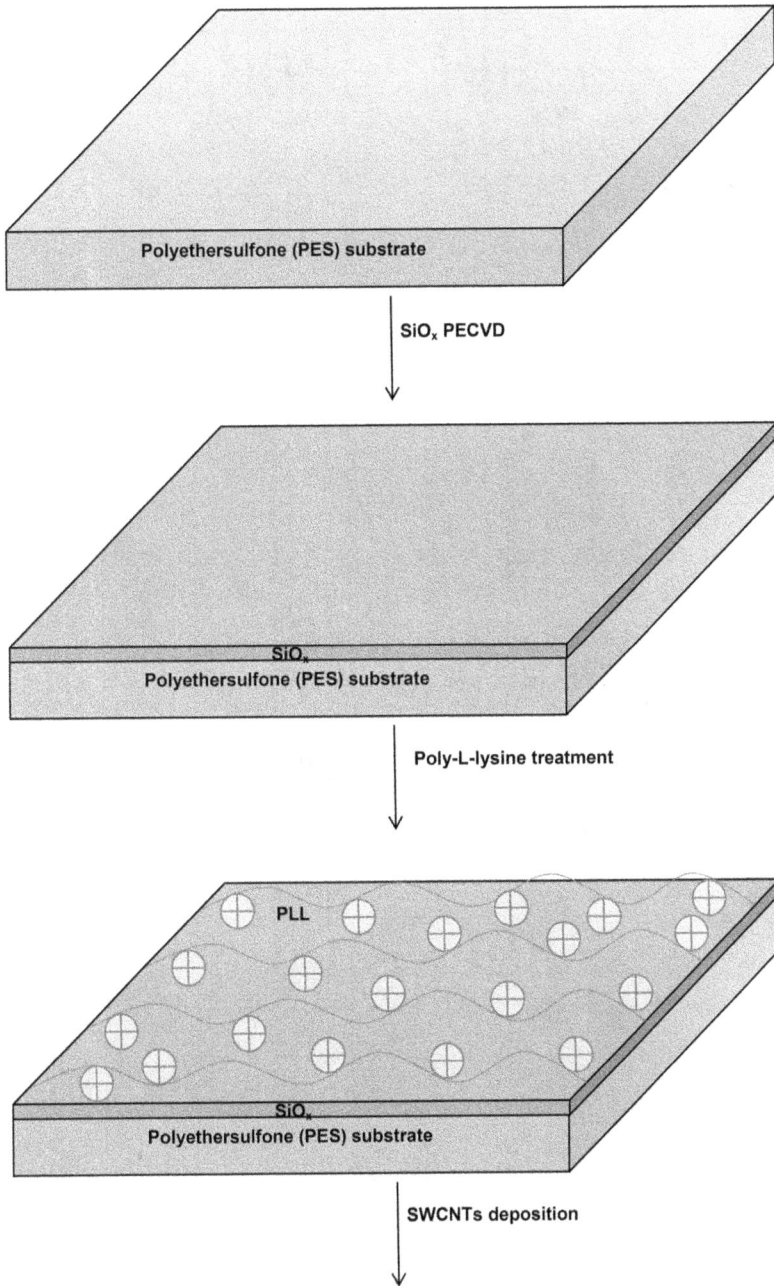

Figure 10.4. Inkjet printed CNT TFT.

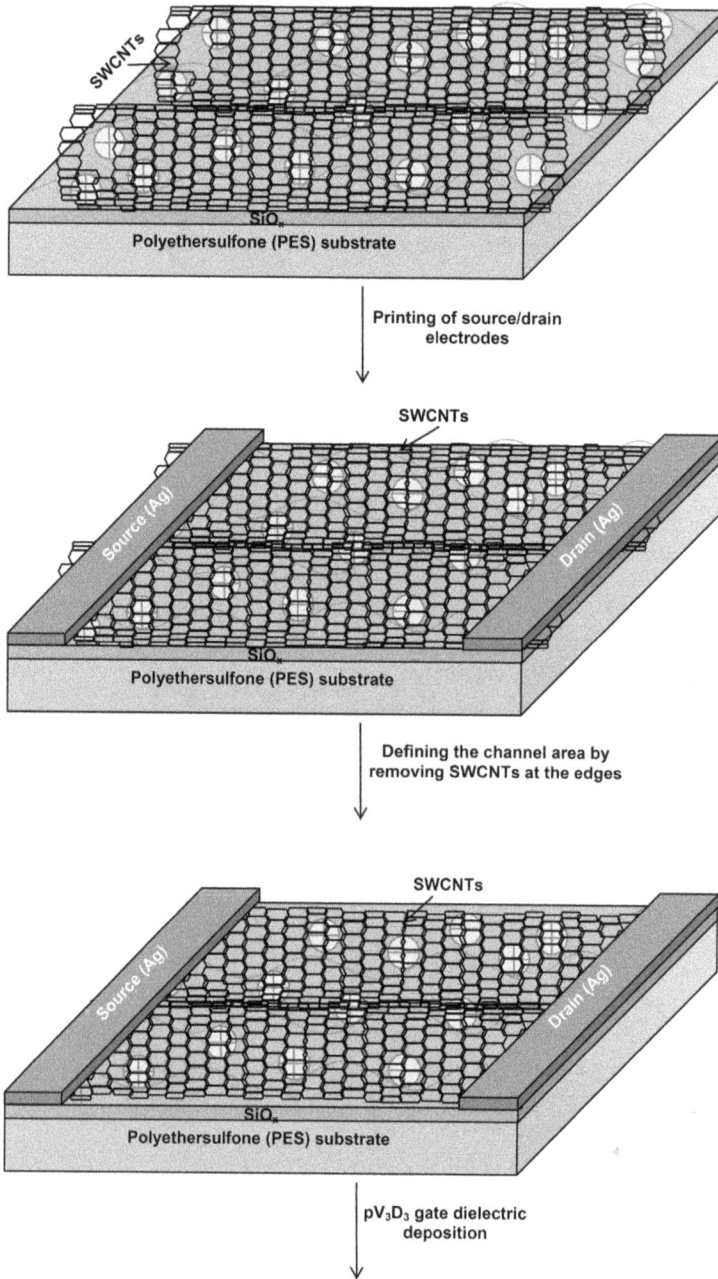

Printing of source/drain electrodes

Defining the channel area by removing SWCNTs at the edges

pV_3D_3 gate dielectric deposition

Figure 10.4. (Continued.)

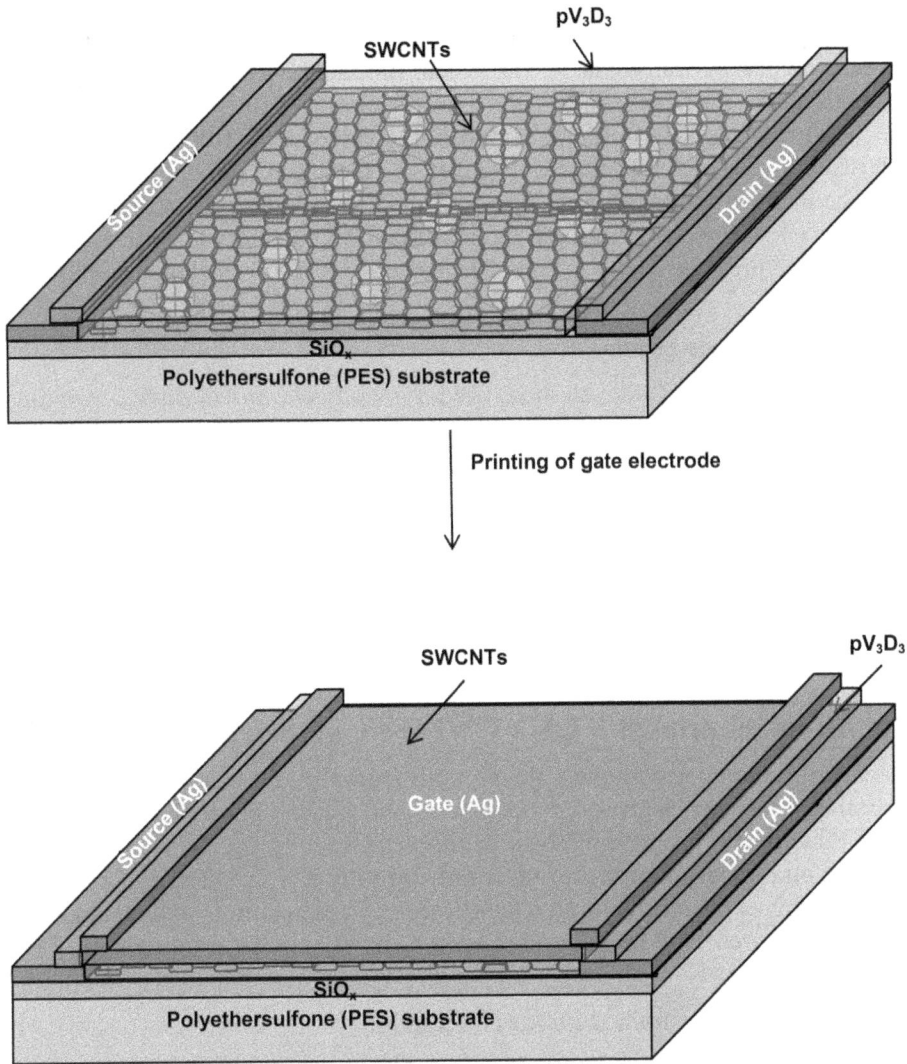

Figure 10.4. (Continued.)

10.5.3 Source–drain electrode printing

A conducting silver nanoparticle ink is used. Filtration through a 5 μm PTFE syringe filter avoids clustering of nanoparticles. The inkjet printer has 50 μm orifice nozzles. The droplet diameter is 100 μm. The volume of the droplet is 40–50 pL. The drop pitch is 80–90 μm at 300 Hz.

10.5.4 Channel area definition

This is done by printing polyvinylpyrrolidone (PVP), $(C_6H_9NO)_n$. It acts as a mask against oxygen plasma etching. All the SWCNTs outside the channel area are removed by etching in oxygen plasma.

10.5.5 Gate dielectric deposition

This is done in a custom-made iCVD system (initiated chemical vapor deposition). The monomer is 1,3,5-trimethyl-1,3,5-trivinyl cyclotrisiloxane (V3D3), $C_9H_{18}O_3Si_3$. It is polymerized. The initiator is tert-butyl peroxide (TBPO), linear formula $(CH_3)_3COOC(CH_3)_3$, molecular weight 146.23. Both V3D3 and TBPO are vaporized. The filament temperature is 200 °C. V3D3 (flow rate = 2.5 sccm) and TBPO (flow rate = 1 sccm) vapors are injected into the iCVD reactor. The substrate is kept at 40 °C. The pressure is 300 mTorr. The obtained pV3D3 thickness is 40 nm.

10.5.6 Gate electrode printing

Silver ink is used and the method used for source/drain electrode printing is followed.

10.5.7 TFT parameters

The operating voltage of TFT is <4 V. The current on–off ratio is 3×10^4 and the mobility is 9.76 cm^2 V^{-1} s^{-1}. Inverter, NAND and NOR gates are realized with these TFTs. Their output characteristics show the viability of execution of logic functions (Lee *et al* 2016).

10.6 All-inkjet printed 5 GHz CNT TFT on Kapton polyimide film

No vacuum coating techniques, no photolithography, no surface pre-treatment/functionalization, no clean room operations and no complicated expensive processes! Technology is progressing towards simplistic solutions solely utilizing printing for device and circuit manufacturing. Inkjet printing is a maskless process. It does not require any exposure at elevated temperature. Vaillancourt *et al* (2008) described top gate configured TFT fabrication process using only inkjet printing (figure 10.5).

10.6.1 Printing source/drain electrodes on Kapton

Silver nanoink is used for printing the source/drain electrodes. The ink is baked at 130 °C for 30 min. The electrodes are 50 µm wide. Their separation distance, i.e., the channel length, is 100 µm.

10.6.2 Printing active carrier transport layer

An ultra-pure, electronic grade CNT solution is used to obtain a high density >1000 CNTs µm^{-2} thin film having extremely low amorphous carbon content. The printing is done several times so that a uniform film giving a source-to-substrate resistance of 200 kΩ is formed. The CNT film is dried in air at room temperature.

10.6.3 Printing gate dielectric

A thin film of ion gel (ionic liquid plus triblock copolymer) is printed.

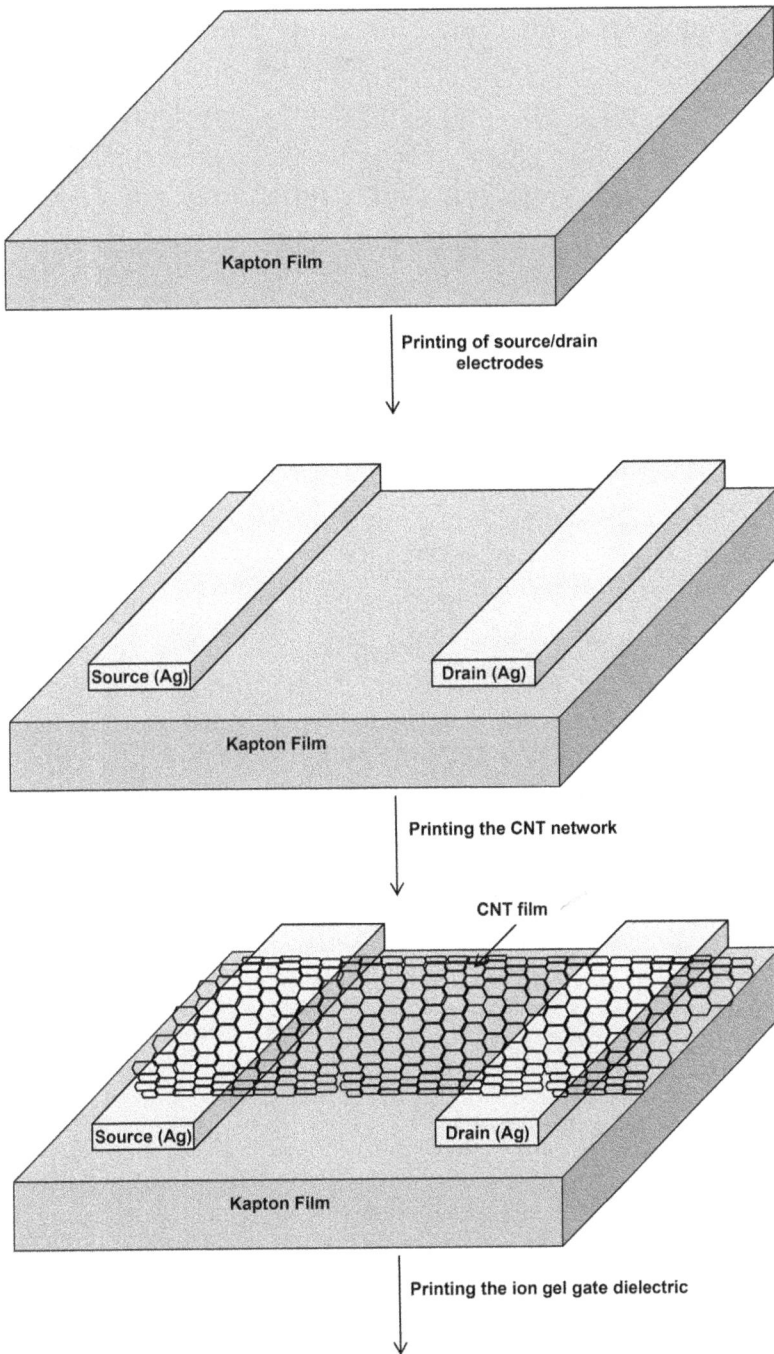

Figure 10.5. All-inkjet printed CNT TFT.

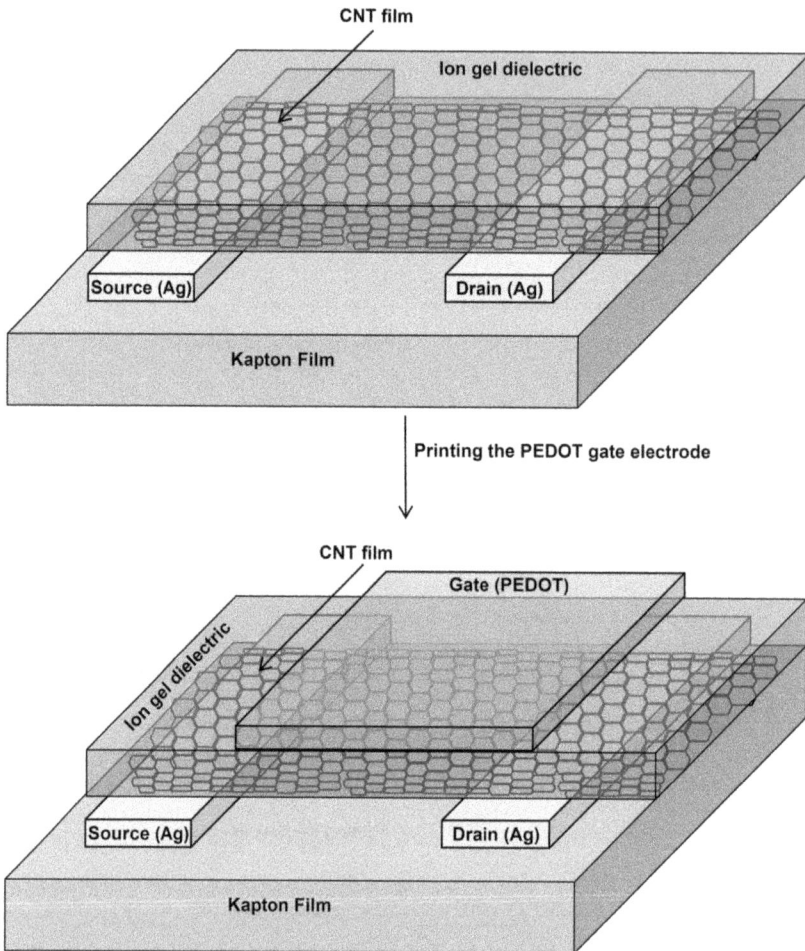

Figure 10.5. (Continued.)

10.6.4 Printing the top gate electrode

The gate electrode is printed with a conducting polymer poly(3,4-ethylenedioxy-thiophene) (PEDOT), linear formula $(C_6H_4O_2S)_n$.

10.6.5 TFT performance

The CNT network works as a P-type material. The source/drain current on–off ratio of the TFT is 138. Gate–source leakage current is in pA range. The operating frequency is 5 GHz (Vaillancourt *et al* 2008).

Gu *et al* (2011) achieved a high current on–off ratio of 10^3 using high-purity, additive-free aqueous SWCNT ink with trace metal impurity content of 500 ppb.

10.7 Gravure printed SWCNT-based TFT for D flip–flop, half-adder and full-adder on PET foil

As a first step towards realization of printed flexible IC, Noh *et al* (2011a) developed a D flip–flop which is fabricated fully by gravure printing: R2R (roll-to-roll) gravure and R2P (roll-to-plate) gravure. Three color units of the printer are employed for printing the gate electrode, the gate dielectric and source/drain electrodes. For printing the SWCNT ink to make the active layer, R2P gravure printer is used.

10.7.1 D flip–flop

The process steps for TFT are (Noh *et al* 2011a) (figure 10.6):

   (i) The substrate: The substrate is a PET foil of thickness 75 μm.

   (ii) Printing the gate electrode: Nanoparticle-based Ag gravure ink is used. The required quantities of ethylene glycol, $HOCH_2CH_2OH$ or $C_2H_6O_2$ and hexanol, $CH_3(CH_2)_5OH$ or $C_6H_{14}O$ are added to obtain a surface tension of 36 mN m^{-1} and viscosity of 200 cP. The gate electrode is 200 μm wide. After R2R gravure printing (roll pressure = 0.8 MPa, web speed = 10 m min^{-1}), the PET foil is passed through the first heating chamber maintained at a temperature of 150 °C for 5 s. Thus the ink is cured.

   (iii) Printing the gate dielectric: The ink is a hybrid of barium titanate, $BaTiO_3$ powder (particle diameter 50 nm) with poly(methyl methacrylate), $[CH_2C(CH_3)(CO_2CH_3)]_n$. Its surface tension is 30 mN m^{-1} and viscosity is 200 cP. It has a relative permittivity of 14. After R2R gravure printing with roll pressure and web speed the same as for the gate electrode, passage through the second inline heating chamber at 150 °C for 5 s cures the ink. The dielectric layer is 3 μm thick. It has a surface roughness of 70 nm.

   (iv) Printing the source/drain electrodes: The same silver-based ink is used as for the gate electrode. After R2R gravure printing of the source/drain electrodes with roll pressure and web speed the same as for the gate electrode, the PET foil is passed through the third heating chamber at 150 °C for 5 s to cure the ink. Thickness of these electrodes is 680 nm. Their surface roughness is 200 nm.

   (v) Printing the active semiconductor layer: SWCNT ink is used for printing the active layer of the load TFT. For printing the active layer of the drive TFT, this SWCNT ink is diluted tenfold. Here R2P gravure printing is done at roll pressure = 0.5 MPa, web speed = 10 m min^{-1}, instead of R2R gravure printing. For R2P gravure printing, the SWCNT ink is filled into the engraved region of the plate mold. The ink is conveyed directly from the chemically etched engraved printing plate to the targeted active area of the device on the PET foil. The foil is fixed on the impression roller covered with rubber. Curing is done at 150 °C for 1 min in an oven.

By using different concentrations of SWCNT inks for drive and load TFTs, the R2P printer alters the density of networks of SWCNTs and thereby achieves the desired electrical performance of these TFTs.

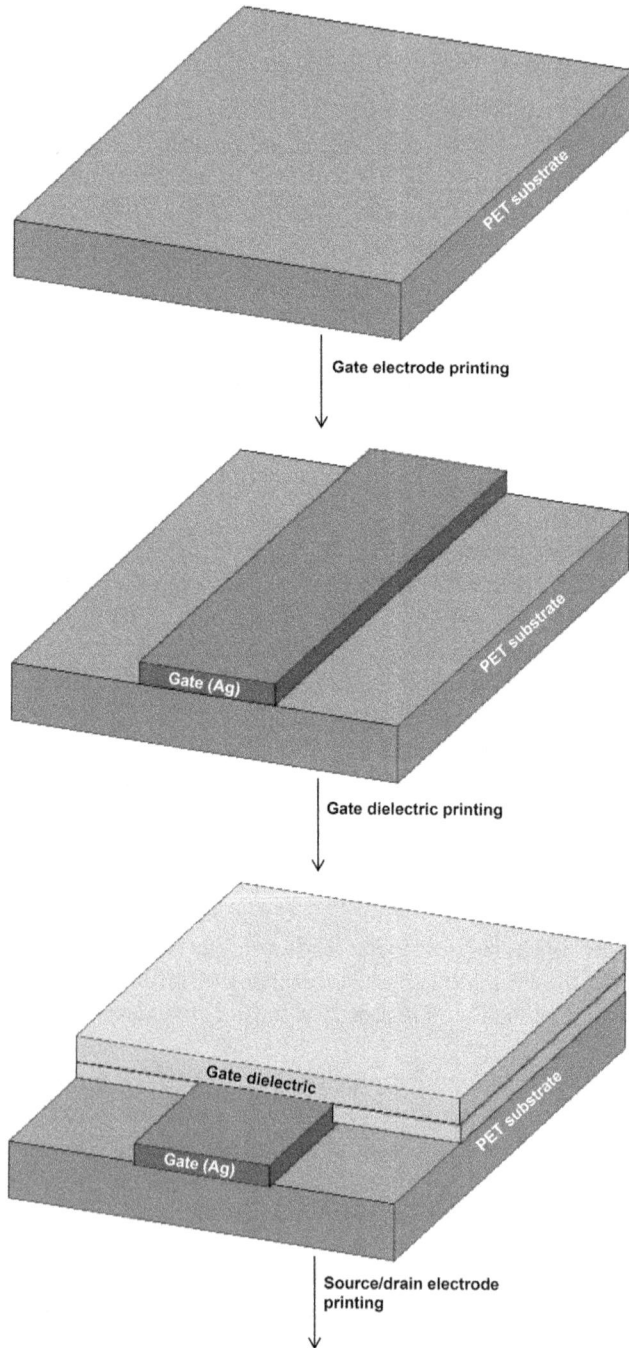

Figure 10.6. Gravure-printed SWCNT TFT.

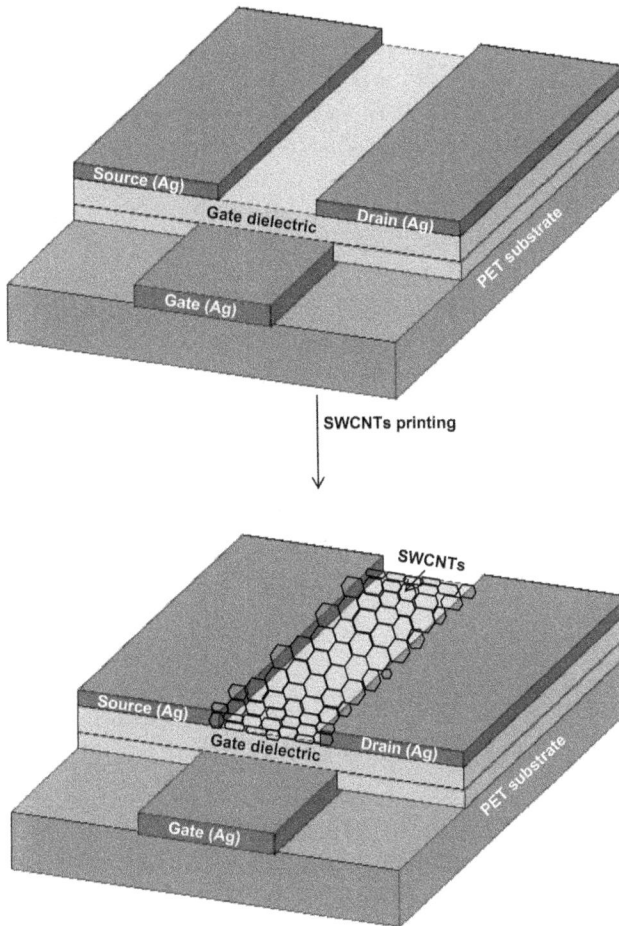

Figure 10.6. (Continued.)

Load and drive TFT, and D flip–flop behavior: The TFTs work as P-channel devices. For the drive TFT, the current on–off ratio is 10^4 and the field-effect mobility is $\sim$0.2 cm^2 V^{-1} s^{-1} at $V_{DS} = -20$ V; the threshold voltage is -5.4 V.

For the load TFT, the current on–off ratio is 10^2 and the field-effect mobility is $\sim$0.5 cm^2 V^{-1} s^{-1} at $V_{DS} = -20$ V; the threshold voltage is -1.5 V.

The D flip–flop comprises two latches with each latch containing two P-channel transmission switches and two inverters. At a clock frequency of 20 Hz, the clock-to-output delay is 23 ms (Noh *et al* 2011a).

10.7.2 Half- and full-adders

To open the doorway for a printed flexible arithmetic logic unit, Noh *et al* (2011b) realized a half-adder by gravure printing process using two color units of an R2R gravure printer at the printing speed of 10 m min^{-1} and an R2P printer at the printing speed of 18 m min^{-1}. For gate electrode printing, the surface tension of the Ag ink is 44 mN m^{-1} and its viscosity is 300 cP. For gate dielectric printing, the surface tension of the dielectric ink is 30 mN m^{-1} and its surface tension is 200 cP. For printing the active layer, the surface tension of the SWCNT ink is 30 mN m^{-1} and its surface tension is 10 cP. The active layer of load TFT is printed with 0.01 wt% content of SWCNTs while that of the drive TFT is printed with 0.001 wt% of SWCNTs. For printing the source/drain electrodes, the surface tension of the silver ink is 44 mN m^{-1} and its viscosity is 500 cP.

The half-adder performance is dependent on variations of edge curliness of source/drain electrodes. For drive TFTs, the on–off current ratio is 10^5 at $V_{DS} = -20$ V; the same for load TFTs is 10^3. For drive TFTs, the field-effect mobility is 0.04–0.07 cm^2 V^{-1} s^{-1} at $V_{DS} = -20$ V, whereas the mobility for load TFTs is 0.1–0.17 cm^2 V^{-1} s^{-1}. For drive TFTs, the threshold voltage is −4.5 V while for load TFTs, it is −2.0 V.

Continuing their efforts towards printed flexible arithmetic logic unit, Noh *et al* (2012) demonstrated a 1-b full-adder circuit containing 27 SWCNT TFTs made fully by R2P gravure printing at a roll pressure of 5 kgf cm^{-2} and web speed of 18 m min^{-1}. The maximum delay of the full-adder is 13.7 ms at 50 Hz with a supply voltage of −20 V.

10.8 Inverse gravure-printed CNT TFT on a PET substrate with solution-deposited SWCNTs

Lau *et al* (2013) deposit the SWCNTs in the first step by immersion in SWCNT solution and thereafter complete the remaining process wholly by inverse gravure printing to achieve higher mobility and current on–off ratio (figure 10.7). A custom-built inverse gravure printer is used. In this printer, Cr-plated flat Cu sheets etched with regularly spaced cells act as the gravure plates. The cells form the print patterns. Pipette-dropped inks are uniformly spread in the cells by doctor blading. A dual camera system enables fine adjustment of stage position for alignment between successive layers. The barrel with the attached PET substrate is lowered to contact the gravure plate. On releasing the barrel rotation clutch, the stage acquires control over the barrel through friction. Movement of the stage is accompanied by rotation of the barrel, transferring the ink from the gravure plate to the PET substrate. Printing of each layer is followed by drying of the ink in an oven at 150 °C for 1 min

10.8.1 Substrate cleaning and surface functionalization

The PET substrate has the dimensions: 15 cm × 25 cm × 100 μm. It is cleaned in oxygen plasma at 120 W for 2 min. After cleaning, it is immersed in 0.1% w/v poly-L-lysine solution in water for 30 min. Poly-L-lysine is a positively charged synthetic linear polymer chain of amino acids with varying molecular weight. A less viscous solution has lower molecular weight. It has one hydrobromide (HBr) per unit of lysine residue. It facilitates adhesion of SWCNTs to the PET surface. After poly-L-lysine treatment, the PET substrate is rinsed with DI water.

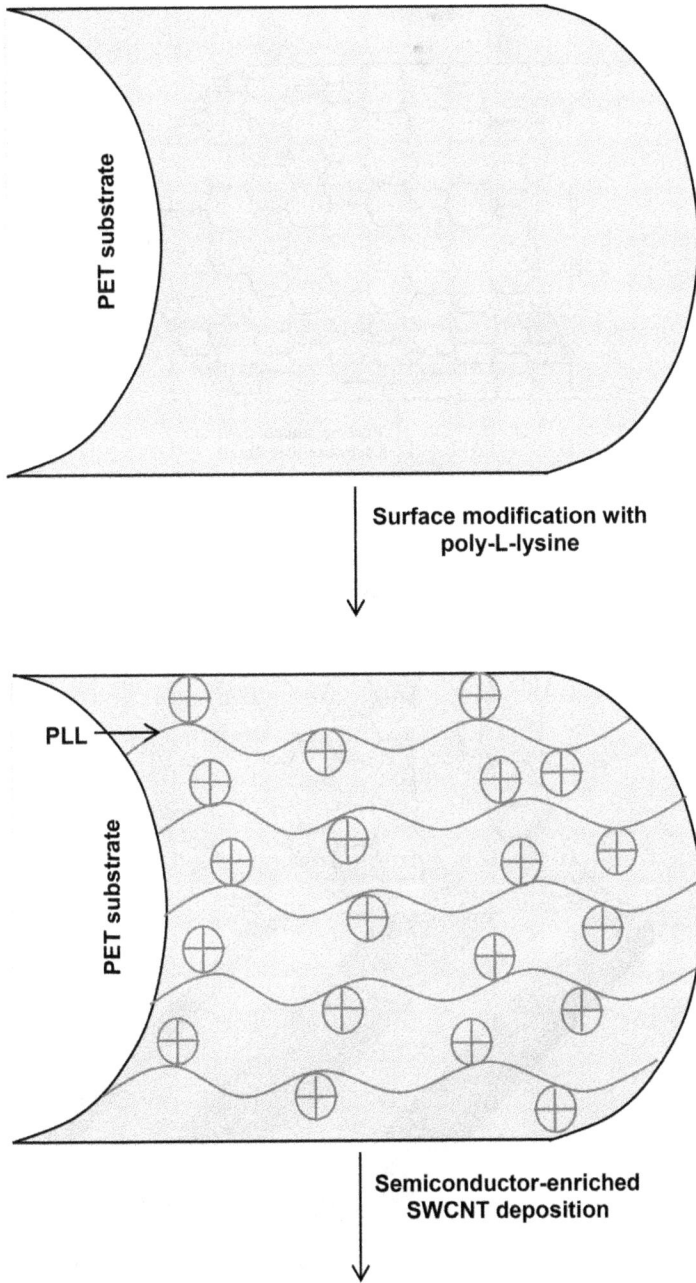

Figure 10.7. Inverse gravure-printed CNT TFT.

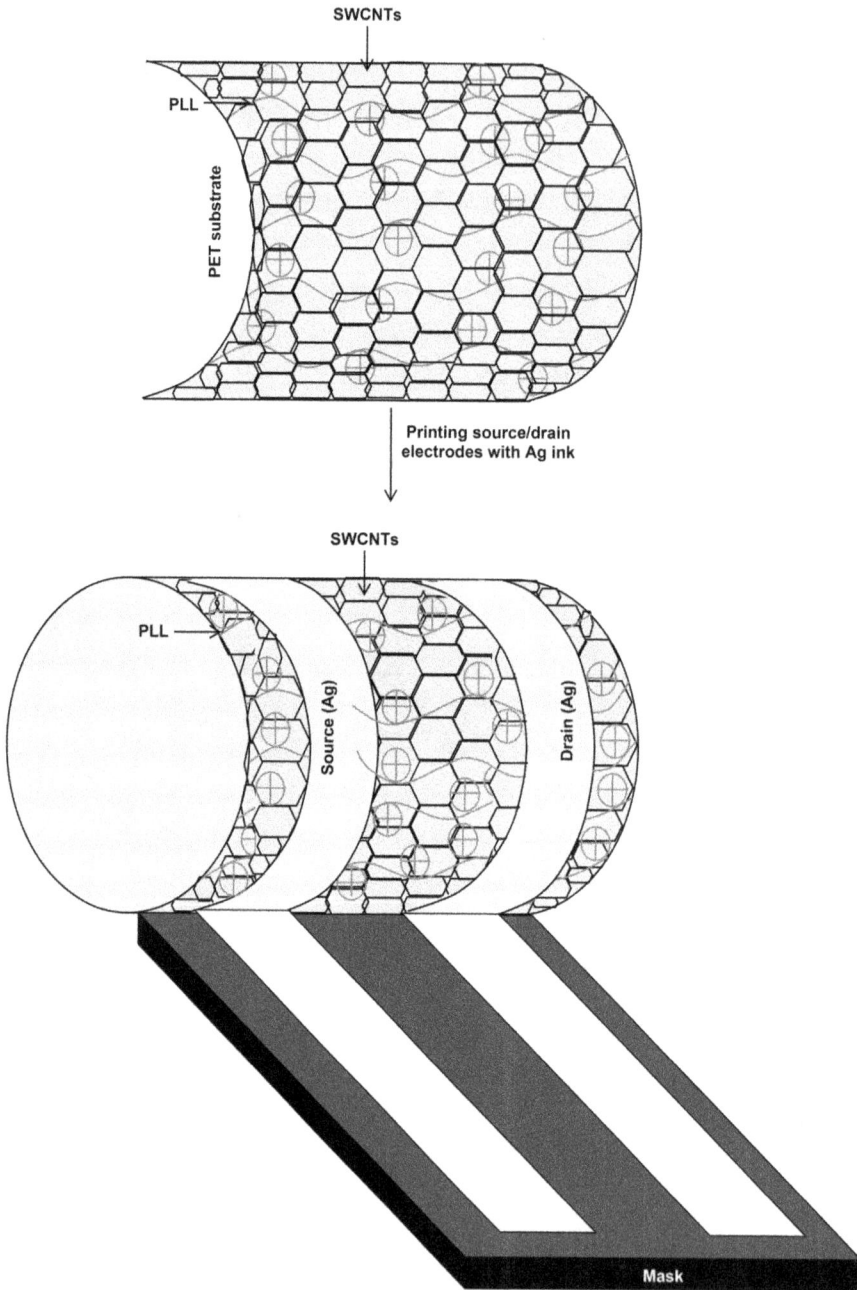

Figure 10.7. (Continued.)

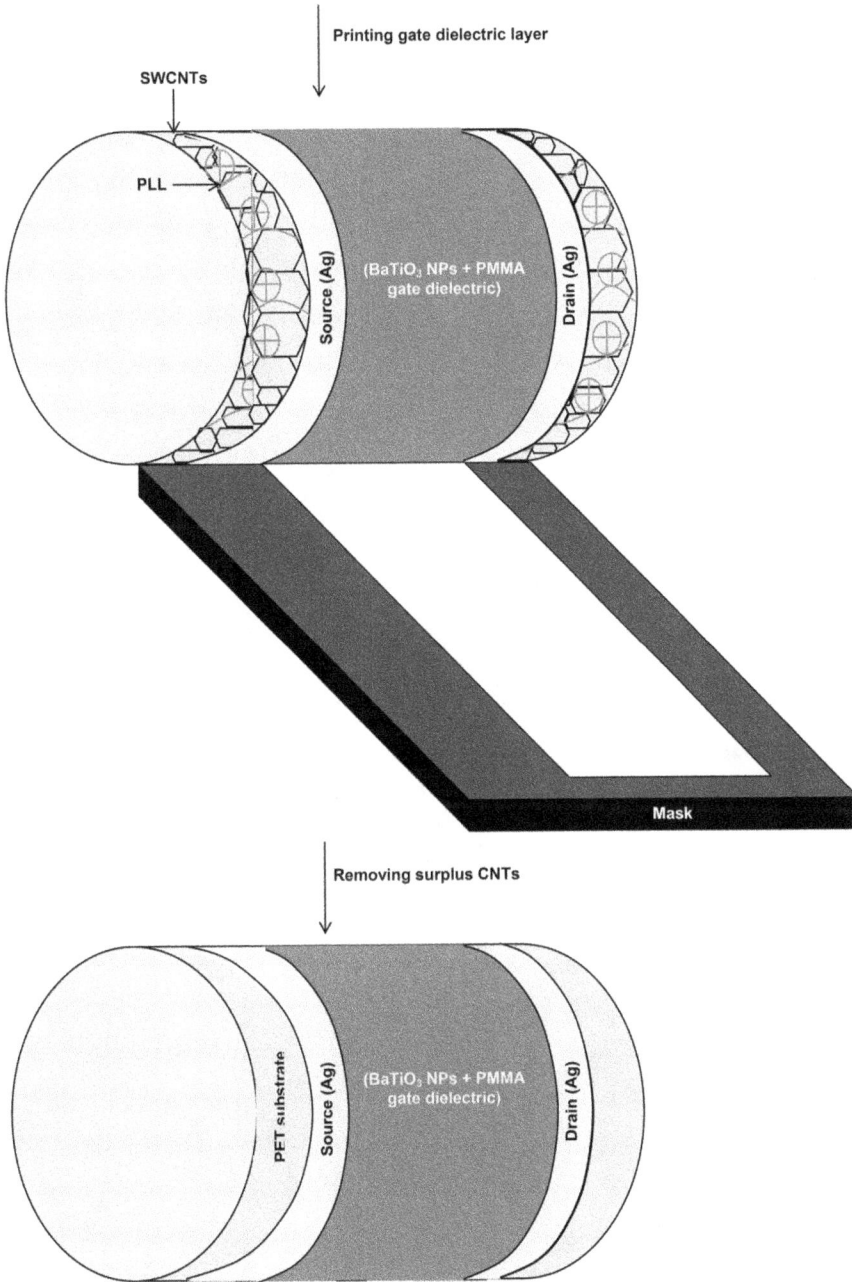

Figure 10.7. (Continued.)

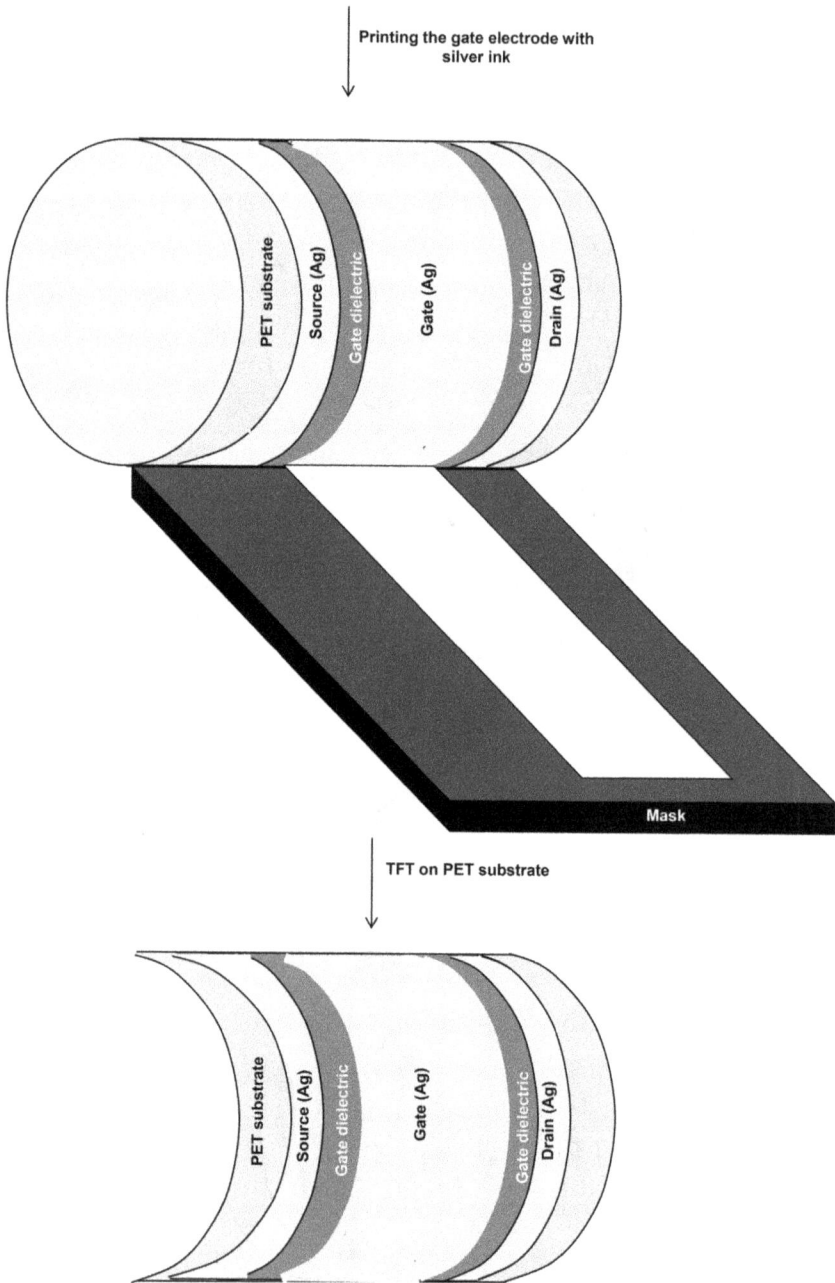

Figure 10.7. (Continued.)

10.8.2 Deposition of active channel material

The functionalized PET substrate is immersed in 99% semiconductor-enriched SWCNT solution. It is kept in this solution for 2 h. After removing it from the solution, the substrate is thoroughly rinsed with DI water and dried in a nitrogen stream. About 60 SWCNTs μm^{-2} density is observed by scanning electron microscopy.

10.8.3 Source–drain electrode printing

Silver nanoparticle ink is used with a film thicness of 2.5 μm.

10.8.4 Gate dielectric printing

This is done with the hybrid ink of high-κ barium titanate nanoparticles and poly (methyl methacrylate). The relative permittivity of the printed film is 17. The dielectric film thickness is 1.5 μm.

10.8.5 Removal of excess SWCNTs

The gate dielectric layer is used as a hard mask to define the gate area. As the SWCNTs are coated on the full PET substrate, the SWCNTs lying outside the gate area are removed by etching in oxygen plasma at 120 W for 2 min.

10.8.6 Gate electrode printing

This is done with the same silver nanoparticle ink as used for source/drain electrodes. The electrode film is 1.25 μm thick.

10.8.7 TFT properties

The W/L ratio of the top-gated TFT is 1250 μm/85 μm = 14.7. The best performance value of on-current density per unit gate width is 32.2 $\mu A\ mm^{-1}$ while transconductance per unit gate width is 5.69 $\mu S\ mm^{-1}$. The current on–off ratio is 5.7×10^5. The field-effect mobility is 9.13 $cm^2\ V^{-1}\ s^{-1}$. The electrical characteristics remain practically constant after 1000 measurement cycles. A threshold voltage shift of 1 V is recorded after 60 days of exposure to atmosphere. No noticeable change in transfer characteristics with respect to the flat condition is observed upon bending the TFT to a radius of curvature of 1 mm (Lau *et al* 2013).

10.9 All-CNT TFT on PEN substrate using a photosensitive dry film

In place of conventional liquid photoresists, Chen *et al* (2018) used a photosensitive dry film. It is a photopolymer film, which can be laminated on a substrate surface instead of spin coating, as done with a photoresist (figure 10.8). Its function is similar to a negative photoresist in the respect that portions irradiated with UV become insoluble so that patterning and etching can be done. Besides the use of photosensitive dry film, a salient feature of this TFT is that the gate/source/drain electrodes as well as the active channel are all made of CNTs only (figure 10.9). The gate dielectric material is poly(methyl methacrylate).

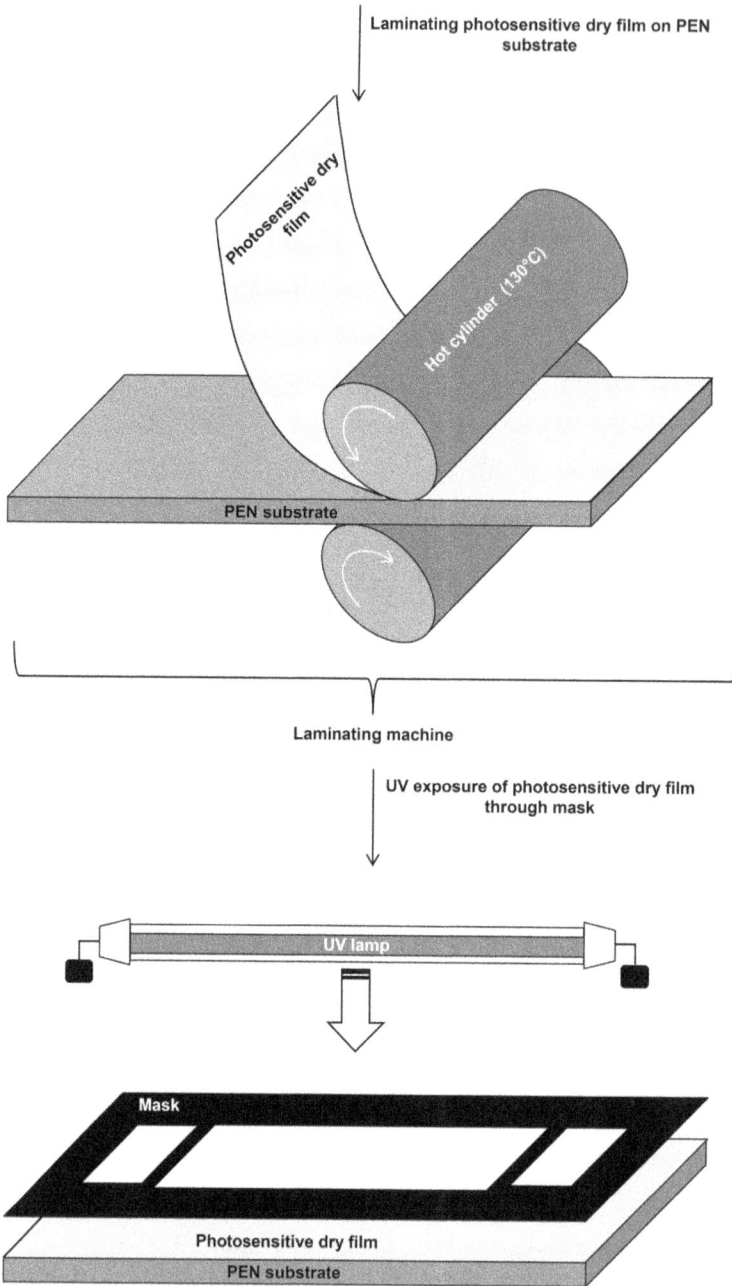

Figure 10.8. Laminating the photosensitive dry film on PEN substrate and creating a geometrical design on it by UV exposure through mask and developing.

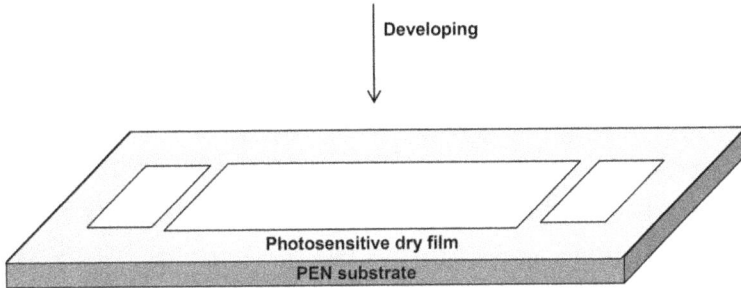

Figure 10.8. (Continued.)

10.9.1 CNT synthesis for gate/source/drain electrodes

Floating catalyst-based CVD (FC-CVD) method is employed. It involves pyrolysis of an organometallic precursor, e.g. ferrocene, $C_{10}H_{10}Fe$ or $Fe(C_5H_5)_2$, molecular weight 186.035 g mol^{-1}, mixed with a hydrocarbon such as ethylene (C_2H_4) as a source of carbon, and a carrier/inert gas H_2/Ar; the temperature is 1100 °C. Dissociation of ferrocene releases iron atoms. These atoms agglomerate into nanoparticles on the substrate. They act as catalytic agents for CNT growth.

For collection of the CNTs, filtration is done through a membrane filter. The filter is made of cellulose acetate (the acetate ester of the plant substance cellulose derived by acetylation of cellulose) mixed with nitrocellulose, $[C_6H_7(NO_2)_3O_5]_n$. The appropriate CNT density for their use as an electrode material is adjusted by varying the time of CNT collection.

10.9.2 Preparation of high-purity semiconducting CNT suspension for channel region

Semiconducting CNTs are gathered from bulk CNTs using a conjugated polymer poly{9-(1-octylonoyl)-9H-carbazole-2,7-diyl} (PcZ). 100 mg CNTs are mixed with 100 mg PcZ in 100 mL xylene, C_8H_{10}, ultrasonically agitated for 30 min at 30% amplitude level and centrifuged at 45 000 g for 1 h. After centrifuging, the upper 90% supernatant is collected for TFT fabrication work.

10.9.3 Creation of an alignment mark on the PEN substrate

This mark is made of titanium (5 nm)/gold (50 nm). Its purpose is to aid in the alignment of successive layers during processing. Three techniques are used to produce this mark: photolithography, electron-beam evaporation and lift-off process.

10.9.4 Transfer of CNT film for source/drain electrodes to PEN substrate

A roll-to-roll process is used to transfer this film from a membrane filter to the substrate.

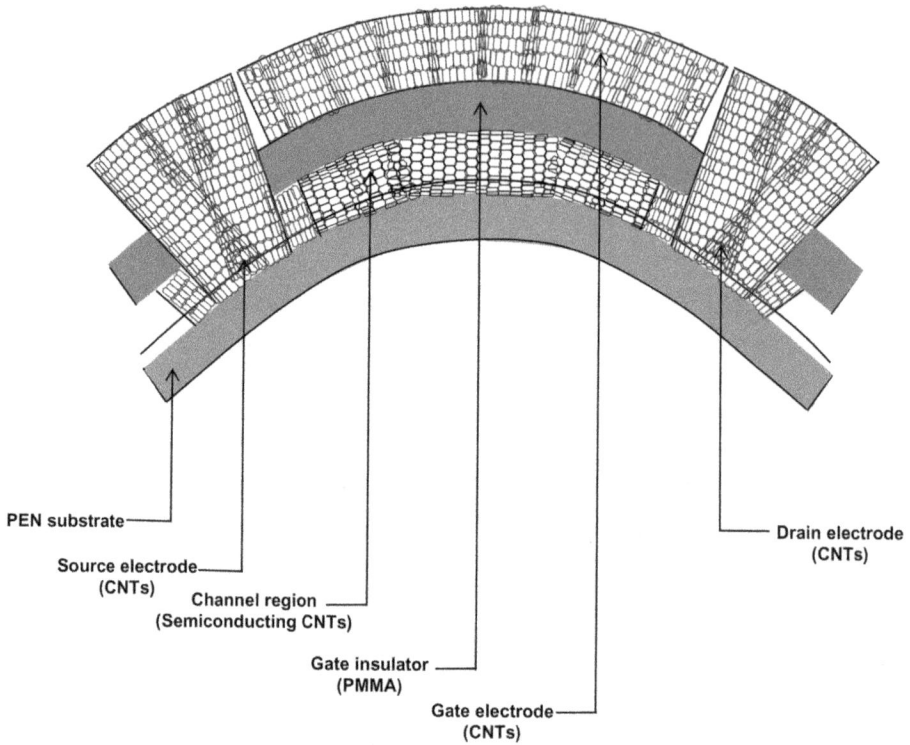

PEN substrate

Source electrode
(CNTs)

Channel region
(Semiconducting CNTs)

Gate insulator
(PMMA)

Gate electrode
(CNTs)

Drain electrode
(CNTs)

Figure 10.9. All-CNT TFT.

10.9.5 Lamination of substrate with photosensitive dry film

The 10 μm thick photosensitive dry film is laminated on the PEN substrate by a roll-to-roll process at 130 °C under 750 Torr pressure at a speed of 1.5 m min^{-1}.

10.9.6 Patterning and etching of source/drain electrodes and interconnections

The photosensitive film is exposed to UV for 10 s through a mask pattern of source/drain electrodes and interconnections. It is then kept at 120 °C for 1.5 min for pattern stabilization. The pattern is developed with 1.5% NAOH solution. The temperature is 70 °C and the developing time is 5 min. Dry etching is performed in oxygen plasma.

10.9.7 Deposition of semiconducting CNTs

The substrate from the preceding step is coated with hexamethyldisilazane (HMDS), linear formula $(CH_3)_3SiNHSi(CH_3)_3$, molecular weight 161.39 to ensure wettability of the surface for CNTs. Then the substrate is immersed in the CNT solution kept in a water bath at 60 °C for 2 h. It is washed in toluene for 5 min, then acetone for 5 min and finally isopropyl alcohol for 5 min. After these washings, it is kept at 120 °C for 10 min and dried with nitrogen.

10.9.8 Patterning of the semiconducting CNT film

This film is patterned using the method for dry photosensitive film as outlined in the case of source/drain electrodes and interconnections.

10.9.9 Deposition of gate insulator

PMMA film of thickness 600 nm is formed by spin coating and baking at 180 °C for 30 min.

10.9.10 Formation of contact windows in PMMA

As PMMA is soluble in the alkaline solution used in developing the photosensitive dry film, acetone along with a resist based on polydimethylglutarimide (PMGI) is used for opening these windows.

10.9.11 Transfer of gate electrodes

This step follows the same process as used for the formation of source/drain electrodes and interconnections.

10.9.12 Electrical performance of TFT

The TFT is a P-channel device. The electrodes have a sheet resistance of 359 Ω/sq. The electrode-to-channel contact resistance is 330 Ω. The current on–off ratio is $>10^5$. The carrier mobility is 33 cm^2 V^{-1} s^{-1}. The all-CNT TFT is transparent with a transmittance of 81%. An inverter circuit with a large noise margin is fabricated. It yields full rail-to-rail output. It has a voltage gain of 30 (Chen *et al* 2018).

10.10 Discussion and conclusions

Various substrate materials are used for fabrication of TFTs, viz, polyimide spin-coated film, PET, PES and PEN. Different methods of CNT film deposition are attempted, notably by printing and solution deposition. The SWCNTs used vary in purity content. It is clear that ultra-pure semiconductor-enriched SWCNTs must be used. The challenge of a low current on–off ratio is satisfactorily dealt with by using high-purity semiconductor-enriched SWCNTs. The presence of even small amounts of metallic CNTs appreciably increases the off-current. Another noteworthy observation is that a better overall performance of TFT in terms of mobility and current on–off ratio is presently achieved using solution deposition process for the CNT film formation.

Review exercises

10.1 Why is the nanoelectronics problem of correct placement of CNTs at prescribed locations a non-issue in macroelectronics?

10.2 How is the SiO$_2$ gate dielectric surface modified for SWCNT deposition? How is the SWCNT solution prepared and deposited on SiO$_2$?

10.3 What is the reason for the high off-current in an SWCNT TFT? How can this current be reduced?

10.4 Write the chemical formulae of 3-aminopropyl trimethoxysilane and Sodium dodecyl sulfate.

10.5 How is the PI film on the surface of Si/SiO$_2$ wafer removed after process completion?

10.6 Explain the function of each layer in the sandwiched gate dielectric: SiO$_x$/Al$_2$O$_3$/SiO$_x$.

10.7 How is the SiO$_x$ surface made ready for semiconductor-enriched SWCNT deposition? Why is vacuum annealing necessary after SWCNTs are solution-casted?

10.8 How is the PI substrate made stretchable by cutting it into a honeycomb mesh structure? How are the active devices positioned on the mesh? Can this device be used for detection and mapping of spatial pressure?

10.9 Why is aluminum oxide selectively deposited on titanium dioxide and not on Kapton?

10.10 How are source/drain electrodes made with conducting nanoparticle ink?

10.11 Which material is used as a mask against oxygen plasma to remove SWCNTs lying outside the channel region?

10.12 How is pV3D3 gate insulator deposited by initiated chemical vapor deposition (iCVD)?

10.13 How is TFT fabricated by an all-inkjet printing process? Explain the formation of different layers constituting the TFT. What is the operating frequency of this TFT?

10.14 With reference to the gravure-printed D flip–flop, describe how the gate electrode, gate dielectric, source/drain electrodes and the active semi-conductor layer are printed to realize the flip–flop.

10.15 What are the current on–off ratio and field-effect mobility for the load TFT and drive TFT?

10.16 In the fabrication of half- and full-adders, what SWCNT concentrations are used for printing the active layers of the load and drive TFTs?

10.17 How is the ink transferred from the gravure plate to the PET substrate in the inverse gravure printer?

10.18 How is the PET substrate functionalized with poly-L-lysine and how are the SWCNTs deposited over the functionalized substrate?

10.19 What material is used for the gate dielectric of the inverse gravure-printed CNT TFT? How is the gate electrode formed?

10.20 How does the inverse gravure-printed CNT TFT behave upon bending?

10.21 What is photosensitive dry film? In what respect is it comparable to a negative photoresist?

10.22 How are the CNTs synthesized by FC-CVD method for the gate/source/drain electrodes of the TFT? What is the synthesis temperature?

10.23 How is the high-purity CNT suspension prepared for forming the channel region of the TFT?

10.24 How is the photosensitive dry film laminated on the substrate? What is the substrate used?

10.25 Highlight the salient features of the all-CNT TFT.

References

Chandra B, Park H, Maarouf A, Martyna G J and Tulevsk G S 2011 Carbon nanotube thin film transistors on flexible substrates *Appl. Phys. Lett.* **99** 072110-1–3

Chen Y-Y, Sun Y, Zhu Q-B, Wang B-W, Yan X, Qiu S, Li Q-W, Hou P-X, Liu C, Sun D-M and Cheng H-M 2018 High-throughput fabrication of flexible and transparent all-carbon nano-tube electronics *Adv. Sci.* **5** 1700965

Gu G, Ling Y, Liu R, Vasinajindakaw P, Lu X, Jones C S, Shih W-S, Kayastha V, Downing N L, Berger U and Renn M 2011 All-printed CNT transistors with a high on-off ratio and bias-invariant transconductance *Carbon Nanotubes, Graphene, and Associated Devices IV* ed D Pribat, Y-H Lee and M Razeghi *Proc. SPIE* **8101** 81010B-1–7

Lau P H, Takei K, Wang C, Ju Y, Kim J, Yu Z, Takahashi T, Cho G and Jave A 2013 Fully printed, high performance carbon nanotube thin-film transistors on flexible substrates *Nano Lett.* **13** 3864–9

Lee D, Yoon J, Lee J, Lee B-H, Seol M-L, Bae H, Jeon S-B, Seong H, Im S G, Choi S-J and Choi Y-K 2016 Logic circuits composed of flexible carbon nanotube thin-film transistor and ultra-thin polymer gate dielectric *Sci. Rep.* **6** 1–7

Noh J, Jung M, Jung K, Lee G, Kim J, Lim S, Kim D, Choi Y, Kim Y, Subramanian V and Cho G 2011a Fully gravure-printed D flip-flop on plastic foils using single-walled carbon-nanotube-based TFTs *IEEE Electron Device Lett.* **32** 638–40

Noh J, Kim S, Jung K, Kim J, Cho S and Cho G 2011b Fully gravure printed half adder on plastic foils *IEEE Electron Device Lett.* **32** 1555–7

Noh J, Jung K, Kim J, Kim S, Cho S and Cho G 2012 Fully gravure-printed flexible full adder using SWNT-based TFTs *IEEE Electron Device Lett.* **33** 1574–6

Snow E S, Campbell P M, Ancona M G and Novak J P 2005 High-mobility carbon-nanotube thin-film transistors on a polymeric substrate *Appl. Phys. Lett.* **86** 033105-1–3

Takahashi T, Takei K, Gillies A G, Fearing R S and Javey 2011 Carbon nanotube active-matrix backplanes for conformal electronics and sensors *Nano Lett.* **11** 5408–13

Vaillancourt J, Zhang H, Vasinajindakaw P, Xia H, Lu X, Han X, Janzen D C, Shih W-S, Jones C S, Stroder M, Chen M Y, Subbaraman H, Chen R T, Berger U and Renn M 2008 All ink-jet-printed carbon nanotube thin-film transistor on a polyimide substrate with an ultrahigh operating frequency of over 5 GHz *Appl. Phys. Lett.* **93** 243301-1–3

IOP Publishing

Flexible Electronics, Volume 2
Thin-film transistors
Vinod Kumar Khanna

Chapter 11

Nanowire FET

Ge/Si core/shell nanowires grown by the vapor–liquid–solid method at temperatures of 270 °C for Ge and 450 °C for Si, on a Si/SiO$_2$ substrate carrying dispersed gold colloids, are contact printed to a PI film using octane/mineral oil as a lubricant and by applying a gentle pressure as the donor substrate slides over the PI. Then TFT fabrication steps are completed, achieving a carrier mobility of 20 cm^2 V^{-1} s^{-1} in the TFT. The device shows minimal change in conductance up to a bending radius of 2.5 mm. When bent to a radius <10 mm, the characteristics remain unruffled up to 2000 bending cycles (Takei $et\ al$ 2010). In another study, gold nanocluster-mediated growth of P-type Si/SiO$_2$ nanowires on a poly-L-lysine-modified Si/SiO$_2$ substrate in a furnace and by an alternative method by laser ablation using Si$_{0.9}$Fe$_{0.1}$ target without any substrate, are described. The randomly oriented nanowires are assembled into parallel arrays by passage through fluidic channels. These NWs are deposited on an AlO$_x$ gate dielectric to form a TFT on a PEEK substrate, which shows a high mobility of 123 cm^2 V^{-1} s^{-1} with a current on–off ratio of 10^5 (Duan $et\ al$ 2003). Further ahead, 20 nm diameter P-type Si NWs grown using gold nanoclusters are deposited on the SiO$_2$ insulator to make a TFT on a Mylar substrate; the carrier mobility in this TFT is 200 cm^2 V^{-1} s^{-1}. The current decreases by 10% when the TFT is bent to 3 mm radius (Mcalpine $et\ al$ 2005). Keeping in view the limitations of forming silicon nanowires of diameters up to 20 nm with 60 nm pitch by electron-beam lithography, a superlattice nanowire pattern (SNAP) transfer technique is developed in which a platinum-coated GaAs/Al$_{1-x}$Ga$_x$As (34 nm/17 nm) superlattice edge is obtained from a superlattice comprising 800 layers of alternating GaAs and Al$_x$Ga$_{1-x}$As. The platinum nanowires are transferred to a doped SOI wafer with the help of PMMA. After these platinum nanowires have been used as a mask for reactive ion etching of silicon, they are dissolved. Then the lower ends of silicon nanowires are detached from buried oxide of SOI wafer by HF etching and the Si nanowires are retrieved by a PDMS slab from which they are harvested by an ITO-coated PET substrate on which SU-8 film is deposited. Using the SU-8 film as the gate insulator, the bottom-gate TFT is fabricated. A top-gate TFT is also fabricated for better transconductance (Mcalpine $et\ al$ 2007).

doi:10.1088/2053-2563/ab0d18ch11

11.1 Introduction

Carbon nanotubes have unique and outstanding properties for device applications. But there are also impediments to progress. One primary limitation is the non-availability of a process to grow specifically metallic and semiconducting nanotubes. The metallic or semiconducting nature of the CNT depends on its diameter and helicity. As we have seen, despite semiconductor enrichment, the presence of even a small quantity of metallic nanotubes has a profound degrading effect on TFT performance. Controlled doping of nanotubes is impossible although it is highly desired for device fabrication (Cui *et al* 2000).

Silicon nanowires have a distinct advantage over CNTs in this regard because their property of being semiconducting or metallic in character is diameter-independent. In addition, the vast knowledge of silicon technology is a valuable backup to resort to in the event of any difficulty.

11.2 Ge/Si NW FET on PI film

Takei *et al* (2010) convincingly showed integration of parallel nanowire arrays functioning as artificial skin. Each pixel of the e-skin is connected to a TFT of the NW array which maintains the state of the pixel.

11.2.1 Substrate

The substrate is a polyimide film of thickness 24 μm (figure 11.1). To carry out the fabrication process, the polyimide film is spin coated twice on a 100 mm diameter handle wafer. The handle wafer is a silicon/silicon dioxide wafer. Curing of the PI film is done in two steps, firstly over the hot plate by increasing the temperature to 300 °C at 5 °C per min and secondly by baking at 300 °C for 1 h. Over the PI film, 25 nm thick SiO_x film is grown by electron-beam evaporation. The areas for nanowire transfer to the substrate are defined in the photoresist by lithography. Hence, the substrate is coated with photoresist with open windows where nanowires are to be deposited as a preparatory step for the lift-off process for selectively depositing the NWs. The substrate is subjected to mild oxygen plasma and then poly-L-lysine treatment.

11.2.2 Growth of Ge/Si core/shell NWs

The technique used for NW growth is called the vapor–liquid–solid process (Lu *et al* 2005, Xiang *et al* 2006). In this process, gold colloids of 10 nm diameter are immobilized on poly-L-lysine on an oxidized Si wafer (SiO_2 thickness 50 nm), which serves as the growth substrate for the NWs (figure 11.2). These Au colloids on Si wafer are placed in a quartz tube furnace. The Au colloids serve as a catalytic agent for growth of Ge NWs.

    (i) Growth of Ge core of the nanowires: The Ge core growth parameters are: temperature = 270 °C; pressure = 45 Torr; gas used = 10% GeH_4 in H_2; flow rate = 12 sccm; time = 30 min. Consequently, 30 μm long Ge nanowires are formed.

Taking Si/SiO$_2$ handle wafer

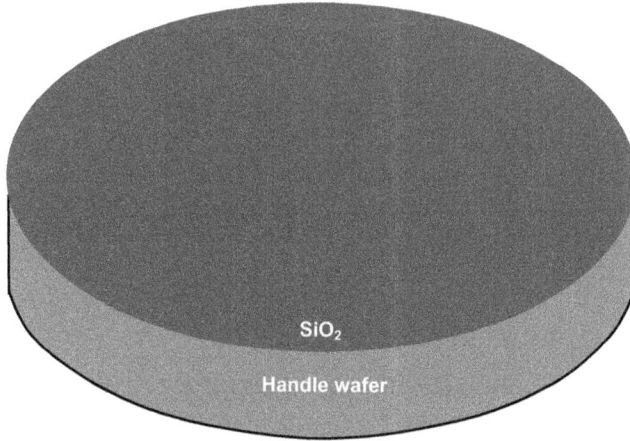

SiO$_2$

Handle wafer

Spin coating and curing
polyimide film

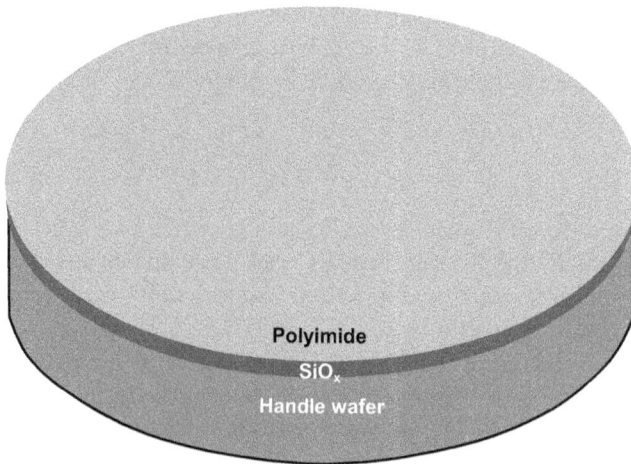

Polyimide

SiO$_x$

Handle wafer

Figure 11.1. Substrate preparation for nanowire transfer.

SiO$_x$ deposition by electron beam evaporation

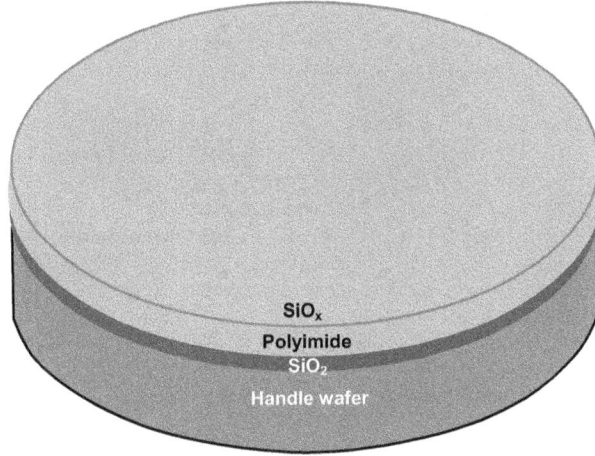

Opening areas in photoresist for nanowire transfer by lift-off photolithography

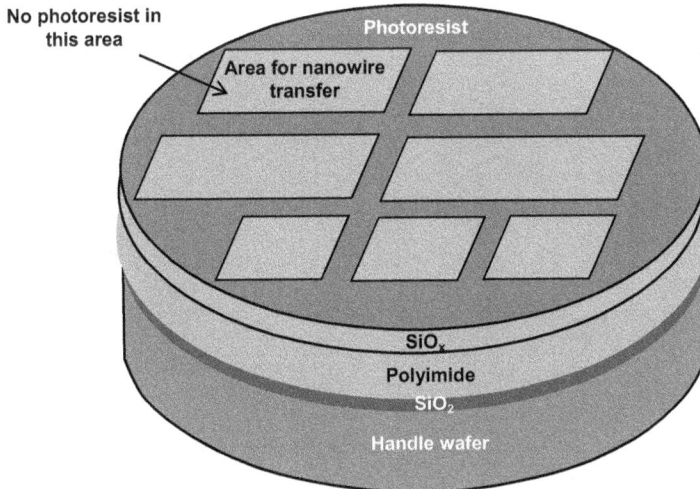

Figure 11.1. (Continued.)

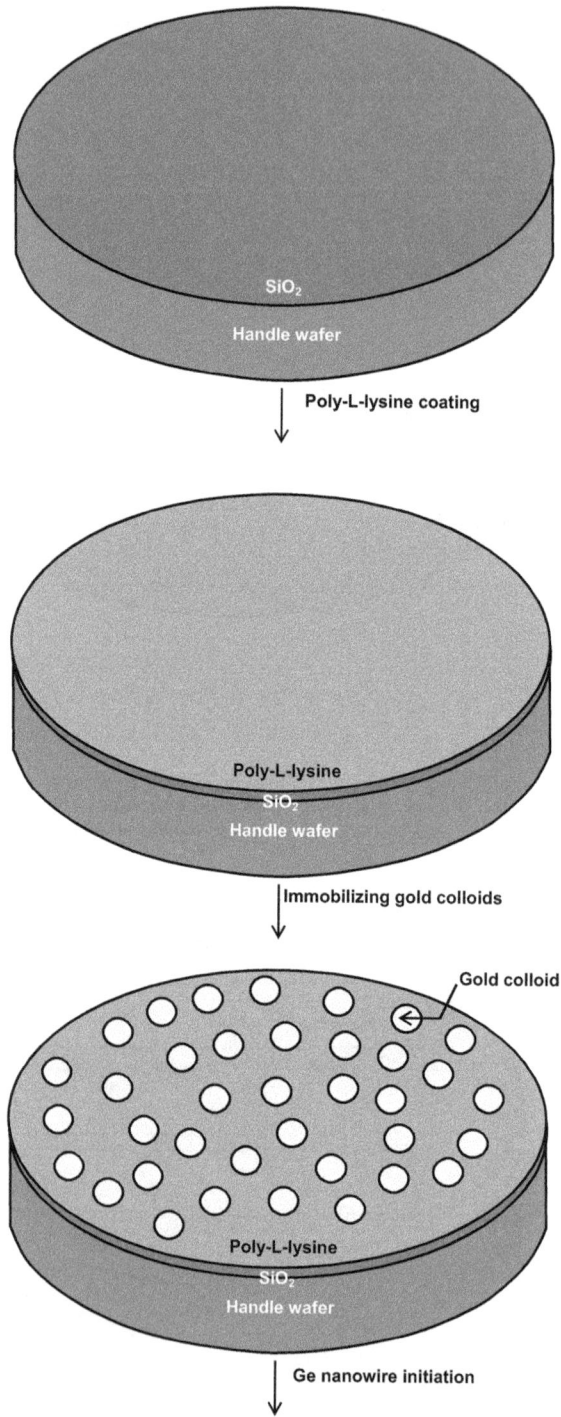

Figure 11.2. Growth and contact printing of silicon nanowires on polyimide.

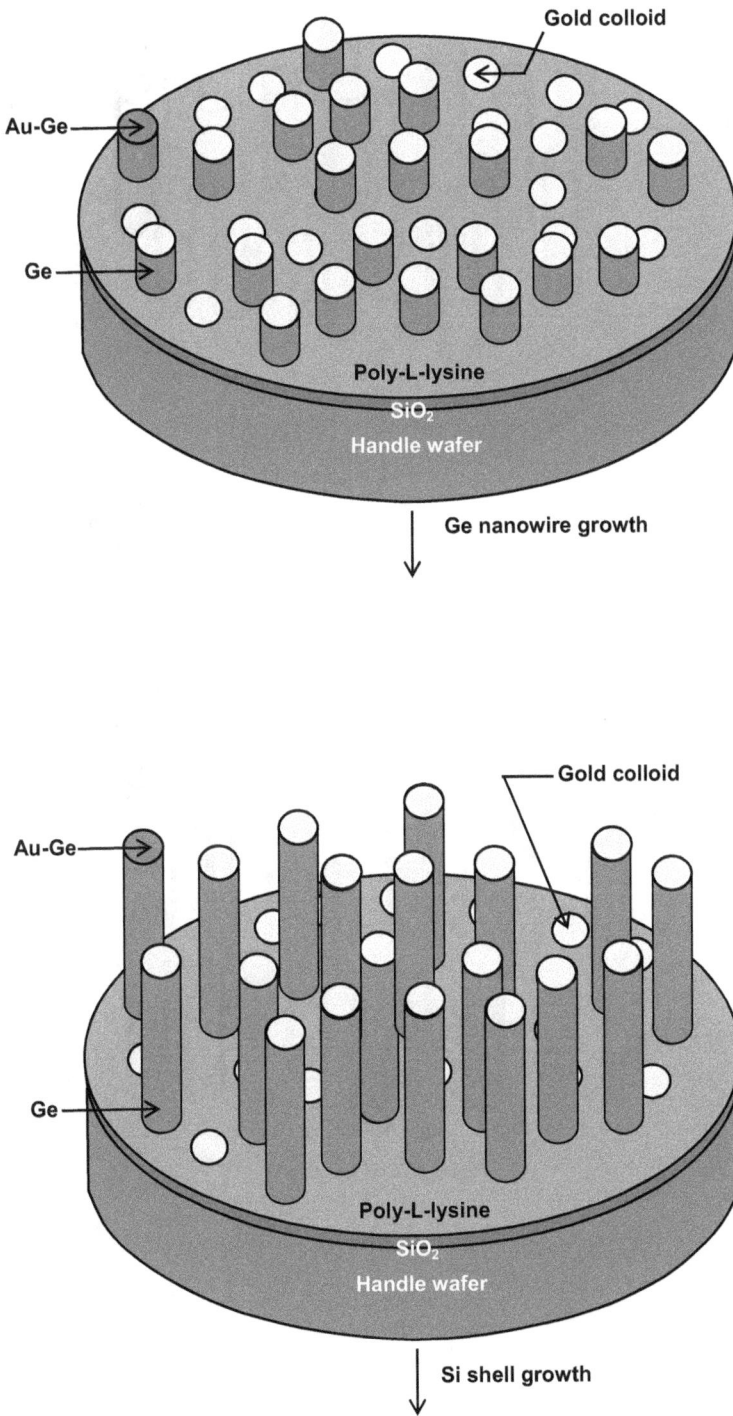

Figure 11.2. (Continued.)

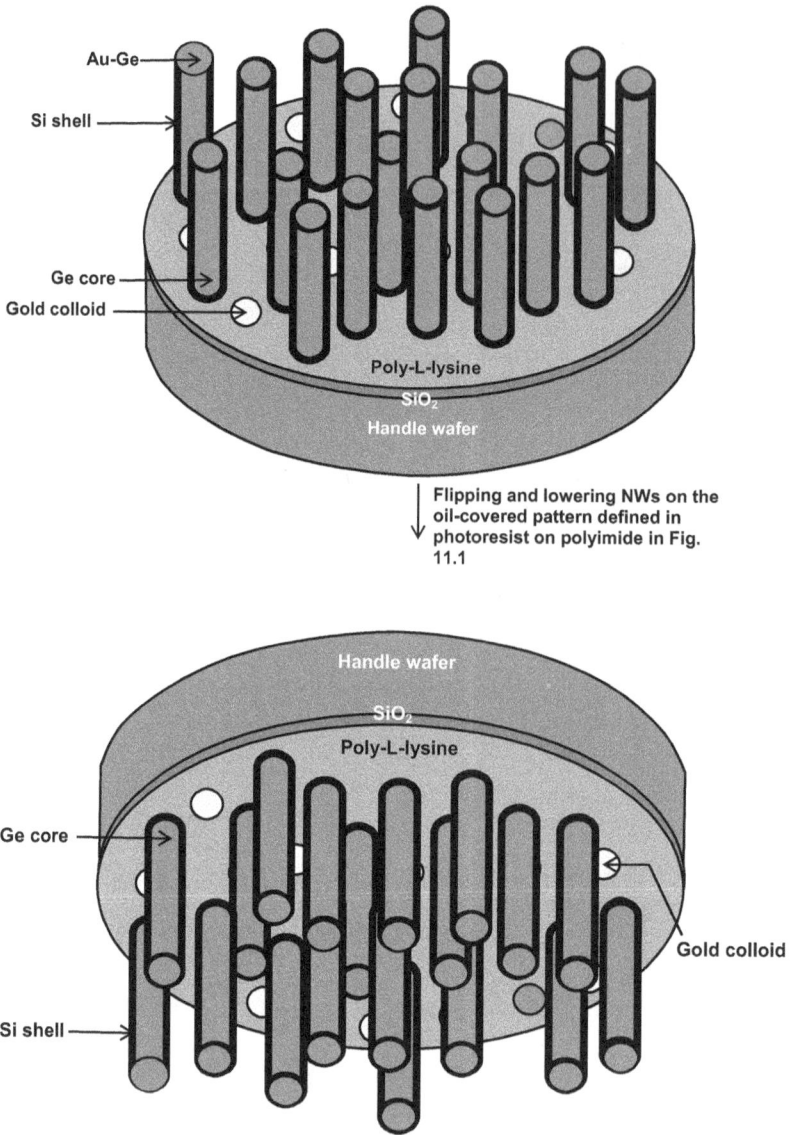

Au-Ge

Si shell

Ge core

Gold colloid

Poly-L-lysine

SiO$_2$

Handle wafer

Flipping and lowering NWs on the
oil-covered pattern defined in
photoresist on polyimide in Fig.
11.1

Handle wafer

SiO$_2$

Poly-L-lysine

Ge core

Gold colloid

Si shell

Figure 11.2. (Continued.)

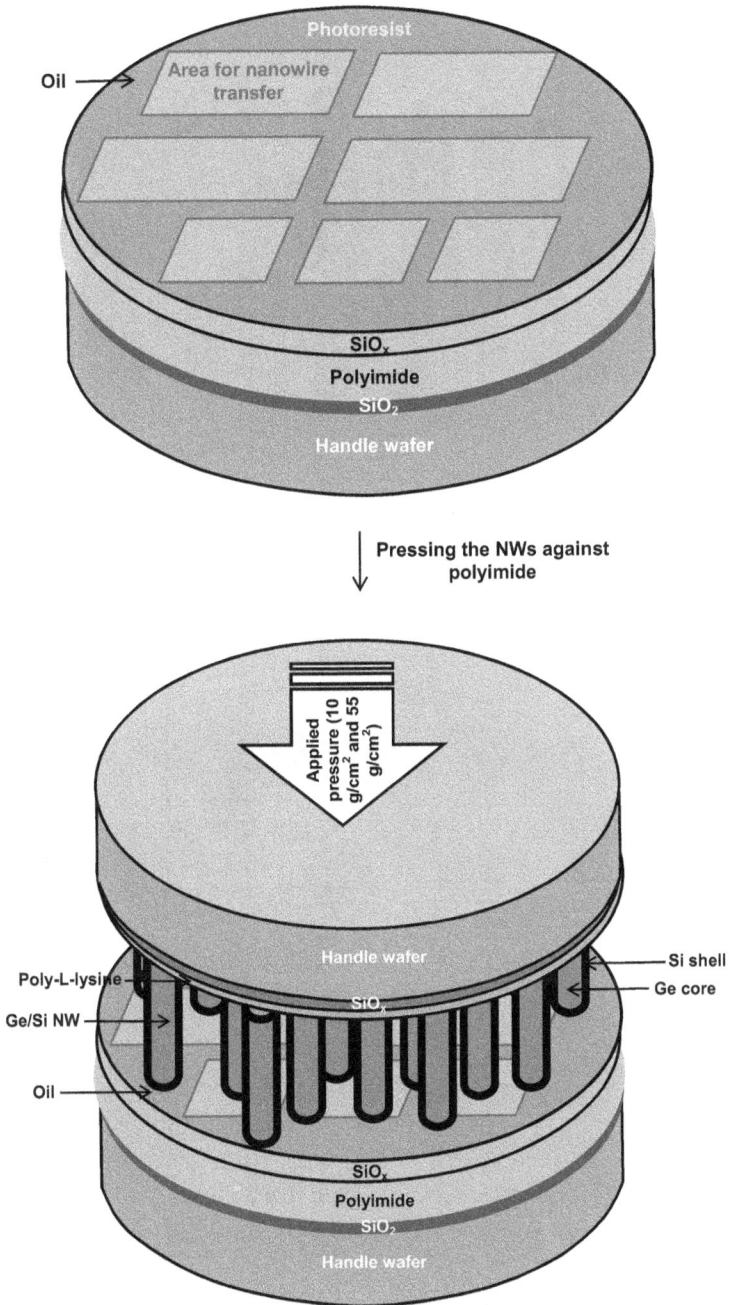

Figure 11.2. (Continued.)

Removing the handle wafer

Ge/Si NW

Oil

SiO$_x$

Polyimide

Represented as

Ge/Si NWs   Ge/Si NWs

Ge/Si NWs   Ge/Si NWs

Ge/Si NWs Ge/Si NWs Ge/Si NWs

SiO$_x$

Polyimide

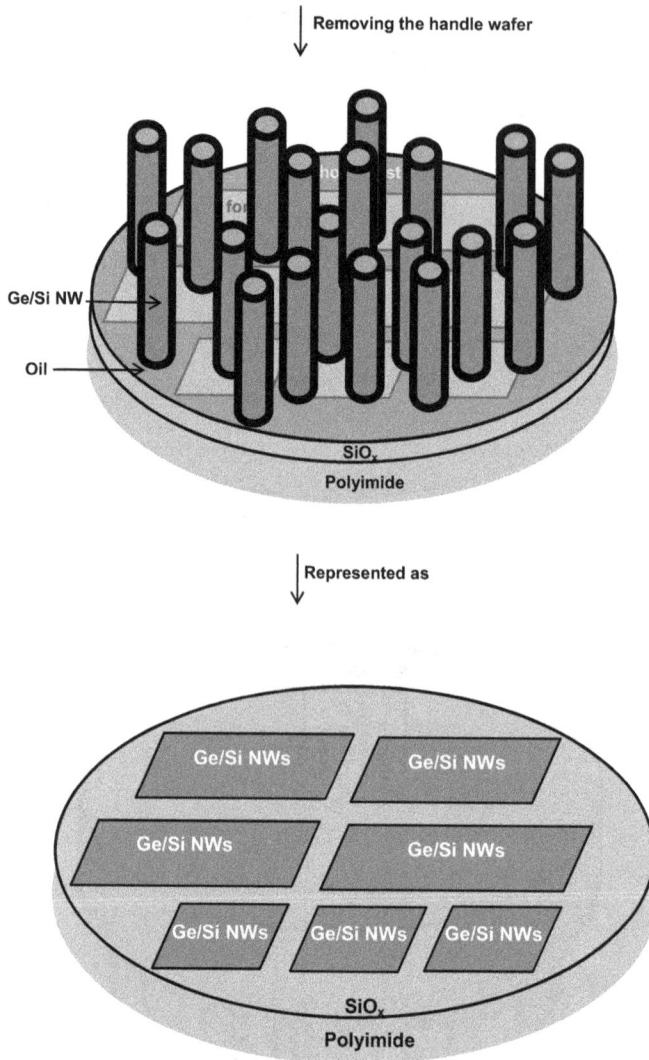

Figure 11.2. (Continued.)

(ii) Deposition of polycrystalline Si shell: Immediately after Ge core growth without exposure to ambient atmosphere, the Si shells are deposited over the Ge cores in the same reactor. The Si shell deposition conditions are: temperature = 450 °C; pressure = 5 torr; gas used = 100% SiH$_4$; flow rate = 5 sccm; time = 5 min. The resulting Si shell is 5 nm thick.

11.2.3 Contact printing of NWs on the substrate

The SiO$_x$/PI/SiO$_2$/Si substrate (acceptor substrate) is mounted on a stage and the Ge/Si NWs/SiO$_2$/Si structure (donor substrate) is placed over it (Fan *et al* 2008a).

Octane/mineral oil as 2:1 v/v mixture is used for lubrication. Applying a gentle pressure of 10 g cm^{-2}, the donor substrate is directionally slid over the acceptor substrate at a speed of 20 mm min^{-1}. After sliding for a distance of 1 mm, an extra pressure of 55 g cm^{-2} is applied. The donor substrate is slowly removed from the acceptor substrate. The nanowires anchored by Van der Waals forces to the receptor substrate remain fixed to it while the untransferred NWs are retained by the donor substrate. Then the acceptor substrate is dried in nitrogen. The acceptor substrate is immersed in acetone to lift off the resist. This is a lift-off process in which the nanowires remain attached to acceptor substrate areas which are not coated with photoresist and are detached from resist-covered areas of acceptor substrate. Thus NWs are deposited in the designated areas. The nanowire density is 5 NWs µm^{-1}.

The octane/mineral oil lubricant plays a crucial role in the NW transfer process. It reduces the NW:NW friction when the two substrates slide against each other. This friction breaks and grinds the nanowires and disturbs their alignment. The lubricant has a low dielectric constant ~2, and, therefore, does not influence the interactions taking place between the chemically modified polar surface and the sliding NWs. Preservation of interactions between the nanowires and the substrate is an essential pre-requisite for the NW assembly process. Dynamic friction is an important parameter affecting the alignment of NWs as well as their detachment from the donor substrate.

The transferred nanowires are highly ordered and aligned. The process is reproducible.

11.2.4 Source/drain electrode formation

Photolithography is done to define source/drain regions (figure 11.3). 50 nm thick nickel film is deposited by thermal evaporation.

11.2.5 Gate dielectric and electrode deposition

The gate dielectric is Al$_2$O$_3$ film of thickness 50 nm. It is formed by atomic layer deposition comprising 500 cycles at 200 °C. Each cycle is made of four pulses. These include: a trimethylaluminum (C$_6$H$_{18}$Al$_2$) pulse for 0.1 s, a nitrogen purging pulse 20 s long, a water pulse of duration 0.2 s and again a nitrogen purging pulse lasting for 20 s.

In a typical process for making FETs with HfO$_2$ dielectric, tetrakis(dimethylamino) hafnium (IV) (TDMAH), C$_8$H$_{24}$HfN$_4$ or [(CH$_3$)$_2$N]$_4$Hf kept at 110 °C and water are used as the precursors (Fan *et al* 2008b). For 9 nm thick film, HfO$_2$ is deposited for 85 cycles at 140 °C. Each cycle consists of a tetrakis(dimethylamino) hafnium precursor pulse, a nitrogen purging pulse, a water vapor pulse and again a nitrogen purging pulse with durations of 1 s, 20 s, 1 s and 40 s respectively.

The gate metal is 100 nm thick aluminum deposited by evaporation and patterned by lift-off lithography.

11.2.6 Passivation and via hole etching

A 500 nm thick insulating parylene-C layer is coated over the device. Plasma etching is done in oxygen to make via holes for contacting the source electrode of each TFT. The vias are electrically connected to nickel pads on the top surface of parylene-C

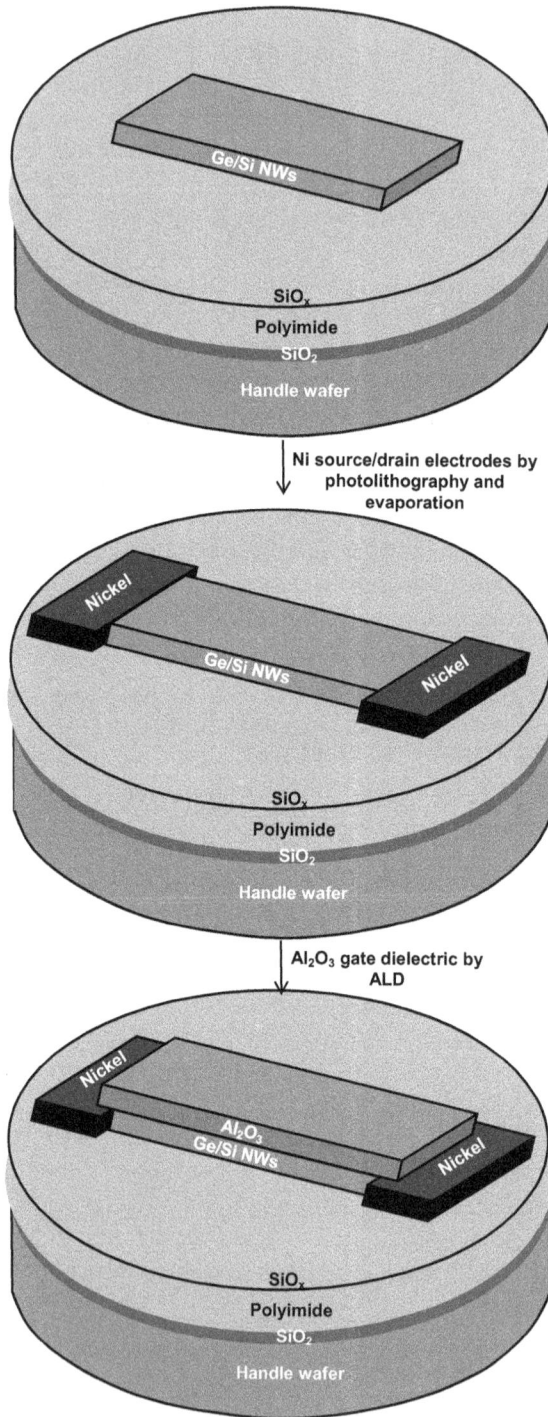

Figure 11.3. Nanowire-based flexible FET.

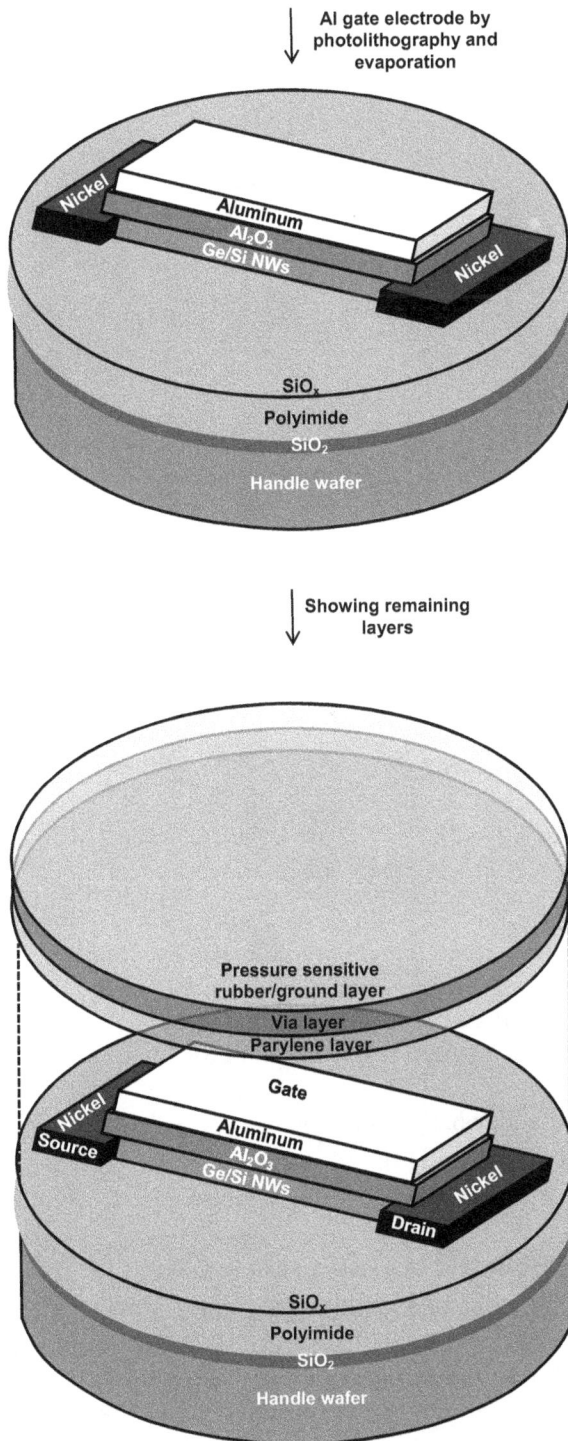

Figure 11.3. (Continued.)

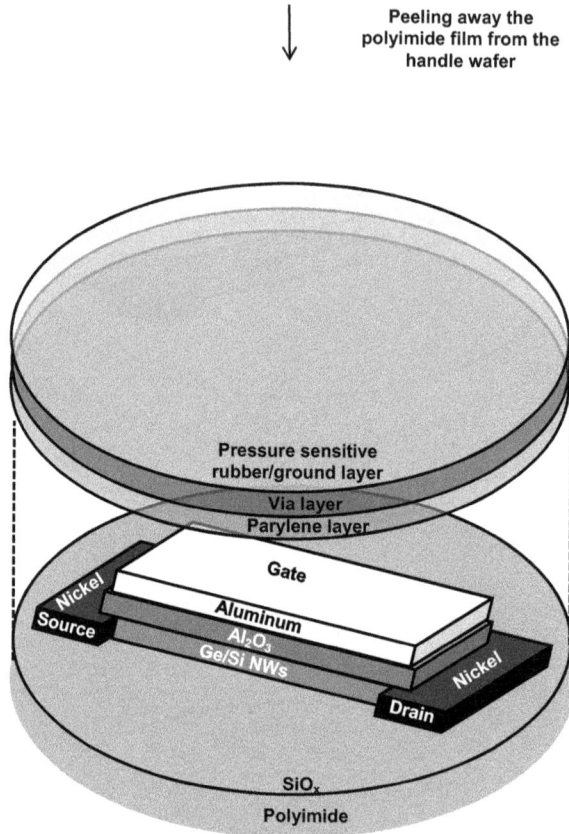

Figure 11.3. (Continued.)

layer. These pads are defined by photolithography. They perform the task of interfacing the TFTs with the pressure sensitive rubber (PSR). The PSR is laminated over the top surface of the device to act as a pressure sensor. The ground electrode is an Al film deposited on the top surface of the PSR.

11.2.7 Peeling off the flexible circuit

After all the fabrication process steps have been completed, the polyimide film with the circuitry upon it is peeled off from the handle wafer to obtain a flexible device.

11.2.8 Electrical performance and mechanical flexibility

The TFTs behave as P-channel devices. They show a high on-state current $I_{ON} =$ 1mA at $V_{DS} = 3$ V. The peak field-effect mobility is 20 cm^2 V^{-1} s^{-1}. The transconductance is measured as 55 µS at $V_{DS} = 0.5$ V (Takei *et al* 2010).

Application of an external pressure changes the conductance of the PSR. The resulting modulation of TFT characteristics alters the output signal from the pixel. The mechanical flexibility, robustness and reliability are evaluated by examining the output conductance of a single pixel in two conditions, firstly when the array is bent and secondly when the array passes through repetitive bending cycles. Negligible change in output conductance is observed upon bending with the ratio (Takei *et al* 2010)

$$g = \frac{\text{Difference in conductance between bent } (G_1) \text{ and relaxed states } (G_0)}{\text{Conductance in relaxed state } (G_0)}$$

$$= \frac{G_1 - G_0}{G_0} = \frac{\Delta G}{G_0} \approx 6\%$$

(11.1)

for a radius of curvature = 2.5 mm. For strain cycling test, the substrate is bent up to a radius <10 mm several times, relaxed after each bending and electrically tested after each bending. Up to 2000 bending cycles, the device showed little variation from its normal behavior. Reliability is principally the result of the small diameter of the nanowires of ~30 nm together with the extremely thin metal layers in the range of 50–100 nm used in the device fabrication. As a result, the chances of cracking of the films as well peeling off from the substrate are reduced.

11.2.9 Upscaling of the process

The process is scalable as evinced from the realization of 7×7 cm^2 arrays which functioned as the active-matrix backplane of an 18×19 pixel pressure sensor array. The backplane circuit operates at <5 V without any impairment in functionality up to a radius of curvature of 2.5 mm for >2000 bending cycles (Takei *et al* 2010).

11.3 P-type Si/SiO$_2$ NW TFT on PEEK

Duan *et al* (2003) showed the fabrication of TFTs on different substrates, including plastic, for which they devised an assembly process which is carried out at a low temperature.

11.3.1 Synthesis of P-type single-crystal silicon NWs

High-quality single-crystal Si nanowires are grown by a vapor–liquid–solid (VLS) mechanism. In this mechanism, metal nanoclusters act as mediators. The nano-clusters mediate the growth mechanism by acting as catalytic agents. Therefore, by using nanocluster catalysts of well-defined size distribution, one can grow nanowires having diameters confined within a narrow dimensional range. The vapor-phase reactant for NW growth is silane (SiH$_4$).

11.3.2 NW growth apparatus

The NW growth apparatus consists of a quartz reactor tube enclosed within a tubular furnace (Morales and Lieber 1998). The quartz tube is maintained at a given temperature by setting the furnace at the chosen temperature. The reactant gas is introduced into the quartz tube from one end and exits the tube at the opposite end into a pumping system.

11.3.3 Preparation of NW growth substrates

Thermally oxidized silicon wafers are used as the growth substrates (Cui *et al* 2001). On the Si/SiO$_2$ substrates 0.1% poly-L-lysine is deposited (figure 11.4). The poly-L-lysine is a synthetic chain of amino acid. It forms a positively charged layer with one hydrobromide per unit of lysine. On the poly-L-lysine-coated surface, gold nano-clusters are dispersed. These nanoclusters have diameters of 5, 10, 20, 30 nm (typically) at a concentration of 10^{11}–10^{12} particles mL^{-1}. They are negatively charged. Hence, they are attracted and bound to the positively charge poly-L-lysine by electrostatic force. Examination by AFM reveals the dispersion of nanoclusters on the substrates. Before loading into the quartz reactor tube, the substrates are cleaned by exposing to oxygen plasma at 100 W power and 0.7 Torr pressure with oxygen flow rate of 250 sccm. The plasma cleaning is done for 5 min.

11.3.4 NW growth parameters

Cui *et al* (2001) create a vacuum <100 mTorr in the quartz reactor. The reactor temperature is 440 °C. Argon is made to flow in the reactor. Silane (10% in helium) is passed through the reactor at a flow rate of 10–80 sccm and the nanowires are grown for 5–10 min.

Duan *et al* (2003) grew nanowires of diameter 20 or 40 nm in an 8″ furnace tube. As they grew P-type nanowires, they used diborane (B$_2$H$_6$) for doping the NWs with boron, along with silane (SiH$_4$) for NW formation. The dopant concentration is controlled by varying the silane–diborane ratio. They used a silane–diborane ratio of 6400:1 for which the boron concentration is ~4 × 10^{17} cm^{-3}. The high boron concentration is necessary to form ohmic contacts with the source/drain electrodes without the requirement of high-temperature metal sintering step. The growth temperature is 420 °C–480 °C. The total pressure in the quartz reactor is 30 Torr and the partial pressure of silane is 2 Torr. The NW growth period is 40 min.

11.3.5 Alternative laser ablation method of NW synthesis

Although not used in this study, this method is an interesting alternative procedure to grow nanowires of different materials. Using an identical growth apparatus, Morales and Lieber (1998) have described a method of NW growth which does not need the Si/SiO$_2$ substrate carrying the dispersed Au nanoclusters. In this setup (figure 11.5), a target is placed inside the quartz tube in the zone at the reaction temperature. Besides the temperature, the pressure and residence time of the target in the quartz tube can be varied as desired. The target consists of the element from which the NWs are to be grown along with the catalyst. A Nd:YAG (yttrium–aluminum–garnet) laser is used to ablate the target. It is a pulsed, frequency-doubled laser with a wavelength of 532 nm. By laser ablation of an Si$_{0.9}$Fe$_{0.1}$ target at 1200 °C, nanowires having a diameter of 10 nm and lengths >1 μm and up to 30 μm are formed. Similarly, laser ablation of a Ge$_{0.9}$Fe$_{0.1}$ target at 820 °C produced Ge NWs of diameters 3–9 nm.

The grown NWs are carried away by the gas entering the quartz tube through a flow controller from the inlet end and flowing towards its exit end where a pump

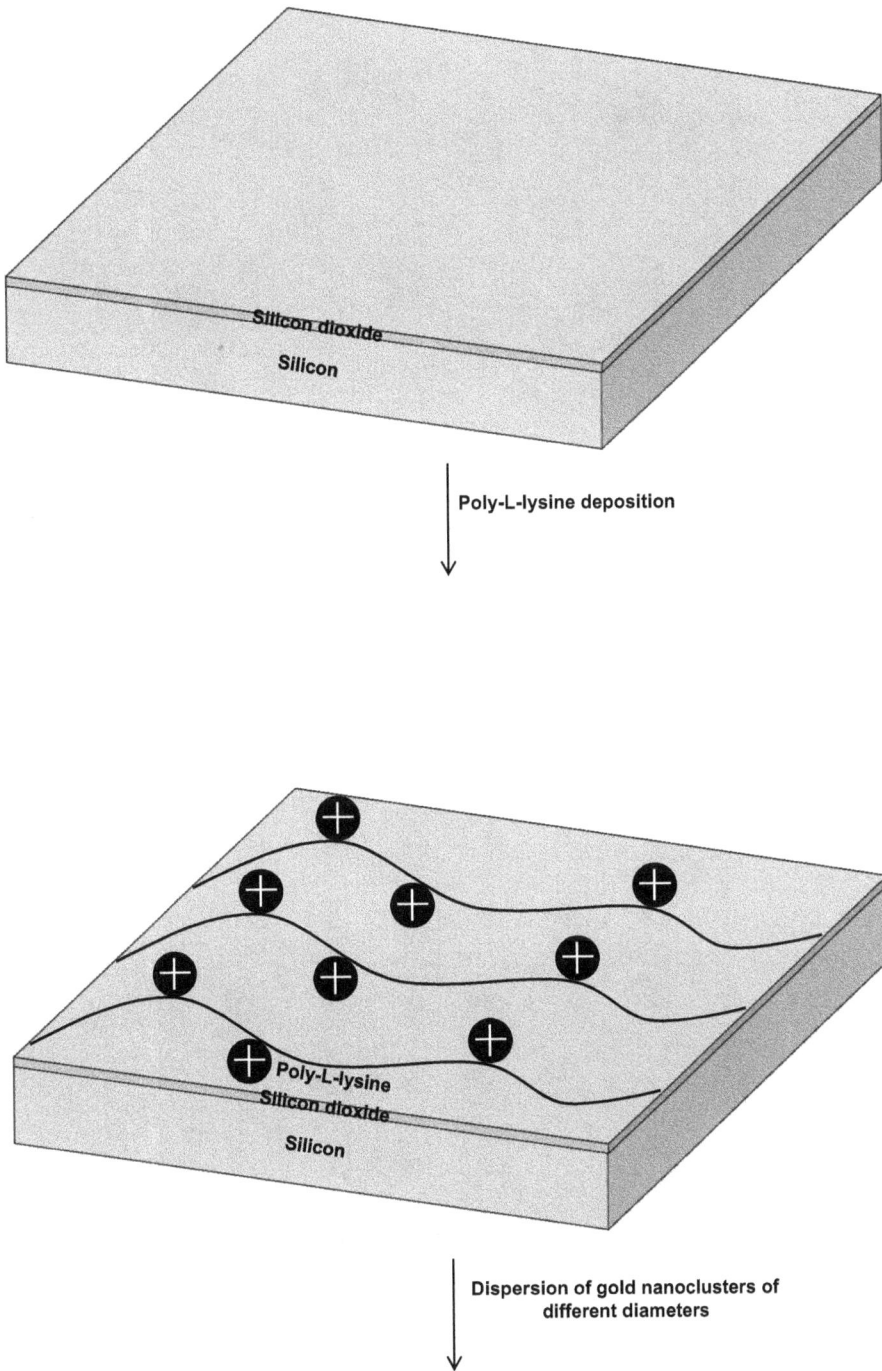

Figure 11.4. Substrate preparation and size-controlled CVD growth of silicon nanowires from gold nanoclusters.

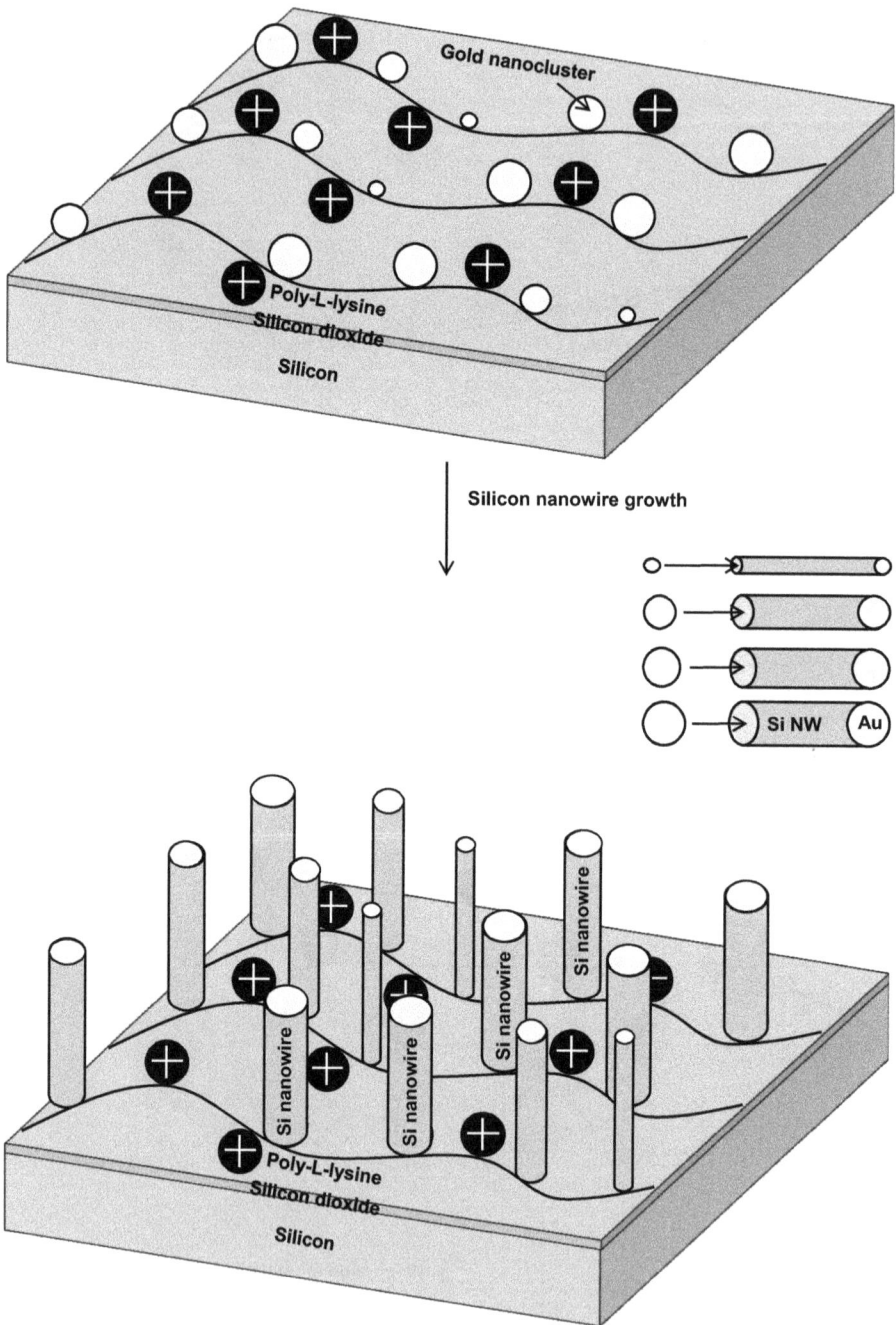

Figure 11.4. (Continued.)

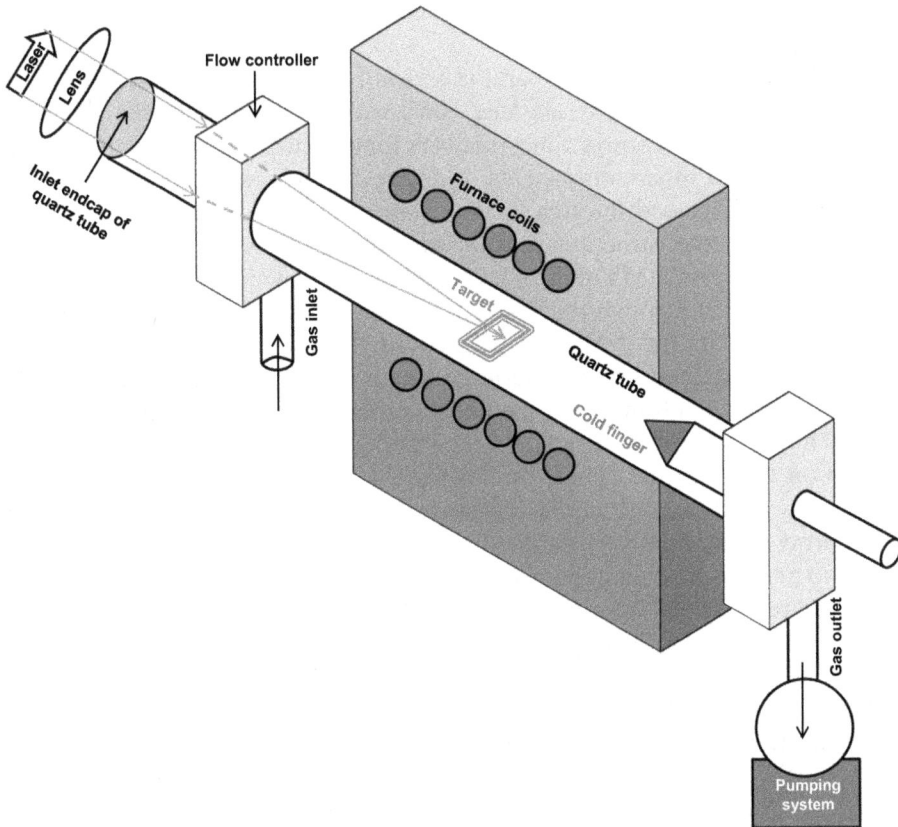

Figure 11.5. Crystalline semiconductor nanowire synthesis by laser ablation.

pulls it out of the quartz tube. The growth of a nanowire halts as soon as it moves out from the hot zone of the furnace towards a cooler zone. All the NWs grown in the tube and carried away by the gas flow collect on a cold finger located at the exit end of the tube.

11.3.6 Directed flow assembly of Si NWs

Back to P-type Si NWs on Si/SiO$_2$ substrate, the grown NWs having lengths of 20–50 μm are harvested from the substrate. They are dispersed in ethanol by ultrasonication of the Si/SiO$_2$ growth substrate in ethanol for 5–10 min. The NWs have a core/shell structure. The core is made of single-crystal silicon while the shell is an amorphous silicon dioxide film of thickness 1–3 nm. In fact, the shell is the native oxide that is formed on a silicon surface when exposed to atmosphere.

The difficulty now is that the NWs are randomly arranged. They are pointing along arbitrary directions. These haphazardly arranged NWs must be aligned parallel to each other along a particular direction on the flexible substrate to obtain a well-defined functional network or thin film in which the constituent NWs are oriented parallely in a given direction. This direction will be the source-to-drain direction of the FET.

The NWs are assembled into parallel arrays by a simple method (Huang *et al* 2001). This method works by aligning NWs with the help of fluidic flow and is generally applicable to various nanostructures for organization of NWs for wiring or interconnections to other components of devices. In this method (figure 11.6), the NW suspension is passed through fluidic channels created between a PDMS mold with engraved channel structures and the underlying flat substrate. These channels are formed when the PDMS mold is brought in contact with the substrate. The NW suspension is passed through the channels at a controlled flow rate. After a certain time, practically all the NWs are aligned in the direction of fluid flow. This alignment is maintained up to quite large length scales, ranging from hundreds of microns to mm, as confirmed by experiments using channels of diameter 50–500 μm and 6–20 mm length. After removal of the PDMS mold, parallel arrays of NWs are observed on the substrate. Thus, by flowing the NWs via the PDMS mold at an adjusted flow rate for a given time, the random arrangement NW bunch transforms into an array of parallel NWs in the direction of fluid flow. Hence, this method is referred to as flow-directed alignment or fluid-directed assembly of NWs.

The flow rate of fluid is a vital parameter determining the degree of alignment of the NWs. The width of the angular distribution of NWs relative to the direction of fluid flow shrinks as the rate of fluid flow increases. From experiments in the flow rate range 4 mm s^{-1} to 10 mm s^{-1}, Huang *et al* (2001) found that the width of angular distribution of NWs decreases with increasing flow rate and becomes approximately constant at the flow rate of 10 mm s^{-1} at which 80% of the NWs are aligned in the direction of fluid flow. Basically, the flow of fluid in the vicinity of the substrate surface is a shear flow which aligns the NWs. A high flow rate produces a larger shear force and consequently, is more suitable for alignment of NWs. Flow rate remaining constant, the duration of flow decides the density of nanowires. The NW density is 250 NWs per 100 μm for a flow duration of 30 min; the corresponding NW-to-NW separation is 400 nm. Apart from fluid flow rate and duration, the chemical functionality of the substrate surface cannot be ignored as a deterministic parameter controlling NW assembly.

The densely packed thin film containing parallel NWs is processed using the standard TFT fabrication technology. The channel of the TFT is kept parallel to the axis of the NW assembly, i.e. the NW length direction.

11.3.7 TFT fabrication with NWs

Having described the methods of growing nanowires and their assembly by directed fluid flow, let us see how these methods are integrated into the TFT fabrication process on plastic (Duan *et al* 2003). The TFT fabrication (figure 11.7) begins with a polyetheretherketone (PEEK) sheet, a colorless thermoplastic with glass transition temperature of 143 °C. The PEEK sheet is 125 μm thick. To obtain a microscopically smooth surface, the PEEK sheet is coated with SU-8 photoresist and baked.

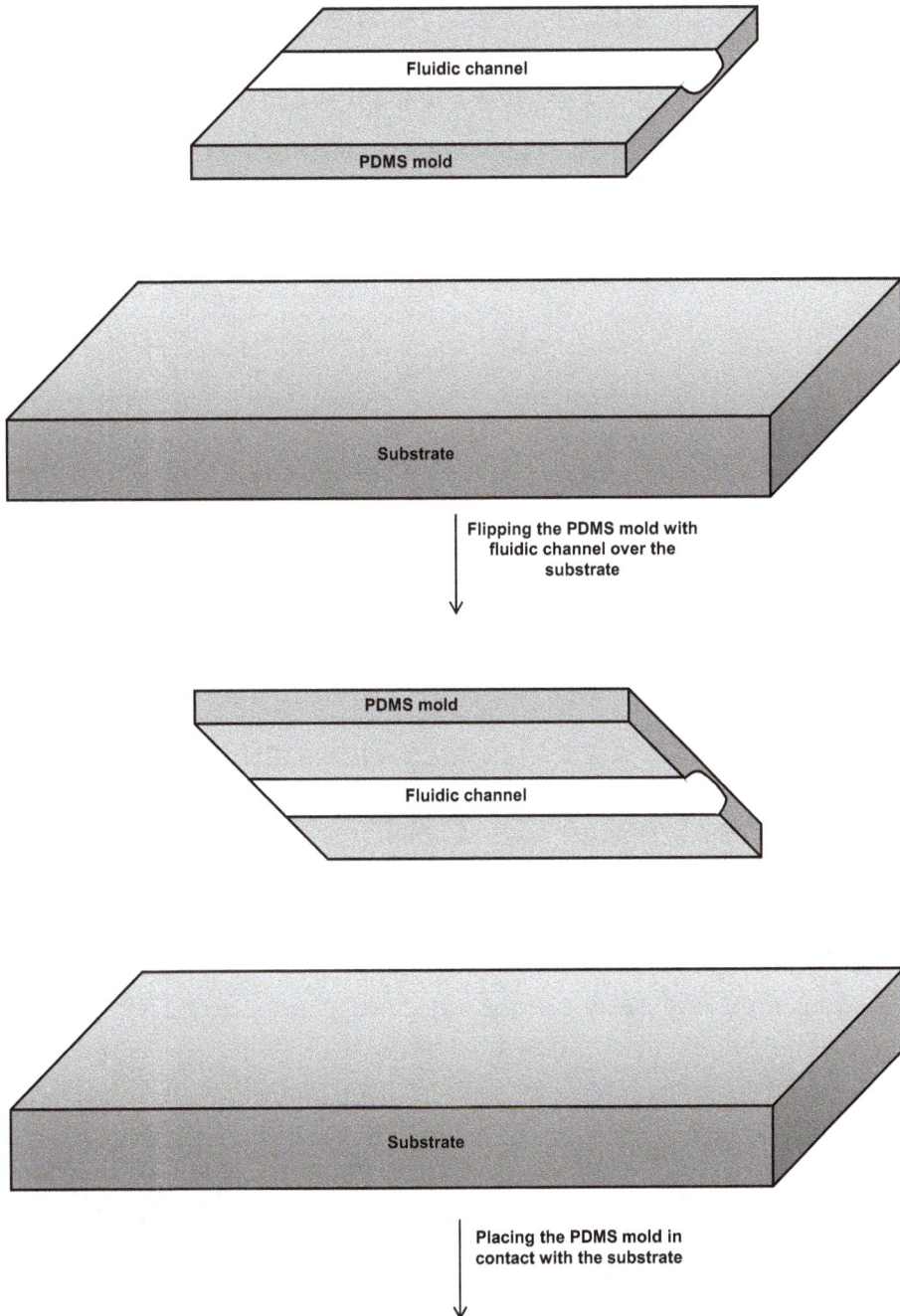

Fluidic channel

PDMS mold

Substrate

Flipping the PDMS mold with
fluidic channel over the
substrate

PDMS mold

Fluidic channel

Substrate

Placing the PDMS mold in
contact with the substrate

Figure 11.6. Fluid flow-directed alignment of silicon nanowires into a parallel array.

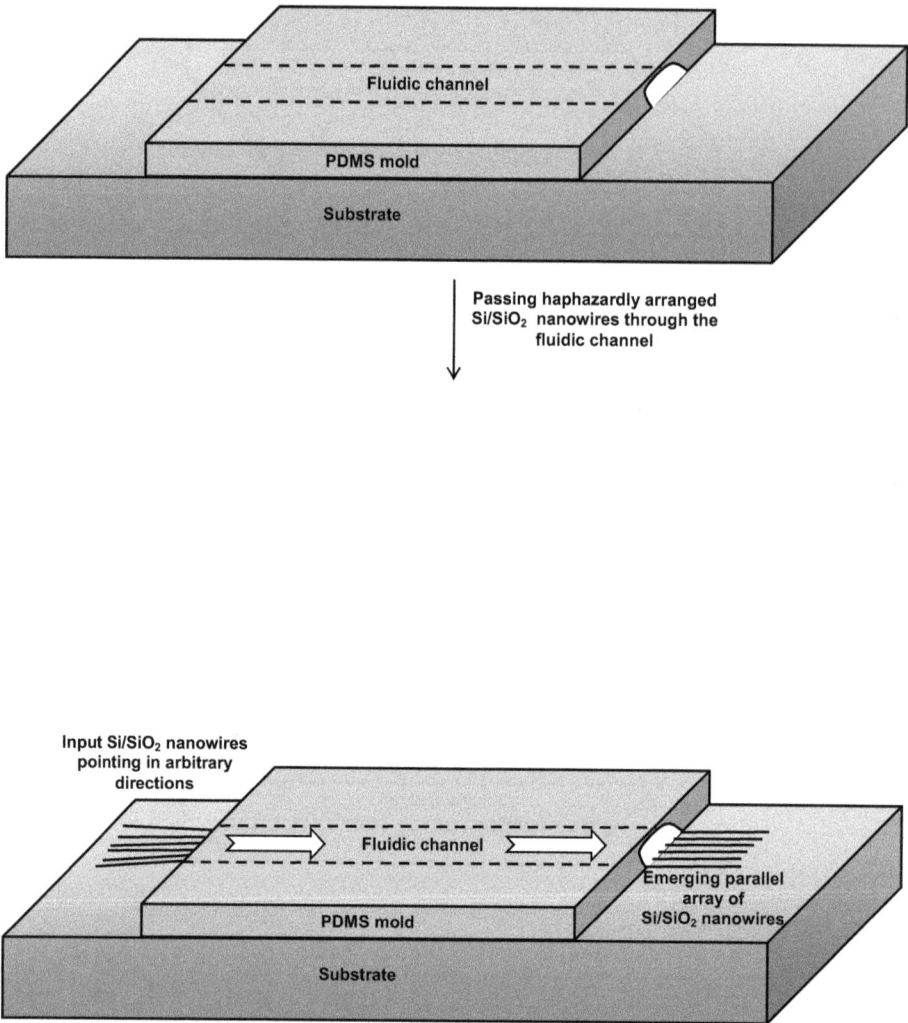

Figure 11.6. (Continued.)

The photoresist thickness is 1–2 μm. For gate electrodes, chromium/gold strips are deposited and defined. The Cr film thickness is 10 nm and gold layer thickness is 30 nm. Over the gate electrode strips, the gate dielectric film is deposited. This film is aluminum oxide AlO_x and the deposition technique used is electron-beam evaporation. The aluminum oxide film is 30 nm thick. Now the aligned Si NW film is deposited over the aluminum oxide layer by the method described in section 11.2.3. Lastly, titanium/gold are deposited and patterned to form the source/drain electrodes. The titanium layer is 60 nm thick and the gold film is 80 nm thick.

The TFT contains 17 NWs having diameters of 40 nm arranged parallel to each other. They form the channel of the TFT; the channel length is 6 μm. The threshold voltage is 3.0 V and the subthreshold swing is 500–800 mV/decade. The current on/

Taking PEEK sheet

PEEK sheet

SU-8 coating

SU-8
PEEK sheet

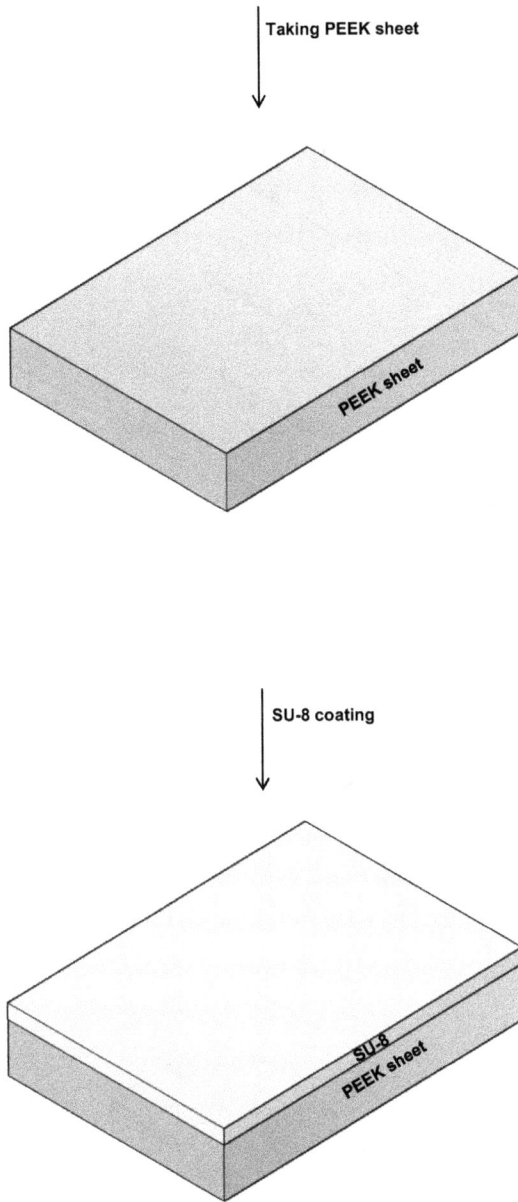

Figure 11.7. Si/SiO$_2$ NW TFT on PEEK.

Cr/Au gate electrode
deposition/patterning

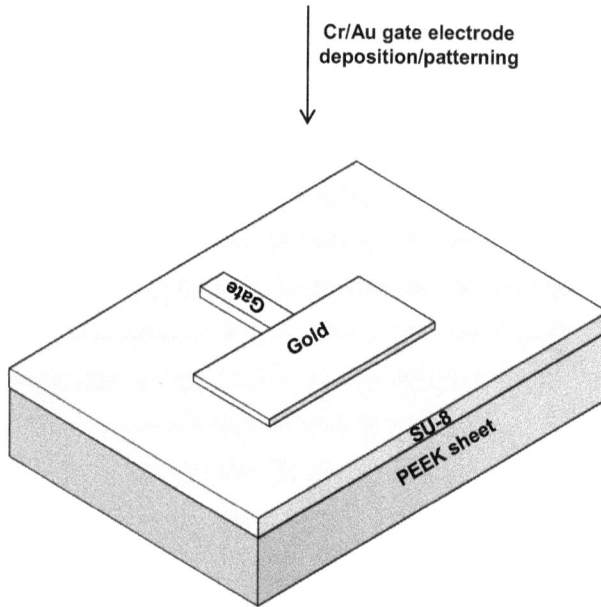

Electron-beam evaporation of
AlO$_x$ gate dielectric and its
patterning

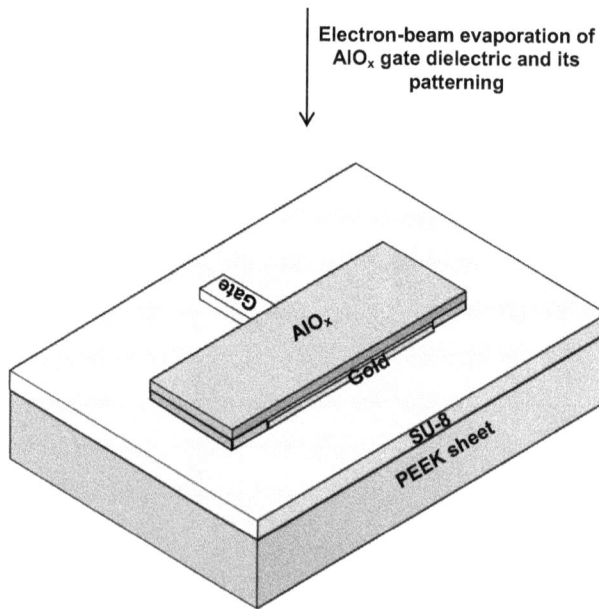

Figure 11.7. (Continued.)

Aligned Si NWs deposition

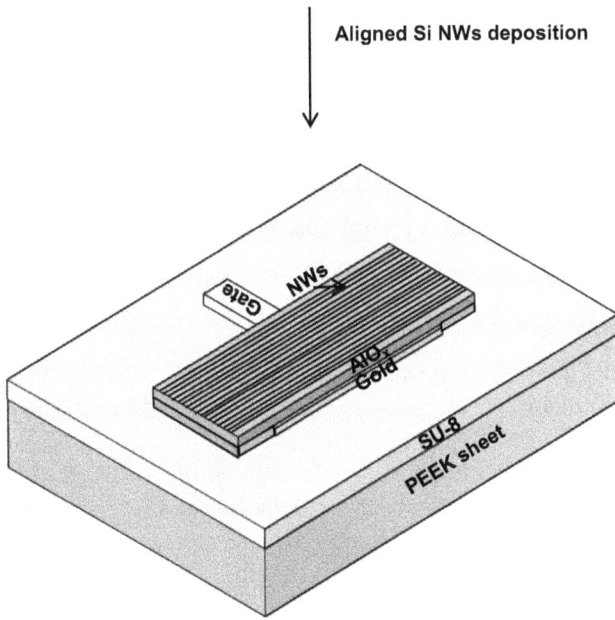

**Ti/Au source/drain electrodes
formation**

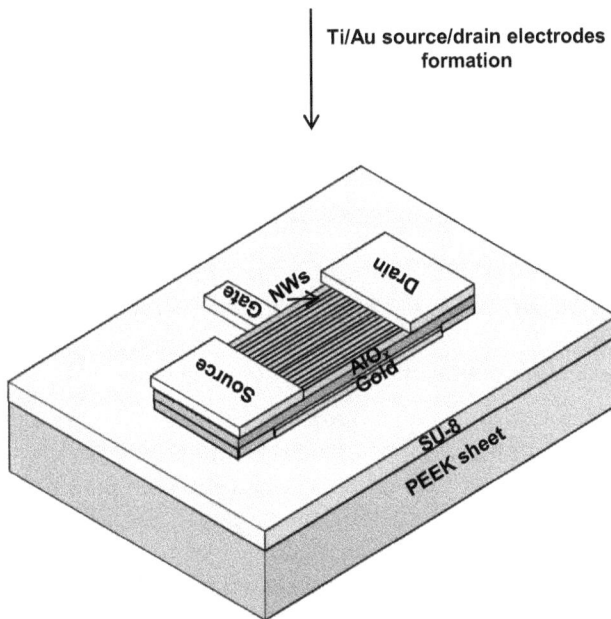

Figure 11.7. (Continued.)

off ratio of the TFT is 10^5. The transconductance is 0.45 μS at $V_{DS} = -1$ V. From measurements with an electrolytic gate, the hole mobility is extracted as 123 cm^2 V^{-1} s^{-1}. When the PEEK substrate is slightly bent with a radius of curvature of 55 mm, the TFT characteristics are insignificantly affected (Duan *et al* 2003).

11.4 P-type Si/SiO$_2$ NW TFT on Mylar

Mcalpine *et al* (2005) fabricated 20 nm-diameter Si NW FET on a Mylar substrate (figure 11.8).

11.4.1 TFT fabrication

Gate electrodes of gold are patterned on Mylar. On the gate electrodes, they deposit SiO$_2$ insulator of thickness 30 nm to function as gate dielectric. They grow P-type Si NWs from 20 nm diameter gold nanoclusters using silane: diborane as reactants in the ratio 8000:1. Then the P-type Si NWs are assembled by fluid-directed alignment on the gate dielectric after preparing stable solution suspensions of nanowires by sonication of growth substrate in ethanol. The source/drain electrodes are made of palladium. They are defined by photolithography. Despite the fact that the source/drain contacts are not annealed, they show practically ohmic behavior.

11.4.2 Electrical and bending characteristics of TFT

The TFT has a threshold voltage of 0.5–1.5 V with a subthreshold slope of 340 mV/decade. The current on–off ratio is 10^5. The TFT has a transconductance of 340 nA V^{-1} from which the hole mobility is 200 cm^2 V^{-1} s^{-1}.

When the TFT is flexed to a radius of curvature of 3 mm and the current is measured in bent condition, it shows a 10% decrease in current. The small change in current indicates the robustness of the TFT towards flexing. The high mobility values are in no way inferior to polysilicon TFTs on non-alkali glass substrates and are much larger than for amorphous silicon TFTs on glass/plastic substrates (Mcalpine *et al* 2005).

11.5 TFT on a PET substrate by the SNAP NW transfer approach

Mcalpine *et al* (2007) fabricate TFT by utilizing a technique for growth of ultrahigh density arrays of aligned NWs of metals and semiconductors (Melosh *et al* 2003).

11.5.1 Limitations of photolithographic methods of making NWs

Two important parameters of the nanowire pattern are the nanowire diameter and pitch; the pitch is the center-to-center spacing between one nanowire and the next. To create an accurate pattern of nanowires of a material containing nanowires of particular diameters at precise locations, the pattern is defined in a photoresist. The pattern in nanoscale dimensions can be made by electron-beam lithography. In lift-off photolithography, the material is deposited over the developed pattern. On removal of the photoresist, the material is pulled out from areas where the material lies over the photoresist. However, it is retained in regions where the material is in direct contact with the substrate. When the nanowire diameters become extremely

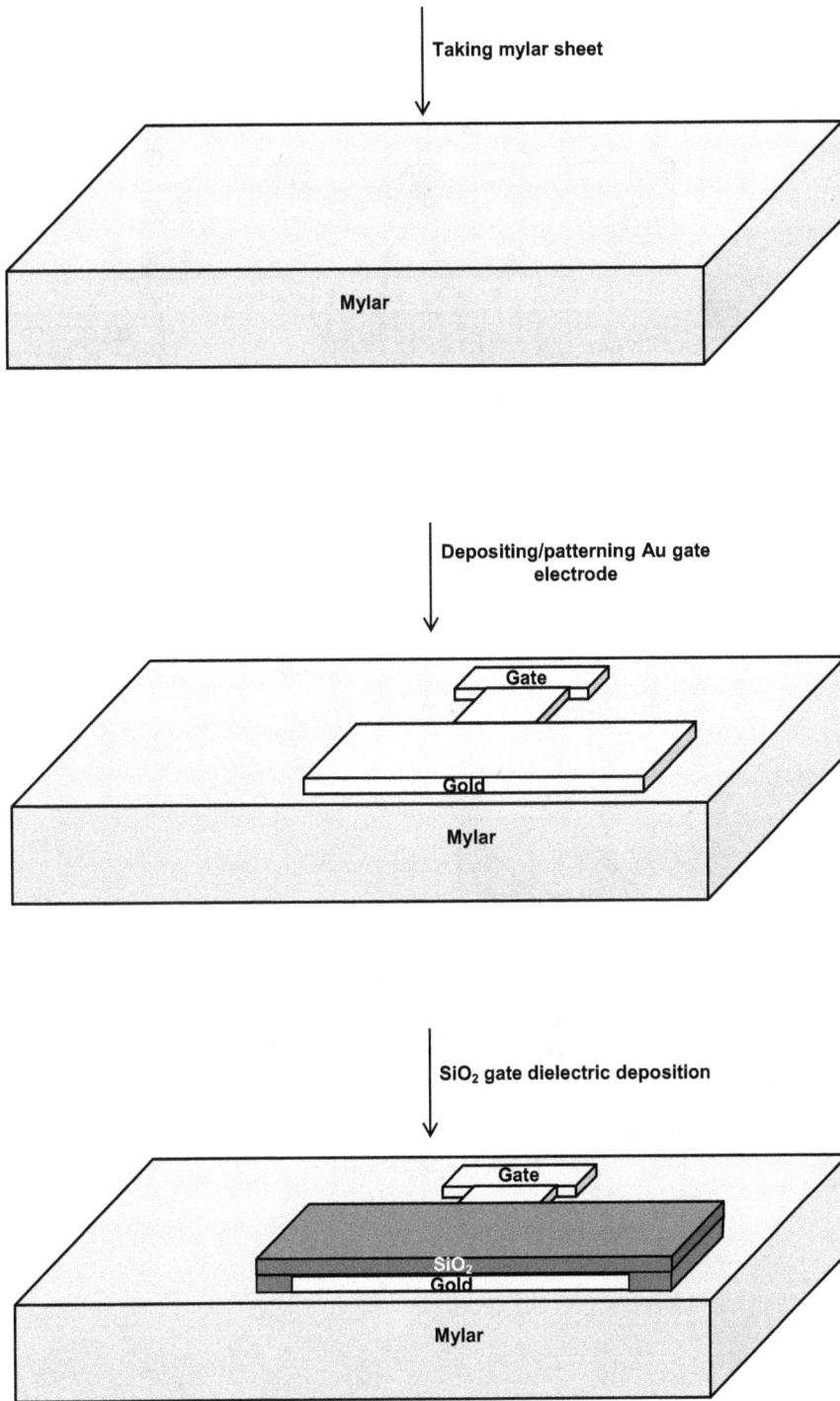

Taking mylar sheet

Mylar

Depositing/patterning Au gate electrode

Gate

Gold

Mylar

SiO₂ gate dielectric deposition

Gate

SiO₂

Gold

Mylar

Figure 11.8. Nanowire TFT on Mylar.

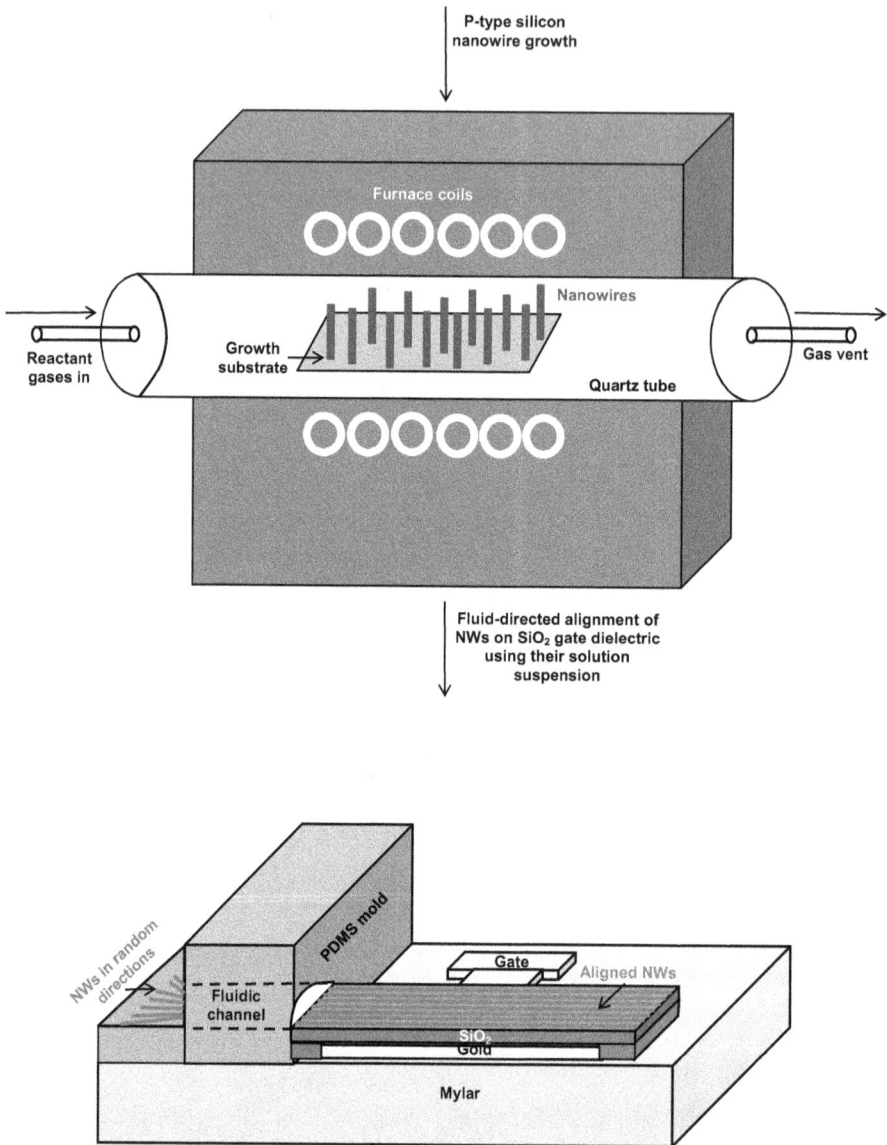

**P-type silicon
nanowire growth**

Furnace coils

Nanowires

Growth
substrate

Quartz tube

Reactant
gases in

Gas vent

**Fluid-directed alignment of
NWs on SiO$_2$ gate dielectric
using their solution
suspension**

NWs in random
directions

PDMS mold

Gate

Aligned NWs

Fluidic
channel

SiO$_2$

Gold

Mylar

Figure 11.8. (Continued.)

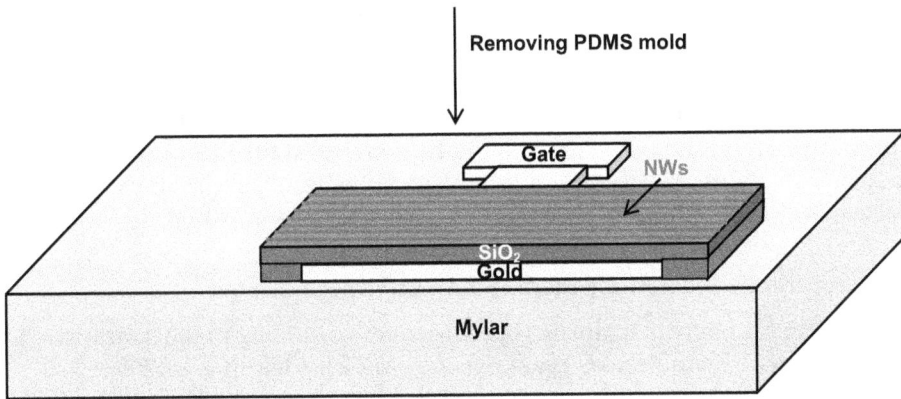

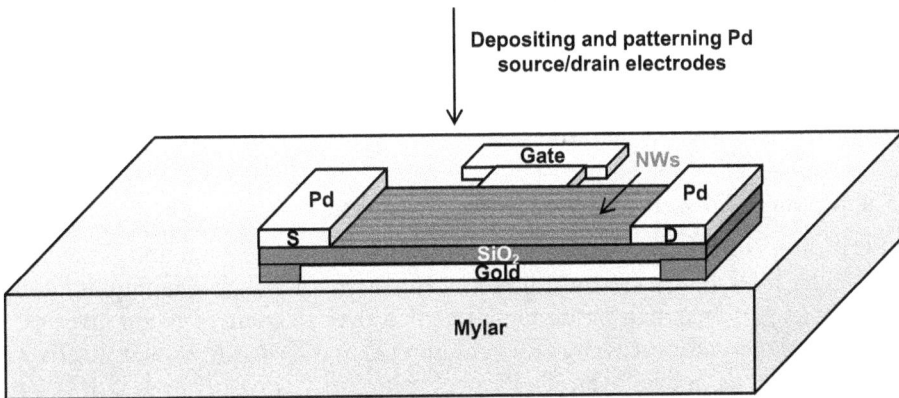

Figure 11.8. (Continued.)

small, it is likely that the peeled off photoresist takes away some neighboring material causing errors in nanowire dimensions. This happens because the area for adhesion of material with the substrate is very small. Conversely, when the pitch is very small, the cohesion of the material hinders the lift-off process. For these reasons, patterns comprising nanowires of diameters up to 20 nm with pitch values up to 60 nm have been successfully realized with e-beam lithography. Beyond these limits, the noise in the processing steps acts as a formidable obstacle. In addition, the serial nature of e-beam lithography makes the process tedious and cumbersome when this patterning is to be done on a larger scale. The difficulties of the serial process have been overcome by resorting to techniques like nanoimprint lithography or microcontact printing. But the limiting factors due to the lift-off process are still persistent. Therefore, the need for a faster technique free from the above boundaries has been felt time and again.

11.5.2 Superlattice nanowire pattern (SNAP) transfer technique

In this method, a physical template is produced for patterning of the nanowires. This template is made by molecular beam epitaxy (MBE) (Melosh *et al* 2003).

11.5.3 Producing a Pt-coated superlattice edge

A superlattice comprising 800 layers of alternating GaAs and $Al_xGa_{1-x}As$ is prepared (Mcalpine *et al* 2007), figure 11.9. The superlattice is a periodic structure of GaAs and $Al_xGa_{1-x}As$ (34 nm/17 nm). It is cleaved along a single crystallographic plane. After the superlattice has been cleaned by sonication in methanol and by mild mopping, the exposed edge of the superlattice is immersed in GaAs etchant. The etchant used is a mixture of ammonia, hydrogen peroxide and water in the volumetric ratio $NH_3/H_2O_2/H_2O$ = 1:20:750. In 10 s, the GaAs regions are selectively etched up to a depth of 30 nm. Consequent upon GaAs etching, the cleaved edge of the superlattice consists of $Al_{1-x}Ga_xAs$ plateaus or highlands separated by GaAs valleys or hollow regions, 30 nm deep. On the edges of the $Al_{1-x}Ga_xAs$ plateaus or highlands, platinum metal is deposited by electron-beam evaporation. During platinum evaporation, the edges are held at an angle of 45° with the beam of Pt atoms. Thus a platinum-coated $GaAs/Al_{1-x}Ga_xAs$ (34 nm/17 nm) superlattice edge is obtained.

11.5.4 Making SOI wafer ready to receive Pt NWs

An SOI wafer is taken with active silicon layer thickness = 32 nm and buried oxide (BOX) layer thickness = 250 nm (figure 11.10). After thorough cleaning, it is rinsed with DI water. P-type spin-on dopant is coated over the active silicon layer of the wafer. Rapid thermal processing (RTP) is applied at 800 °C for 3 min to diffuse the dopant into silicon active layer. The acceptor impurity concentration is 10^{18} cm^{-3}.

11.5.5 Transferring Pt NWs to doped SOI wafer

The P-impurity doped SOI wafer is coated with a PMMA/epoxy layer (Mcalpine *et al* 2007). This PMMA/epoxy coating contains PMMA and epoxy in the gravimetric ratio of 1:50. It is formed by spinning the PMMA/epoxy mixture at 6000 RPM for 30 s on the wafer. We thus have: PMMA–epoxy/SOI structure. When the Pt-coated $GaAs/Al_{1-x}Ga_xAs$ superlattice is brought into contact with the PMMA–epoxy/SOI structure, the Pt-coated $GaAs/Al_{1-x}Ga_xAs$ superlattice/PMMA–epoxy/SOI sandwich is formed. This sandwich is dried on a hot plate at 150 °C for 40 min. The GaAs/Al$_1$

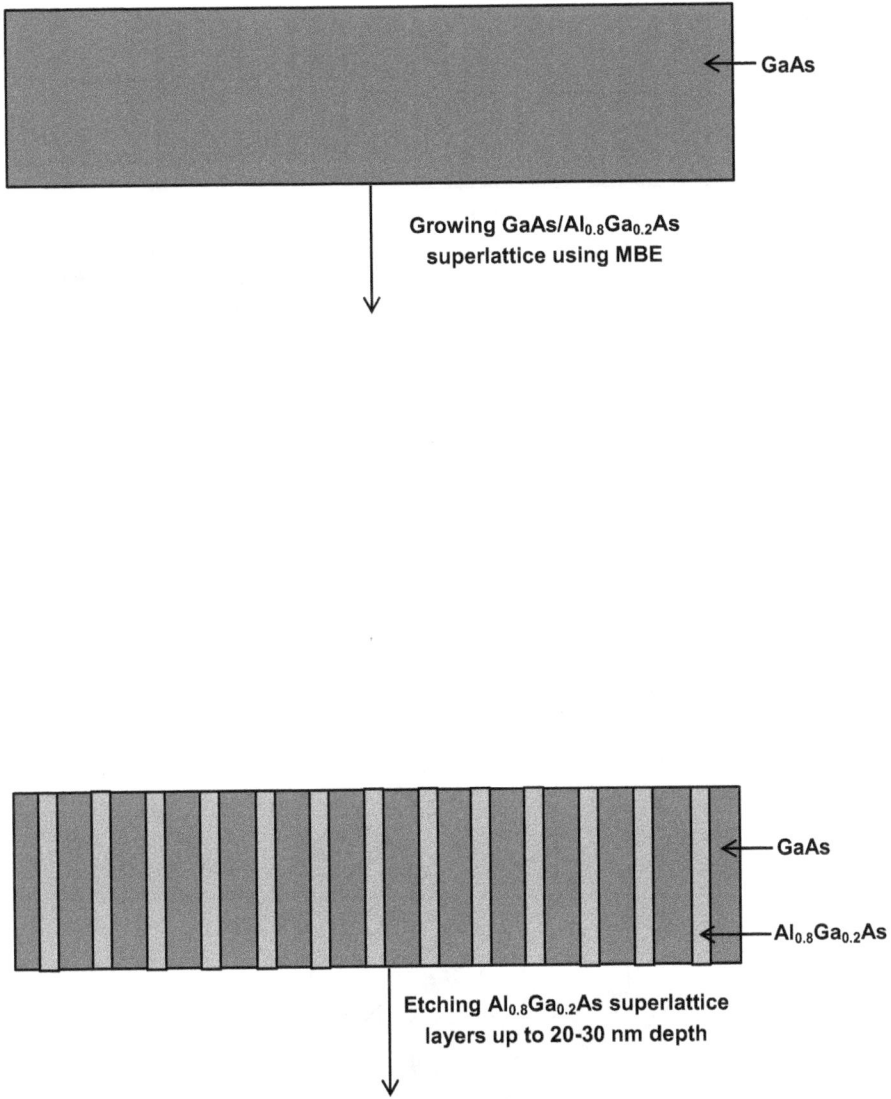

Figure 11.9. Creating platinum wires precisely located at a small distance apart.

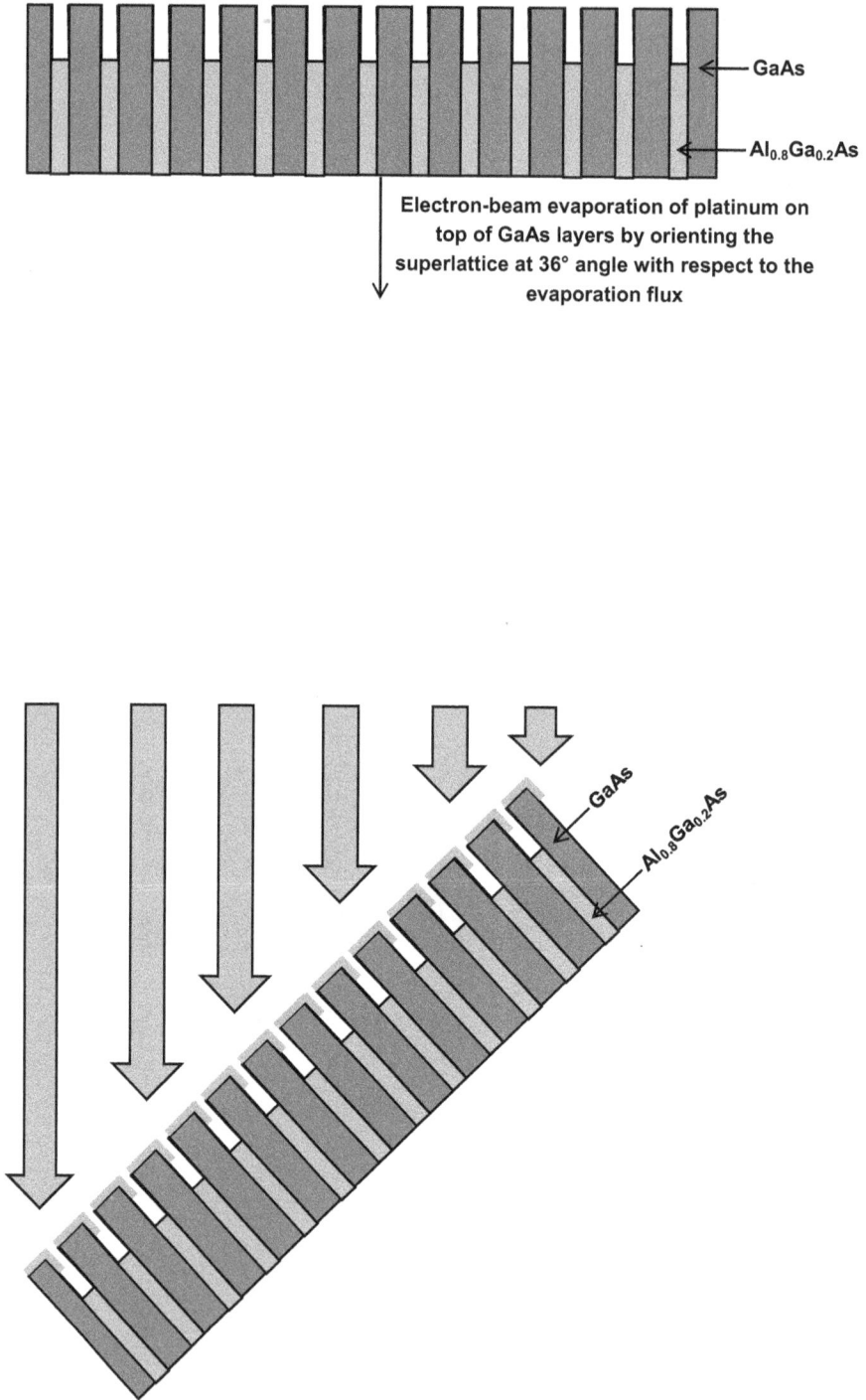

GaAs

Al$_{0.8}$Ga$_{0.2}$As

Electron-beam evaporation of platinum on top of GaAs layers by orienting the superlattice at 36° angle with respect to the evaporation flux

GaAs

Al$_{0.8}$Ga$_{0.2}$As

Figure 11.9. (Continued.)

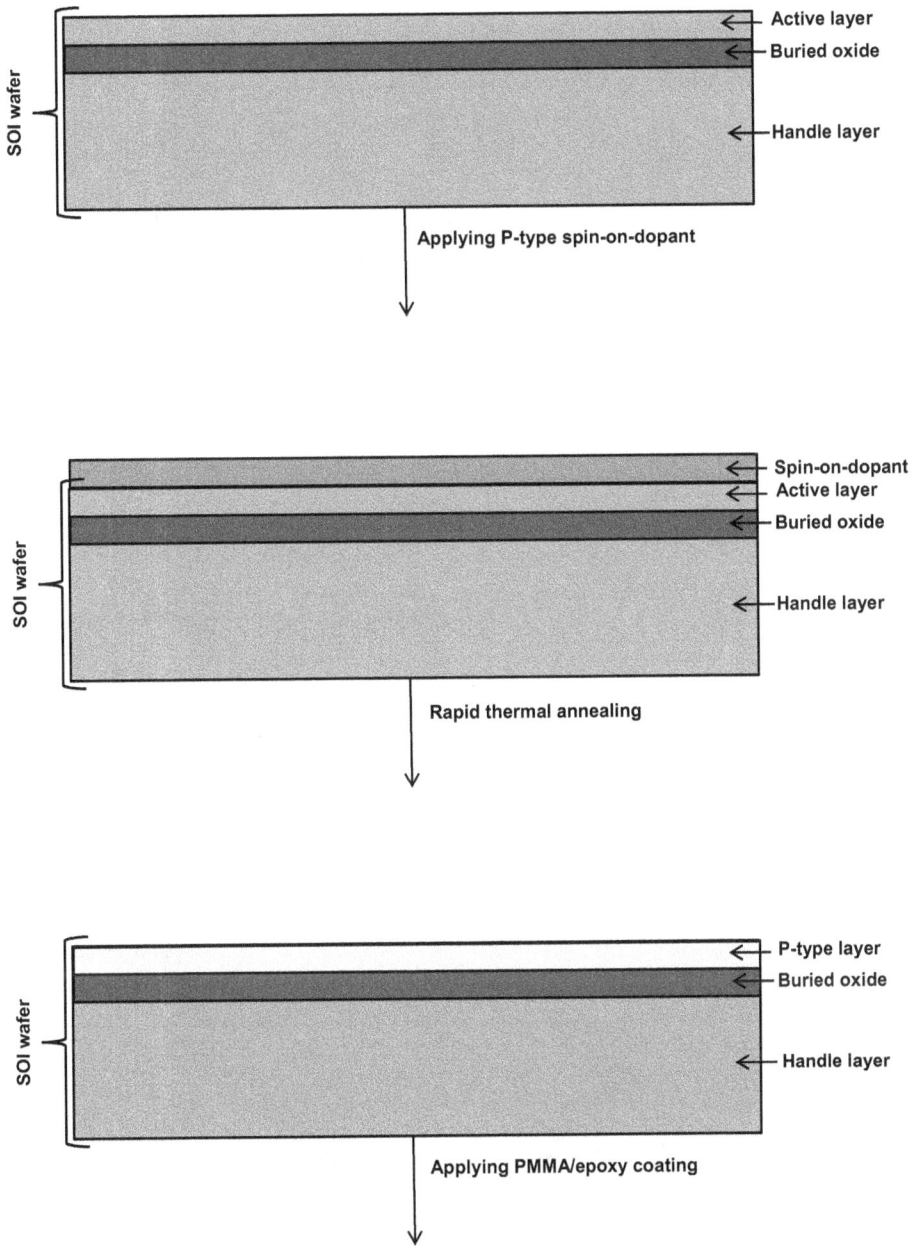

Figure 11.10. Nanowire formation by silicon etching and their retrieval by PDMS.

PMMA/Epoxy
P-type layer
Buried oxide
Handle layer

SOI wafer

Bringing Pt-coated GaAs/Al$_{1-x}$Ga$_x$As superlattice near PMMA/Epoxy-coated SOI wafer

Al$_{0.8}$Ga$_{0.2}$As

GaAs

PMMA/Epoxy
P-type layer
Buried oxide
Handle layer

SOI wafer

Bringing Pt-coated GaAs/Al$_{1-x}$Ga$_x$As superlattice into contact with PMMA/Epoxy-coated SOI wafer

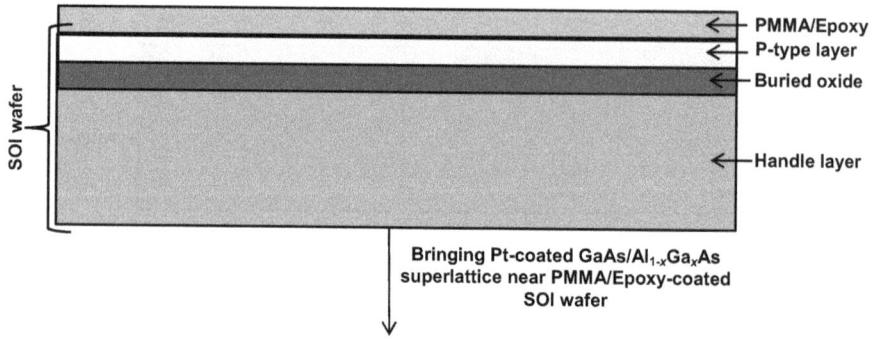

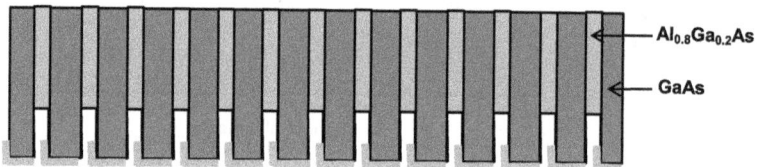

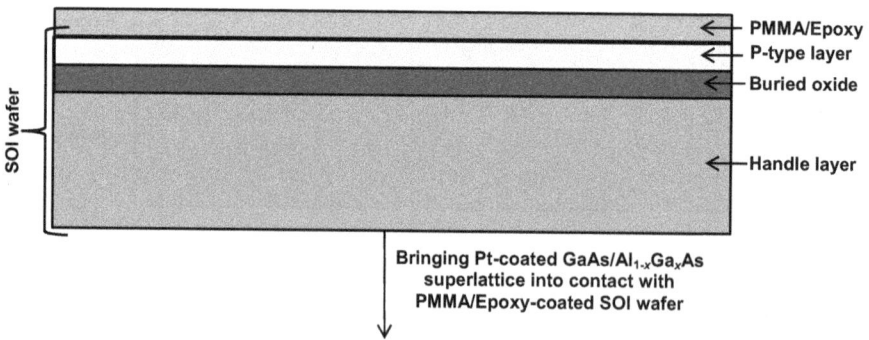

Figure 11.10. (Continued.)

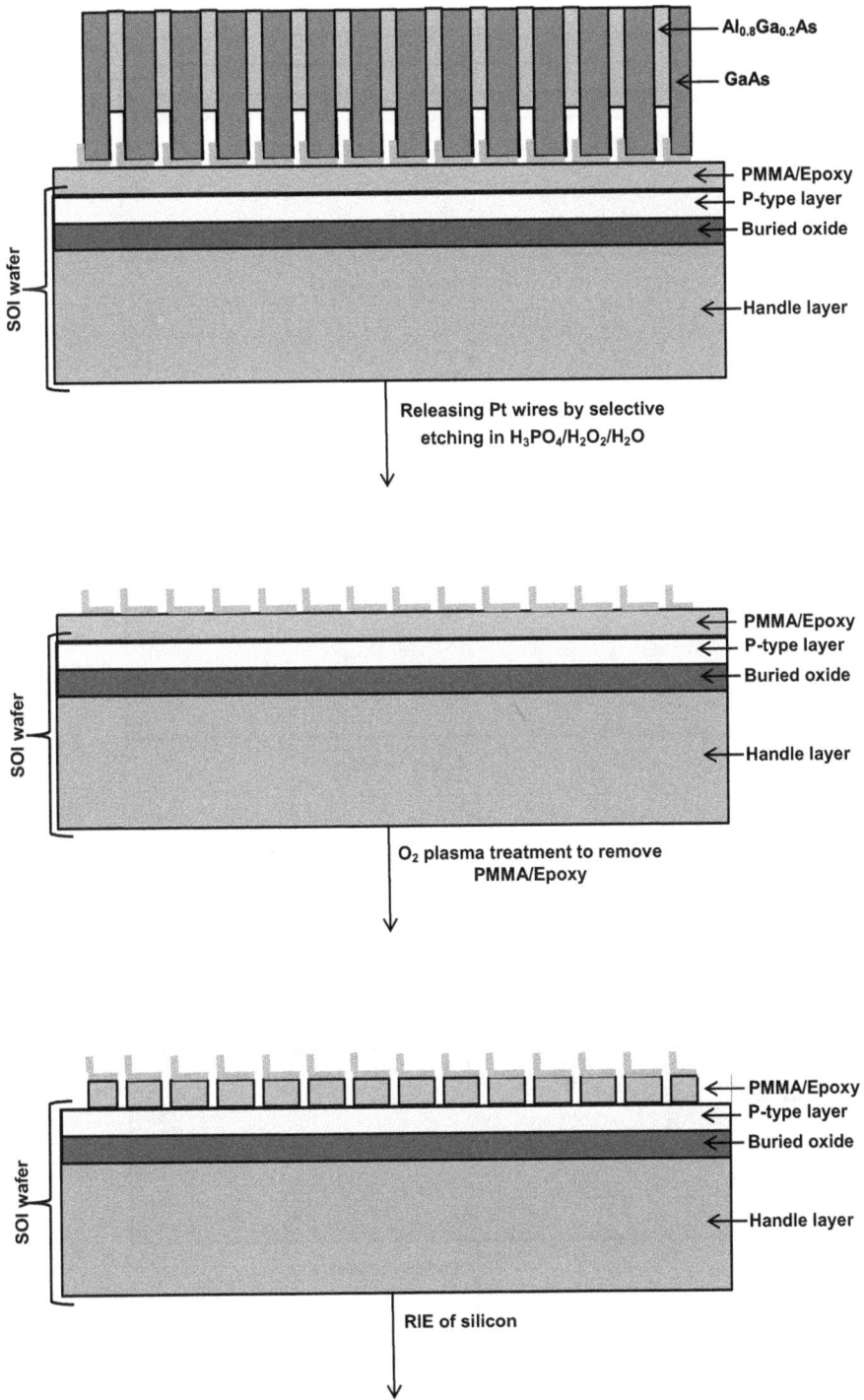

Figure 11.10. (Continued.)

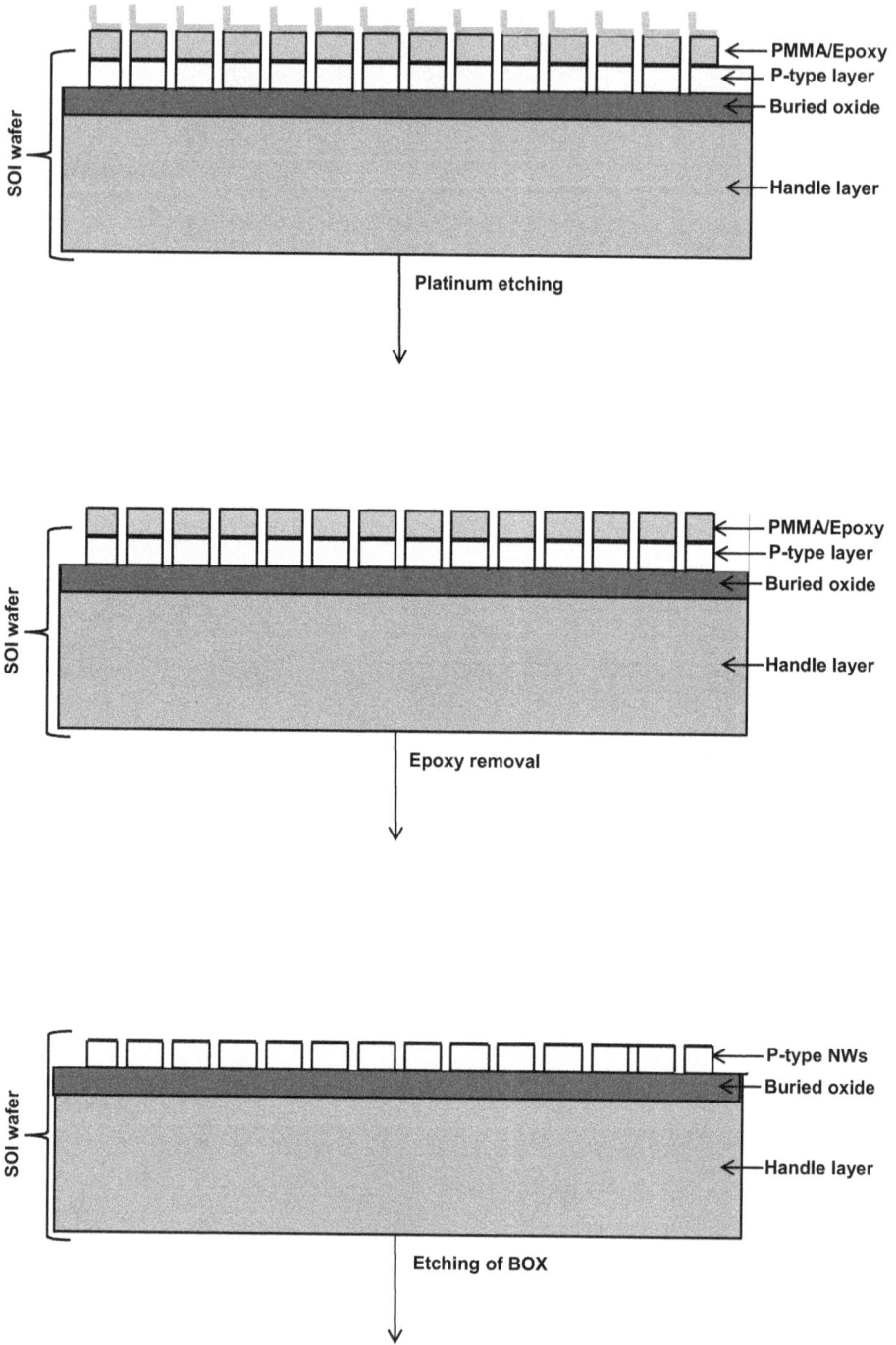

Platinum etching

Epoxy removal

Etching of BOX

Figure 11.10. (Continued.)

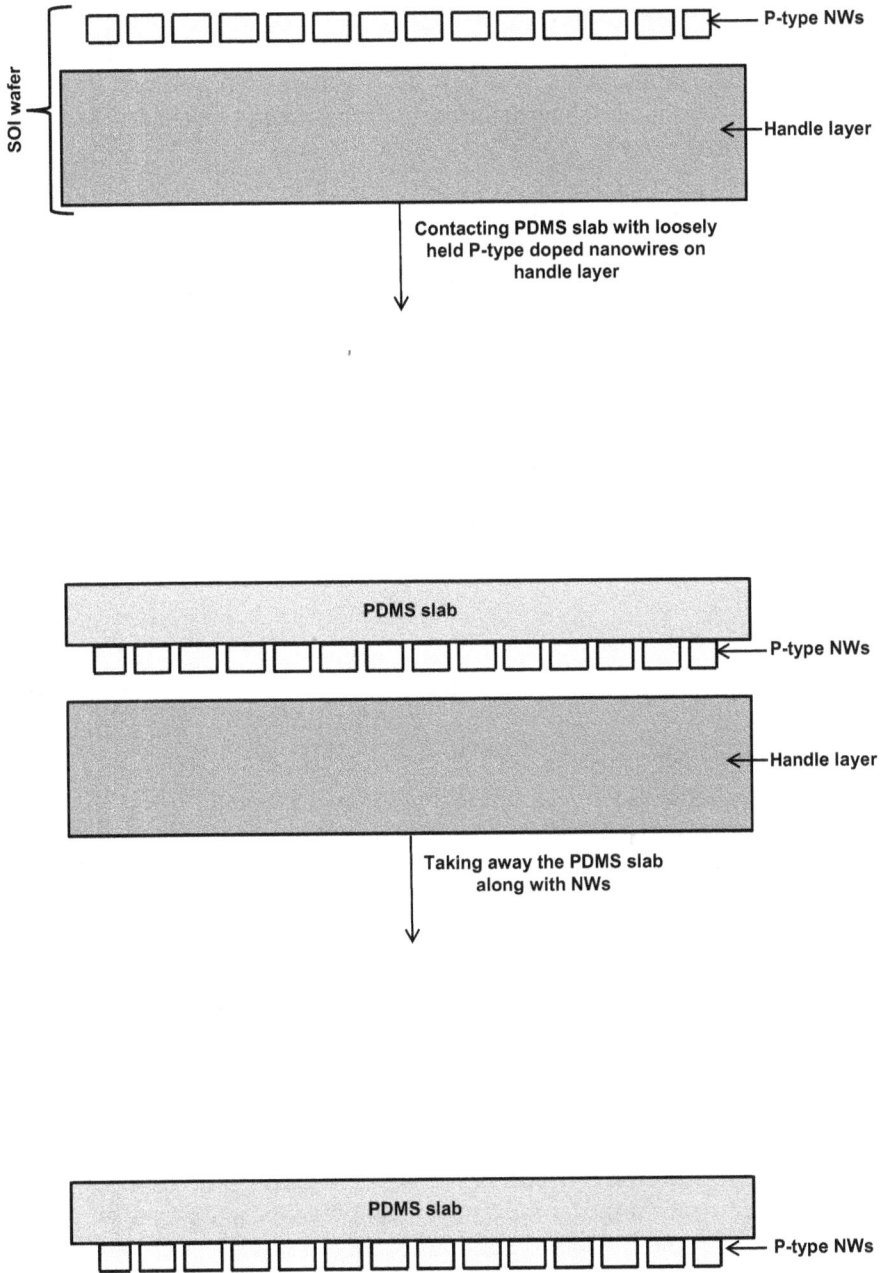

Figure 11.10. (Continued.)

$_{-x}$Ga$_x$As superlattice is released by etching in a solution of phosphoric acid, hydrogen peroxide and water in the ratio $H_3PO_4/H_2O_2/H_2O = 5:1:50$ (v/v). The etching is done for 4.5 h. After the completion of etching, we are left with a highly ordered array of platinum nanowires on the SOI wafer. This array contains 400 platinum nanowires. It is used as a mask for silicon etching in the next step after the oxygen plasma exposure to remove the PMMA/epoxy layer.

11.5.6 Reactive ion etching of silicon

Using the array of platinum NWs as a mask, reactive ion etching of silicon is performed. The RIE parameters are: Gas used: tetrafluoromethane (CF$_4$)/Helium (He); Gas flow rates: CF$_4$:20 sccm, He: 30 sccm; Pressure: 5 mTorr; Power: 40 W; Etching time: 3.5 min. The buried oxide (BOX) of the SOI wafer acts an etch stop. The lower ends of the silicon NWs are bound to the BOX layer of the SOI wafer.

11.5.7 Dissolving platinum NWs to release the Si NWs

The platinum NWs are dissolved by immersion of the structure in aqua regia. In 30 min, the Pt NWs are completely dissolved setting the array of 400 silicon NWs free on the SOI wafer tied at their lower extremeties to the BOX of the SOI wafer.

11.5.8 Residual epoxy removal

The SOI wafer is cleaned in an epoxy remover solution which strips away all epoxy.

11.5.9 Retrieval of NW array by PDMS

The NW array is immersed in concentrated hydrofluoric acid (HF) for 5 s to etch the silicon dioxide holding the NWs at their lower ends with the SOI wafer. Then a slab of PDMS is made to contact conformally the top surface of the SOI wafer loosely holding the NW array. When this slab is pulled away, the NWs are peeled off from the SOI wafer surface and get attached to the PDMS slab. Now, we have the structure: PDMS slab/Si NWs.

11.5.10 Plastic substrate preparation for receiving Si NWs

The substrate is a PET sheet coated with ITO film (Mcalpine *et al* 2007). The thickness of the PET sheet is 100 μm while that of the ITO film is 100 nm (figure 11.11).

The ITO film will act as the gate electrode of the bottom-gate TFT. Over this gate electrode, the gate dielectric is deposited. For this dielectric film, SU-8 photoresist is used. But before depositing the SU-8 film, the ITO film is cleaned by rinsing in acetone, CH_3COCH_3 or C_3H_6O, then isopropanol, $CH_3CHOHCH_3$ or C_3H_8O and finally in DI water. The water traces are dried in a stream of nitrogen.

After drying up the water, the surface of ITO film is activated in oxygen plasma at 300 mTorr pressure with 60 W power for 1 min. Then it is spin coated with SU-8 photoresist at 3000 RPM for 30 s. The SU-8 photoresist is given a soft bake on a hot plate at 65 °C for 1 min. The thickness of the SU-8 coating is 2 μm. Now, the PET sheet/ITO film/SU-8 structure is ready to receive the Si NWs.

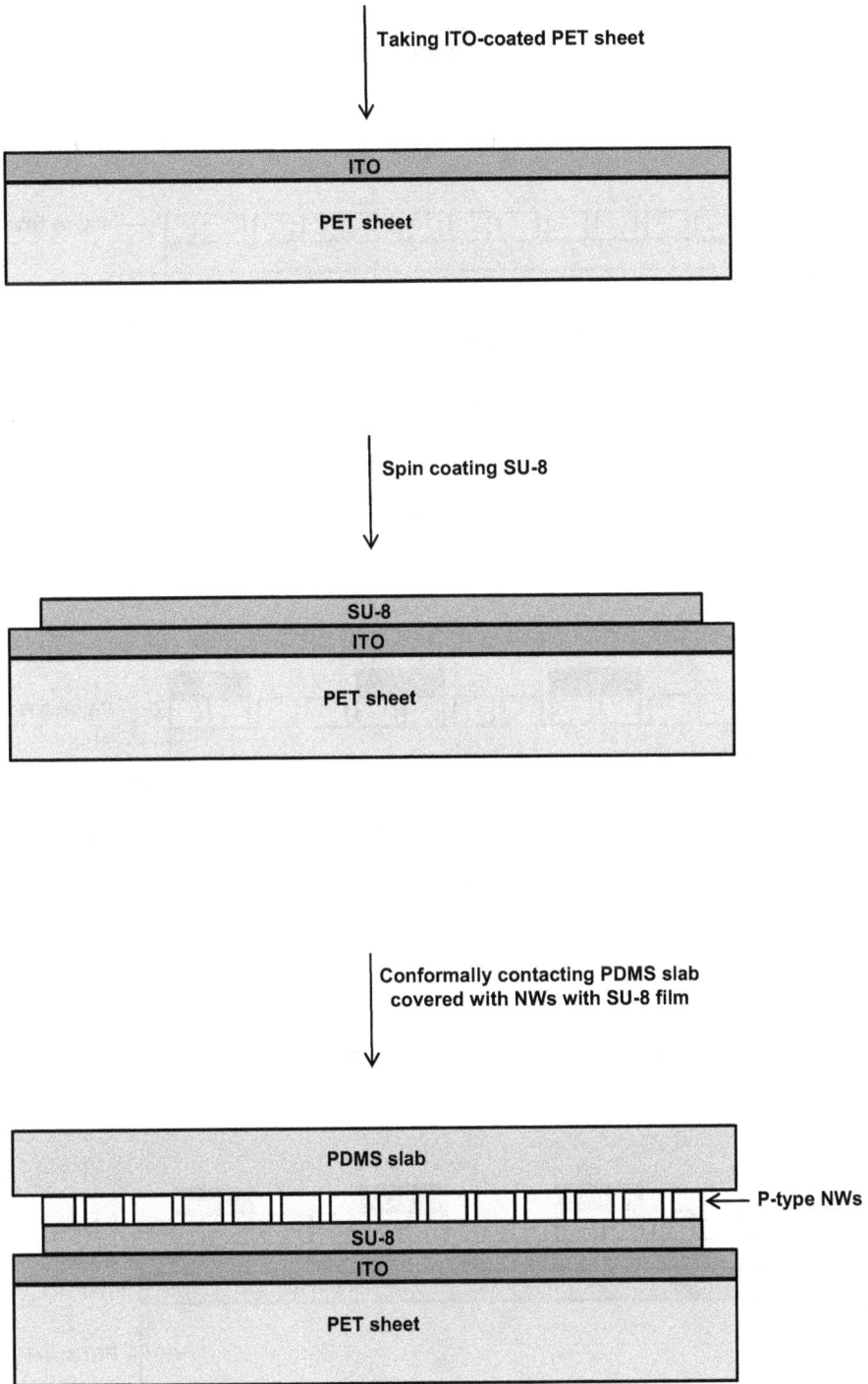

Figure 11.11. Bottom-gate and top-gate NW FET fabrication.

Heating and UV exposure of SU-8
followed by taking away the PDMS
slab

P-type NWs

SU-8

ITO

PET sheet

Bottom-gate FET by forming Ti
source/drain electrodes

S   D   S   D

P-type NWs

SU-8

Gate   Gate   ITO

PET sheet

Subdivision into bottom-gate FET
islands by silicon etching using
RIE

S   D   S   D

P-type NWs

SU-8

Gate   Gate   ITO

PET sheet

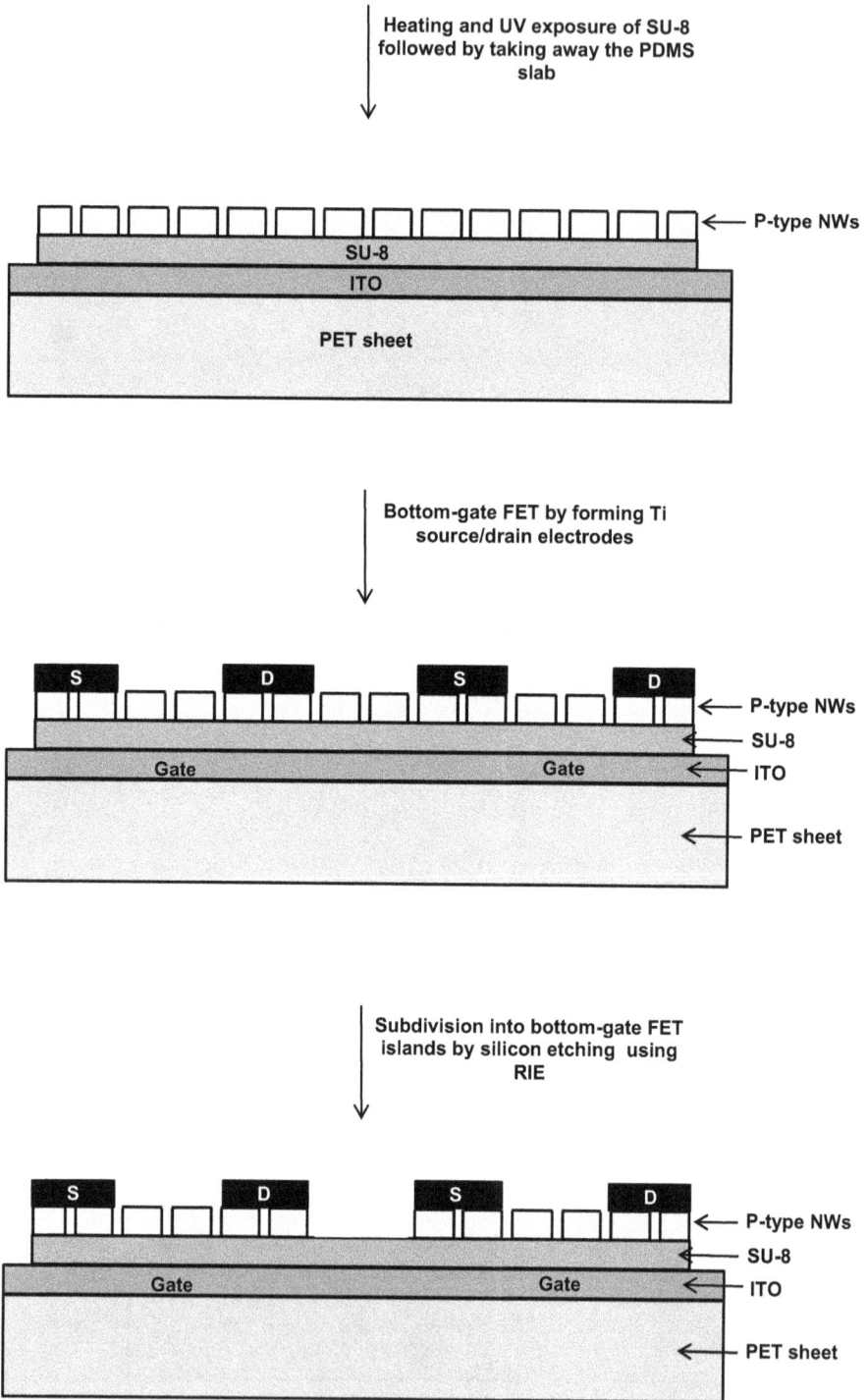

Figure 11.11. (Continued.)

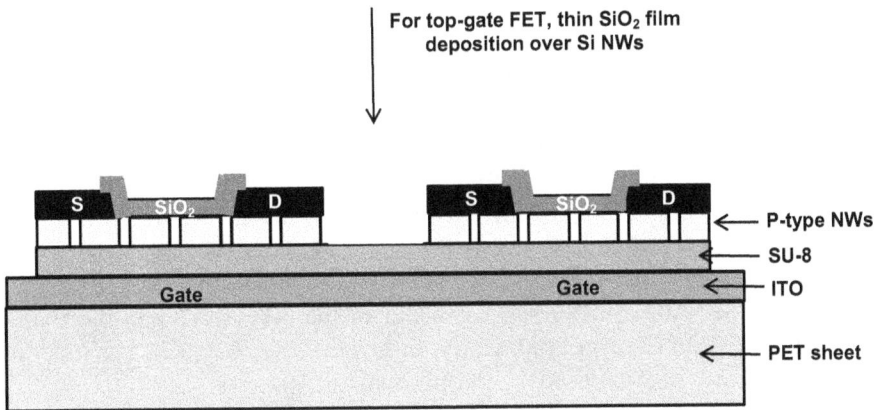

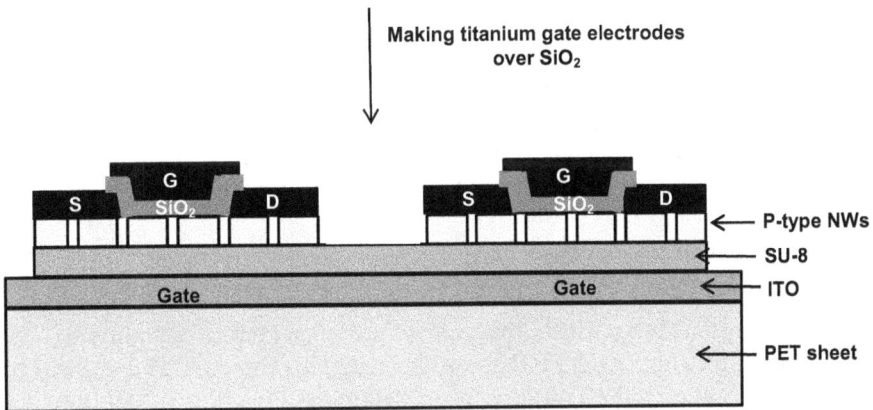

Figure 11.11. (Continued.)

11.5.11 Harvesting Si NWs on the PET substrate from PDMS slab

The PDMS/Si NWs structure is brought into conformal contact with the PET sheet/ ITO film/SU-8 structure (Mcalpine *et al* 2007). The combination on the hot plate is: PET sheet/ITO film/SU-8/Si NWs/PDMS. The hot plate temperature is ramped up to 95 °C. When the temperature reaches 95 °C, the combination is given a medium temperature bake at 95 °C for 5 min. The back side of the PET sheet is exposed to UV radiation for 1 min. The PDMS slab is cautiously withdrawn leaving the structure: PET sheet/ITO film/SU-8/Si NWs.

To make this assembly ready for carrying out the remaining process steps for TFT fabrication, the back side of PET sheet is exposed to UV again for 1 min followed by hardbaking at 115 °C for 15 min whereby the Si NWs are firmly fixed in the SU-8 photoresist. At this stage, the PET sheet carries the back gate electrode (ITO film), the gate dielectric (SU-8) and the P-type semiconductor thin film comprising the Si NWs. The average diameter of the NWs is 18 nm and the mean pitch is 41 nm, which are in conformity with the GaAs/Al$_{1-x}$Ga$_x$As superlattice periodicity within limits imposed by the process.

11.5.12 Forming source/drain electrodes for completing the bottom-gate TFT

The bottom-gate TFT needs only the source/drain electrodes for completion (Mcalpine *et al* 2007). Before making these electrodes, the NWs are cleaned by rinsing in water and treatment with mild oxygen plasma at 300 mTorr with 30 W power for 30 s. Dipping in buffered HF for 3 s removes the thin silicon dioxide film formed on the surfaces of NWs. After native oxide removal, titanium is deposited by electron-beam evaporation over the Si NW thin film. The titanium layer is 100 nm thick. By photolithography, the pattern of source/drain electrodes is defined. Unwanted titanium is removed in 5 s using the etchant: HF/H$_2$O$_2$/DI water in the ratio 1:1:10 (v/v).

11.5.13 Sectioning the geometry into device islands

A mask is used to subdivide the SW NW thin film into device islands. The design for this subdivision is defined in the photoresist. The unwanted silicon Si NW portions are removed by dry etching with SF$_6$ at flow rate of 20 sccm under 20 mTorr pressure. The etching is done at 30 W for 1 min. Thus we have TFT islands with Si NWs between the islands removed. After etching, photoresist is stripped in acetone.

11.5.14 Constructing the top-gate TFT

The gate dielectric of the bottom-gate TFT is the 2 μm thick SU-8 film (Mcalpine *et al* 2007). Due to its large width, this gate will provide poor modulation of channel current. Therefore, a top-gate TFT is constructed. This top-gate TFT utilizes the Si NW thin film along with the Ti source/drain electrodes. Over the Si NW thin film, an SiO$_2$ film is deposited by electron-beam evaporation to act as the gate dielectric of the top-gate TFT. This SiO$_2$ film is 25 nm thick.

Now photolithography is performed to define the gate electrode region on the SiO$_2$ thin film. This gate electrode region is matched with the source/drain gaps of the bottom-gate TFT so that the source/drain electrodes of the bottom-gate TFT can also be used as the source/drain electrodes of the top-gate TFT. The lift-off technique is applied. 50 nm thick titanium is deposited on the photoresist pattern. When the photoresist is removed in acetone, the titanium sticks only at the gate areas forming the gate electrodes. At the remaining regions, titanium comes off with the photoresist. Thus the top-gate TFT comprises the titanium gate electrode, SiO_2 gate dielectric, the semiconductor channel region made of Si NW thin film and the titanium source/drain electrodes. Note that the semiconductor channel region and the source/drain electrodes of the top-gate TFT also serve as the semiconductor layer and source/drain contacts of the bottom-gate TFT.

11.5.15 Top-gate TFT characteristics

This TFT has a threshold voltage $= -2$ to -3 V. It is enhancement-mode type. The subthreshold voltage is 300 mV/decade. The top gate modulates the channel conductance by 5 orders of magnitude. The transconductance of top-gate TFT is 5 μS.

The nanowires are also used as NO_2 sensors by functionalizing via microfluidic channels through which various silanes are injected. They exhibit sensitivity to NO_2 gas in parts per billion range. When exposed to 20 ppm NO_2, there is 3000% increase in current flowing through the NWs after 1.25 min; 20 ppb NO_2 causes 10% current rise after 15 min (Mcalpine *et al* 2007).

11.6 Discussion and conclusions

One method of silicon nanowire growth that has been extensively utilized is the vapor–liquid–solid method mediated by gold nanoclusters as catalysts. This method is used to grow Si nanowires on a Si/SiO$_2$ substrate. A substrate-independent method has also received attention. Here the nanowires are formed by ablating a target of the material of which the nanowires are to be grown with the help of Nd:YAG laser beam. The nanowires are collected on a cold finger on which they deposit with the gas flowing through the quartz tube The haphazard network of nanowires is aligned into parallel arrays in an assigned direction by flowing the nanowires through microfluidic channels formed by placing a PDMS mold with engraved trenches over the substrate. Efforts have been made still further. Dimensional limitations of photolithographically formed nanowires have been understood and a $GaAs/Al_xGa_{1-x}$ As superlattice-based method has been evolved to overcome these bounds.

The mobilities, transconductances and bending characteristics of nanowire TFTs fabricated on various substrates are encouraging enough to enable them to be placed in the frontal leading position amongst 1D nanostructure devices in flexible electronics.

Review exercises

11.1 Point out the advantages of silicon nanowires over carbon nanotubes for TFT fabrication.

11.2 How is the 24 μm thick polyimide film deposited by spin coating on a handle wafer? How is it cured?

11.3 In the vapor–liquid–solid method of nanowire growth, what is the role of the gold colloids? What is the typical diameter of the gold colloids?

11.4 What is the substrate on which the nanowires are grown? What gas is used for growing the germanium core? What is the temperature of growth? What is the length of the Ge nanowire?

11.5 Which gas is used for depositing the polycrystalline silicon shell? How thick is the silicon shell? At what temperature is the shell deposition done?

11.6 Explain how the lubricant plays a crucial part in the transfer printing of nanowires from the growth substrate to polyimide film?

11.7 Explain how the donor substrate is slid over the acceptor substrate during transfer printing of nanowires while applying a gentle pressure and changing this pressure?

11.8 In the Ge/Si NW FET, what is the gate electrode made of? What is the gate dielectric material? Of what material are the source/drain electrodes made?

11.9 How is the Ge/Si NW FET passivated? How is it used as a pressure sensor?

11.10 How much does the device conductance change on bending up to a radius of 2.5 mm?

11.11 How is the strain cycling test done? What happens when the device is bent 2000 cycles up to a radius less than 10 mm?

11.12 What growth substrates are used for nanowires to fabricate P-type Si/SiO$_2$ NW TFT? Why are the gold nanoclusters dispersed on the substrates? What are the diameters of the gold nanoclusters used?

11.13 What apparatus is used for nanowire growth? What is the gas used for nanowire synthesis? Which gas is used for doping the nanowires P-type? What is the growth temperature? What is the total pressure in the growth reactor?

11.14 Can you grow nanowires without using a growth substrate? If yes, how? Where are the nanowires collected?

11.15 What target is used for growing silicon nanowires by laser ablation? What target is used for germanium nanowire growth? What are the typical growth temperatures for silicon and germanium nanowires? What is the wavelength of the laser source used?

11.16 You have grown a silicon nanowire, how does it become a Si/SiO$_2$ core/shell nanowire when placed in the atmosphere?

11.17 The as-grown silicon nanowires are randomly arranged pointing in all directions: describe a simple method by which nanowires can be aligned into parallel arrays in a given direction.

11.18 In the fluidic flow method of nanowire alignment, which is more effective: a high flow rate or a low flow rate of fluid? Why?

11.19 Describe how the nanowires aligned by fluidic flow are used to fabricate a TFT on a PET substrate? Give the materials used for the gate electrode, the source/drain electrodes, and the gate dielectric.

11.20 Nanowire patterns with diameters up to 20 nm and pitches of 60 nm have been fabricated using electron-beam lithography. Discuss the hurdles faced in crossing these limits.

11.21 What is a superlattice? Describe how will you make a platinum-coated GaAs/Al$_{1-x}$Ga$_x$As (34 nm/17 nm) superlattice edge.

11.22 How are the platinum nanowires transferred to a doped SOI wafer? How is RIE of silicon done using platinum nanowires as a mask? How are platinum nanowires dissolved?

11.23 How is the silicon nanowire array retrieved with PDMS? How are the silicon nanowires harvested on the ITO-deposited PET substrate from PDMS slab using SU-8?

11.24 Why is it necessary to fabricate a separate top-gate TFT when a bottom-gate TFT with SU-8 as dielectric has already been fabricated during harvesting of the silicon nanowires from PDMS slab?

References

Cui Y, Duan X, Hu J and Lieber C M 2000 Doping and electrical transport in silicon nanowires *J. Phys. Chem.* B **104** 5213–6

Cui Y, Lauhon L J, Gudiksen M S, Wang J and Lieber C M 2001 Diameter-controlled synthesis of single-crystal silicon nanowire *Appl. Phys. Lett.* **78** 2214–6

Duan X, Niu C, Sahi V, Chen J, Parce J W, Empedocles S and Goldman J L 2003 High-performance thin-film transistors using semiconductor nanowires and nanoribbons *Nature* **425** 274–8

Fan Z, Ho J C, Jacobson Z A, Razavi H and Javey A 2008b Large-scale, heterogeneous integration of nanowire arrays for image sensor circuitry *Proc. Natl. Acad. Sci.* **105** 11066–70

Fan Z, Ho J C, Jacobson Z A, Yerushalmi R, Alley R L, Razavi H and Javey A 2008a Wafer-scale assembly of highly ordered semiconductor nanowire arrays by contact printing *Nano Lett.* **8** 20–5 Supporting Information S1–S7.

Huang Y, Duan X, Wei Q and Lieber C M 2001 Directed assembly of one-dimensional nanostructures into functional networks *Science* **291** 630–3

Lu W, Xiang J, Timko B P, Wu Y and Lieber C M 2005 One-dimensional hole gas in germanium silicon nanowire heterostructures *Proc. Natl. Acad. Sci.* **02** 10046–51

Mcalpine M C, Ahmad H, Wang D and Heath J R 2007 Highly ordered nanowire arrays on plastic substrates for ultrasensitive flexible chemical sensors *Nat. Mater.* **6** 379–84

Mcalpine M C, Friedman R S and Lieber C M 2005 High-performance nanowire electronics and photonics and nanoscale patterning on flexible plastic substrates *Proc. IEEE* **93** 1357–63

Melosh N A, Boukai A, Diana F, Gerardot B, Badolato A, Petroff P M and Heath J R 2003 Ultrahigh-density nanowire lattices and circuits *Science* **300** 112–5

Morales A M and Lieber C M 1998 A Laser ablation method for the synthesis of crystalline semiconductor nanowires *Science* **279** 208–11

Takei K, Takahashi T, Ho J C, Ko H, Gillies A G, Leu P W, Fearing R S and Javey A 2010 Nanowire active-matrix circuitry for low-voltage macroscale artificial skin *Nat. Mater.* **9** 1–6

Xiang J, Lu W, Hu Y, Wu Y, Yan H and Lieber C M 2006 Ge/Si nanowire heterostructures as high- performance field-effect transistors *Nature* **441** 489–93

www.ingramcontent.com/pod-product-compliance
Lightning Source LLC
Chambersburg PA
CBHW082130210326
41599CB00031B/5927